P9-AGF-482

College Trigonometry

College Trigonometry

RICHARD N. AUFMANN

VERNON C. BARKER

RICHARD D. NATION, JR.

Palomar College

HOUGHTON MIFFLIN COMPANY BOSTON

Dallas Geneva, Illinois Palo Alto Princeton, New Jersey

Cover photograph by Edward Slaman

Library of Congress Number 89-080912

ISBN: Examination copy 0-395-52624-8
Text 0-395-38098-7

Printed in the U.S.A.

ABCDEFGHIJ-VH-9543210

94

Contents

Preface

This text provides a comprehensive and mathematically sound treatment of the topics considered essential for a college trigonometry course. It is intended for the student who has successfully completed an intermediate algebra course.

To help the student master the concepts in this text, we have tried to maintain a balance among theory, application, and drill. Each definition and theorem is precisely stated and many theorems are proved. Carefully developed mathematics is complemented by abundant, creative applications that are both contemporary and representative of a wide range of disciplines.

Extensive exercise sets ranging from routine exercises to thought-provoking problems are provided at the end of each section. An ample selection of review exercises can be found at the end of each chapter.

Features

Interactive Presentation *College Trigonometry* is written in a style that encourages the student to interact with the textbook. At various places throughout the text, a question in the form of (Why?) is asked of the reader. This question encourages the reader to pause and think about the current discussion and to answer the question. To make sure the student does not miss important information, the answer to the question is provided as a footnote on the same page.

Each section contains a variety of worked examples. Each example is given a name so that a student can see at a glance the type of problem being illustrated. Each example is accompanied by annotations that assist the student in moving from step to step. Following the worked example is a suggested exercise from the exercise set for the student to work. The exercises are color coded by number in the exercise set and the complete solution of that exercise can be found in an appendix in the text.

Extensive Exercise Sets The exercise sets of *College Trigonometry* were carefully developed to provide the student with a variety of exercises. The exercises range from drill and practice to interesting challenges and were chosen to illustrate the many facets of topics discussed in the text. Besides the regular exercise sets, there is a set of supplemental problems that includes material from previous chapters, present extensions of topics, or are of the form "prove or disprove."

Applications One way to motivate a student to an interest in mathematics is through applications. The applications in *College Trigonometry* have been taken from agriculture, architecture, biology, business, chemistry, earth science, economics, engineering, medicine, and physics. Besides providing motivation to study mathematics, the applications provide an avenue to problem solving. The applications problems require the student to organize and implement a problem solving scheme.

Supplements for the Student

Two computerized study aids, the Computer Tutor and the Math Assistant, accompany this text.

The Computer Tutor The Computer Tutor is an interactive instructional microcomputer program for student use. Each section in the text is supported by a lesson on the Computer Tutor. Lessons on the tutor provide additional instruction and practice and can be used in several ways: (1) to cover material the student missed because of absence from class; (2) to reinforce instruction on a concept that the student has not yet mastered; (3) to review material in preparation for examinations. This tutorial is available for the IBM PC and compatible microcomputers.

The Math Assistant The Math Assistant is a collection of programs that can be used by both the instructor and the student. Some programs are instructional and allow the student to practice a skill like finding the inverse of a matrix. Other programs are computational routines that perform numerical calculations. In addition, there is a function grapher that graphs elementary functions and polar equations. The Math Assistant is available for the IBM PC and compatible microcomputers.

Supplements for the Instructor

College Trigonometry has an unusually complete set of teaching aids for the instructor.

Solutions Manual The Solutions Manual contains worked-out solutions for all end-of-section, supplemental, challenge and review exercises.

Instructor's Manual with Testing Program The Instructor's Manual contains the printed testing program, which is the first of three sources of testing material available to the user. Four printed tests (in two formats—free response and multiple choice) are provided for each chapter. In addition, the Instructor's Manual includes documentation of all the software ancillaries—the Math Assistant, the Computer Tutor, and the Computerized Test Generator. Finally, it contains answers to all the even-numbered exercises in the text.

Computerized Test Generator The Computerized Test Generator is the second source of testing material. The data base contains more than 1500 test items. These questions are unique to the test generator and do not repeat items provided in the Instructor's Manual testing program. The Test Generator is designed to produce an unlimited number of tests for each chapter of the text, including cumulative tests and final exams. It is available for the IBM PC and compatible microcomputers, the Macintosh microcomputer and the Apple II family of microcomputers.

Printed Test Bank The Printed Test Bank, the third component of the testing material, is a printout of all items in the Computerized Test Generator. Instructors using the Test Generator can use the test bank to select specific items from the data base. Instructors who do not have access to a computer can use the test bank to select items to be included on a test being prepared by hand.

Acknowledgments

We would like to express our gratitude to Linda Murphy of Northern Essex Community College who read the entire book for accuracy. In addition, we sincerely wish to thank the following reviewers who reviewed the manuscript at various stages of development for their valuable contributions.

Frank Battles, Massachusetts Maritime Academy
Diane Johnson, University of Rhode Island
Alberto L. Delgado, Kansas State University
Judith L. Willoughby, Minneapolis Community College
Mel Hamburger, Laramie County Community College
Janice McFatter, Gulf Coast Community College
John W. Burns, Mt. San Antonio College
Cynthia Moody, City College of San Francisco
Archie L. Ritchie, Peace College
W. R. Wilson, Delta State University
Pauline Lowman, Western Kentucky University
F. P. Mathur, California State Polytechnic University
James Younglove, University of Houston
Jim Delany, California Polytechnic State University
David C. Kay, University of North Carolina
Beth D. Layton, Appalachian State University
Gillian C. Raw, University of Missouri–St. Louis
Burdette C. Wheaton, Mankato State University
Ann Thorne, College of DuPage
Gerald Schrag, Central Missouri State University

1 Introduction to Functions

So that we may say the door is now opened, for the first time, to a new method fraught with numerous and wonderful results which in future years will command the attention of other minds.

GALILEO GALILEI (1564–1642)

TRAP-DOOR FUNCTIONS, CRYPTOGRAPHY, AND SMART CARDS

Some activities are naturally easier to *reverse* than others. Activities that are not easy to reverse include squeezing the toothpaste out of the tube and cooking fresh eggs to make an omelet. These activities can be thought of as *one-way activities.*

The activity of using your car to pull a trailer is a *trap-door activity.* It is said to be a trap-door activity, because although you can drive either forward or backward, the activity of backing up is very difficult until you have mastered the necessary skills.

There are mathematical functions called *trap-door functions.* For example, the function of cubing a number is a trap-door function. It is easy to cube a number, but more difficult to discover the method of determining what number has been cubed, by examining the cube.

Ideally, a mathematical trap-door function would be such that it and its inverse can be evaluated in a matter of seconds. However, just to discover how to evaluate the inverse function requires several years of intense effort.

Mathematicians have found interesting and useful applications of trap-door functions. One of these applications is the art of writing and deciphering secret codes. This art, called cryptography, is vital to the military and commerce of every large nation.

Adi Shamir, Amos Fiat, and Uriel Feige, of the Weizmann Institute have recently used trap-door functions to develop the concept of zero-knowledge proofs. This has lead to the development of credit cards called "smart" cards. These credit cards contain electronic chips that use zero-knowledge proofs. The cards can identify the user of the card as the owner of the card, without giving a sales clerk the card number or any other information that could be used to make unauthorized purchases.

1.1 Preliminaries

The real numbers are used extensively in mathematics. The set of real numbers is quite comprehensive and contains several unique sets of numbers.

The **integers** is the set of numbers

$$\{\ldots, -4, -3, -2, -1, 0, 1, 2, 3, 4, \ldots\}.$$

Recall that the brace symbols, { }, are used to identify a set. The positive integers are called **natural numbers.**

The **rational numbers** is the set of numbers of the form a/b, where a and b are integers, and $b \neq 0$. Thus, the rational numbers include $\frac{-3}{4}$ and $\frac{5}{2}$. Because each integer can be expressed in the form a/b with denominator $b = 1$, the integers are included in the set of rational numbers. Every rational number can be written as either a terminating or repeating decimal.

A number written in decimal form that does not repeat or terminate is called an **irrational number.** Some examples of irrational numbers are $0.141141114\ldots$, $\sqrt{2}$ and π. These numbers cannot be expressed as quotients of integers. The set of real numbers is the union of the sets of rational and irrational numbers.

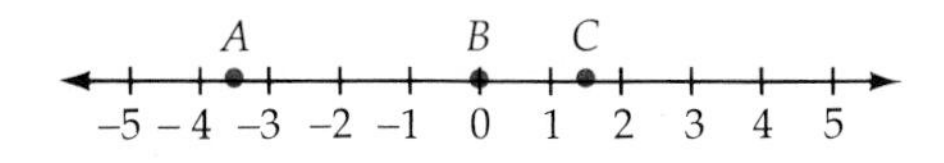

Figure 1.1
A coordinate line

A real number can be represented geometrically on a **coordinate line.** Each point on this line is associated with a real number called the **coordinate** of the point. Conversely, each real number can be associated with a point on a coordinate line. In Figure 1.1, the coordinate of A is $-\frac{7}{2}$, the coordinate of B is 0, and the coordinate of C is $\sqrt{2}$.

Given any two real numbers a and b, we say that a is **less than** b, denoted by $a < b$, if $a - b$ is a negative number. Similarly, we say that a is **greater than** b, denoted by $a > b$, if $a - b$ is a positive number. When a **equals** b, $a - b$ is zero. The symbols $<$ and $>$ are called **inequality symbols.** Two other inequality symbols, $\leq$ (less than or equal) and $\geq$ (greater than or equal) are also used. The inequality symbols can be used to designate sets of real numbers.

If $a < b$, the notation (a, b) is used to indicate the set of real numbers between a and b. This set of numbers can be described using **set builder notation:**

$$(a, b) = \{x \,|\, a < x < b\}.$$

When reading a set written in set builder notation, read "$\{x\,|$" as "the set of x such that." The expression that follows the vertical bar designates the elements in the set.

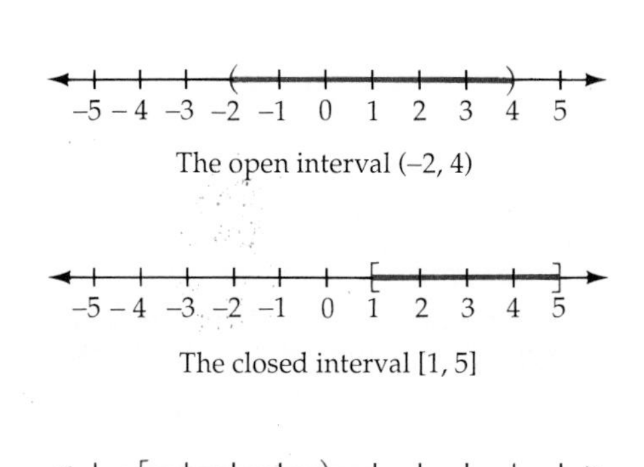

The open interval (−2, 4)

The closed interval [1, 5]

The half-open interval [−4, 0)

The set (a, b) is called a **open interval.** The graph of the open interval consists of all the points on the coordinate line between a and b, not including a and b. A **closed interval,** denoted by $[a, b]$, consists of all points between a and b including a and b. We can also discuss **half-open intervals.** An example of each of these is shown in Figure 1.2.

$(-2, 4) = \{x \,|\, -2 < x < 4\}$	An open interval
$[1, 5] = \{x \,|\, 1 \leq x \leq 5\}$	A closed interval
$[-4, 0) = \{x \,|\, -4 \leq x < 0\}$	A half-open interval
$(-6, -2] = \{x \,|\, -6 < x \leq -2\}$	A half-open interval

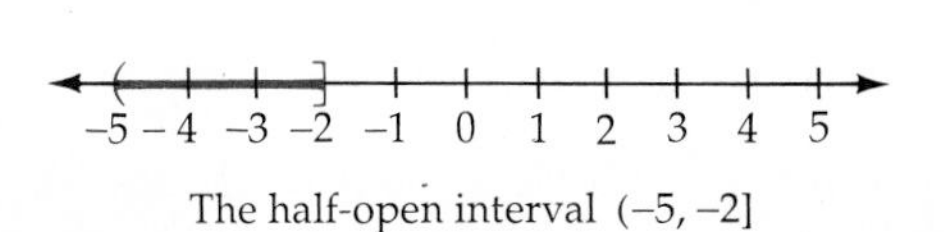

The half-open interval (−5, −2]

Figure 1.2

The *absolute value* of a number is a measure of the distance from zero to the point associated with the number on a coordinate line. Therefore, the absolute value of a number is always positive or zero. We now give a more formal definition of absolute value.

Absolute Value

For a real number a, the **absolute value** of a, denoted by $|a|$, is

$$|a| = \begin{cases} a & \text{if } a \geq 0. \\ -a & \text{if } a < 0. \end{cases}$$

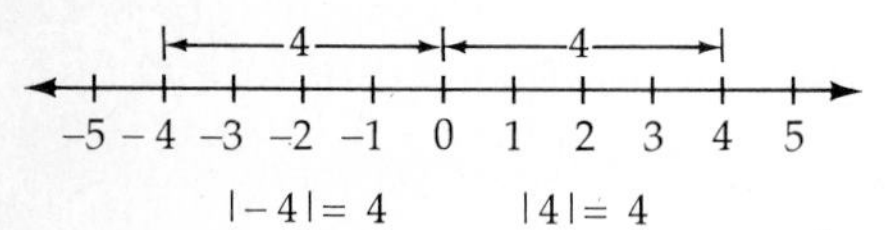

Figure 1.3

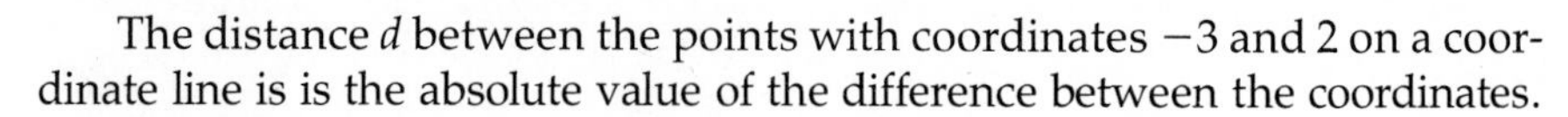

The distance d between the points with coordinates -3 and 2 on a coordinate line is is the absolute value of the difference between the coordinates.

$$d = |2 - (-3)| = 5$$

Because the absolute value is used, we could also write

$$d = |(-3) - 2| = 5$$

In general, we define the distance between any two points A and B on a coordinate line as the absolute value of the difference between the coordinates of the points.

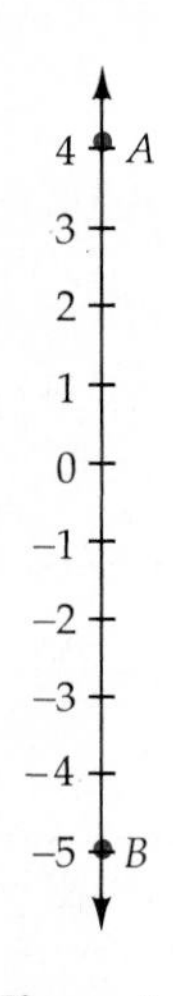

Figure 1.4

Distance Between Two Points on a Coordinate Line

Let a and b be the coordinates of the points A and B, respectively, on a coordinate line. Then the distance between A and B, denoted $d(A, B)$ is

$$d(A, B) = |a - b|.$$

Remark This formula applies to any coordinate line. Thus, it can be used to find the distance between two points on a vertical coordinate line as shown in Figure 1.4.

An **equation** is a statement about the equality of two expressions. Examples of equations are

$$7 = 2 + 5, \qquad x^2 = 4x + 5, \qquad 3x - 2 = 2(x + 1) + 3$$

The values of the variable that make an equation a true statement are the **roots** or **solutions** of the equation. To **solve** an equation means to find the solutions of the equation. The **solution set** of an equation is the set of all solutions of the equation.

First Degree Equation

A **first degree** or **linear equation** in one variable is an equation of the form $ax + b = c$, where $a \neq 0$.

To solve a first degree equation, isolate the variable on one side of the equal sign.

EXAMPLE 1 Solve a First Degree Equation

Solve the equation $3x - 5 = 2$.

Solution

$$3x - 5 = 2$$

$$3x - 5 + 5 = 2 + 5 \quad \text{Add 5 to each side of the equation.}$$

$$3x = 7$$

$$\frac{3x}{3} = \frac{7}{3} \quad \text{Divide each side of the equation by 3.}$$

$$x = \frac{7}{3}$$

The solution is $\frac{7}{3}$.

■ *Try Exercise* **6**, *page 9.*

Quadratic Equation

An equation of the form $ax^2 + bx + c = 0$, $a \neq 0$ is a **quadratic equation** or **second degree equation** in one variable.

A quadratic equation can be solved using the quadratic formula.

Quadratic Formula

The solution of the quadratic equation $ax^2 + bx + c = 0$ is given by

$$x = \frac{-b \pm \sqrt{b^2 - 4ac}}{2a}.$$

EXAMPLE 2 Solve a Quadratic Equation

Solve the equation $2x^2 - 4x + 1 = 0$.

Solution We have $a = 2$, $b = -4$, and $c = 1$.

$$x = \frac{-(-4) \pm \sqrt{(-4)^2 - 4(2)(1)}}{2(2)} = \frac{4 \pm \sqrt{16 - 8}}{4}$$

$$= \frac{4 \pm \sqrt{8}}{4} = \frac{4 \pm 2\sqrt{2}}{4} = \frac{2 \pm \sqrt{2}}{2}$$

The solutions are $\frac{2 + \sqrt{2}}{2}$ and $\frac{2 - \sqrt{2}}{2}$.

■ *Try Exercise* **16**, *page 9.*

Although every quadratic equation can be solved using the quadratic formula, sometimes it is easier to use another method. Factoring and using the Principle of Zero Products, is one such method.

Principle of Zero Products

If a and b are real numbers and $ab = 0$, then $a = 0$ or $b = 0$.

For example, to solve $2x^2 + x - 6 = 0$, first factor the polynomial.

$$2x^2 + x - 6 = 0$$

$$(2x - 3)(x + 2) = 0$$

Now let each factor equal zero and solve for x.

$$2x - 3 = 0 \qquad x + 2 = 0$$

$$x = \frac{3}{2} \qquad x = -2.$$

The solutions are $\frac{3}{2}$ and -2.

Rectangular Coordinate System

Each point on a flat two-dimensional surface, called a **plane**, can be associated with a pair of numbers. To do this, we use a coordinate system.

A **rectangular coordinate system** in a plane is formed by two coordinate lines, one horizontal and one vertical, that intersect at the origin of each line. The two coordinate lines are called **coordinate axes** or simply **axes**, and the point of intersection is called the **origin** of the coordinate system. The horizontal axis is usually referred to as the ***x*-axis**, the vertical axis as the ***y*-axis**, and the plane as the ***xy*-plane**. The coordinate axes divide a plane into four regions called **quadrants**. The quadrants are numbered counterclockwise from I to IV. See Figure 1.5.

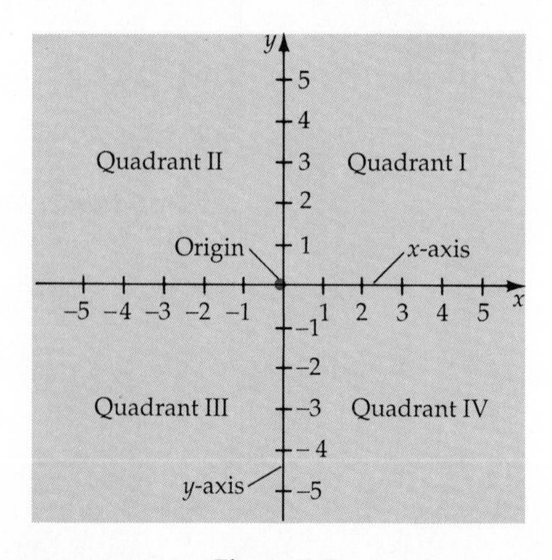

Figure 1.5

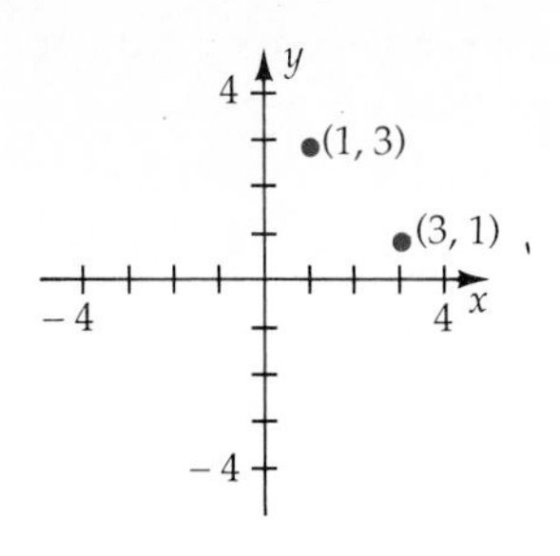

Figure 1.6

In the xy-plane, each point P can be associated with an **ordered pair** of numbers called **coordinates** of the point. Ordered pairs are denoted by (a, b), where a is the **x-coordinate** or **abscissa** and b is the **y-coordinate** or **ordinate**. The first number a of the pair locates a point on the line parallel to the y-axis passing through a on the x-axis. The second number b of the pair locates a point on the line parallel to the x-axis passing through b on the y-axis. To **plot** a point means to place a dot at the coordinates of the point.

The order in which the coordinates of an ordered pair are listed is important. Figure 1.6 shows that $(1, 3)$ and $(3, 1)$ do not denote the same point.

Equality of Ordered Pairs

Two ordered pairs (a, b) and (c, d) are equal if and only if

$$a = c \quad \text{and} \quad b = d.$$

The combination of algebra and geometry with a coordinate system is the branch of mathematics called **analytic geometry.** This branch of mathematics extends the techniques available to us to analyze not only problems in mathematics but also problems in such areas as science, engineering, and business.

The distance between any two points in a coordinate plane can be found by using the Pythagorean Theorem. Recall that a right triangle contains one 90° angle called a **right angle.** The side opposite the 90° angle is the hypotenuse and the two other sides are the legs of the triangle.

Pythagorean Theorem

If a and b denote the lengths of the legs of a right triangle and c is the length of the hypotenuse, then

$$a^2 + b^2 = c^2.$$

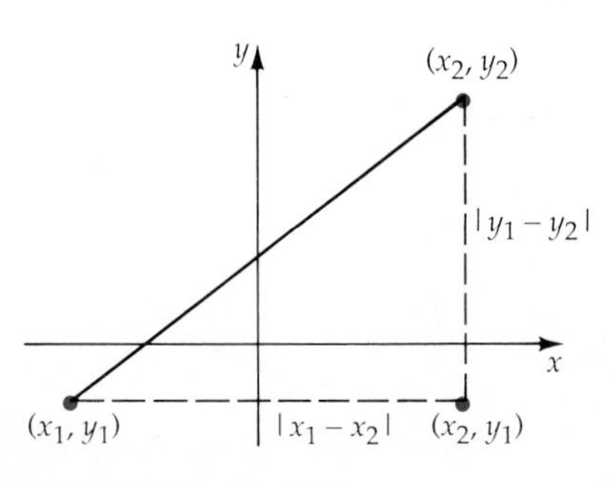

Figure 1.7

Let (x_1, y_1) and (x_2, y_2) be two points in an xy-plane. Notice in Figure 1.7 that the distance between the two points is the hypotenuse of a right triangle whose legs are horizontal and vertical line segments. The length of the horizontal segment is $|x_1 - x_2|$, and the length of the vertical segment is $|y_1 - y_2|$. Applying the Pythagorean Theorem to the triangle and using the fact that $|a|^2 = a^2$, we have

$$d^2 = |x_1 - x_2|^2 + |y_1 - y_2|^2$$

$$d^2 = (x_1 - x_2)^2 + (y_1 - y_2)^2$$

$$d = \sqrt{(x_1 - x_2)^2 + (y_1 - y_2)^2}$$

Distance Formula

Let (x_1, y_1) and (x_2, y_2) be two points in a plane. The distance d between the two points is given by

$$d = \sqrt{(x_1 - x_2)^2 + (y_1 - y_2)^2}.$$

Remark This formula also is valid for points that lie on a vertical or horizontal line.

EXAMPLE 3 Find the Distance Between Two Points

Find the distance d between the points

a. $(-4, 3)$ and $(2, 7)$ b. $(5, -1)$ and $(-2, -4)$

Solution

a. $d = \sqrt{[(-4) - 2)]^2 + [3 - 7]^2} = \sqrt{36 + 16} = \sqrt{52} = 2\sqrt{13}$

b. $d = \sqrt{[5 - (-2)]^2 + [(-1) - (-4)]^2} = \sqrt{49 + 9} = \sqrt{58}$

■ *Try Exercise* **28**, *page 9.*

The midpoint of the line segment with endpoints $P_1(x_1, y_1)$ and $P_2(x_2, y_2)$ is the point on the line equidistant from P_1 and P_2. The coordinates of this point are given by the *Midpoint Formula.*

Midpoint Formula

The midpoint M of the line segment joining P_1 and P_2 is given by

$$\left(\frac{x_1 + x_2}{2}, \frac{y_1 + y_2}{2}\right).$$

To prove this formula, let $M(x_3, y_3)$ be the midpoint of the line segment joining P_1 and P_2 where $x_1 < x_2$. Now consider the two triangles P_1QP_2 and P_1RM in Figure 1.8. Because these are similar triangles, the ratios of corresponding sides are equal. Thus,

$$\frac{d(P_1, R)}{d(P_1, Q)} = \frac{d(P_1, M)}{d(P_1, P_2)}$$

$$\frac{x_3 - x_1}{x_2 - x_1} = \frac{1}{2}$$

$$x_3 - x_1 = \frac{1}{2}(x_2 - x_1)$$

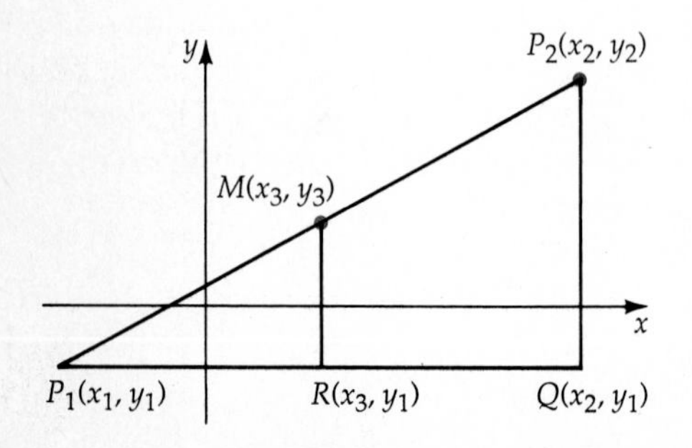

Figure 1.8

Solving for x_3, we have

$$x_3 = \frac{x_1 + x_2}{2}.$$

We have proved the equation of the x-coordinate of the midpoint. In a similar manner, we can verify the formula for the y-coordinate.

EXAMPLE 4 Find the Midpoint of a Line Segment

Find the midpoint of the line segment connecting
a. $(6, -2)$ and $(1, 3)$ b. $(3, -2)$ and $(-5, -4)$

Solution

a. $$\left(\frac{6+1}{2}, \frac{(-2)+3}{2}\right) = \left(\frac{7}{2}, \frac{1}{2}\right)$$

b. $$\left(\frac{3+(-5)}{2}, \frac{-2+(-4)}{2}\right) = (-1, -3)$$

■ *Try Exercise* **38,** *page 9.*

EXERCISE SET 1.1

In Exercises 1 to 10, solve the following first degree equations.

1. $2x - 1 = 7$
2. $3x + 2 = 11$
3. $4 - 5x = 19$
4. $1 - 7x = 15$
5. $2x + 4 = x - 5$
6. $5x - 2 = 3x + 4$
7. $2 - 3(2x + 1) = 5$
8. $3x - 2(2x - 3) = 4(2x + 1)$
9. $5x - 3(1 - 2x) = 2 - 5(2x + 1)$
10. $4 - 3(2 - 2x) = 3x + 2(3 - 2x)$

In Exercises 11 to 20, solve the following quadratic equations. Solve by factoring when possible.

11. $x^2 - x - 6 = 0$
12. $2x^2 - x - 1 = 0$
13. $x^2 - 3x = 10$
14. $6x^2 - 5x = 6$
15. $4x^2 - 4x + 1 = 0$
16. $x^2 - 6x - 4 = 0$
17. $x^2 + 6x = 2$
18. $2x^2 - x = 3$
19. $3x^2 = 2x + 1$
20. $4x^2 = 7x - 1$

In Exercises 21 to 30, find the distance between the given points.

21. $(2, 3)$ and $(-3, 4)$
22. $(4, 0)$ and $(8, -2)$
23. $(0, -4)$ and $(6, 3)$
24. $(-2, -4)$ and $(5, 4)$
25. $(-5, 6)$ and $(5, -6)$
26. $(7, 0)$ and $(3, -3)$
27. $(1, -5)$ and $(-4, -1)$
28. $(5, -2)$ and $(2, -4)$
29. $(5, -2)$ and $(4, -2)$
30. $(-1, 5)$ and $(-1, 3)$

In Exercises 31 to 40, find the midpoint of the line segment connecting the two points.

31. $(4, 1)$ and $(-2, 3)$
32. $(3, 2)$ and $(6, -4)$
33. $(5, -2)$ and $(0, 4)$
34. $(-1, -3)$ and $(2, 6)$
35. $(6, -6)$ and $(-4, 5)$
36. $(0, 3)$ and $(2, -3)$
37. $(-1, 4)$ and $(-3, -5)$
38. $(1, -4)$ and $(3, -5)$
39. $(9, -4)$ and $(5, -4)$
40. $(-8, 1)$ and $(-8, 3)$

41. **a.** Plot the points $A(1, 2)$, $B(5, 6)$, $C(10, 1)$, and $D(6, -3)$.
b. Draw line segments AB, BC, CD, and AD. Show that $d(A, B) = d(C, D)$ and $d(B, C) = d(A, D)$.
c. Show that $d(B, D) = d(A, C)$ for the diagonals BD and AC.
d. Name the figure $ABCD$.

42. **a.** Plot the points $A(-1, -1)$, $B(5, 2)$, $C(10, -3)$, and $D(4, -6)$.
b. Draw line segments AB, BC, CD, and AD. Show that $d(A, B) = d(C, D)$ and $d(B, C) = d(A, D)$.
c. Show that for $d(B, D) \neq d(A, C)$ the diagonals BD and AC.
d. Name the figure $ABCD$.

43. The points $A(0, 2)$, $B(2, 6)$, $C(10, 2)$, and $D(8, -2)$ are the vertices of a rectangle. Find the area of the rectangle.

44. The points $A(1, 4)$, $B(7, 7)$ and $C(4, 1)$ form an isosceles triangle. Find the area of the triangle. (*Hint:* Find the midpoint M of the side AC. Then BM is the altitude of the triangle.)

Supplemental Exercises

45. Complete the proof of the midpoint formula by showing that $y_3 = \dfrac{y_1 + y_2}{2}$.

46. Prove that the lengths of the diagonals of a rectangle are equal. (*Hint:* Construct a rectangle with coordinates $A(0, 0)$, $B(0, a)$, $C(b, a)$, and $D(b, 0)$. Show that $d(A, C) = d(B, D)$.

In Exercises 47 to 50, use the square with vertices $A(-1, 2)$, $B(5, 8)$, $C(11, 2)$ and $D(5, -4)$.

47. Find the area of the square.

48. Form a second square within the first square by connecting the midpoints of each line segment.

49. Find the area of the square in Exercise 48.

50. Find the ratio of the area of the inscribed square to the area of the square.

1.2 Graphs of Equations

Analytic geometry also can be used to represent an equation in two variables by means of a graph. Rene Descartes was one of the first mathematicians to use analytic geometry in his study of mathematics.

Graph of an Equation

> The **graph of an equation** in two variables x and y is the set of all points whose coordinates are solutions of the equation.

To graph an equation, plot the points whose coordinates are solutions of the equation. It is often convenient to solve the equation for y and then substitute values for x. The resulting ordered pairs can be listed in a table as shown in the following example.

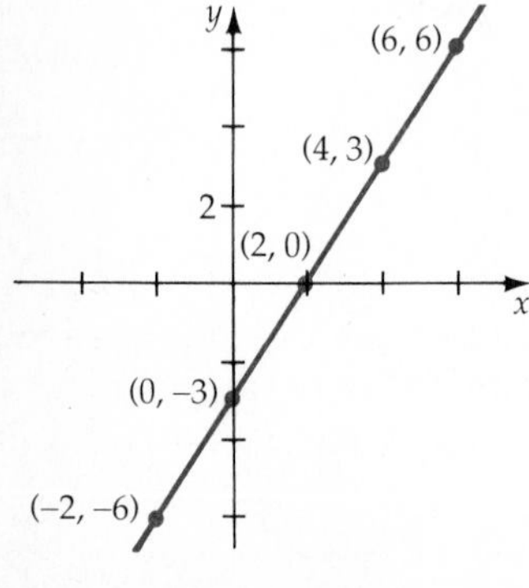

Figure 1.9
$3x - 2y = 6$

EXAMPLE 1 Graph an Equation in Two Variables

Graph each of the following equations by plotting points.

a. $3x - 2y = 6$ b. $y = |x + 3|$

Solution

a. First we solve for y.

$$3x - 2y = 6$$
$$-2y = -3x + 6$$
$$y = \frac{3}{2}x - 3$$

When x is	-2	0	2	4	6
y is	-6	-3	0	3	6

Plot the points corresponding to the ordered pairs $(-2, -6)$, $(0, -3)$, $(2, 0)$, $(4, 3)$ and $(6, 6)$. Now connect the points with a smooth curve.

We chose the values of x arbitrarily. Multiples of 2 were chosen to obtain points with integer coordinates.

b. The equation is already solved for y.

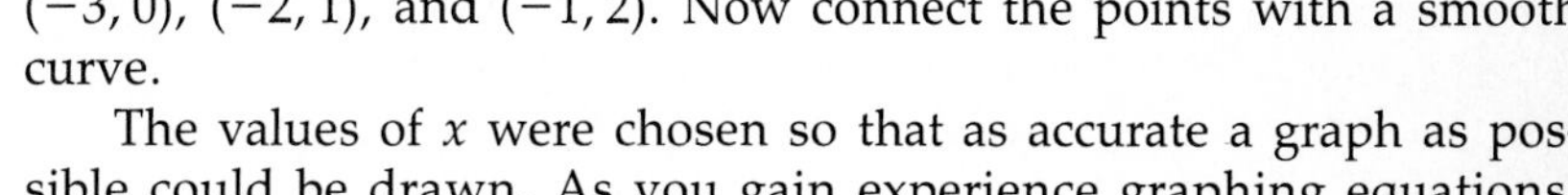

$y = |x + 3|$

When x is	-5	-4	-3	-2	-1
y is	2	1	0	1	2

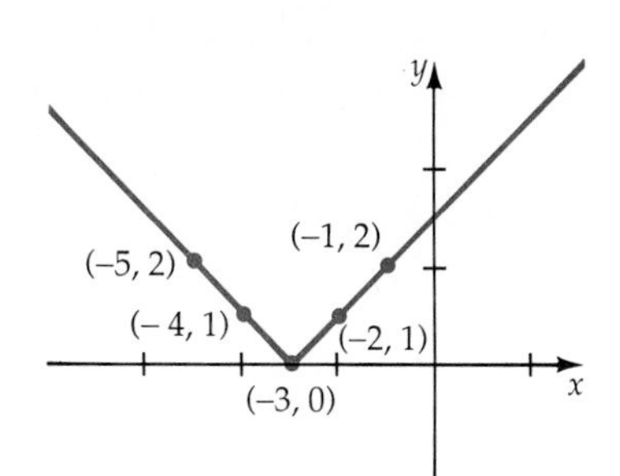

Figure 1.10
$y = |x + 3|$

Plot the points corresponding to the ordered pairs $(-5, 2)$, $(-4, 1)$, $(-3, 0)$, $(-2, 1)$, and $(-1, 2)$. Now connect the points with a smooth curve.

The values of x were chosen so that as accurate a graph as possible could be drawn. As you gain experience graphing equations, you will recognize the appropriate values of x to use.

■ *Try Exercise* **16**, *page 15.*

You may recall that (as in Example 1a) an equation of the form $Ax + By = C$ is a **linear equation in two variables.** The graph of a linear equation is a straight line. The graph of a linear absolute value equation (as in Example 1b) is frequently V shaped.

If the graph of an equation is not a straight line, it may be necessary to plot quite a few points before the graph of the curve can be sketched. For these *nonlinear* equations, it is helpful to have an idea of the general shape of the graph.

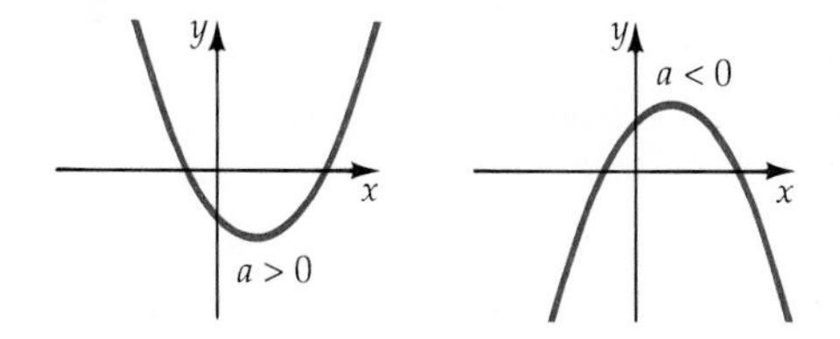

Figure 1.11
$y = ax^2 + bx + c$

Graph of a Quadratic Equation in Two Variables

The graph of the **quadratic equation in two variables,**

$$y = ax^2 + bx + c, \quad a \neq 0$$

is a **parabola**. The parabola opens up when $a > 0$ and opens down when $a < 0$.

EXAMPLE 2 Graph a Parabola

Graph $y = x^2 + 4x - 1$.

Solution Comparing this equation with the general quadratic equation in two variables, we see that $a = 1$. Therefore, the graph will open up. Choose enough values of x to obtain the accurate location of the graph. See Figure 1.12

$y = x^2 + 4x - 1$

When x is	-4	-3	-2	-1	0
y is	-1	-4	-5	-4	-1

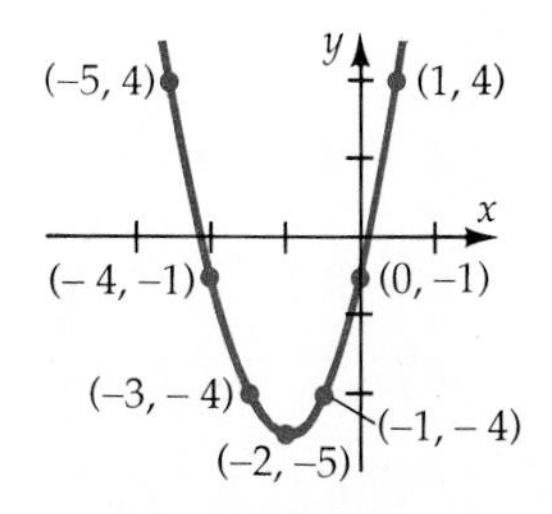

Figure 1.12
$y = x^2 + 4x - 1$

■ *Try Exercise* **26**, *page 15.*

Any point whose x- or y-coordinate is zero is called an *intercept* of the graph of the equation. At these points the graph intersects the x-axis or y-axis.

Definition of x- and y-intercepts

> The point $(x_1, 0)$ is an **x-intercept** of the graph of an equation if and only if it is a solution of the equation.
>
> The point $(0, y_1)$ is a **y-intercept** of the graph of an equation if and only if it is a solution of the equation.

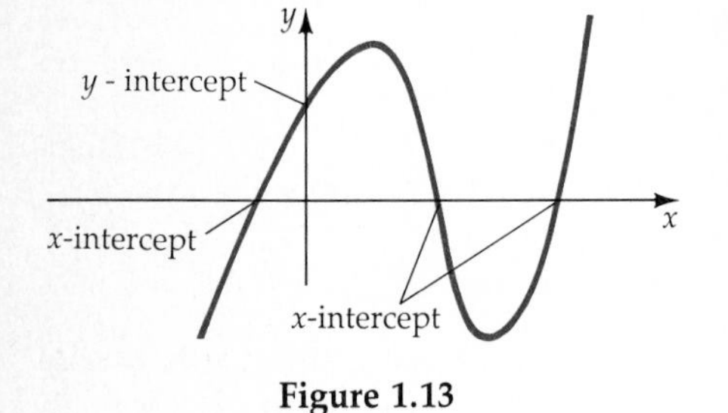

Figure 1.13

Remark To find the x-intercept(s) of the graph of an equation, let $y = 0$ and solve for x. To find the y-intercept(s) of the graph of an equation, let $x = 0$ and solve for y.

EXAMPLE 3 Find the Intercepts of the Graph of a Linear Equation in Two Variables

Find the intercepts and graph the equation $3x + 5y = 15$.

Solution To find the x-intercept, let $y = 0$ and solve for x.

$$3x + 5(0) = 15$$
$$3x = 15$$
$$x = 5$$

The x-intercept is $(5, 0)$.

To find the y-intercept, let $x = 0$ and solve for y.

$$3(0) + 5y = 15$$
$$5y = 15$$
$$y = 3$$

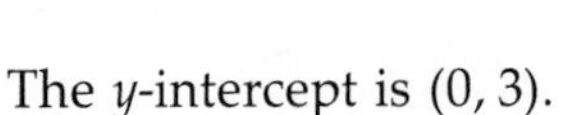

The y-intercept is $(0, 3)$.

Because the equation is a linear equation in two variables, the graph will be a straight line. Plot the two intercepts and draw a line through the two points.

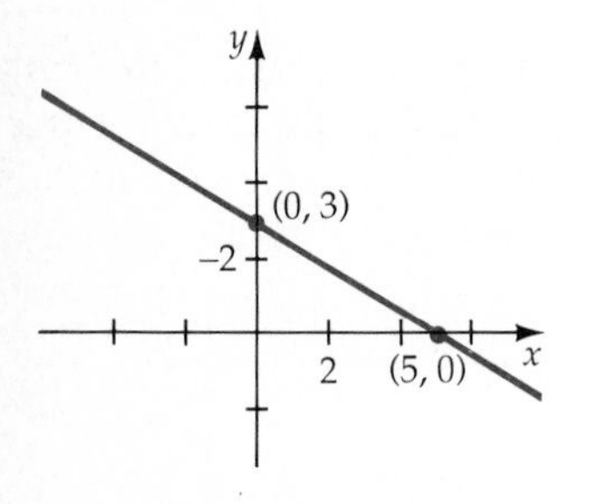

Figure 1.14
$3x + 5y = 15$

■ *Try Exercise* **32,** *page 15.*

EXAMPLE 4 Find the Intercepts of the Graph of a Parabola

Find the intercepts and graph the equation $y = 2x^2 - x - 3$.

Solution To find the x-intercepts, let $y = 0$ and solve for x.

$$y = 2x^2 - x - 3$$
$$0 = 2x^2 - x - 3 \qquad \text{This is a quadratic equation. Solve by factoring.}$$

$$0 = (2x - 3)(x + 1)$$

$$2x - 3 = 0 \qquad x + 1 = 0$$

$$x = \frac{3}{2} \qquad\qquad x = -1$$

The x-intercepts are $\left(\frac{3}{2}, 0\right)$ and $(-1, 0)$.

To find the y-intercept, let $x = 0$ and solve for y.

$$y = 2x^2 - x - 3$$

$$y = 2(0)^2 - 0 - 3 \quad \text{and thus} \quad y = -3.$$

The y-intercept is $(0, -3)$.

Now we find a few more points so that an accurate graph can be drawn. This is a parabola with $a > 0$. Thus, the parabola opens up.

$y = 2x^2 - x - 3$

When x is	-2	1	2
y is	7	-2	3

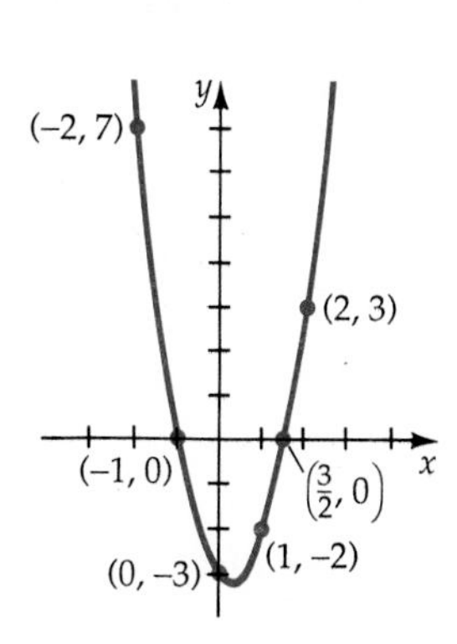

Figure 1.15
$y = 2x^2 - x - 3$

■ *Try Exercise* **40**, *page 15.*

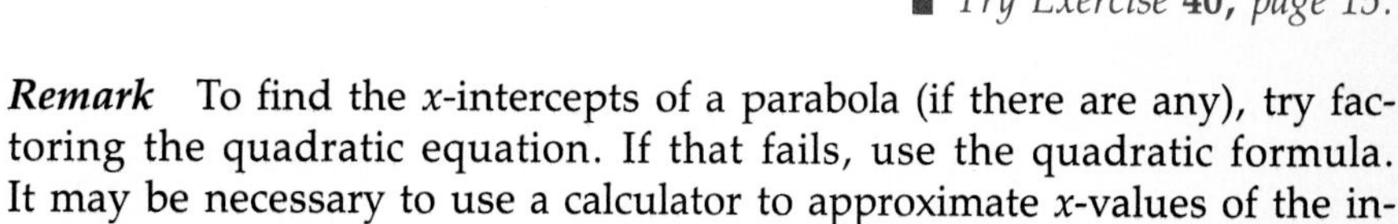

Remark To find the x-intercepts of a parabola (if there are any), try factoring the quadratic equation. If that fails, use the quadratic formula. It may be necessary to use a calculator to approximate x-values of the intercepts.

A **circle** is the set of points in a plane that are a constant distance from a specified point. The distance is the **radius** of the circle, and the specified point is the **center** of the circle. Using the distance formula, we can derive the equation of a circle.

Let (h, k) be the center of a circle of radius r. By the definition of a circle, the point (x, y) is on the circle if and only if the distance from the center (h, k) to the point (x, y) is r units. Thus, (x, y) is on the circle if and only if

$$\sqrt{(x - h)^2 + (y - k)^2} = r.$$

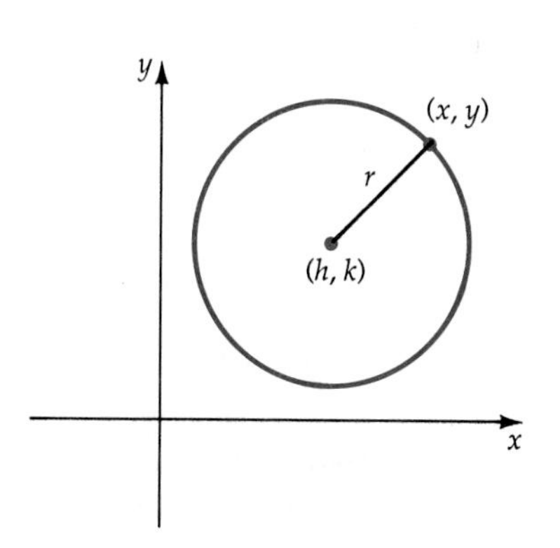

Figure 1.16

Squaring each side, we have

$$(x - h)^2 + (y - k)^2 = r^2.$$

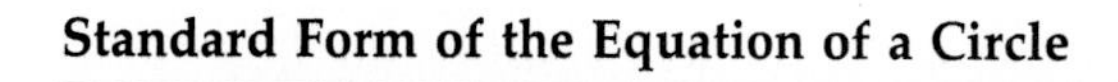

Standard Form of the Equation of a Circle

Let (h, k) be the equation of a circle with center (h, k) and radius r. The **standard form of the equation of the circle** is

$$(x - h)^2 + (y - k)^2 = r^2.$$

Remark If the center of the circle is at the origin $(0, 0)$, then the equation of the circle simplifies to

$$x^2 + y^2 = r^2.$$

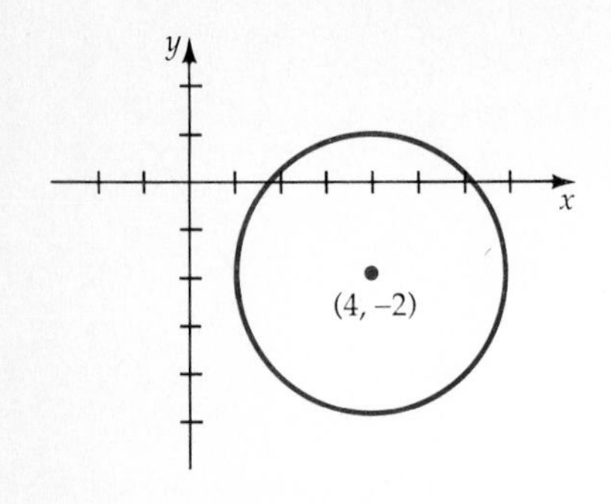

Figure 1.17
$(x - 4)^2 + (y + 2)^2 = 9$

The circle of radius 1 with center at the origin is called the **unit circle.** The equation of the unit circle is

$$x^2 + y^2 = 1.$$

The unit circle plays an important role in trigonometry. The graph of the circle $(x - 4)^2 + (y + 2)^2 = 9$ is shown in Figure 1.17. In standard form, the equation is

$$(x - 4)^2 + (y - (-2))^2 = 3^2,$$

from which we can determine that $h = 4$, $k = -2$, and $r = 3$. The center of the circle is $(4, -2)$ and the radius is 3.

EXAMPLE 5 **Find the Standard Form of the Equation of a Circle**

Find the standard form of the equation of a circle that has center $C(-3, 2)$ and contains the point $P(2, -2)$.

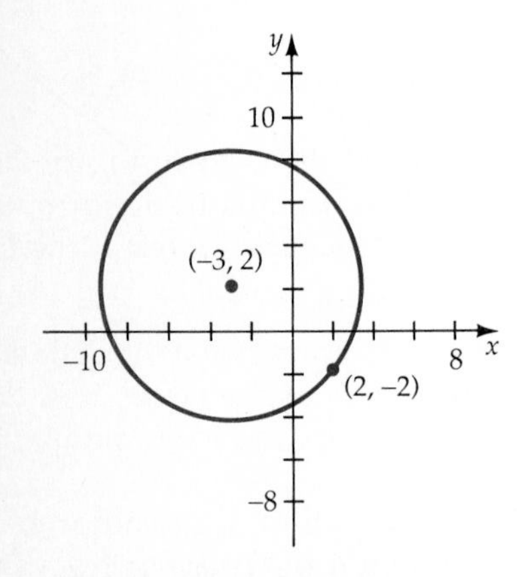

Figure 1.18
$(x + 3)^2 + (y - 2)^2 = 41$

Solution Because the point $P(2, -2)$ is on the circle, the radius of the circle must equal the distance from C to P. Thus,

$$\begin{aligned} r &= \sqrt{(2 - (-3))^2 + (-2 - 2)^2} \\ &= \sqrt{25 + 16} \\ &= \sqrt{41} \end{aligned}$$

Using the equation of a circle in standard form with $h = -3$, $k = 2$, and $r = \sqrt{41}$, we have

$$\begin{aligned} (x - (-3))^2 + (y - 2)^2 &= (\sqrt{41})^2 \\ (x + 3)^2 + (y - 2)^2 &= 41. \end{aligned}$$

■ *Try Exercise* **60,** *page 16.*

If we rewrite the answer to Example 5 by squaring the binomials and combining like terms, we have

$$\begin{aligned} (x + 3)^2 + (y - 2)^2 &= 41 \\ x^2 + 6x + 9 + y^2 - 4y + 4 &= 41 \end{aligned}$$

or,

$$x^2 + y^2 + 6x - 4y - 28 = 0.$$

This form of the equation of a circle is known as the *general form of the equation of a circle.*

General Form of the Equation of a Circle

> The **general form of the equation of a circle** is
>
> $$x^2 + y^2 + ax + by + c = 0.$$

In the next example, we use the technique of *completing the square* to change the equation of a circle from the general form to the standard form.

EXAMPLE 6 **Find the Equation of a Circle in Standard Form by Completing the Square**

Find the center and radius of the circle with equation

$$x^2 + y^2 + 8x - 3y + 12 = 0.$$

Solution First, we arrange and group the terms as shown.

$$(x^2 + 8x) + (y^2 - 3y) = -12.$$

Now we can complete the square of $(x^2 + 8x)$ by adding $\left(\frac{1}{2} \cdot 8\right)^2 = 16$ to each side of the equation. To complete the square $(y^2 - 3y)$, add $\left(\frac{1}{2} \cdot (-3)\right)^2 = \frac{9}{4}$ to each side of the equation.

$$(x^2 + 8x + 16) + \left(y^2 - 3y + \frac{9}{4}\right) = -12 + 16 + \frac{9}{4}$$

Note that, to produce a new equivalent equation, 16 and $\frac{9}{4}$ is also added to the right side of the equation. Now factor and simplify.

$$(x + 4)^2 + \left(y - \frac{3}{2}\right)^2 = \frac{25}{4}$$

The center is $\left(-4, \frac{3}{2}\right)$ and the radius is $\frac{5}{2}$.

■ *Try Exercise* **52,** *page 16.*

EXERCISE SET 1.2

In Exercises 1 to 26, graph each equation.

1. $y = 3x - 2$

2. $y = -2x + 3$

3. $y = 4 - 2x$

4. $y = 5 + 3x$

5. $y = \frac{3}{4}x - 2$

6. $y = \frac{5}{2}x + 3$

7. $3x - 2y = 6$

8. $2x + 5y = 10$

9. $4x - 3y = 12$

10. $3x - 5y = 15$

11. $y = |2x + 1|$

12. $y = |x - 4|$

13. $y = |-3x + 2|$

14. $y = |-2x - 1|$

15. $y = |2x| - 2$

16. $y = -|x| + 3$

17. $y = -|3x| + 2$

18. $y = |3x| - 4$

19. $y = x^2 - 1$

20. $y = x^2 + 4$

21. $y = x^2 - 2x - 3$

22. $y = x^2 + 4x - 3$

23. $y = 2x^2 - 6x + 1$

24. $y = 2x^2 + 8x - 5$

25. $y = -x^2 + 5x - 6$

26. $y = -2x^2 + 4x - 5$

In Exercises 27 to 40, find the intercepts and graph the equation.

27. $5x + 2y = 10$

28. $x - 4y = 8$

29. $2x - 3y = 12$

30. $4x + 5y = 20$

31. $3x + 5y = 9$

32. $2x + 3y = -9$

33. $4x - 3y = 0$

34. $2x + 3y = 0$

35. $y = x^2 - x - 6$

36. $y = x^2 + 4x - 12$

37. $y = 2x^2 + 5x - 3$

38. $y = 3x^2 - 10x - 8$

39. $y = x^2 - 2x - 1$

40. $y = x^2 - 4x + 1$

In Exercises 41 to 58, find the center and radius of each equation whose graph is a circle.

41. $x^2 + y^2 = 1$

42. $x^2 + y^2 = 36$

43. $(x - 3)^2 + (y + 1)^2 = 16$

44. $(x + 2)^2 + (y - 1)^2 = 9$

45. $(x + 4)^2 + (y - 2)^2 = 25$

46. $(x - 3)^2 + (y + 3)^2 = 49$

47. $(x - 5)^2 + (y - 2)^2 = 1$

48. $(x + 3)^2 + (y + 4)^2 = 4$

49. $x^2 + y^2 - 6x + 8y - 11 = 0$

50. $x^2 + y^2 + 8x - 10y - 8 = 0$

51. $x^2 + y^2 + 4x - 4y - 8 = 0$

52. $x^2 + y^2 - 6x - 4y - 12 = 0$

53. $x^2 + y^2 + 5x - 6y + 1 = 0$

54. $x^2 + y^2 - 8x + 3y + 2 = 0$

55. $x^2 + y^2 - 4x - 6y + 13 = 0$

56. $x^2 + y^2 - 10x + 8y + 41 = 0$

57. $x^2 + y^2 + 6x - 8y + 36 = 0$

58. $x^2 + y^2 + 12x - 6y + 50 = 0$

59. Find the equation of the circle passing through the point $(2, -3)$ and having center $(-1, 1)$.

60. Find the equation of the circle passing through the point $(-1, -2)$ and having center $(2, -1)$.

61. A circle has center $(3, 2)$ and passes through the point $(0,0)$. Find the equation of the circle.

62. A circle has center $(-4, 1)$ and passes through the point $(-1, 3)$. Find the equation of the circle.

Supplemental Exercises

In Exercises 63 to 70, sketch the graph of each of the following.

63. $|x + y| = 1$

64. $|x - y| = 1$

65. $|x| + |y| = 1$

66. $|x| - |y| = 1$

67. $y = |x + 3| + |x|$

68. $y = |x - 4| - |x|$

69. $y = |x^2 - 4|$

70. $y = |x^2 - 9|$

71. A circle is circumscribed around a square that has vertices $(0,0)$, $(0, 2)$, $(2, 2)$, and $(2, 0)$. Find the equation of the circle.

72. A circle is inscribed in a square that has vertices $(0, 0)$, $(0, 2)$, $(2, 2)$, and $(2, 0)$. Find the equation of the circle.

1.3 Functions

The relationship between two variables is frequently described by an equation. For example, the time t it takes a computer program to alphabetize a list of n names is given by the equation

$$t = \frac{n^2}{1000}.$$

The ordered pair $(10, 0.\overline{1})$ indicates that it takes 0.1 seconds to alphabetize 10 names. Other ordered pairs determined by this equation are $(0, 0)$, $(5, 0.025)$, and $(100, 10)$. This is one example of a mathematical relationship called a *function*.

Definition of a Function

> A **function** f from a set D (the **domain**) to a set R (the **range**) is a correspondence, or rule, that pairs each element of D with exactly one element of R.

The equation $t = n^2/1000$ defines t as a function of n. For each choice of n, there corresponds exactly one value for t. Since the value of t *depends* on n, t is called the **dependent variable** and n is the **independent variable** of the function.

The set D is the domain of the function and is the set of all allowable values for the independent variable. The range of the function is the set of all corresponding values of the dependent variable.

The correspondence or rule can be an equation (as shown above) or can be defined by a graph, a table, or by any method that determines a relationship between two sets.

Not all equations, however, define a function. Consider the equation

$$y^2 = 25 - x^2.$$

This equation does not define a function because if x is 3, y can be 4 or -4. Thus, there is an x that is not paired with *exactly* one element of R. The value 3 is paired with 4 and -4. To determine if an equation defines a function,

1. solve (if possible) for the dependent variable, and
2. determine if each single value of the independent variable produces exactly one value of the dependent variable.

The specific letters used for the independent and dependent variables are not important. For example, the equation $s = t^2$ represents the same function as the equation $y = x^2$. Traditionally, the letter x is used for the independent variable and y for the dependent variable.

Remark Anytime we use the phrase "y is a function of x", or a similiar phrase with different letters, the variable that follows "function of" is the independent variable. The definition of function requires that for each value of the independent variable there is exactly one value of the dependent variable.

EXAMPLE 1 Determine If y Is a Function of x

Identify which of the following equations define y as a function of x.

a. $3x + y = 2$ b. $y = x^2$ c. $y^2 = x$

Solution

a. $$3x + y = 2 \qquad \text{Solve for } y.$$
$$y = -3x + 2$$

Because each value of x produces exactly one value of y, the equation $3x + y = 2$ defines y as a function of x.

b. $y = x^2$ is already solved for y.

Because each value of x produces exactly one value of y, the equation $y = x^2$ defines y as a function of x. Note, for this function, that two *different* x's can have the *same* y. For example, when x is 2, y is 4. When x is -2, y is 4. This does not contradict the definition of a function.

c. $$y^2 = x \qquad \text{Solve for } y.$$
$$y = \pm\sqrt{x}$$

Because each nonzero value of x produces two values of y, the equation $y^2 = x$ does not define y as a function of x. For example, when x is 9, y is 3 or -3.

■ *Try Exercise* **4**, *page 22.*

Often functions are represented in **functional notation** by letters or a combination of letters such as f, g, f_1, log, or tan.

If y is a specific range value corresponding to x of the function f, we write

$$y = f(x).$$

This is read "y equals f of x" or "y is a function of x." Writing the equation

$$2x - 3y = 6 \quad \text{as} \quad y = \frac{2}{3}x - 2$$

indicates that y is the dependent variable. Using functional notation, this function would be written as

$$f(x) = \frac{2}{3}x - 2.$$

Functional notation indicates which variable is the independent variable, in this case x, and it provides a name for the function, namely f.

Caution $f(x)$ is *not* a function; $f(x)$ is the value of the function at x. The function is denoted by f.

The symbol $f(6)$ is used to denote the value of the dependent variable when the independent variable is 6. For the function defined by $f(x) = \frac{2}{3}x - 2$, we have

$$f(6) = \frac{2}{3}(6) - 2 = 4 - 2 = 2.$$

We say that the **value of the function** is 2 when x is 6. The process of finding the value of a function is called **evaluating the function.**

EXAMPLE 2 Evaluate a Function

For the function g defined by $g(x) = 2x^2 - x - 4$, evaluate each of the following: a. $g(0)$ b. $g(-2)$

Solution

a. $$g(0) = 2(0)^2 - 0 - 4 = 0 - 0 - 4 = -4$$

b. $$g(-2) = 2(-2)^2 - (-2) - 4 = 8 + 2 - 4 = 6$$

■ *Try Exercise* **24**, *page 22.*

Functions can be evaluated at other variables. Considering the function defined by $f(x) = x^2 - 1$, we have

$$\begin{aligned} f(w + 2) &= (w + 2)^2 - 1 \\ &= w^2 + 4w + 4 - 1 \\ &= w^2 + 4w + 3 \end{aligned}$$

If the domain of a function is not specifically stated, then it is the largest set of real numbers for which the defining expression in x produces a value in the range that is a real number. If the defining expression is an equation, then the domain of the function is restricted to avoid

- division by zero, and
- even roots of negative numbers

For example, the domain of $h(x) = \sqrt{2x - 6}$ is all real numbers x where $x \geq 3$. If x is less than 3, then $\sqrt{2x - 6}$ is not a real number. The domain of

$$r(x) = \frac{x}{x - 1}$$

is the set of real numbers except 1. If $x = 1$, $r(x)$ is not defined.

For any element b in the range of a function f, there is at least one value of c in the domain of f so that $f(c) = b$.

EXAMPLE 3 Find an Element in the Domain Corresponding to a Given Value in the Range

Given that -2 is in the range of the function defined by $f(x) = x^2 - x - 8$, find two values of c in the domain of f such that $f(c) = -2$.

Solution We must find c so that $f(c) = -2$.

$$f(c) = c^2 - c - 8 \quad \text{Evaluate } f \text{ at } c.$$

and

$$f(c) = -2$$

Therefore,

$$c^2 - c - 8 = -2.$$

Solve the quadratic equation.

$$\begin{aligned} c^2 - c - 8 &= -2 \\ c^2 - c - 6 &= 0 \\ (c - 3)(c + 2) &= 0 \end{aligned}$$

Thus, $c = 3$ or $c = -2$. As a check,

$$f(3) = (3)^2 - (3) - 8 = 9 - 3 - 8 = -2$$

$$f(-2) = (-2)^2 - (-2) - 8 = 4 + 2 - 8 = -2$$

■ *Try Exercise* **34**, *page 22.*

Algebra of Functions

It is possible to define addition, subtraction, multiplication, division, and powers of functions.

Operations on Functions

Addition	$(f + g)(x) = f(x) + g(x)$
Subtraction	$(f - g)(x) = f(x) - g(x)$
Multiplication	$(fg)(x) = f(x) \cdot g(x)$
Division	$\left(\dfrac{f}{g}\right)(x) = \dfrac{f(x)}{g(x)}, \quad g(x) \neq 0$
Powers	$f^n(x) = [f(x)]^n$

EXAMPLE 4 **Operate with Functions**

For the functions defined by $f(x) = 3x - 4$ and $g(x) = x^2 - 1$, find

a. $(f + g)(x)$ b. $(f - g)(x)$ c. $(f \cdot g)(x)$ d. $\left(\dfrac{f}{g}\right)(x)$
e. $f^3(x)$

Solution

a. $(f + g)(x) = f(x) + g(x) = (3x - 4) + (x^2 - 1)$
$= x^2 + 3x - 5$

b. $(f - g)(x) = f(x) - g(x) = (3x - 4) - (x^2 - 1)$
$= -x^2 + 3x - 3$

c. $(f \cdot g)(x) = f(x) \cdot g(x) = (3x - 4)(x^2 - 1)$
$= 3x^3 - 4x^2 - 3x + 4$

d. $\left(\dfrac{f}{g}\right)(x) = \dfrac{f(x)}{g(x)} = \dfrac{3x - 4}{x^2 - 1}, \quad x \neq 1, -1$

e. $f^3(x) = [f(x)]^3 = (3x - 4)^3$
$= 27x^3 - 108x^2 + 144x - 64$

■ *Try Exercise* **30**, *page 22.*

Applications of Functions

Suppose that studies of peach trees show that planting twenty-five peach trees per acre will yield two hundred seventy-five quarts of peaches per tree. For each additional tree planted per acre, the yield of a tree decreases by five quarts. To find the function that describes the relationship between the yield per acre and the number of trees planted, let

$$x = \text{the number of additional trees planted over twenty-five}.$$

Since yield per tree decreases by five quarts for each additional tree planted,

$$275 - 5x \text{ is the yield per tree}$$

$$25 + x \text{ is the number of trees per acre}$$

The function Y that represents the relationship between the yield and the number of trees is given by

$$Y(x) = (275 - 5x)(25 + x) \quad \text{$Y(x)$ is the product of the yield per tree and the number of trees.}$$

$$= -5x^2 + 150x + 6875$$

If $x = 11$, then 36 trees are planted per acre. The yield per acre would be

$$Y(x) = -5x^2 + 150x + 6875$$

$$Y(11) = -5(11)^2 + 150(11) + 6875$$

$$= 7920$$

If $x = 20$, then forty-five trees are planted per acre. The yield per acre would be

$$Y(x) = -5x^2 + 150x + 6875$$

$$Y(20) = -5(20)^2 + 150(20) + 6875$$

$$= 7875$$

Notice that thirty-six trees per acre yield more fruit than forty-five trees per acre. Using some methods of calculus, we can determine the optimal number of trees to plant per acre so that the yield will be as large as possible.

EXAMPLE 5 Solve an Application

An underground telephone cable is being laid between two homes on opposite sides of a river. The cost for laying the cable is \$45 per meter on land and \$70 per meter under the river. If the cable is laid to a point x meters from home A and then under the river to home B, find the function that represents the cost of the cable as a function of x.

Solution If x is the distance in meters from A to P, then the cost C_{AP} to lay the cable from A to P, is

$$C_{AP} = 45x \quad \text{(Cost per meter)(number of meters)}$$

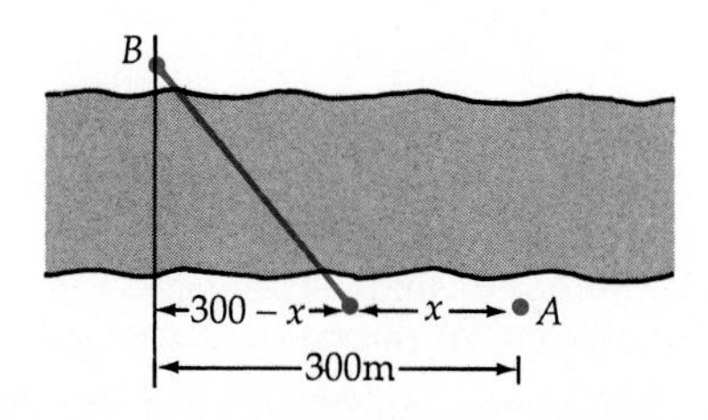

Figure 1.19

The distance $d(P, B)$ from P to B is

$$d(P, B) = \sqrt{(300 - x)^2 + 100^2}.$$

The cost C_{PB} to lay the cable from P to B, is

$$C_{PB} = 70\sqrt{(300 - x)^2 + 100^2}.$$

The total cost $C(x)$ to lay the cable is the sum of C_{AP} and C_{PB}.

$$C(x) = 45x + 70\sqrt{(300 - x)^2 + 100^2}.$$

■ *Try Exercise* **38**, *page 22.*

EXERCISE SET 1.3

In Exercises 1 to 6, determine whether the equation defines y as a function of x.

1. $-2x + y = 3$ **2.** $4x + 2y = 1$

3. $x^2 - 2y = 1$ **4.** $2x^2 + y = 3$

5. $x^2 - y^2 = 9$ **6.** $y = x^3$

In Exercises 7 to 20, find the domains of the functions defined by each equation.

7. $f(x) = x^2 + 4$ **8.** $f(x) = 3x - 2$

9. $g(u) = 3u^2 - u + 4$ **10.** $p(q) = q^3 + 4q - 3$

11. $\text{ABS}(x) = |x|$ **12.** $\text{SQR}(x) = x^2$

13. $y(x) = \sqrt{3x - 12}$ **14.** $s(t) = \sqrt{2t + 6}$

15. $r(x) = \dfrac{x}{x^2 - 9}$ **16.** $u(v) = \dfrac{6}{v + 4}$

17. $H(r) = \sqrt{r^2 + 4}$ **18.** $A(t) = \sqrt{4 - 5t}$

19. $C(z) = \dfrac{z^2 - 4}{z^2}$ **20.** $F(x) = \dfrac{x^2 - 9}{x^2 + 2x + 5}$

For Exercises 21 to 30, using the functions

$f(x) = 3x^2 - 2x + 1$, $g(x) = \dfrac{x}{x^2 - 1}$, $h(x) = 4 - 7x$, and $k(x) = 3x + 1$, find each of the following.

21. $g(0)$ **22.** $f(-2)$

23. $h(4)$ **24.** $k\left(-\dfrac{1}{2}\right)$

25. $f(a - 1)$ **26.** $g(\sqrt{2})$

27. $(f + h)(x)$ **28.** $(h \cdot k)(x)$

29. $\left(\dfrac{f}{h}\right)(1)$ **30.** $(h - k)(4)$

31. Given that 4 is in the range of the function defined by $f(x) = 3x - 2$, find c in the domain of f such that $f(c) = 4$.

32. Given that 5 is in the range of the function defined by $f(x) = 7 - 2x$, find c in the domain of f such that $f(c) = 5$.

33. Given that -1 is in the range of the function defined by $f(x) = x^2 - 5$, find two values of c in the domain of f such that $f(c) = -1$.

34. Given that 2 is in the range of the function defined by $f(x) = x^3 - 6$, find c in the domain of f such that $f(c) = 2$.

35. A pebble is dropped into a still pond. The outer ripple is a circle whose radius is $3/\sqrt{t}$ centimeters where t is measured in seconds. Find the area of a circle formed by the outer ripple as a function of t. (*Hint:* $A = \pi r^2$.)

36. The radius of a spherical balloon from which air is escaping is given by $10 - 0.5t$ centimeters where t is measured in seconds. Find the volume of the balloon after t seconds as a function of time. (*Hint:* $V = \frac{4}{3}\pi r^3$.)

37. A transmitter is located one kilometer from a straight road. A truck is travelling along the road away from the transmitter at seventy-five kilometers per hour. Find the distance the truck is from the transmitter as a function of time.

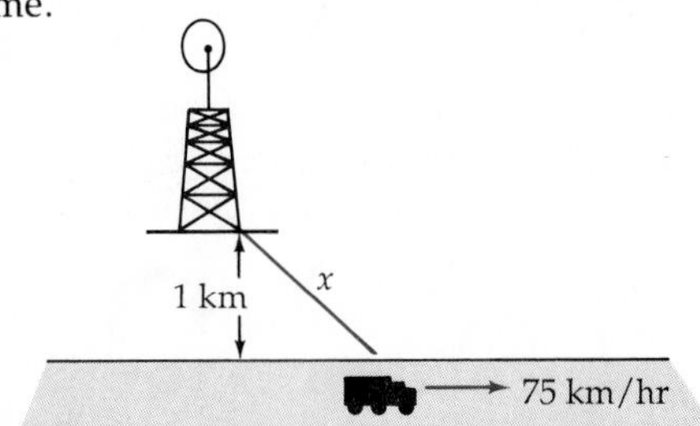

38. An observer is 50 meters from a weather balloon that is about to be released. When the balloon is released, it rises vertically at a rate of 3 meters per second. Find the distance from the observer to the balloon as a function of time.

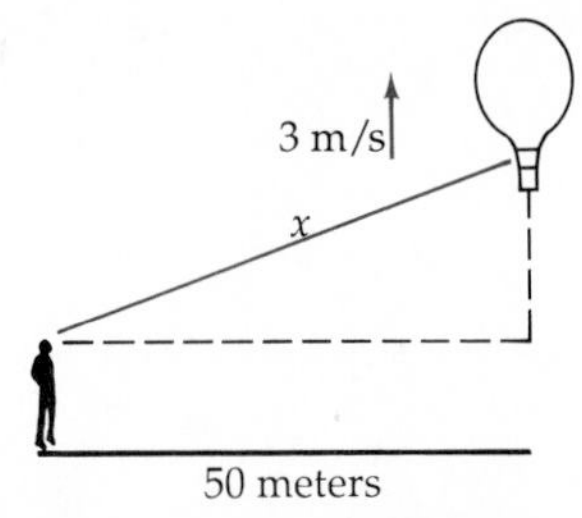

39. When an orange grove is planted with thirty trees per acre, the yield per tree is 300 quarts of oranges. For each additional tree planted per acre, the yield per tree decreases by 4 quarts. Find the yield per tree as a function of the number of trees over 30 planted per acre.

40. The management of a hotel knows that two hundred rooms will be occupied at a rate of \$50 per night. For each \$2 decrease in the price of a room, one more room will be occupied. Find the nightly revenue of the hotel as a function of the number of \$2 decreases in the price of a room. (*Hint:* revenue = (price per room) (number of rooms occupied))

Supplemental Exercises

41. Does the equation $x^2 - y^3 = 0$ define y as a function of x?

42. If an equation defines y as a function of x, is it also true that the equation defines x as a function of y? Give an example.

43. Is there a value of c in the domain of the function defined by $f(x) = \dfrac{x}{x - 4}$ so that $f(c) = 1$? Why or why not?

44. Is there a value of c in the domain of the function defined by $f(x) = \dfrac{2x^2 + 1}{x^2 - 8}$ so that $f(c) = 2$? Why or why not?

45. Consider a function f and two real numbers a and b. If $f(a) = f(b)$, does this mean that $a = b$? Prove your answer.

In Exercises 46 to 49, use the functions defined by the following: $f(x) = 3x - 2$ and $g(x) = x^2$.

46. Find $f(a + h)$.

47. Find $\dfrac{f(a + h) - f(a)}{h}$.

48. Find $g(2 + h)$.

49. Find $\dfrac{g(2 + h) - g(2)}{h}$.

1.4 Graphs of Functions

There is an alternate definition of function that suggests a way to represent a function geometrically. We can use this definition to graph a function.

Alternate Definition of a Function

> A function f is a set of ordered pairs (x, y) such that no two ordered pairs with the same first coordinate have different second coordinates.

Remark Using the notation $y = f(x)$, the ordered pairs of the function are written $(x, f(x))$.

The **graph** of a function is obtained by plotting the ordered pairs $(x, f(x))$ of the function f.

To graph the function defined by $f(x) = x^2 + 1$, think of it as the equation $y = x^2 + 1$. Recall from Section 2, that the graph of this equation is a parabola that opens up. Make a table of values of x and $f(x) = y$, and plot the ordered pairs.

$f(x) = x^2 + 1$

When x is	−2	−1	0	1	2
$f(x)$ is	5	2	1	2	5

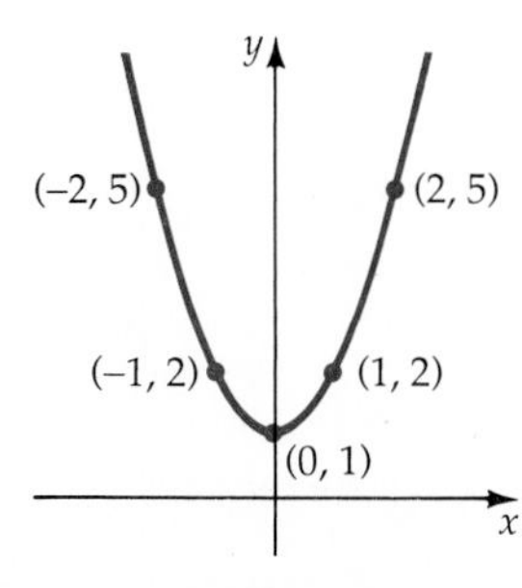

Figure 1.20
$f(x) = x^2 + 1$

Plot the ordered pairs $(x, f(x))$ and connect the points with a smooth curve.

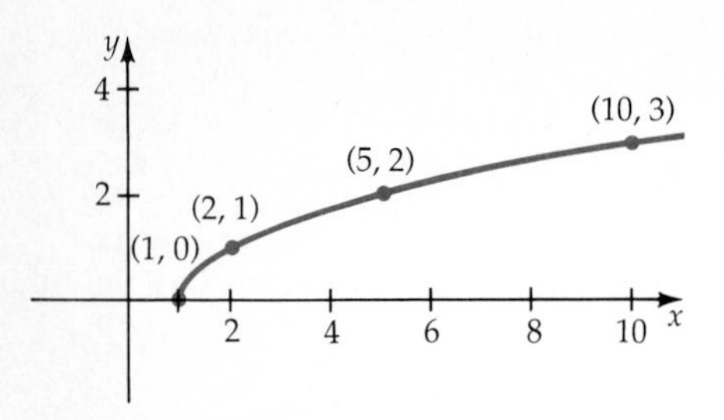

Figure 1.21
$f(x) = \sqrt{x - 1}$

EXAMPLE 1 Graph a Function

Sketch a graph of $f(x) = \sqrt{x - 1}$, $x \geq 1$.

Solution Recall that the symbol $\sqrt{}$ represents the positive square root of a number. Also, note that the domain of f is restricted to $x \geq 1$.

$f(x) = \sqrt{x - 1}$

When x is	1	2	5	10
$f(x)$ is	0	1	2	3

Plot the ordered pairs $(1, 0)$, $(2, 1)$, $(5, 2)$, and $(10, 3)$, and connect the points with a smooth curve.

■ *Try Exercise* **6**, *page 31.*

According to the definition of a function, there is only one point $(a, f(a))$ on the graph of f for each a in the domain of f. Geometrically this means that every vertical line passes through the graph of a function in at most one point. This property is known as the *vertical line test.*

Vertical Line Test

> A graph is the graph of a function if and only if no vertical line intersects the graph at more than one point.

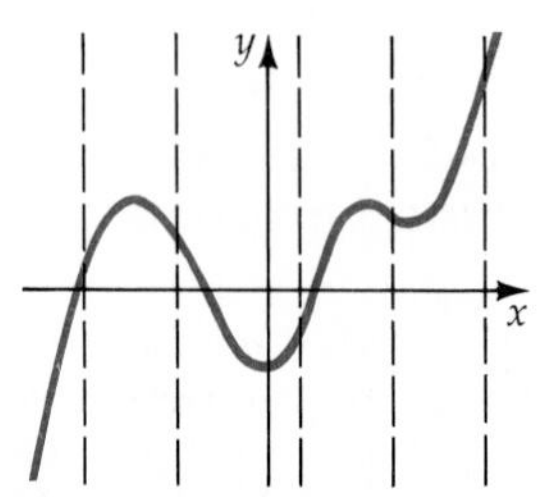

The graph of a function. Any vertical line intersects the graph only once.

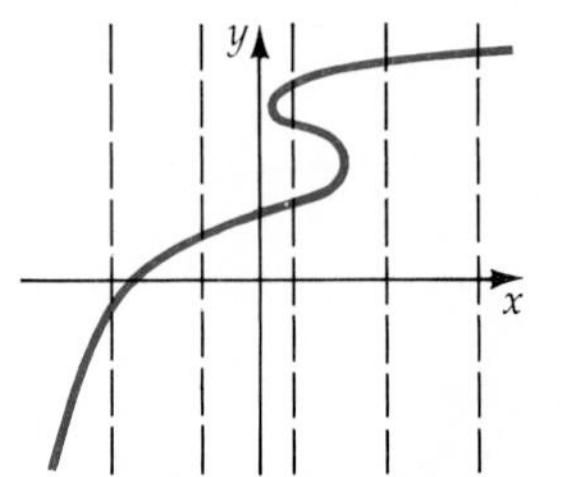

Not the graph of a function. A vertical line intersects the graph at more than one point.

Figure 1.22

EXAMPLE 2 Use the Vertical Line Test for a Function

Which of the graphs in Figure 1.23 on page 125 represent the graph of a function?

Solution

a. This is the graph of a function because every vertical line intersects the graph in exactly one point.
b. This is not the graph of a function because some vertical lines intersect the graph at more than one point.

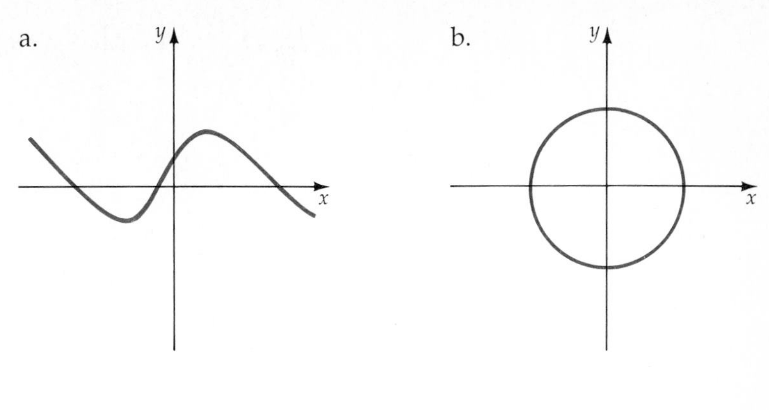

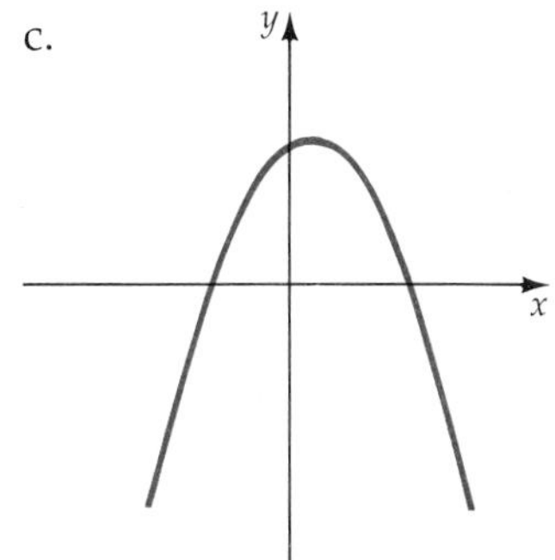

Figure 1.23

c. This is the graph of a function because every vertical line intersects the graph in exactly one point.

■ *Try Exercise* **16**, *page 31.*

Properties of Functions

A function is said to be decreasing if the values of f decrease as x increases. If the values of f increase as x increases, then the function is increasing. When the values of f remain constant as x increases, the function is said to be constant. These ideas are expressed in the next definition.

Decreasing, Increasing, and Constant Function

Let f be a function defined on the interval $a \leq x \leq b$, and let x_1 and x_2 be two points in the interval.

1. f is **increasing** on the interval if $f(x_1) < f(x_2)$ whenever $x_1 < x_2$.
2. f is **decreasing** on the interval if $f(x_1) > f(x_2)$ whenever $x_1 < x_2$.
3. f is **constant** on the interval if $f(x_1) = f(x_2)$ for every x_1 and x_2 in.

A function that decreases (increases) on the domain of the function is a **decreasing (increasing) function.**

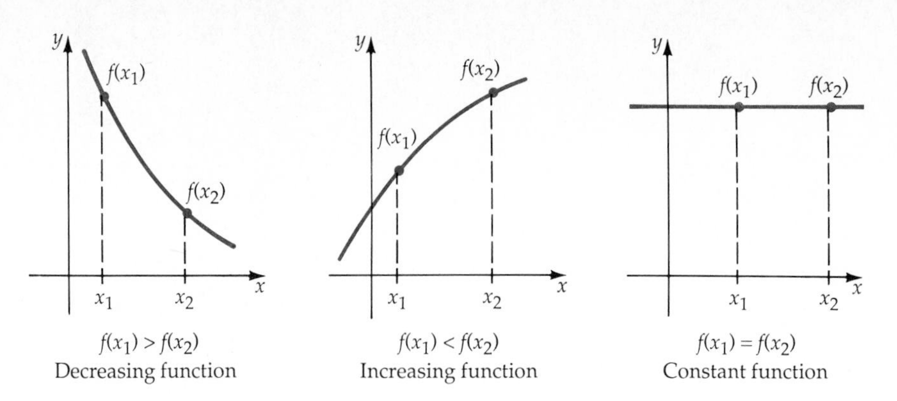

Figure 1.24

The graphs in Figure 1.24 show examples of decreasing, increasing and constant functions.

Any horizontal line intersects the graph of an increasing or decreasing function at exactly one point. This means that given any value b in the range of the function f, there is exactly one c in the domain of f such that $f(c) = b$. A function that satisfies this condition is called a *one-to-one function.*

Definition of a One-To-One Function

> Let f be a function with domain D and range R. Then f is a **one-to-one function** if given any element b in R, there is exactly one element c in D such that $f(c) = b$.

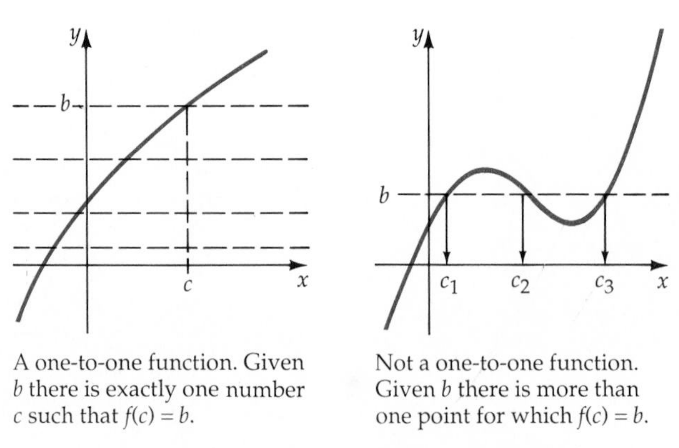

A one-to-one function. Given b there is exactly one number c such that $f(c) = b$.

Not a one-to-one function. Given b there is more than one point for which $f(c) = b$.

Figure 1.25

Remark All increasing functions and all decreasing functions are one-to-one functions.

In a manner similar to the vertical line test, we can state a horizontal line test for one-to-one functions.

Horizontal Line Test for a One-To-One Function

> If any horizontal line intersects a graph of a function at most once, then the graph is the graph of a one-to-one function.

EXAMPLE 3 **Use the Horizontal Line Test**

Which of the graphs represent a one-to-one function?

a.

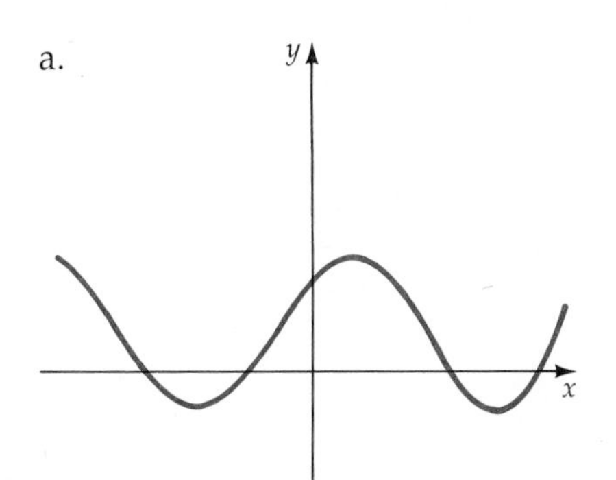

b.

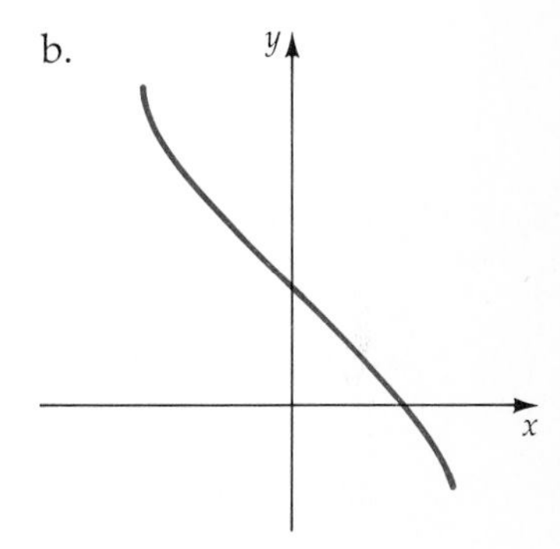

Figure 1.26

Solution

a. Because some horizontal lines intersect the graph at more than one point, the graph is not the graph of a one-to-one function.
b. Because any horizontal line intersects the graph at exactly one point, the graph is the graph of a one-to-one function.

■ *Try Exercise* **28,** *page 32*

Two additional classes of functions are called *even* or *odd* functions.

Even and Odd Functions

> A function f is an **even function** if $f(-x) = f(x)$ for all x in the domain of the function. A function f is an **odd function** if $f(-x) = -f(x)$ for all x in the domain of the function.

EXAMPLE 4 **Determine Even and Odd Functions**

Determine if the given functions are even, odd, or neither.

a. $f(x) = 2x^4 - x^2$ b. $f(x) = 2x^3 + 3x^2$ c. $f(x) = \dfrac{x}{x^2 + 1}$

Solution To determine if a function is even or odd, replace x by $-x$ and then simplify.

a. $$f(x) = 2x^4 - x^2$$
$$f(-x) = 2(-x)^4 - (-x)^2 = 2x^4 - x^2 = f(x)$$

The function is an even function.

b. $$f(x) = 2x^3 + 3x^2$$
$$f(-x) = 2(-x)^3 + 3(-x)^2 = -2x^3 + 3x^2$$

$-2x^3 + 3x^2 \neq f(x)$ and $-2x^3 + 3x^2 \neq -f(x)$, therefore the function is neither even nor odd.

c. $$f(x) = \frac{x}{x^2 + 1}$$
$$f(-x) = \frac{(-x)}{(-x)^2 + 1} = \frac{-x}{x^2 + 1} = -\left(\frac{x}{x^2 + 1}\right) = -f(x)$$

The function is an odd function.

■ *Try Exercise* **36**, *page 32.*

The graph of an even function is **symmetric with respect to the y-axis.** That is, if the coordinate plane were folded along the y-axis, then the graphs on each side would coincide.

The graph of an odd function is **symmetric with respect to the origin.** This means that for all x in the domain of the function, the points (x, y) and $(-x, -y)$ are on the graph.

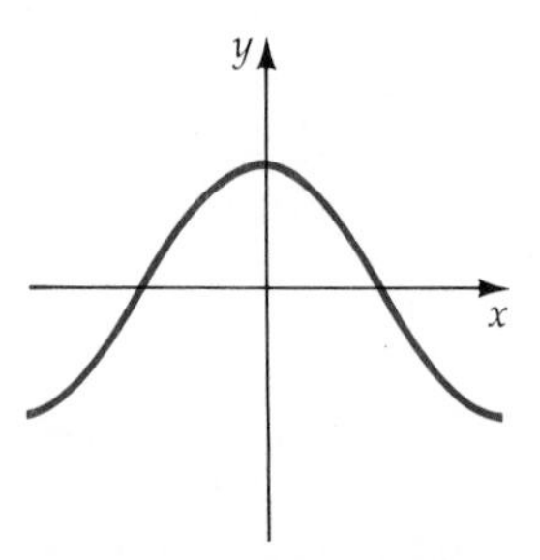

The graph of an even function. The graph is symmetric with respect to the y-axis.

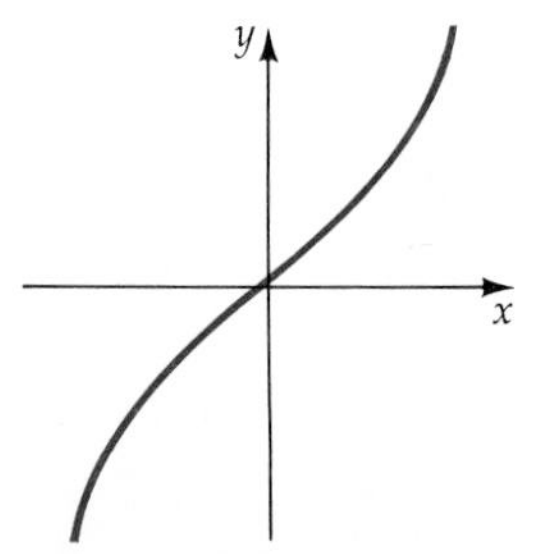

The graph of an odd function. The graph is symmetric with respect to the origin.

Figure 1.27

A graph can also exhibit **symmetry with respect to the x-axis.** That is, if the coordinate plane were folded along the x-axis, then the graphs on each side would coincide. See Figure 1.28. By the vertical line test, the graph would not be the graph of a function.

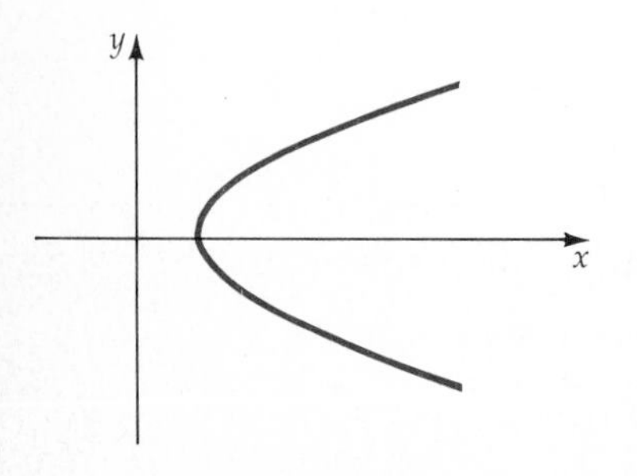

Figure 1.28
x-axis symmetry

Translations of the Graph of a Function

A graph may be exactly the same as another graph except for its position relative to the origin of the coordinate plane. For example, compare

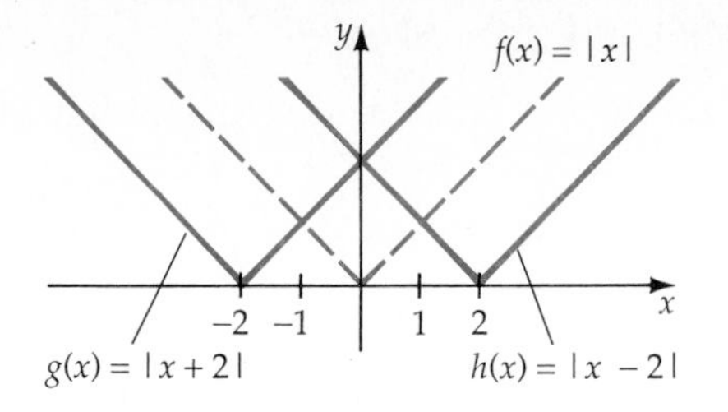

Figure 1.29
Horizontal Translations of $f(x) = |x|$

the graph of $f(x) = |x|$ with that of $g(x) = |x + 2|$ and $h(x) = |x - 2|$ in Figure 1.29. Note that the graphs are exactly the same except that each point of the graph of g is two units to the left of the graph of f. Each point on the graph of h is two units to the right of the graph of f. We say that the graphs of g and h are *horizontal translations* or *horizontal shifts* of the graph of f.

Horizontal Translation of a Graph

> Given a function defined by $y = f(x)$ and a positive constant c, then the graph of the function defined by:
>
> - $y = f(x - c)$ is the graph of $y = f(x)$ shifted horizontally c units to the right;
> - $y = f(x + c)$ is the graph of $y = f(x)$ shifted horizontally c units to the left.

Now compare the graph of $f(x) = x^2$ with that of $g(x) = x^2 - 2$ and $h(x) = x^2 + 2$ in Figure 1.30. The graphs are exactly the same except that each point of the graph of g is two units below the graph of f and each point on the graph of h is two units above the graph of f. The graphs of g and h are *vertical translations* or *vertical shifts* of the graph of f.

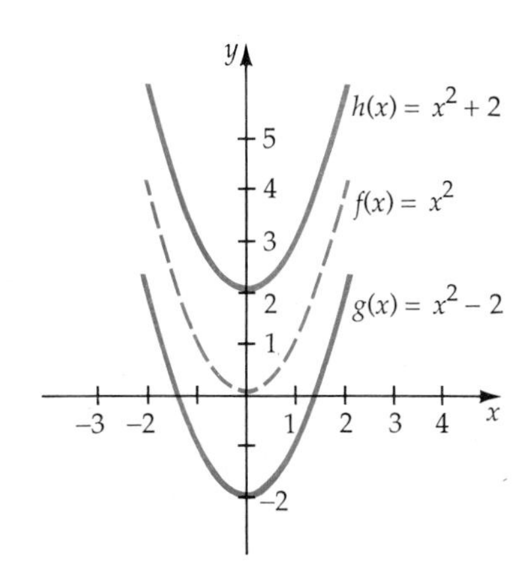

Figure 1.30
Vertical Translations of $f(x) = x^2$

Vertical Translation of a Graph

> Given a function defined by $y = f(x)$ and a positive constant c, the graph of the function defined by
>
> - $y = f(x) - c$ is the graph of $y = f(x)$ shifted vertically c units down;
> - $y = f(x) + c$ is the graph of $y = f(x)$ shifted vertically c units up.

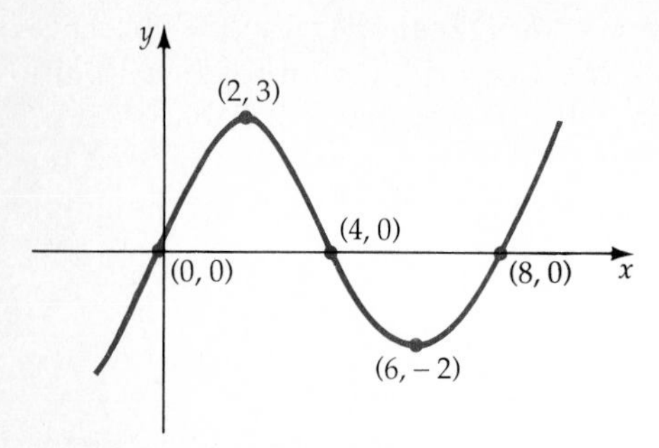

Figure 1.31

EXAMPLE 5 **Sketch a Graph Using Translations**

Use the graph of $y = f(x)$ shown in Figure 1.31 to sketch the graph of each of the following.

a. $y = f(x + 1)$ b. $y = f(x) - 1$ c. $y = f(x - 3) + 1$

Solution

a.

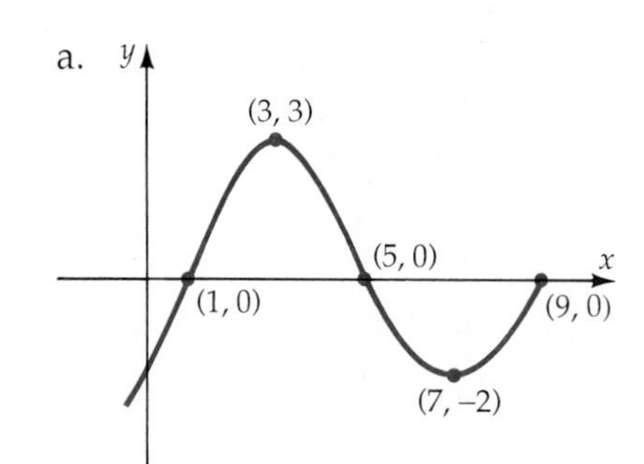

b.

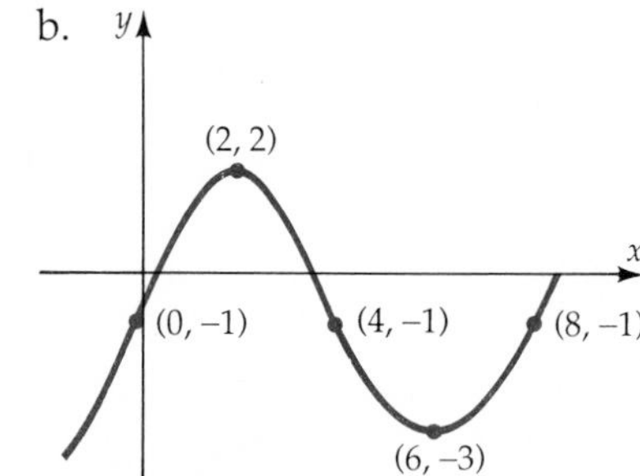

c.

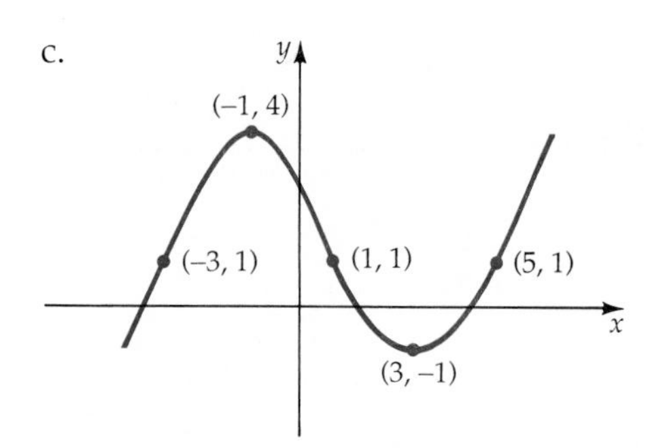

Figure 1.32

■ *Try Exercise* **46,** *page 32.*

The graph of a function gives us a way to picture the range of a function. Consider the graph of the function defined by $f(x) = 1/(x - 1)$ in Figure 1.33. To find the range of the function, think of all the values $f(x)$ being projected onto the y-axis. The only point excluded as a value of y is $y = 0$. The range of the function is all real numbers except 0. This can be proved algebraically by asking "Is there a value of c in the domain D of f such that $f(c) = 0$?" That is, is there a c in D such that $f(c) = 0$? Since

$$f(c) = \frac{1}{c - 1},$$

we have

$$\frac{1}{c - 1} = 0.$$

This equation has no solution. Thus, 0 is not in the range of f.

Figure 1.33

From the graph of each function in Figure 1.34, we can determine the range of the function. The range of the graphs of the functions in are, respectively, $y \geq -2$, $0 \leq y \leq 2$, and the set of real numbers.

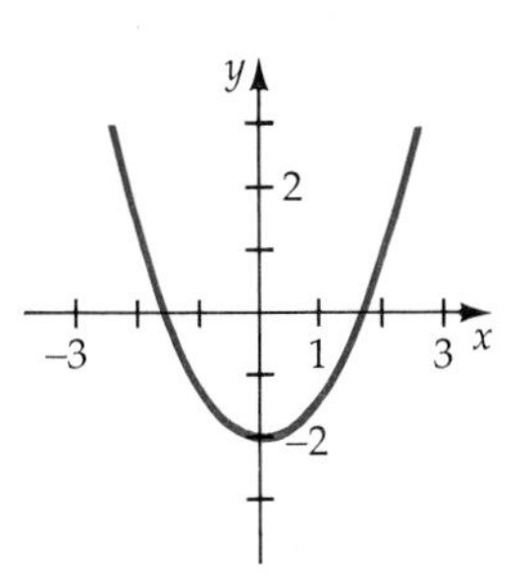

The range is all real numbers y where $y \geq -2$.

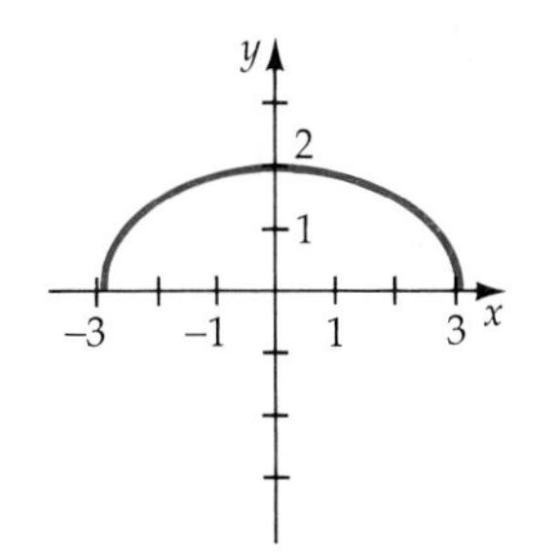

The range is all real numbers y where $0 \leq y \leq 4$.

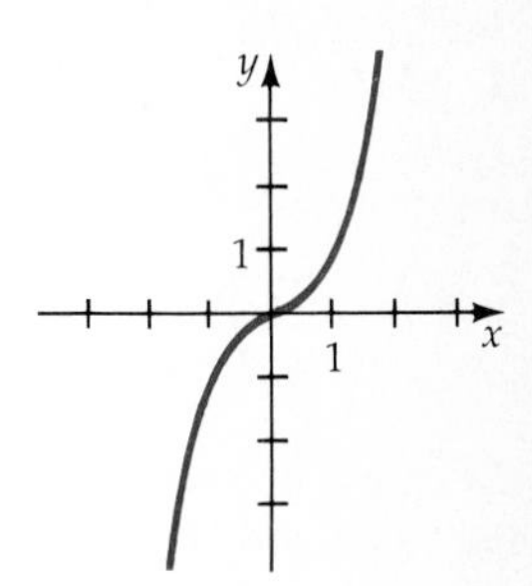

The range is all real numbers.

Figure 1.34

EXERCISE SET 1.4

In Exercises 1 to 10, sketch a graph of each function.

1. $f(x) = 4 - 2x$

2. $f(x) = 5x - 4$

3. $f(x) = -x^2$

4. $f(x) = 2x^2$

5. $f(x) = x^2 - x$

6. $f(x) = x^2 + 4x$

7. $f(x) = |2x - 8|$

8. $f(x) = |3x + 6| - 1$

9. $f(x) = -2|x + 1|$

10. $f(x) = -2x^2 - 4x + 3$

In Exercises 11 to 20, determine if the graph represents a function.

11.

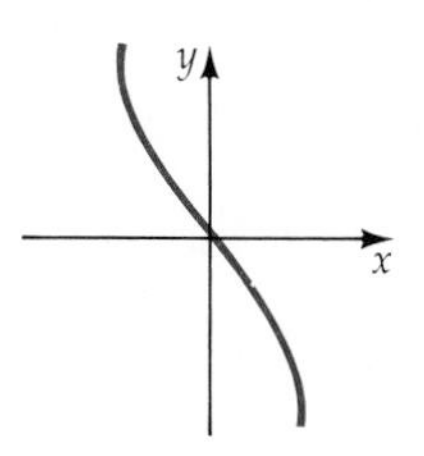

12.

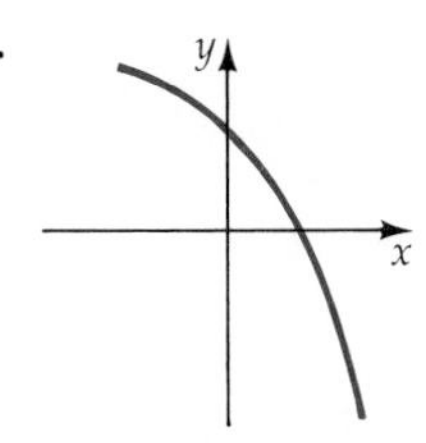

13.

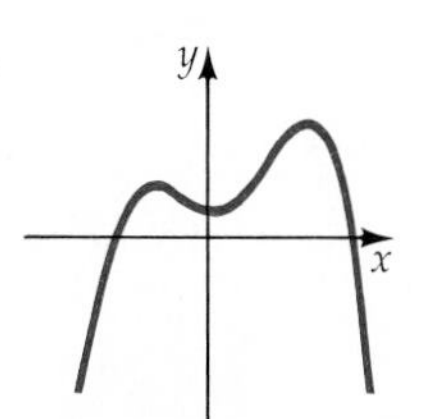

14.

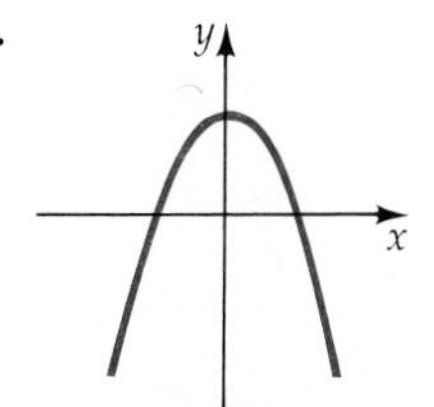

15.

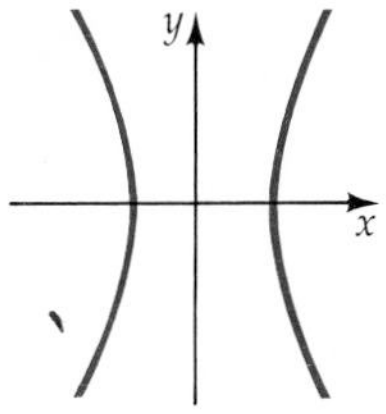

16.

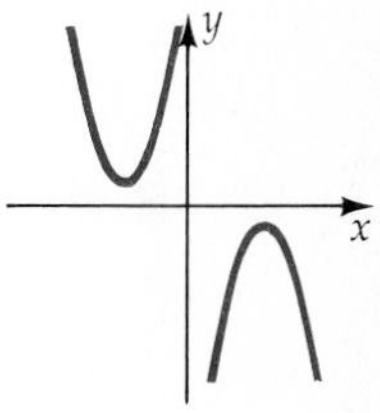

17.

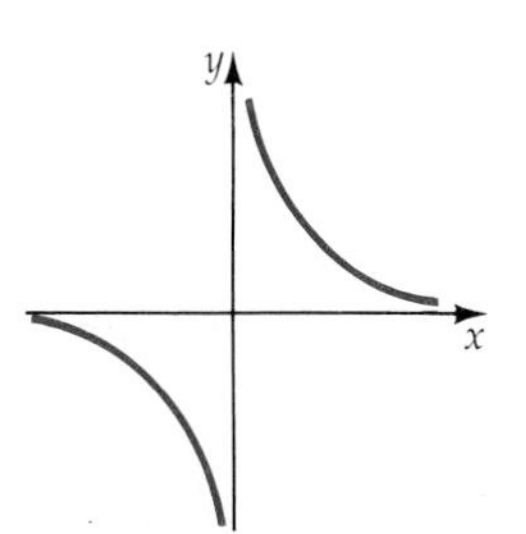

18.

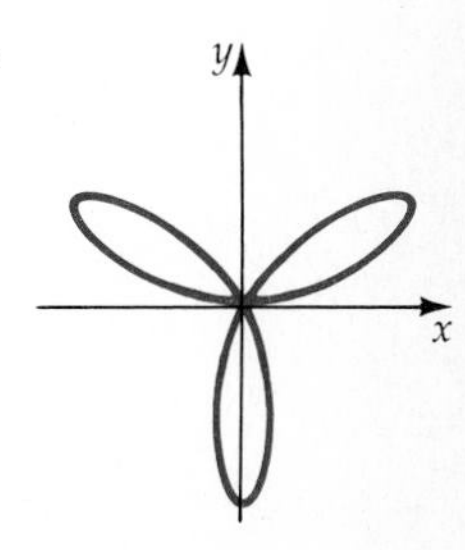

19.

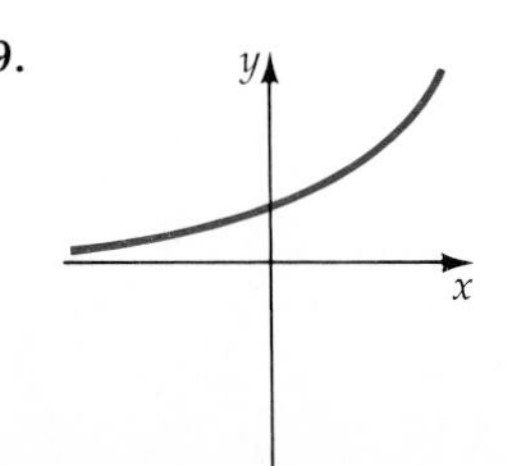

20.

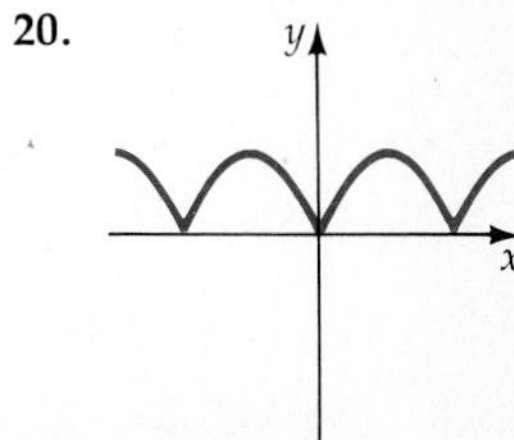

21–30. Determine which of the graphs in Exercises 11 to 20 represent a one-to-one function.

In Exercises 31 to 40, determine if the function is even, odd, or neither.

31. $f(x) = |x|$
32. $f(x) = x + 1$
33. $f(x) = x^3 - x$
34. $f(x) = x^2$
35. $f(x) = 3x - 5$
36. $f(x) = |x - 2|$
37. $f(x) = \dfrac{x^2}{x^3 + x}$
38. $f(x) = \dfrac{x^2}{x^2 + 1}$
39. $f(x) = 2|x| - 1$
40. $f(x) = x^3 + 3$

The graph of a function f is shown below. In Exercises 41 to 46, sketch the graph.

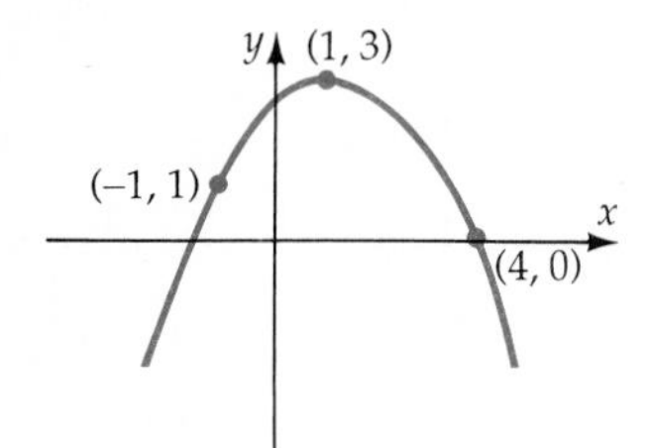

41. $f(x + 1)$
42. $f(x - 3)$
43. $f(x) - 2$
44. $f(x) + 1$
45. $f(x - 2) - 1$
46. $f(x + 3) + 2$

The graph of a function f is shown below. In Exercises 47 to 52, sketch the graph.

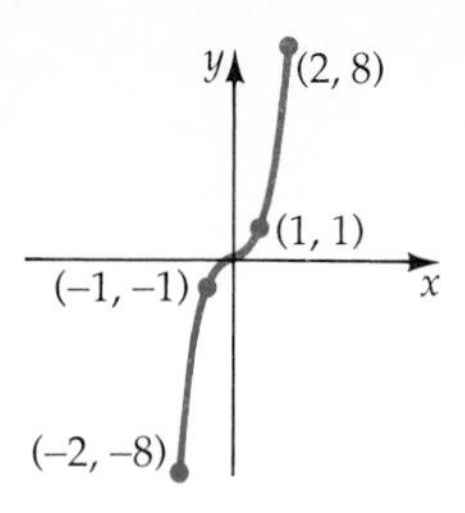

47. $f(x - 1)$
48. $f(x + 2)$
49. $f(x) + 3$
50. $f(x) - 2$
51. $f(x + 2) - 1$
52. $f(x - 3) - 3$

Supplemental Exercises

In Exercises 53 to 56, determine if the given function is even, odd, or neither.

53. $f(x) = x + |x|$
54. $f(x) = x - |x|$
55. $f(x) = x \cdot |x|$
56. $f(x) = \dfrac{x}{|x|}, x \neq 0$

57–60. Graph the functions defined in Exercises 53 to 56.

1.5 Composition of Functions and Inverse Functions

Another way to combine functions is by forming the *composition* of two functions.

Consider the two functions $f(x) = x^2 + 4$ and $g(x) = 2x - 3$ shown in Figure 1.35.

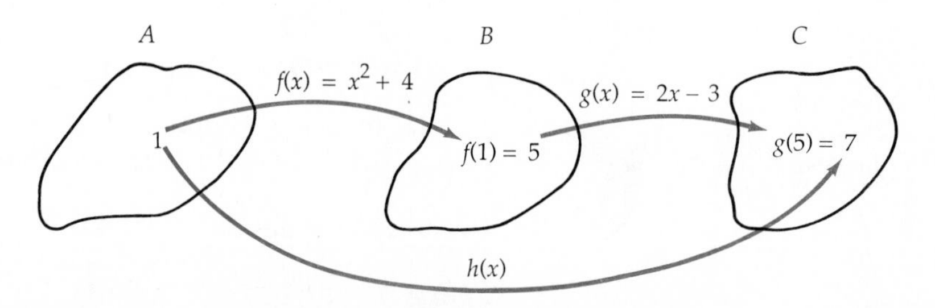

Figure 1.35

The function f pairs the number 1 in A with the number 5 in B. The function g pairs the number 5 in B with the number 7 in C. The composition of the two functions is the single function h that would pair 1 in A with 7 in C.

To find the function h, we will consider the element a in A.

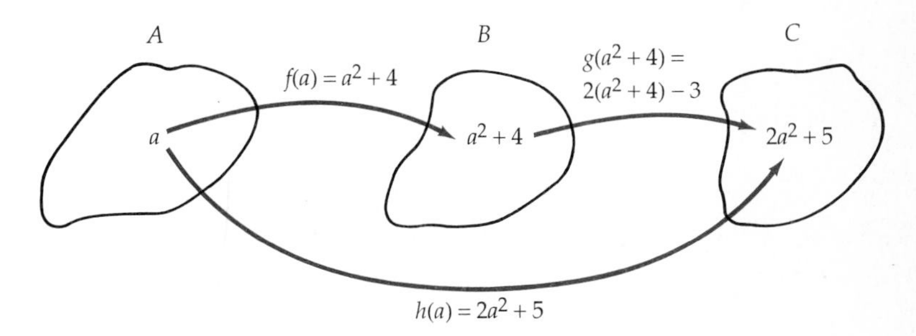

Figure 1.36

In general, we can find the composition function as follows:

$$\begin{aligned} h(x) &= g[f(x)] \\ &= g[x^2 + 4] \\ &= 2(x^2 + 4) - 3 \\ &= 2x^2 + 8 - 3 \\ &= 2x^2 + 5 \end{aligned}$$

Composition of Functions

> The **composition** of two functions g and f, written $(g \circ f)(x)$, is given by
>
> $$(g \circ f)(x) = g(f(x))$$
>
> for all x in the domain of f such that $f(x)$ is in the domain of g.

Remark Do not confuse composition of functions with multiplication of functions. For composition an open circle $\circ$ is used. For multiplication a dot $\cdot$ is used.

EXAMPLE 1 Find the Composition of Two Functions

Given the functions defined by $f(x) = 2x + 1$ and $g(x) = x^2 - x$ find
a. $f(g(3))$ b. $g(f(-2))$ c. $(f \circ g)(x)$ d. $(g \circ f)(x)$

Solution

a. $g(3) = 3^2 - 3 = 9 - 3 = 6$

$f(g(3)) = f(6) = 2(6) + 1 = 13$

b. $f(-2) = 2(-2) + 1 = -3$

$$g(f(-2)) = g(-3) = (-3)^2 - (-3) = 12$$

c. $(f \circ g)(x) = f(g(x)) = 2(g(x)) + 1 = 2(x^2 - x) + 1$

$$= 2x^2 - 2x + 1$$

d. $(g \circ f)(x) = g(f(x)) = [f(x)]^2 - [f(x)]$

$$= (2x + 1)^2 - (2x + 1) = 4x^2 + 2x$$

■ *Try Exercise* **4,** *page 39.*

Remark Note that in Example 1c and 1d, $(f \circ g)(x) \neq (g \circ f)(x)$. In general, the composition of two functions is not a commutative operation.

Inverse Functions

For many operations in mathematics, there are inverse operations. The inverse of addition is subtraction, and the inverse of multiplication is division.

It also is possible to define the inverse of a function. We do this relying on the second or alternate definition of a function stated in Section 4.

Definition of a Function

A **function** f is a set of ordered pairs (x, y) such that no two ordered pairs with the same first component have different second components.

We can define an inverse function by interchanging the coordinates of f as follows.

Definition of the Inverse Function

Let f be a function with an inverse. Then the **inverse function** is the set of ordered pairs of f with the coordinates interchanged.

Consider the functions defined by $f(x) = x - 1$. Some of the ordered pairs of f are $(-1, -2)$, $(0, -1)$, $(1, 0)$, and $(3, 2)$. The inverse of f contains the ordered pairs $(-2, -1)$, $(-1, 0)$, $(0, 1)$ and $(2, 3)$.

Now consider the function $g(x) = |x| - 1$. Some of the ordered pairs of g are $(-2, 1)$, $(-1, 0)$, $(0, -1)$, $(1, 0)$ and $(2, 1)$. By interchanging the coordinates of the ordered pairs, we have $(1, -2)$, $(0, -1)$, $(-1, 0)$, $(0, 1)$, and $(1, 2)$. These ordered pairs do *not* satisfy the definition of a function, because there are ordered pairs with the same first coordinates and *different* second coordinates. For example, the ordered pairs $(1, -2)$ and $(1, 2)$.

This example illustrates that not all functions have an inverse function. Figure 1.37 shows the graphs of the functions f and g. Note that f is

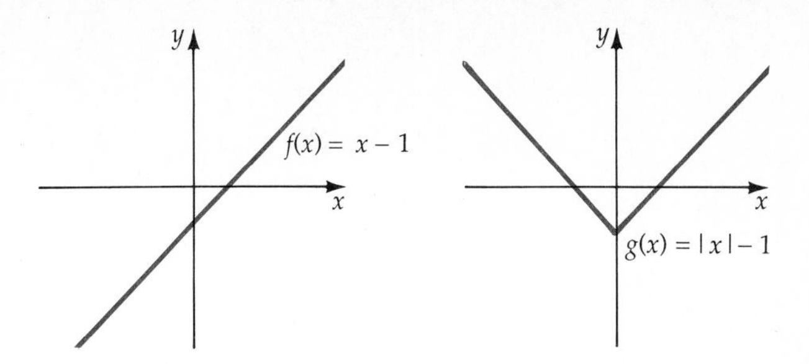

Figure 1.37

the graph of a one-to-one function, but g is not the graph of a one-to-one function.

Condition for a Function to Have an Inverse

A function f will have an inverse function if and only if f is a one-to-one function. The inverse of the function f is denoted f^{-1}.

Caution The notation for an inverse function is sometimes confusing. The function f^{-1} is the inverse of the function f and is not the same as $1/f$. The function $1/f$ is called the **reciprocal function** and is an entirely different function.

To find the inverse of the function defined by $f(x) = 2x - 6$, begin by replacing $f(x)$ by y.

$$y = 2x - 6$$

Now interchange x and y.

$$x = 2y - 6$$

Solve for y.

$$x + 6 = 2y$$

$$\frac{x + 6}{2} = y$$

or,

$$y = \frac{1}{2}x + 3$$

The inverse function is defined by

$$f^{-1}(x) = \frac{1}{2}x + 3.$$

Figure 1.38 shows the graphs of a function f and the inverse function

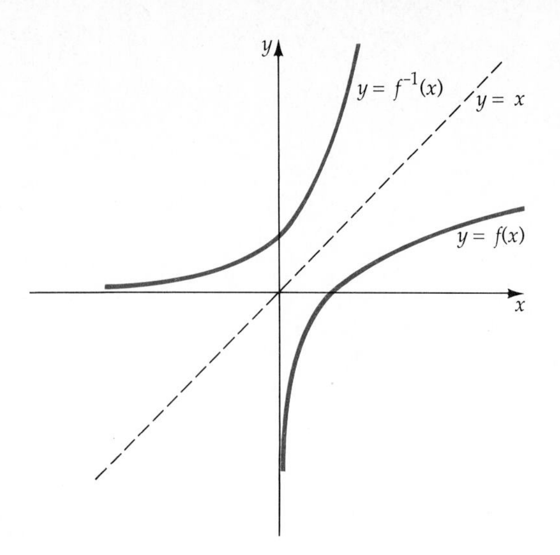

Figure 1.38

f^{-1}. Note that the graphs are symmetric with respect to the line $y = x$. This is true for a function and its inverse.

Symmetry Property of f and f^{-1}

> The graph of a function f and the graph of the inverse function f^{-1} are symmetric with respect to the line $y = x$.

EXAMPLE 2 **Find the Inverse of a Function**

Find the inverse of the function defined by $f(x) = 3x + 9$.

Solution

$$f(x) = 3x + 9$$

$$y = 3x + 9$$

$$x = 3y + 9 \quad \text{Interchange } x \text{ and } y.$$

$$x - 9 = 3y \quad \text{Solve for } y.$$

$$\frac{x - 9}{3} = y$$

$$f^{-1}(x) = \frac{1}{3}x - 3$$

■ *Try Exercise* **14,** *page 39.*

EXAMPLE 3 Find the Inverse of a Function

Find the inverse of the function defined by $f(x) = \dfrac{2x}{x+3}$, $x \neq -3$.

Solution

$$f(x) = \frac{2x}{x+3}$$

$$y = \frac{2x}{x+3}$$

$$x = \frac{2y}{y+3} \quad \text{Interchange } x \text{ and } y.$$

$$x(y+3) = 2y$$

$$xy + 3x = 2y$$

$$3x = 2y - xy = (2-x)y$$

$$\frac{3x}{2-x} = y$$

$$f^{-1}(x) = \frac{3x}{2-x}$$

■ *Try Exercise* **18,** *page 39.*

Consider the function f and f^{-1} in Example 2. By forming the composition of the two functions we have

$$\begin{aligned}(f \circ f^{-1})(x) &= f[f^{-1}(x)] \\ &= f\left(\frac{1}{3}x - 3\right) \\ &= 3\left(\frac{1}{3}x - 3\right) + 9 \\ &= x - 9 + 9 \\ &= x\end{aligned}$$

Similarly,

$$\begin{aligned}(f^{-1} \circ f)(x) &= f^{-1}[f(x)] \\ &= f^{-1}(3x + 9) \\ &= \frac{1}{3}(3x + 9) - 3 \\ &= x + 3 - 3 \\ &= x\end{aligned}$$

This result can be generalized to any function that has an inverse function.

Condition for an Inverse Function

> A function f has an inverse function f^{-1} if and only if
> $$(f \circ f^{-1})(x) = (f^{-1} \circ f)(x) = x.$$

The function defined by

$$f(x) = x^2 + 1$$

is not a one-to-one function, and therefore does not have an inverse function. It is possible, however, to restrict the domain of f so that an inverse of f with the restricted domain exists.

Define a function f as

$$f(x) = x^2 + 1, \quad x \geq 0.$$

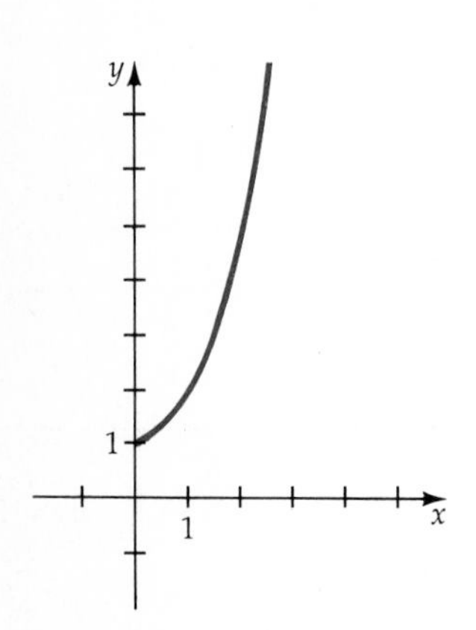

Figure 1.39

Thus,

domain of the function f: $x \geq 0$
range of the function f: $y \geq 1$

Because the coordinates of the ordered pairs of the inverse function are reversed, the domain of the inverse function is the range of the function. The range of the inverse function is the domain of the function. Thus, for the inverse function f^{-1},

domain of the inverse function f^{-1}: $x \geq 1$
range of the inverse function f^{-1}: $y \geq 0$

To find the inverse of this function, we proceed as before.

$$f(x) = x^2 + 1$$
$$y = x^2 + 1$$
$$x = y^2 + 1 \quad \text{Interchange } x \text{ and } y.$$
$$x - 1 = y^2$$
$$\pm\sqrt{x - 1} = y$$

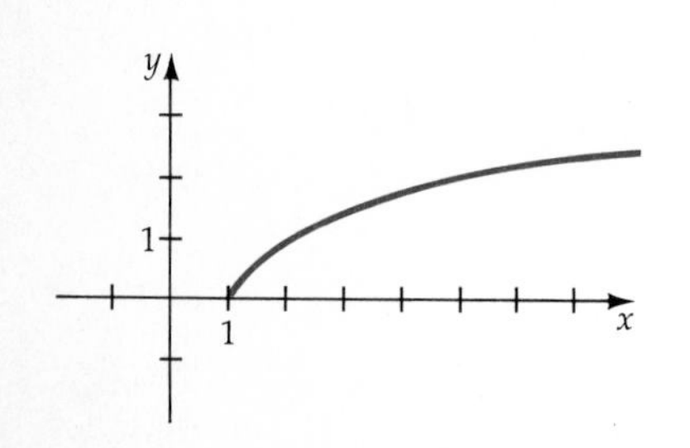

Figure 1.40

Because the range of the inverse function is all nonnegative real numbers ($y \geq 0$), we can remove the $-$ sign from the $\pm$. The inverse function is then defined by

$$f^{-1}(x) = \sqrt{x - 1}.$$

This method of restricting the domain of a function so that an inverse function exists is used often in trigonometry.

EXAMPLE 4 Form the Composition of a Function and Its Inverse Function

Show that the functions f and f^{-1} defined as $f(x) = x^2 + 1$, $x \geq 0$ and $f^{-1}(x) = \sqrt{x - 1}$, $x \geq 1$, are inverses of each other.

Solution

$$\begin{aligned}(f \circ f^{-1})(x) &= f[f^{-1}(x)] \\ &= f(\sqrt{x-1}) \\ &= (\sqrt{x-1})^2 + 1 \\ &= x - 1 + 1 \\ &= x\end{aligned} \qquad \begin{aligned}(f^{-1} \circ f)(x) &= f^{-1}[f(x)] \\ &= f^{-1}(x^2+1) \\ &= \sqrt{(x^2+1) - 1} \\ &= \sqrt{x^2} \\ &= x\end{aligned}$$

■ *Try Exercise* **30,** *page 39.*

EXERCISE SET 1.5

In Exercises 1 to 10, use the functions defined by the equations $f(x) = x^2 - x$, $g(x) = 1/x$, and $k(x) = 2x - 3$ to find each of the following.

1. $f(k(3))$ **2.** $k(f(3))$ **3.** $g(k(2))$ **4.** $k(g(2))$

5. $k(k(-1))$ **6.** $g(g(2))$ **7.** $f(g(a))$ **8.** $g(f(a))$

9. $f(k(x))$ **10.** $k(f(x))$

In Exercises 11 to 26, find the inverse of the given function.

11. $f(x) = 2x - 4$ **12.** $g(t) = 3t$

13. $u(v) = 1 - 3v$ **14.** $H(r) = 5r + 10$

15. $f(z) = z^2,\ z \geq 0$ **16.** $y(x) = x^3$

17. $p(q) = \dfrac{1}{q}$ **18.** $G(s) = \dfrac{3}{s-2},\ s \neq 2$

19. $r(t) = \sqrt{t+1},\ t \geq -1$ **20.** $F(z) = \sqrt{1-z},\ z \leq 1$

21. $H(s) = s^2 - 4,\ s \geq 0$ **22.** $f(x) = 1 - x^2,\ x \geq 0$

23. $P(v) = \sqrt[3]{v} - 1$ **24.** $u(t) = \sqrt[3]{t-2}$

25. $f(x) = \dfrac{x}{x-1},\ x \neq 1$ **26.** $f(x) = \dfrac{x+1}{x},\ x \neq 0$

In Exercises 27 to 32, show that the given pair of functions are inverses of each other.

27. $g(t) = 2t - 4,\ g^{-1}(t) = \dfrac{1}{2}t + 2$

28. $h(v) = \dfrac{1}{3}v + 2,\ h^{-1}(v) = 3v - 6$

29. $f(x) = x^2 + 2,\ x \geq 0$
$f^{-1}(x) = \sqrt{x-2},\ x \geq 2$

30. $G(s) = s^2 - 4,\ s \geq 0$
$G^{-1}(s) = \sqrt{s+4},\ s \geq -4$

31. $F(x) = \dfrac{1}{x+1},\ x \neq -1$
$F^{-1}(x) = \dfrac{1-x}{x},\ x \neq 0$

32. $H(x) = \dfrac{1}{x} + 1,\ x \neq 0$
$H^{-1}(x) = \dfrac{1}{x-1},\ x \neq 0$

In Exercises 33 to 42, graph each pair of functions on the same coordinate plane. Note that the graphs of the functions are symmetric with respect to the line $y = x$.

33. $f(x) = 3x + 3,\ f^{-1}(x) = \dfrac{1}{3}x - 1$

34. $f(x) = x - 4,\ f^{-1}(x) = x + 4$

35. $f(x) = \dfrac{1}{2}x,\ f^{-1}(x) = 2x$

36. $f(x) = 2x - 4,\ f^{-1}(x) = \dfrac{1}{2}x + 2$

37. $f(x) = x^2 + 2,\ x \geq 0$
$f^{-1}(x) = \sqrt{x-2},\ x \geq 2$

38. $f(x) = x^2 - 3,\ x \geq 0$
$f^{-1}(x) = \sqrt{x+3},\ x \geq -3$

39. $f(x) = (x-2)^2,\ x \geq 2$
$f^{-1}(x) = \sqrt{x} + 2,\ x \geq 0$

40. $f(x) = (x+3)^2,\ x \geq -3$
$f^{-1}(x) = \sqrt{x} - 3,\ x \geq 0$

41. $f(x) = \sqrt{x+1},\ x \geq -1$
$f^{-1}(x) = x^2 - 1,\ x \geq 0$

42. $f(x) = \sqrt{x} - 2,\ x \geq 0$
$f^{-1}(x) = (x+2)^2,\ x \geq -2$

Supplemental Exercises

In Exercises 43 to 48, find the inverse of each function.

43. $f(x) = x^2 - 4x + 1,\ x \geq 2$ (*Hint:* Write $y = x^2 - 4x + 1$ and complete the square of $x^2 - 4x$.)

44. $f(x) = x^2 + 6x - 6,\ x \geq -3$ (*Hint:* Write $y = x^2 + 6x - 6$ and complete the square of $x^2 + 6x$.)

45. $f(x) = x^2 + 8x - 9,\ x \geq -4$ (*Hint:* Write $y = x^2 + 8x - 9$ and complete the square of $x^2 + 8x$.)

46. $f(x) = x^2 - 2x - 2,\ x \geq 1$ (*Hint:* Write $y = x^2 - 2x - 2$ and complete the square of $x^2 - 2x$.)

47. $f(x) = \dfrac{x-1}{x+1},\ x \neq -1$

48. $f(x) = \dfrac{2-x}{x+2},\ x \neq -2$

49. If the ordered pair (a, b) is on the graph of the function f and f has an inverse, then (b, a) is on the graph of f^{-1}. Show that the points $P(a, b)$ and $Q(b, a)$ are symmetric with respect to the line $y = x$ by showing that the midpoint of the line segment PQ lies on the line $y = x$.

Just as it is possible to multiply more than two numbers together, it is also possible to find the composition of more than two functions. If f, g and h are functions, we define

$$(f \circ g \circ h)(x) = f[g(h(x))].$$

In Exercises 50 to 55, use this definition and the functions defined by

$$f(x) = 2x + 1, \quad g(x) = x - 3, \quad \text{and} \quad h(x) = 3x + 4.$$

50. Find $f[g(h(1))]$

51. Find $f[g(h(0))]$

52. Find $(f \circ g \circ h)(-1)$

53. Find $(f \circ g \circ h)(-2)$

54. Find $(f \circ g \circ h)(x)$

55. Find $(g \circ h \circ f)(x)$

Chapter Review

1.1 Preliminaries

The Pythagorean Theorem states a relationship between the legs of a right triangle and the hypotenuse of the triangle. If a and b are the measures of the legs of a right triangle and c is the measure of the hypotenuse, then

$$a^2 + b^2 = c^2.$$

The distance between two points $P_1(x_1, y_1)$ and $P_2(x_2, y_2)$ in the xy-plane is given by the distance formula

$$d = \sqrt{(x_1 - x_2)^2 + (y_1 - y_2)^2}.$$

The midpoint M of the line segment joining the points $P_1(x_1, y_1)$ and $P_2(x_2, y_2)$ is given by the midpoint formula:

$$\left(\frac{x_1 + x_2}{2}, \frac{y_1 + y_2}{2}\right).$$

1.2 Graphs of Equations

The graph of an equation in two variables is the set of all points whose coordinates satisfy the equation.

A linear equation in two variables is an equation of the form $Ax + By = C$. The graph of a linear equation in two variables is a straight line.

A quadratic equation in two variables is an equation of the form $y = ax^2 + bx + c$, $a \neq 0$. The graph of a quadratic equation is a parabola that opens up when $a > 0$ and opens down when $a < 0$.

The standard form of the equation of a circle is given by the equation $(x - h)^2 + (y - k)^2 = r^2$, where the center of the circle is the point $C(h, k)$ and the radius is r.

The general form of the equation of a circle is given by the equation $x^2 + y^2 + ax + by + c = 0$.

1.3 Functions

A function f from a set D to a set R is a correspondence, or rule, that pairs each element of D with exactly one element of R. The set D is the domain of the function. The set R is the range of the function.

1.4 Graphs of Functions

The vertical line test for a function states that a graph is the graph of a function if and only if no vertical line intersects the graph at more than one point.

Given an interval $a \leq x \leq b$, a function f is increasing on the interval if $f(x_1) < f(x_2)$ whenever $x_1 < x_2$. A function f is decreasing on the interval if $f(x_1) > f(x_2)$ whenever $x_1 < x_2$. A function f is constant if $f(x_1) = f(x_2)$ for all x_1 and x_2 in the interval.

A function f is a one-to-one function if given any element b in the range of f, there is exactly one element a in the domain of f such that $f(a) = b$.

A function f is an even function if $f(x) = f(-x)$ for all x in the domain of f. A function f is an odd function if $f(-x) = -f(x)$ for all x in the domain of f.

1.5 Composition of Functions and Inverse Functions

The composition of two functions f and g is defined as $(g \circ f)(x) = g[f(x)]$ when $f(x)$ is in the domain of g.

Let f be a function with an inverse. The inverse function is the set of ordered pairs of f with the components reversed.

The graph of the function f and the graph of the function f^{-1} are symmetric with respect to the line $y = x$.

A function f has an inverse function f^{-1} if and only if $(f \circ f^{-1})(x) = x$ and $(f^{-1} \circ f)(x) = x$.

CHALLENGE EXERCISES

In Exercises 1 to 10, answer true or false. If the statement is false, give an example.

1. Every function has an inverse function.
2. Let f be any function. Then $f(a) = f(b)$ implies that $a = b$.
3. If f is a function given by the equation $f(x) = ax$, then $f(u + v) = f(u) + f(v)$ for all real numbers u and v.

4. As a generalization of Exercise 3, $f(u + v) = f(u) + f(v)$ for any function and for all real numbers u and v.
5. If $f(f^{-1}(x)) = f^{-1}(f(x)) = x$, then f and f^{-1} are inverse functions.
6. Suppose that $f(x) = f(x + 4)$ for all real numbers x. If $f(2) = 3$, then $f(18) = 3$.
7. For all functions f, $f^2(x) = f(f(x))$.
8. Let f be any function. Then for $f(b) \neq 0$ and $b \neq 0$, $\dfrac{f(a)}{f(b)} = \dfrac{a}{b}$.
9. If f is the function defined by $f(x) = x$ (this is called the **identity function**), then $f^{-1}(x) = x$.
10. For all functions f and g, $(f \circ g)(x) \neq (g \circ f)(x)$.

REVIEW EXERCISES

1. Find the distance between the points $P(-2, 3)$ and $Q(4, -2)$. Find the midpoint of the line segment connecting P and Q.
2. Find the distance between the points $P(2, -4)$ and $Q(-3, 5)$. Find the midpoint of the line segment connecting P and Q.
3. Let $P(x_1, y_1)$ be a point on the line $y = 2x - 1$. Find the distance between P and the point $(2, 1)$ in terms of x_1.
4. Let $P(x_1, y_1)$ be a point on the line $y = 3 - 3x$. Find the midpoint of the line segment from P to $(1, -1)$ in terms of x_1.

In Exercises 5 to 22, sketch the graph of the equation or function.

5. $3x - 2y = 12$
6. $y = -2x + 3$
7. $|y| = x$
8. $y = |x + 2| - 1$
9. $x^2 - 3x + 4 - y = 0$
10. $2y + x^2 - 4 = 0$
11. $|y| = |x|$
12. $y^2 = x$
13. $(x - 1)^2 + (y - 3)^2 = 4$
14. $(x + 3)^2 + (y + 1)^2 = 9$
15. $(x + 4)^2 + (y - 1)^2 = 0$
16. $(x - 3)^2 + (y + 6)^2 = -1$
17. $x^2 + y^2 - 2x + 2y - 2 = 0$
18. $x^2 + y^2 - 4x + 2y - 4 = 0$
19. $(x + 1)^2 + y^2 = 16$
20. $x^2 + (y - 2)^2 = 25$
21. $x^2 + y^2 - 6x - 4y - 12 = 0$
22. $x^2 + y^2 + 8x - 2y - 8 = 0$

In Exercises 23 to 28, find the x- and y-intercepts for the equations.

23. $5x - 2y = 10$
24. $2x + 3y = 4$
25. $y = x^2 + 4x - 12$
26. $y = x^2 - 3x - 4$
27. $y = x^2 - 2x + 3$
28. $y = x^2 + 4x + 4$

In Exercises 29 to 38, find the domain of each function.

29. $f(x) = (x - 2)^2 + 4$
30. $f(x) = 3x - 4$
31. $f(x) = \dfrac{3x}{x + 1}$
32. $f(x) = \dfrac{-2x}{x - 3}$
33. $f(x) = \dfrac{x - 1}{x^2 + 1}$
34. $f(x) = \dfrac{x^2}{x + 2}$
35. $f(x) = \sqrt{3 - 6x}$
36. $f(x) = \sqrt{2x + 7}$
37. $f(x) = \dfrac{x - 4}{x^2 - x - 6}$
38. $f(x) = \dfrac{2x + 1}{x^2 + 4x + 5}$

In Exercises 39 to 50, use the functions $f(x) = 1 - 2x$ and $g(x) = x^2 - 2x - 3$, and $h(x) = x^2$ to find the indicated value.

39. $f(-2)$
40. $g(0)$
41. $h(a + b)$
42. $(f - g)(4)$
43. $f(3) \cdot h(-2)$
44. $f[g(1)]$
45. $g[h(2)]$
46. $h[f(-1)]$
47. $f[g(x)]$
48. $g[h(x)]$
49. $(f \cdot g)(x)$
50. $f^2(x)$

In Exercises 51 to 62, sketch the graph of the function.

51. $f(x) = -2x^2$
52. $f(x) = 2x^2 + 1$
53. $f(x) = 3 - |x|$
54. $f(x) = -3|x - 1| + 2$
55. $f(x) = \sqrt{x + 1}$
56. $f(x) = \sqrt{2x - 2}$
57. $f(x) = (x + 2)^2$
58. $f(x) = -(x - 1)^2$
59. $f(x) = -2(x - 3)^2$
60. $f(x) = (2x - 1)^2$
61. $f(x) = (x + 2)^2 - 2$
62. $f(x) = -(x - 2)^2 + 1$

In Exercises 63 to 68, use the graph of the function f below to sketch the graph of the function.

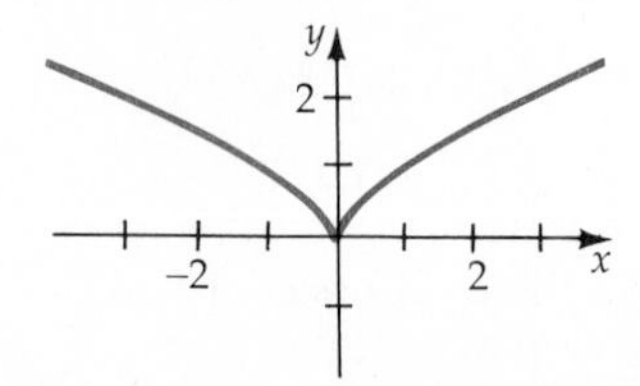

63. $f(x + 2)$

64. $f(x - 3)$

65. $f(x) - 1$

66. $f(x) + 2$

67. $f(x - 1) + 2$

68. $f(x + 2) - 3$

69–80. Use the horizontal line test to determine which of the graphs in Exercises 51 to 62 are the graphs of one-to-one functions.

In Exercises 81 to 86, find the inverse of each function f. Show that $(f \circ f^{-1})(x) = (f^{-1} \circ f)(x) = x$.

81. $f(x) = 2x - 8$

82. $f(x) = 3x + 3$

83. $f(x) = \dfrac{5}{x + 4}$

84. $f(x) = \sqrt{x - 2}, \quad x \geq 2$

85. $f(x) = (x + 1)^2, \quad x \geq -1$

86. $f(x) = 2x^2 - 4, \quad x \geq 0$

87. An oil spill is spreading in a circular pattern whose radius at time t is $2t^2$ meters. Find the area of the oil spill as a function of time.

88. An apple orchard has thirty trees per acre, and the average yield is two hundred fifty quarts of apples per tree. For each additional tree planted per acre, the yield per tree will decrease by five quarts. Write a function to describe the yield per acre as a function of the number of trees over 30.

89. A rectangular play area is to be constructed along the side of a house by building a fence on three sides, using the house wall as the fourth side. If the perimeter of the fence is 30 feet, find a function that describes the area of the play area as a function of its width.

90. A wire 40 centimeters long is cut into two pieces. One piece is bent into a square, and the second piece is shaped as a circle. Write a function that describes the sum of the areas of the two shapes.

2 Trigonometric Functions

A mathematician, like a painter or a poet, is a maker of patterns. If his patterns are more permanent than theirs, it is because they are made of ideas.

G. H. HARDY (1877–1947)

TSUNAMIS AND EARTHQUAKES

Imagine an undersea earthquake. The energy from the earthquake would be translated to the water as water waves. These water waves are called *tsunamis* or *tidal waves*. Although the phrase tidal waves is still used to describe these waves, tsunami is the preferred description because the waves have nothing to do with the tides.

In the open ocean, the distance between crests of a tsunami may be as great as 60 miles and the height of the wave no more than 2 feet. As the depth of the ocean decreases however, the water wave slows down. As it slows, the height of the wave increases. When a tsunami reaches the shore, the wave height can be quite high with crests 100 feet above the normal tide level.

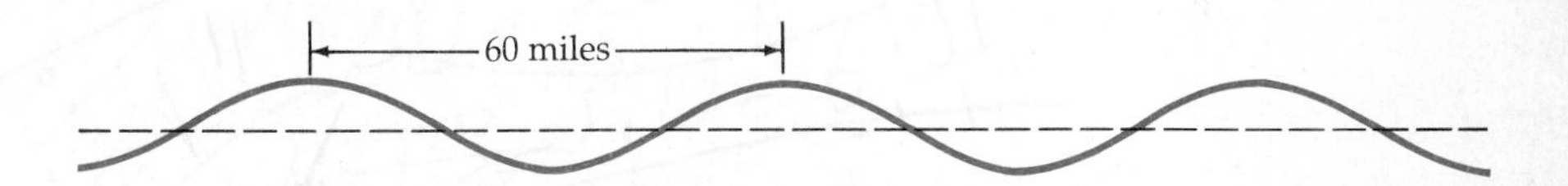

The earthquake that generated the tsunamis also creates waves within the earth. Two of the wave types that are created are the *primary* or *P wave* and the *secondary* or *S wave*. These two waves are quite different. The P wave is very much like a sound wave. It alternately compresses and dilates the substances within the earth. These waves can travel through solid rock and water.

S waves are slower than P waves and are more like water waves. As an S wave travels through the earth, it shears the rock sideways at right angles to the direction of travel. The S wave causes much of the structural damage from an earthquake. Wave phenomena exhibited by tsunami and earthquake waves can be described by trigonometric functions, the subject of this chapter.

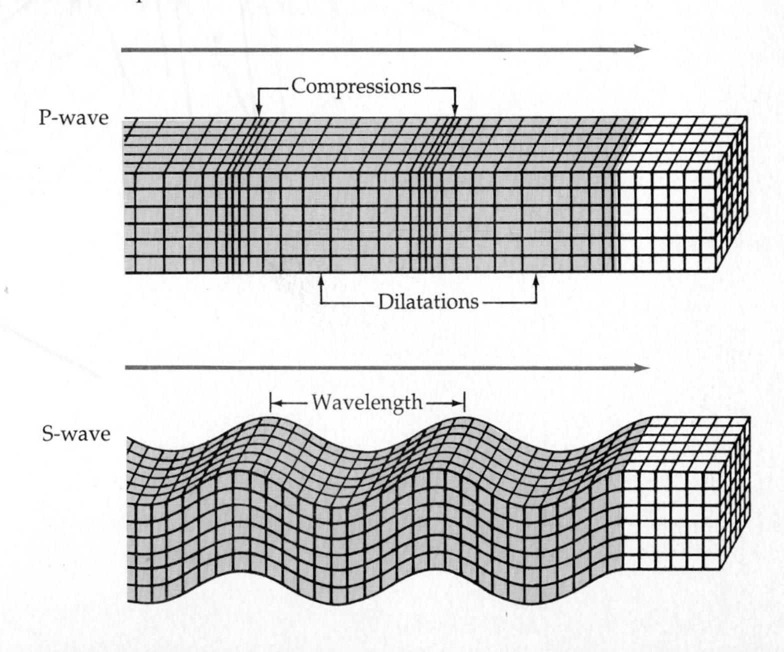

2.1 Measuring Angles

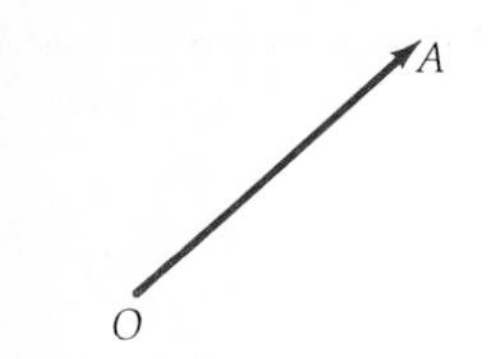

Figure 2.1

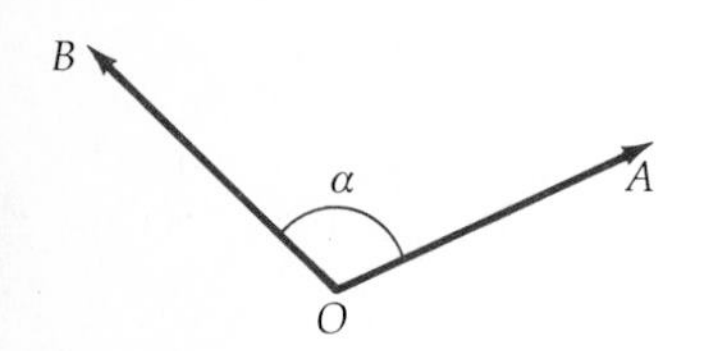

Figure 2.2

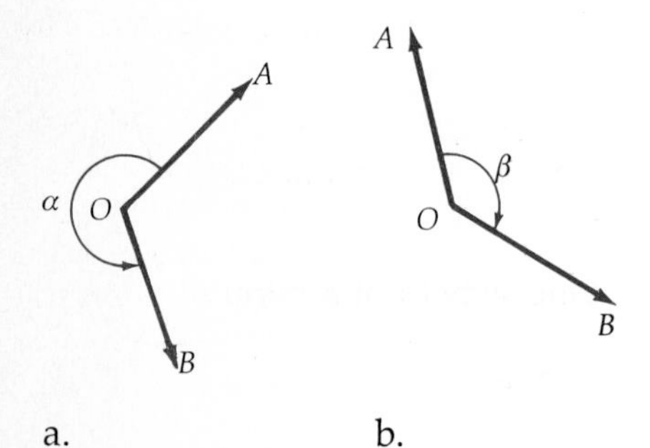

Figure 2.3
a. Positive angle;
b. Negative angle.

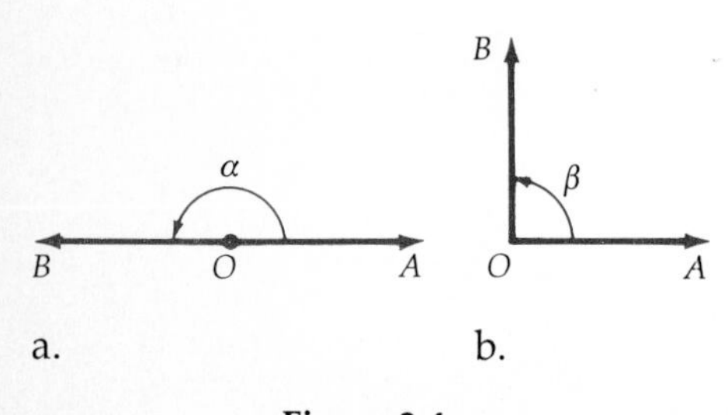

Figure 2.4
a. Straight angle (180°);
b. Right angle (90°).

Early Babylonians noticed that the seasons repeated about every 360 days. Thinking that the earth was at the center of the universe, they assumed the universe made one complete revolution in 360 days. Our concept of measuring angles in degrees is an outgrowth of those early beliefs.

A ray is a part of a line originating at a point and extending infinitely, as Figure 2.1 shows. An **angle** is formed by rotating a ray about its endpoint. The initial position of the ray is called the **initial side** of the angle. The position of the ray after it has been rotated is called the **terminal side** of the angle. The endpoint of the ray is called the vertex of the angle. Figure 2.2 shows an angle with vertex O.

In Figure 2.3 the ray OA is the **initial side** of the angle and OB is the **terminal side** of the angle. Angles formed by a counterclockwise rotation are considered **positive** angles; angles formed by a clockwise rotation are considered **negative** angles.

The measure of an angle is the amount of rotation of the ray. One unit of angle measurement is the degree.

Definition of Degree

An angle formed by rotating a ray $\frac{1}{360}$ of a complete revolution has a measure of one **degree**. The symbol for degree is °.

Using this definition, an angle formed by rotating a ray one complete revolution has a measure of 360°. A **straight angle** is formed by rotating a ray one-half a complete revolution. The measure of a straight angle is 180°. A **right angle** is formed by rotating a ray one-quarter of a complete revolution. The measure of a right angle is 90°. (See Figure 2.4.)

An angle is **acute** if its measure is between 0° and 90°. An angle is **obtuse** if its measure is between 90° and 180°. (See Figure 2.5.)

Two angles are **complementary angles** if the sum of the measures of the angles is 90°. In this case, one angle is the *complement* of the other angle. Two angles are **supplementary angles** if the sum of the measures of the two angles is 180°. One angle is the *supplement* of the other angle. (See Figure 2.6.)

We also can measure angles greater than 360° $[-720° + (-270°)]$ (1 revolution). An angle of 720° is 2 revolutions counterclockwise. An angle of 450° (360° + 90°) is $1\frac{1}{4}$ revolutions counterclockwise. The angle $-990°$ is $2\frac{3}{4}$ revolutions clockwise.

Degrees may be subdivided into smaller units by using decimal degrees or by using the degree, minute, second system (DMS). In the decimal system,

$$8.53° \text{ means } 8° \text{ plus 53 hundredths of } 1°.$$

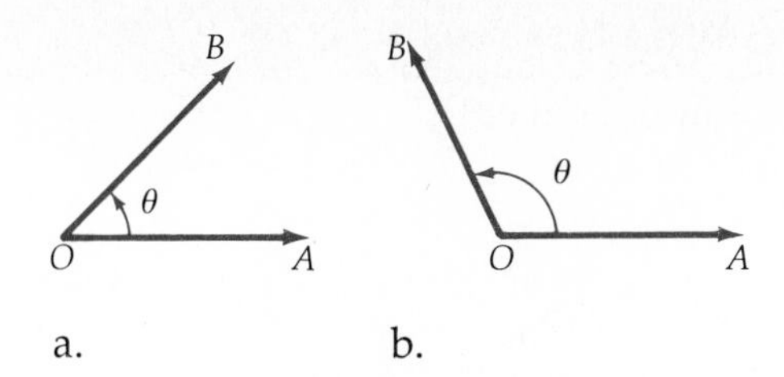

Figure 2.5
a. Acute angle: $0° < \theta < 90°$;
b. Obtuse angle: $90° < \theta < 180°$.

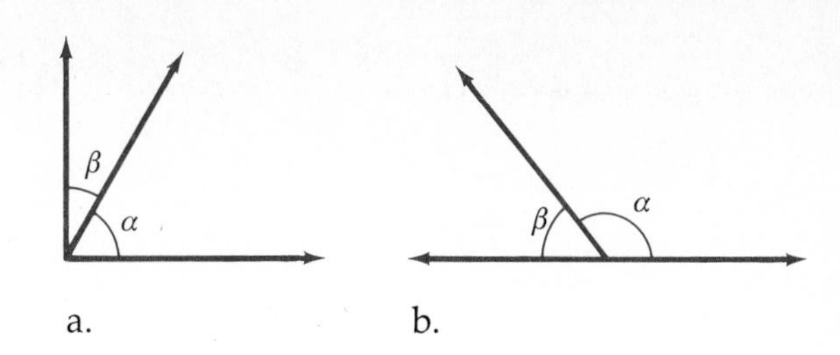

Figure 2.6
a. Complementary angles: $\alpha + \beta = 90°$;
b. Supplementary angles: $\alpha + \beta = 180°$.

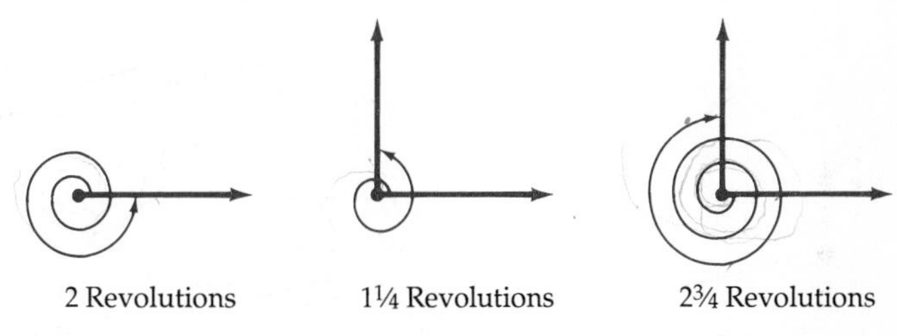

Figure 2.7

In the DMS system, the degree is subdivided into smaller units of minutes, which are further divided into seconds.

$$\text{One minute } (1') = \left(\frac{1}{60}\right)^{\circ} \quad \text{One-sixtieth of a degree}$$

$$\text{One second } (1'') = \left(\frac{1}{60}\right)' \quad \text{One-sixtieth of a minute}$$

Thus we can write $60' = 1°$ and $60'' = 1'$.

Therefore we know that $1° = 3600''$. Why?[1]

EXAMPLE 1 Convert from DMS to Decimal Degrees

Convert 20°14′ to decimal degrees. Round to the nearest thousandth of a degree.

Solution We use the conversion factor $1° = 60'$ to convert minutes to decimal degrees.

$$20°14' = 20° + 14' = 20° + 14'\left(\frac{1°}{60'}\right) \quad 1° = 60', \text{ therefore } 1 = \frac{1°}{60'}.$$

$$\approx 20° + 0.233° = 20.233°$$

■ *Try Exercise* **22,** *page 54.*

[1] $1° = 60' = 60' \cdot \dfrac{60''}{1'} = 3600''$.

EXAMPLE 2 **Convert from Decimal Degrees to DMS**

Convert 42.82° to the DMS system of measurement.

Solution

$$\begin{aligned} 42.82° &= 42° + 0.82° \\ &= 42° + 0.82°\left(\frac{60'}{1°}\right) && 1 = \frac{60'}{1°}. \\ &= 42° + 49.2' \\ &= 42° + 49' + 0.2' \\ &= 42° + 49' + 0.2'\left(\frac{60''}{1'}\right) && 1 = \frac{60''}{1'}. \\ &= 42° + 49' + 12'' \\ &= 42°49'12'' \end{aligned}$$

■ *Try Exercise* **30**, *page 54.*

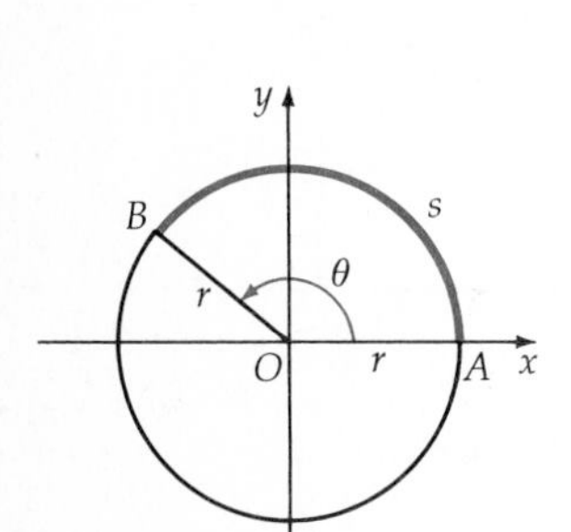

Figure 2.8

Another common angle measurement is the *radian*. To define a radian, first we consider a circle of radius r and two radii OA and OB. The angle θ formed by the two radii is a **central angle.** The portion of the circle between A and B is an **arc** of the circle and is written $\overparen{AB}$. We say that $\overparen{AB}$ *subtends* the angle θ. The length of $\overparen{AB}$ is s (see Figure 2.8).

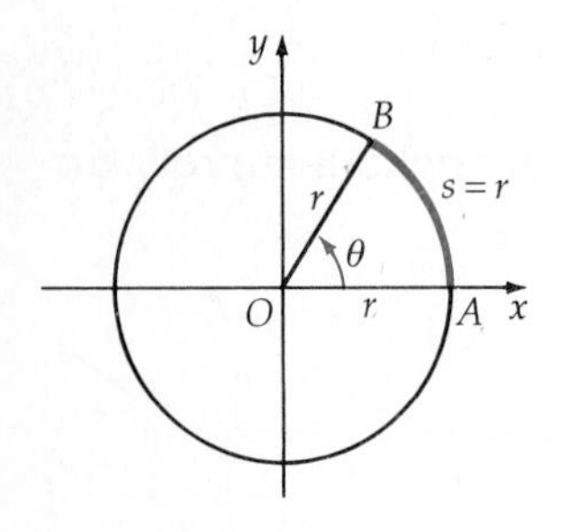

Figure 2.9

Definition of Radian

> One **radian** is the measure of the central angle subtended by an arc of length r on a circle of radius r.

For example, an arc length of 15 centimeters on a circle with a radius of 5 centimeters will subtend an angle of 3 radians, as shown in Figure 2.10. We can obtain the same result by dividing 15 centimeters by 5 centimeters. To find the number of radians in *any central angle* θ, divide the length s of the arc that subtends θ by the radius of the circle.

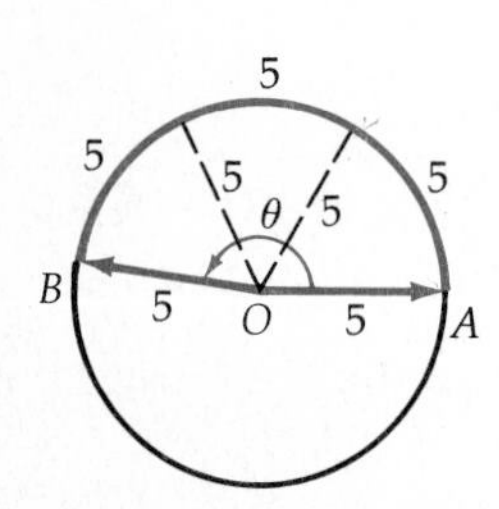

Figure 2.10

Radian Measure

> Given an arc of length s on a circle of radius r, the radian measure of the angle subtended by the arc is $\theta = s/r$.

Figure 2.11 shows the measure of the central angles for different arc lengths. For $\frac{1}{2}$ revolution we see that the central angle measure is π radians, and for a complete revolution the central angle measure is 2π radians.

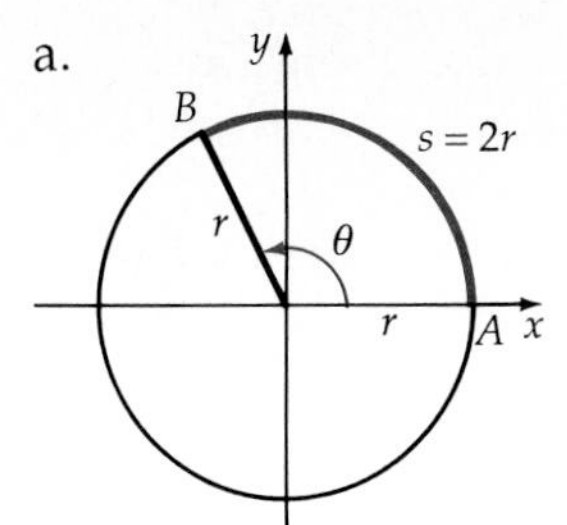

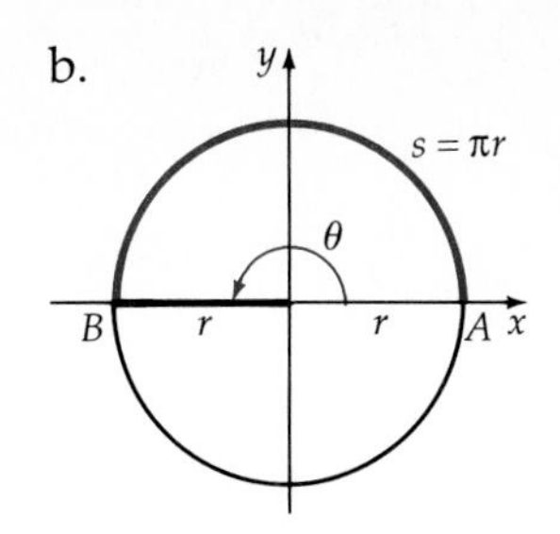

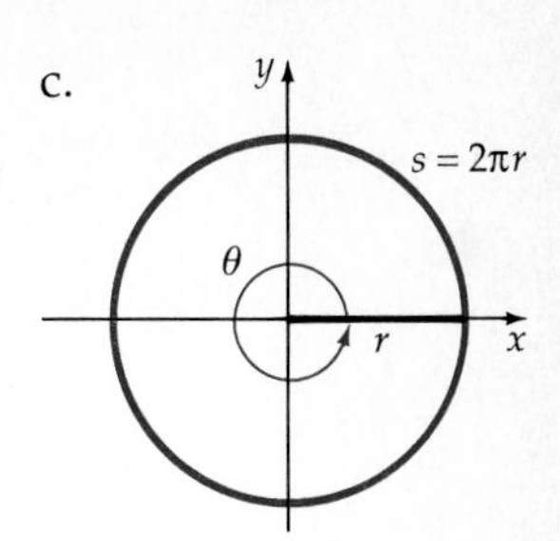

Figure 2.11
a. $\theta = s/r = 2r/r = 2$ radians;
b. $\theta = s/r = \pi r/r = \pi$ radians;
c. $\theta = s/r = 2\pi r/r = 2\pi$ radians.

EXAMPLE 3 Find the Measure of a Central Angle

An arc of length 10 centimeters on a circle with a radius of 4 centimeters subtends an angle θ. Find the measure of θ in radians.

Solution Use the formula $\theta = \frac{s}{r}$ to find the central angle.

$$\theta = \frac{s}{r} = \frac{10\,\text{cm}}{4\,\text{cm}} = 2.5\,\text{rad}$$

■ *Try Exercise* **74,** *page 54.*

Remark From Example 3, note that the magnitude of a radian is the ratio of two length quantities (centimeters, in this case). Thus a radian is a dimensionless quantity, a fact that is used in application problems.

EXAMPLE 4 Convert Revolutions to Radian Measure

Find the measure in radians of an angle formed by rotating the initial side $1\frac{1}{3}$ revolutions in a counterclockwise direction.

Solution From Figure 6.11, one complete counterclockwise revolution corresponds to 2π radians. Thus to find the number of radians in $1\frac{1}{3}$ counterclockwise revolutions, multiply $1\frac{1}{3}$ by 2π.

$$\theta = 1\frac{1}{3} \cdot 2\pi = \frac{8}{3}\pi$$

■ *Try Exercise* **82,** *page 54.*

Referring again to Figure 2.11, one complete revolution has a degree measure of 360°, or a radian measure of 2π. Thus $360° = 2\pi$ radians, and

$180° = \pi$ radians. From the last equation, we can derive the following conversion factors to convert degrees to radians or radians to degrees:

Radian/Degree Conversion Factors

$$1 \text{ radian} = \frac{180°}{\pi} \qquad 1° = \frac{\pi}{180} \text{ radians}$$

Remark Using a calculator, we have

$$1 \text{ radian} \approx 57.29577951° \quad \text{and} \quad 1° \approx 0.017453292 \text{ radian}.$$

EXAMPLE 5 Convert Degree Measure to Radian Measure

Convert 240° to radians.

Solution We use the conversion factor π rad/180° to convert degrees to radians. The exact answer can be given in terms of π. A decimal approximation is obtained when an approximate value of π is used.

$$\begin{aligned} 240° &= 240°\left(\frac{\pi}{180°} \text{ radians}\right) \\ &= \frac{4}{3}\pi \text{ radians} && \text{Exact answer} \\ &\approx 4.19 \text{ radians} && \text{Approximate answer} \end{aligned}$$

■ *Try Exercise* **42,** *page 54.*

Remark In most cases the radian measure of angles will be stated in terms of π and the decimal approximation will not be used.

EXAMPLE 6 Convert Radians to Degrees

Convert $\frac{11\pi}{12}$ radians to degrees.

Solution We use the conversion factor $\frac{180°}{\pi}$ radians to convert radians to degrees.

$$\frac{11\pi}{12} \text{ radians} = \frac{11\pi}{12} \cdot \frac{180°}{\pi} = 165°$$

■ *Try Exercise* **52,** *page 54.*

The following table lists the degree and radian measure of selected angles. Figure 2.12 illustrates the relationships.

Degrees	Radians
0	0
30	$\pi/6$
45	$\pi/4$
60	$\pi/3$
90	$\pi/2$
120	$2\pi/3$
135	$3\pi/4$
150	$5\pi/6$
180	π
210	$7\pi/6$
225	$5\pi/4$
240	$4\pi/3$
270	$3\pi/2$
300	$5\pi/3$
315	$7\pi/4$
330	$11\pi/6$
360	2π

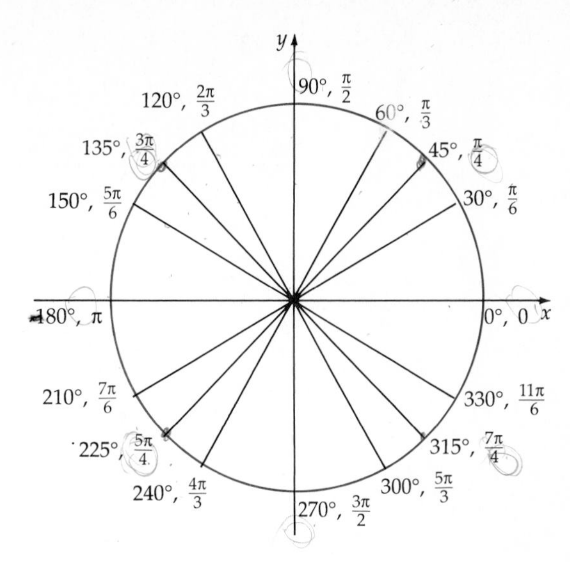

Figure 2.12
Degree and radian measures of selected angles.

Consider a circle of radius r. The use of radians is helpful for finding the measure of the length of an arc s. By solving the formula $\theta = s/r$ for s, we have an equation that gives the length of the arc of a circle.

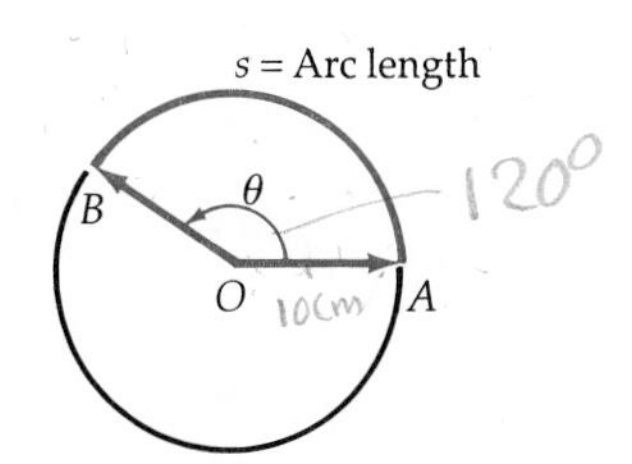

Figure 2.13

Arc Length of a Circle

> Let r be the radius of a circle C and θ the radian measure of a central angle of C. Then the length of the arc s that subtends θ is $s = r\theta$.

Remark When the circle C has a radius $r = 1$, $s = \theta$. This circle, called the unit circle, will be particularly useful in later sections.

EXAMPLE 7 Find Arc Length of a Circle

Find the length of an arc that subtends a central angle of 120° in a circle of radius 10 centimeters.

Solution The formula for the length of a circular arc requires that the angle be measured in radians. Convert 120° to radians and use the formula $s = r\theta$ to find the length of the arc.

$$\theta = 120^\circ = 120 \cdot \frac{\pi}{180^\circ} \text{ radians} = \frac{2\pi}{3} \text{radian}$$

$$s = 10 \text{ cm}\left(\frac{2\pi}{3}\right) = \frac{20\pi}{3} \text{ cm}$$

■ *Try Exercise* **80**, *page 54.*

EXAMPLE 8 **Solve an Application Involving Radians**

A pulley with a radius of 10 inches uses a belt to drive a pulley with a radius of 6 inches. The 10-inch pulley turns through an angle of 2π radians. Find the angle through which the 6-inch pulley turns (see Figure 2.14).

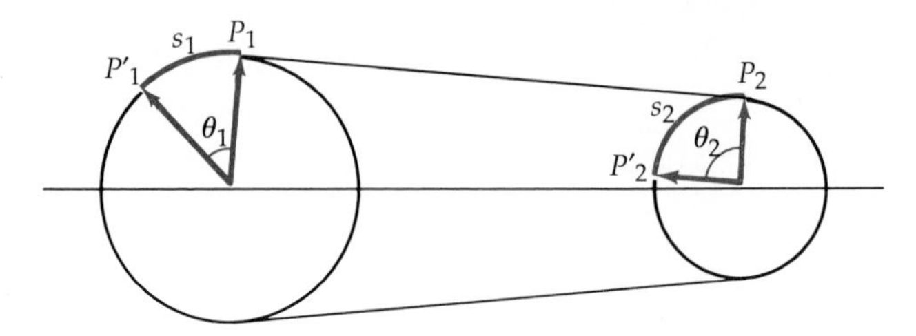

Figure 2.14

Solution

A point P_1 on the 10-inch pulley moves: $s_1 = r_1\theta_1$

A point P_2 on the 6-inch pulley moves: $s_2 = r_2\theta_2$

Since one point on the belt moves the same distance as any other point on the belt, the point P_1 moves through the same distance as P_2. Thus $s_1 = s_2$ and $r_1\theta_1 = r_2\theta_2$. Solve for θ_2. Substitute the given values for r_1, r_2, and θ_1 and simplify.

$$\theta_2 = \left(\frac{r_1}{r_2}\right)\theta_1 = \frac{10}{6}(2\pi) = \frac{10}{3}\pi \quad r_1 = 10,\ r_2 = 6,\ \theta_1 = 2\pi$$

■ *Try Exercise* **84**, *page 54.*

A car traveling at a speed of 55 miles per hour travels 55 miles in 1 hour, 110 miles in 2 hours. **Linear velocity** is defined as the distance traveled per unit time, or, in equation form, $v = d/t$, where v is the velocity, d the distance, and t the time. The units may be expressed in feet per second, miles per hour, meters per second, or in any unit of length per unit of time.

Remark Velocity actually has two characteristics: speed and direction. The distinction between speed and velocity is discussed when we discuss vectors.

The floppy disk in a computer revolving at 300 rpm makes 300 revolutions in 1 minute. A radius OP of the disk would move through or generate an angle θ in time t. **Angular velocity** is defined as the angle generated per unit of time, or $\omega = \theta/t$, where ω is the angular velocity in units of revolutions per second, revolutions per minute, radians per second, or in any unit of angle measure per unit of time. The angle generated is θ and t is the time.

EXAMPLE 9 Solve an Application Involving Angular Velocity

A hard disk in a computer rotates at 3600 revolutions per minute (rpm). Convert the angular velocity to radians per second. Round to the nearest tenth.

Solution Since we must find the angular velocity in radians per second, first we find the total number of radians.

$$3600(2\pi) = 7200\pi$$

Use the formula $\omega = \theta/t$ to find the angular velocity in radians per second.

$$\omega = \frac{\theta}{t} = \frac{7200\pi}{60} = 377.0 \text{ rad/s}$$

■ *Try Exercise* **86,** *page 54.*

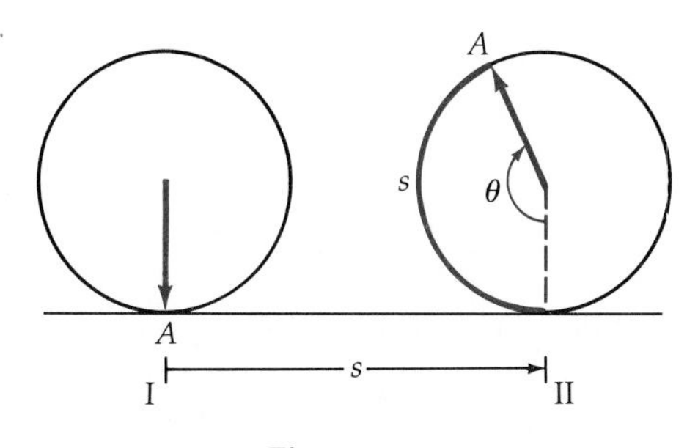

Figure 2.15

A rotating circle may have both linear and angular velocity related by a simple equation. Assume that a wheel is rolling without slipping. Point A on the wheel in position I is in contact with the ground. As the wheel moves a distance s, the point A moves through an angle θ. The arc length subtending angle θ in position II is equal to s. (See Figure 2.15.)

The linear velocity of the wheel is given by $v = s/t$

Substitute $s = r\theta$ for s $= r\theta/t$

Substitute ω for θ/t $= r\omega$

This expression $v = r\omega$ gives the velocity of a point on a rotating body at a distance r from the axis of rotation.

EXAMPLE 10 Solve an Application Involving Linear Velocity

The tires of an automobile are rotating at a rate of 600 revolutions per minute. The radius of the tire is 14 inches. Find the speed of the automobile in miles per hour. Round to the nearest mile per hour.

Solution First convert 600 revolutions per minute to 36,000 revolutions per hour by multiplying by 60. Next multiply by 2π to get $\omega = 72000\pi$. Now use $v = r\omega$ to find the velocity.

$$v = \frac{14}{63{,}360} \cdot 72000\pi \approx 50 \text{ mph} \quad \text{There are 63,368 inches per mile.}$$

■ *Try Exercise* **88,** *page 54.*

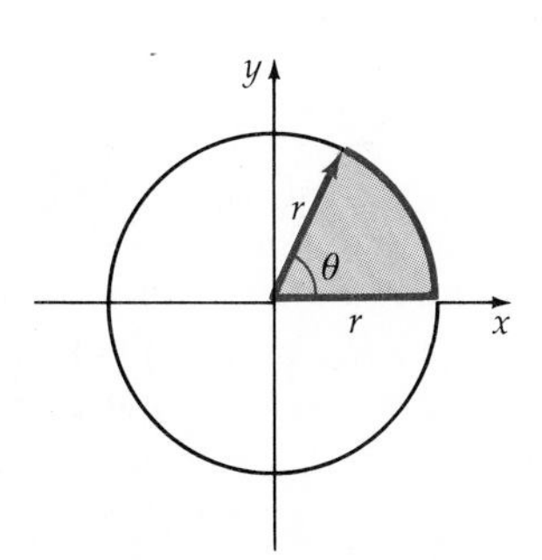

Figure 2.16

A **sector** of a circle is the figure bounded by two radii and the intercepted arc as Figure 2.16 shows. It can be shown that the ratio of the area of the sector (A) to the area of the circle (πr^2) is equal to the ratio of the *central angle* (θ) of the sector to one complete revolution of the circle (2π).

$$\frac{A}{\pi r^2} = \frac{\theta}{2\pi}$$

Solving for A, we have $$A = \frac{1}{2}r^2\theta.$$

EXERCISE SET 2.1

In Exercises 1 to 9, find the complement of the given angle.

1. $15°$ **2.** $72°$ **3.** $86.3°$
4. $66.14°$ **5.** $77.55°$ **6.** $7°4'$
7. $46°9'$ **8.** $54°36'14''$ **9.** $10°55'35''$

In Exercises 10 to 18, find the supplement of the given angle.

10. $37°$ **11.** $67°$ **12.** $101.5°$
13. $172.34°$ **14.** $34.56°$ **15.** $123.8°$
16. $102.4°$ **17.** $45.78°$ **18.** $147.66°$

In Exercises 19 to 27, convert the DMS measure of each angle to decimal degree measure to the nearest thousandth of a degree.

19. $78°8'$ **20.** $5°39'$ **21.** $16°44''$
22. $35°42''$ **23.** $47°20'$ **24.** $20°4'45''$
25. $165°36'54''$ **26.** $68°16'50''$ **27.** $95°28'8''$

In Exercises 28 to 36, convert the decimal degree measure of each angle to the DMS system to the nearest second.

28. $110.4°$ **29.** $36.6°$ **30.** $55.44°$
31. $66.72°$ **32.** $7.05°$ **33.** $6.19°$
34. $12.28°$ **35.** $132.58°$ **36.** $102.76°$

In Exercises 37 to 42, convert the measure of the angle to exact radian measure.

37. $15°$ **38.** $165°$ **39.** $315°$
40. $420°$ **41.** $630°$ **42.** $585°$

In Exercises 43 to 48, convert the degree measure of the angle to radian measure to the nearest hundredth of a radian.

43. $29°$ **44.** $148°$ **45.** $166°$
46. $434°$ **47.** $610°$ **48.** $295°$

In Exercises 49 to 54, convert the radian measure of the angles to exact degree measure.

49. $\pi/6$ **50.** $\pi/9$ **51.** $3\pi/8$
52. $11\pi/18$ **53.** $11\pi/3$ **54.** $6\pi/5$

In Exercises 55 to 60, convert the radian measure of the angles to degree measure to the nearest hundredth of a degree.

55. 1.2 **56.** 0.35 **57.** 0.64
58. 5.66 **59.** 4.38 **60.** 8

In Exercises 61 to 66, find the complement of each angle measured in radians. Express your answer in terms of π.

61. $\pi/6$ **62.** $\pi/4$ **63.** $5\pi/12$
64. 0.75 **65.** 1.22 **66.** 0.30

In Exercises 67 to 72, find the supplement of each angle measured in radians. Express your answer in terms of π.

67. $3\pi/4$ **68.** $7\pi/12$ **69.** $7\pi/8$
70. 2.5 **71.** 1.76 **72.** 0.55

In Exercises 73 to 76, find the measure in radians and in degrees of the central angle of a circle with the given radius and arc length.

73. $r = 2$ in, $s = 8$ in **74.** $r = 7$ ft, $s = 4$ ft
75. $r = 5.2$ cm, $s = 12.4$ cm **76.** $r = 35.8$ m, $s = 84.3$ m

In Exercises 77 to 80, find the measure of the intercepted arc of a circle with the given radius and central angle.

77. $r = 8$ in., $\phi = \pi/4$ radians
78. $r = 3$ ft, $\theta = 7\pi/2$ radians
79. $r = 25$ cm, $\phi = 42°$ **80.** $r = 5$ m, $\theta = 144°$

81. Find the equivalent number of radians in $1\frac{1}{2}$ revolutions.

82. Find the equivalent number of radians in $\frac{3}{8}$ revolution.

83. A pulley with a radius of 14 inches uses a belt to drive a pulley with a radius of 28 inches. The 14-inch pulley turns through an angle of 150°. Find the angle through which the 28-inch pulley turns.

84. A pulley with a diameter of 1.2 meters uses a belt to drive a pulley with a diameter of 0.8 meter. The 1.2-meter pulley turns through an angle of 240°. Find the angle through which the 0.8-meter pulley turns.

85. Find the angular velocity of the second hand of a watch.

86. Find the angular velocity of a point on the equator in radians per second. The radius of the earth is 3960 miles.

87. A wheel is rotating at 50 revolutions per second. Find the angular velocity in radians per second.

88. A wheel is rotating at 200 revolutions per minute. Find the angular velocity in radians per second.

89. A car with a 14-inch wheel is moving with a velocity of 55 mph. Find the angular velocity of the wheel in radians per second.

90. The 15-inch tires of an automobile are rotating at a rate of 450 revolutions per minute. Find the velocity of the automobile in miles per hour.

91. The 18-inch tires of a truck are rotating at a rate of 500 revolutions per minute. Find the velocity of the truck in miles per hour.

Supplemental Exercises

In Exercises 92 to 97 find the area of the sector of a circle with the given radius and central angle.

92. $r = 5$ in, $\theta = \pi/3$ rad **93.** $r = 2.8$ ft, $\theta = 5\pi/2$ rad

94. $r = 120$ cm, $\theta = 0.65$ rad **95.** $r = 30$ ft, $\theta = 62°$

96. $r = 20$ m, $\theta = 125°$ **97.** $r = 25$ cm, $\theta = 220°$

98. The minute hand on the clock atop city hall measures 6 ft 3 inches from the tip to its axle.

a. Through what angle does the minute hand pass between 9:12 A.M. and 9:48 A.M.?

b. What distance does the tip of the minute hand travel during this period?

99. At a time when the earth was 93,000,000 miles from the sun, using a transit you observed through a properly smoked glass that the diameter of the sun occupied an arc of 31′. Calculate the approximate diameter of the sun to 2 significant digits.

100. A merry-go-round horse is 11.6 meters from the center. The merry-go-round makes $14\frac{1}{4}$ revolutions per ride. How many meters does the horse travel? How fast is it moving in meters per second?

101. **a.** A car with 13-inch tires makes an 8-mile trip. Find the number of revolutions the tire makes on the 8-mile trip.

b. A car with 15-inch tires makes an 8-mile trip. Find the number of revolutions the tire makes on the 8-mile trip.

102. A water wheel has a 10-foot radius. When the wheel makes 18 revolutions per minute, what is the speed of the river?

103. A pulley with a 50-centimeter diameter drives a pulley with a 20-centimeter diameter. The larger pulley makes 30 revolutions per minute. What is the linear speed of a point on the surface of the smaller pulley?

104. Find the area of the shaded portion of the graph shown. The radius of the circle is 9 inches.

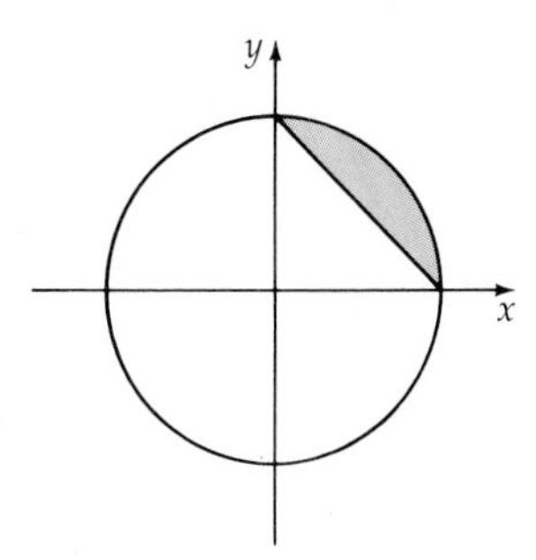

105. Latitude describes the position of a point on the earth's surface in relation to the equator. A point on the equator has a latitude of 0°. The north pole has a latitude of 90°. The radius of the earth is approximately 3960 miles. Assuming that the earth is a perfect sphere, find the distance along the earth's surface that subtends a central angle of latitude (a) 1°, (b) 1′, and (c) 1″. Express your answer to 3 significant digits.

2.2 Trigonometric Functions of Acute Angles

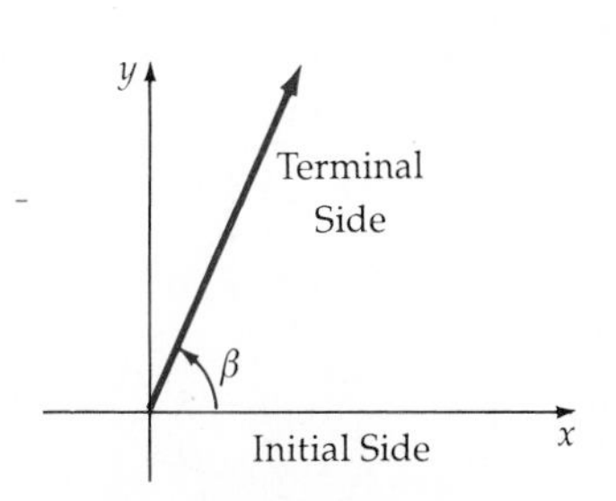

Figure 2.17

The study of trigonometry, which means "triangle measurement," began more than 2000 years ago, partially as a means to solving surveying problems. Early trigonometry used the length of a chord of a circle as the value of a *trigonometric function.* In the sixteenth century, right triangles were used to define a trigonometric function. We will use a modifed right triangle approach to define the trigonometric functions by placing one of the acute angles of a right triangle on a coordinate grid.

An angle is in **standard position** when the initial side coincides with the x-axis and the vertex is at the origin of the coordinate axes. The angle β in Figure 2.17 is in standard position.

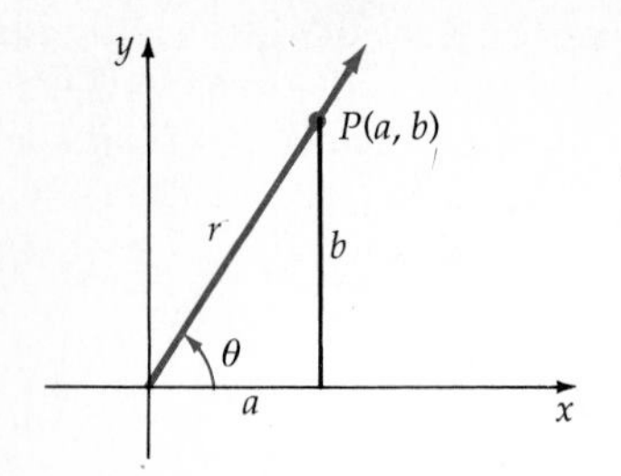

Figure 2.18

Now consider angle θ in Figure 2.18 in standard position and a point $P(a, b)$ on the terminal side of the angle. A line perpendicular to the x-axis has been drawn. The values a, b and r represent the measure of the legs and hypotenuse of the right triangle formed. Six possible ratios can be formed:

$$\frac{a}{r}, \frac{b}{r}, \frac{r}{a}, \frac{r}{b}, \frac{a}{b}, \frac{b}{a}.$$

Each ratio defines a value of a trigonometric function; the functions are the sine (sin), cosine (cos), tangent (tan), cosecant (csc), secant (sec), and cotangent (cot).

Trigonometric Functions of an Acute Angle

Let θ be an acute angle in standard position and $P(a, b)$ a point on the terminal side of the angle. The six trigonometric functions of θ are

$$\sin\theta = \frac{b}{r} \qquad \cos\theta = \frac{a}{r} \qquad \tan\theta = \frac{b}{a}$$

$$\csc\theta = \frac{r}{b} \qquad \sec\theta = \frac{r}{a} \qquad \cot\theta = \frac{a}{b}$$

where $r = \sqrt{a^2 + b^2}$.

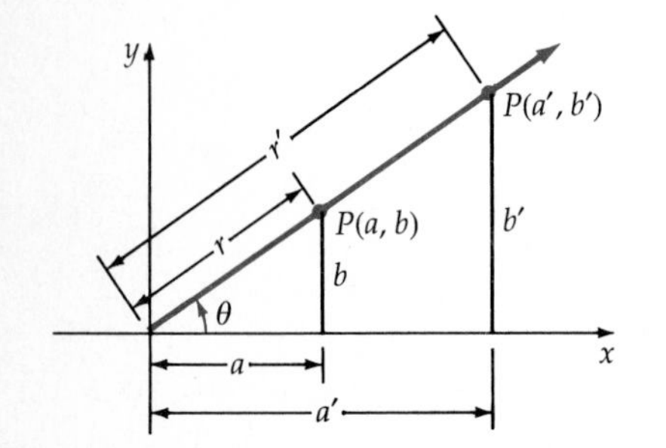

Figure 2.19

Consider any two points on the terminal side of the angle in standard position and drawing the perpendicular lines from the two points to the x-axis. The first triangle formed has legs of lengths a and b, and the second triangle formed has legs of lengths a' and b'. The right triangles formed are similar triangles, and the corresponding sides are proportional. Thus,

$$\frac{b}{r} = \frac{b'}{r'} \quad \text{and} \quad \sin\theta = \frac{b}{r} = \frac{b'}{r'}.$$

Thus the value of the sine function is independent of the length of the sides of a triangle. The values of the remaining trigonometric functions can also be shown to be independent of the lengths of the sides.

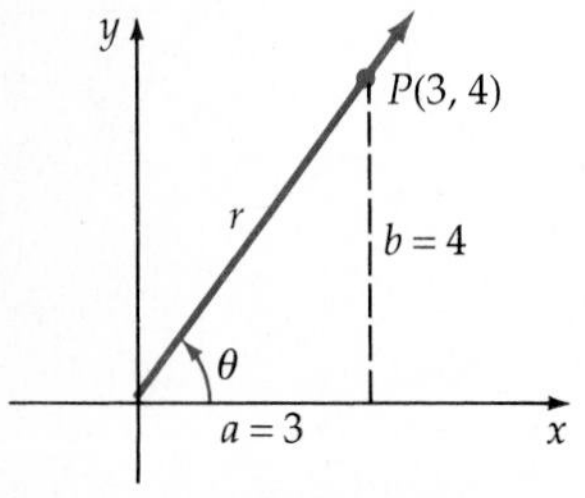

Figure 2.20

EXAMPLE 1 Evaluate Trigonometric Functions

Find the values of the six trigonometric functions of the acute angle in standard position with the terminal side passing through the point $P(3, 4)$.

Solution Sketch the angle θ on the coordinate axes. (See Figure 2.20.) Find the length of r by using the Pythagorean Theorem.

$$r = \sqrt{3^2 + 4^2} = 5$$

Use the definitions of the trigonometric functions to evaluate the trigonometric functions.

$$\sin\theta = \frac{b}{r} = \frac{4}{5} \qquad \csc\theta = \frac{r}{b} = \frac{5}{4} \qquad a = 3,\ b = 4,\ r = 5$$

$$\cos\theta = \frac{a}{r} = \frac{3}{5} \qquad \sec\theta = \frac{r}{a} = \frac{5}{3}$$

$$\tan\theta = \frac{b}{a} = \frac{4}{3} \qquad \cot\theta = \frac{a}{b} = \frac{3}{4}$$

■ *Try Exercise* **8**, *page 61.*

EXAMPLE 2 **Evaluate Trigonometric Functions**

If ϕ is an acute angle in standard position and $\sin\phi = \frac{5}{13}$, find the values of the other five trigonometric functions of ϕ.

Solution Sketch the angle in standard position (see Figure 6.21). Because $\sin\phi = \frac{b}{r} = \frac{5}{13}$, we use $r = 13$, $b = 5$.

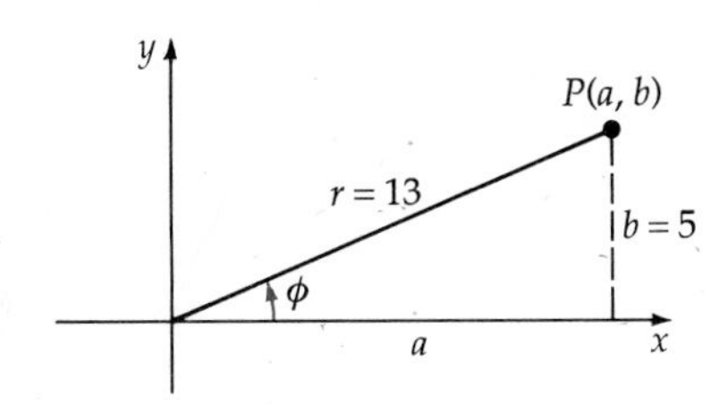

Figure 2.21

$$r^2 = a^2 + b^2 \quad \text{Use the Pythagorean Theorem to find } a.$$

$$13^2 = a^2 + 5^2$$

$$a = 12$$

$$\cos\phi = \frac{a}{r} = \frac{12}{13} \qquad \tan\phi = \frac{b}{a} = \frac{5}{12}$$

$$\csc\phi = \frac{r}{b} = \frac{13}{5} \qquad \sec\phi = \frac{r}{a} = \frac{13}{12} \qquad \cot\phi = \frac{a}{b} = \frac{12}{5}$$

■ *Try Exercise* **18**, *page 61.*

For most angles, advanced mathematical methods are required to find the value of a trigonometric function. However, the value of a trigonometric function for some *special angles* can be found by geometric methods. These special angles are 0° (0), 30° ($\pi/6$), 45° ($\pi/4$), 60° ($\pi/3$), and 90° ($\pi/2$).

First we will find the six trigonometric functions of 45°. (The discussion is based on angles measured in degrees. However, we could have used radian measure without changing the results.) Figure 2.22 shows a 45° angle in standard position with a perpendicular line drawn from $P(a, b)$ to the x-axis. The measure of the legs of the triangle are equal. Let the length of each leg be equal to a. By the Pythagorean Theorem,

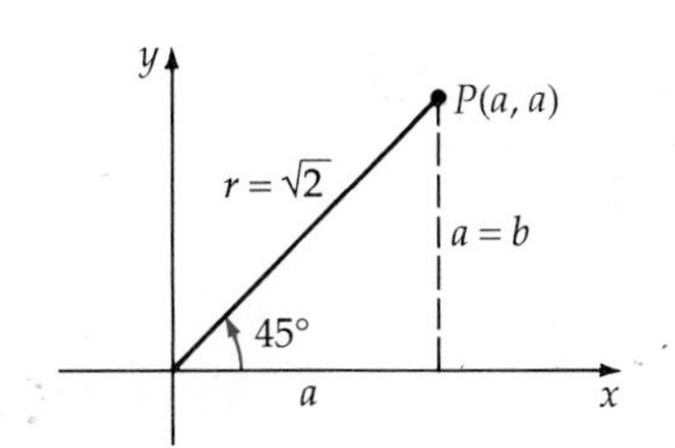

Figure 2.22

$$r^2 = a^2 + a^2 = 2a^2$$

$$r = a\sqrt{2}.$$

The values of the six trigonometric functions of 45° are

$$\sin 45° = \frac{a}{a\sqrt{2}} = \frac{1}{\sqrt{2}} = \frac{\sqrt{2}}{2} \qquad \csc 45° = \frac{a\sqrt{2}}{a} = \sqrt{2}$$

$$\cos 45° = \frac{a}{a\sqrt{2}} = \frac{1}{\sqrt{2}} = \frac{\sqrt{2}}{2} \qquad \sec 45° = \frac{a\sqrt{2}}{a} = \sqrt{2}$$

$$\tan 45° = \frac{a}{a} = 1 \qquad \cot 45° = \frac{a}{a} = 1$$

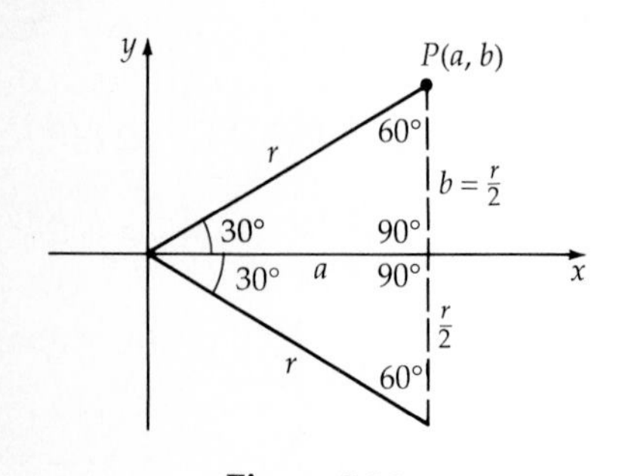

Figure 2.23

The trigonometric values of the special angles 30° and 60° can be found by drawing an equilateral triangle and bisecting one of the angles, as Figure 2.23 shows. The angle bisector also bisects one of the sides. Thus $\alpha = 30°$ and the measure of the side opposite the 30° angle is one-half the hypotenuse.

Let r be the measure of the hypotenuse. Then the measure of the side opposite 30° is $r/2$. The length of the side adjacent to the 30° angle (a) is found by using the Pythagorean Theorem.

$$r^2 = \left(\frac{r}{2}\right)^2 + a^2$$

$$\frac{3r^2}{4} = a^2$$

$$\frac{\sqrt{3}r}{2} = a$$

Note that in a 30°–60° right triangle, the side opposite the 30° angle is one-half the hypotenuse and the side opposite the 60° angle is $\sqrt{3}/2$ times the hypotenuse. The point $P(\sqrt{3}r/2, r/2)$ is on the terminal side of a 30° angle. (See Figure 2.24.) The values of the six trigonometric functions of 30° are

$$\sin 30° = \frac{r/2}{r} = \frac{1}{2} \qquad \cos 30° = \frac{\sqrt{3}r/2}{r} = \frac{\sqrt{3}}{2}$$

$$\tan 30° = \frac{r/2}{\sqrt{3}r/2} = \frac{\sqrt{3}}{3} \qquad \csc 30° = \frac{r}{r/2} = 2$$

$$\sec 30° = \frac{r}{\sqrt{3}r/2} = \frac{2\sqrt{3}}{3} \qquad \cot 30° = \frac{\sqrt{3}r/2}{r/2} = \sqrt{3}$$

The trigonometric functions of 60° can be found by placing a 60° angle in standard position and using the information in Figure 2.25. The mea-

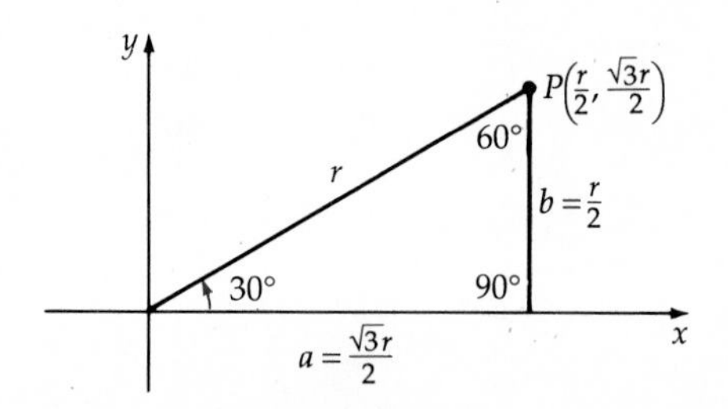

Figure 2.24

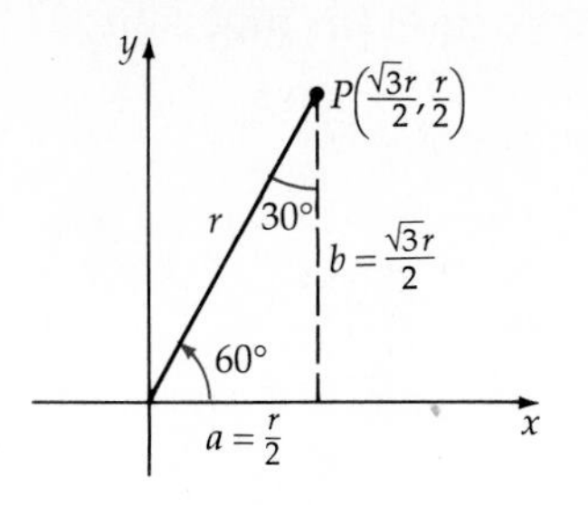

Figure 2.25

sure of the hypotenuse is r. The measure of the side *opp*
is $\sqrt{3}r/2$, and the measure of the side *adjacent* to the 60° ang
point $P(r/2, \sqrt{3}r/2)$ is on the terminal side of the 60° angle. The
the six trigonometric functions of 60° are

$$\sin 60° = \frac{\sqrt{3}r/2}{r} = \frac{\sqrt{3}}{2} \qquad \cos 60° = \frac{r/2}{r} = \frac{1}{2}$$

$$\tan 60° = \frac{\sqrt{3}r/2}{r/2} = \sqrt{3} \qquad \csc 60° = \frac{r}{\sqrt{3}r/2} = \frac{2\sqrt{3}}{3}$$

$$\sec 60° = \frac{r}{r/2} = 2 \qquad \cot 60° = \frac{r/2}{\sqrt{3}r/2} = \frac{\sqrt{3}}{3}$$

Table 2.1 summarizes the values of the trigonometric functions for the special angles 30° ($\pi/6$), 45° ($\pi/4$), and 60° ($\pi/3$).

TABLE 2.1 Values of Trigonometric Functions for 30°, 45° and 60°

	$\sin \phi$	$\cos \phi$	$\tan \phi$	$\csc \phi$	$\sec \phi$	$\cot \phi$
$\phi = 30° = \frac{\pi}{6}$	$\frac{1}{2}$	$\frac{\sqrt{3}}{2}$	$\frac{\sqrt{3}}{3}$	2	$\frac{2\sqrt{3}}{3}$	$\sqrt{3}$
$\phi = 45° = \frac{\pi}{4}$	$\frac{\sqrt{2}}{2}$	$\frac{\sqrt{2}}{2}$	1	$\sqrt{2}$	$\sqrt{2}$	1
$\phi = 60° = \frac{\pi}{3}$	$\frac{\sqrt{3}}{2}$	$\frac{1}{2}$	$\sqrt{3}$	$\frac{2\sqrt{3}}{3}$	2	$\frac{\sqrt{3}}{3}$

EXAMPLE 3 Evaluate Trigonometric Expressions with Special Angles

Find the exact value of $\sin^2 \pi/6 + \cos^2 \pi/6$.

Solution Substitute the values for $\sin \pi/6$ and $\cos \pi/6$ into the expression and simplify.

$$\sin^2 \frac{\pi}{6} + \cos^2 \frac{\pi}{6} = \left(\frac{1}{2}\right)^2 + \left(\frac{\sqrt{3}}{2}\right)^2 = \frac{1}{4} + \frac{3}{4} = 1 \qquad \begin{aligned} \sin \pi/6 &= 1/2 \\ \cos \pi/6 &= \sqrt{3}/2 \end{aligned}$$

■ *Try Exercise* **70,** *page 62.*

Appendix III includes a table of values of the trigonometric functions to four decimal places in increments of 10′. For more accuracy, a calculator or interpolation can be used. (Interpolation is discussed in Appendix I.)

To find the values of trigonometric functions using a scientific calculator, first notice that the calculator has only the function keys sin, cos, and tan; the values of the cosecant, secant, and cotangent are found by using *reciprocal* keys.

From the definitions of the sine and the cosecant functions,

$$(\sin \theta)(\csc \theta) = \frac{b}{r} \cdot \frac{r}{b} = 1$$

Thus

$$(\sin \theta)(\csc \theta) = 1 .$$

By rewriting the above equation, the sine and cosecant functions can be written in the following forms:

$$\sin \theta = \frac{1}{\csc \theta} \qquad \text{or} \qquad \csc \theta = \frac{1}{\sin \theta}$$

The sine and cosecant functions are called **reciprocal functions.**

The secant is the reciprocal of the cosine and the cotangent is the reciprocal of the tangent. Why?[2]

Table 2. shows each trigonometric function and its reciprocal. These relationships ar true for all values of the variable θ for which the functions are defined.

TABLE 2.2 Trigonometric Functions and Their Reciprocals.

$\sin \theta = \frac{1}{\csc \theta}$	$\cos \theta = \frac{1}{\sec \theta}$	$\tan \theta = \frac{1}{\cot \theta}$
$\csc \theta = \frac{1}{\sin \theta}$	$\sec \theta = \frac{1}{\cos \theta}$	$\cot \theta = \frac{1}{\tan \theta}$

For example, to calculate cos 33°, cos 58°, and sec 58°, perform the following sequence of keystrokes. (Be sure the calculator is in *degree* mode.)

Trigonometric Function	Key Sequence	Calculator Display
cos 33°	33 [cos]	0.83867057
cos 58°	58 [cos]	0.52991926
sec 58°	58 [cos] [1/x]	1.88707992

Remark The [1/x] key is used to calculate the reciprocal of the number in the display. By pressing this key when the value of cos 58° is shown in the display, we are performing the mathematical equivalent of

$$\frac{1}{\cos 58^\circ} = \sec 58^\circ .$$

As a second *important* point, many calculator errors are a result of not placing the calculator in the correct degree or radian mode before beginning a calculation. Always ensure that the calculator is in the correct mode before starting a calculation.

[2] $\sec \theta = r/a$ and $\cos \theta = a/r$, therefore $(\sec \theta)(\cos \theta) = 1$; $\cot \theta = a/b$, $\tan \theta = b/a$, therefore $(\cot \theta)(\tan \theta) = 1$.

EXAMPLE 4 Evaluate Trigonometric Function Using a Calculator

Use a scientific calculator to find:

a. $\sin 23.33°$ b. $\csc 79.35°$ c. $\tan \frac{\pi}{12}$

Solution

Trigonometric Function	Key Sequence	Calculator Display
a. $\sin 23.33°$	23.33 [sin]	0.39602635
b. $\csc 79.35°$	79.35 [sin] [1/x]	1.01752747
c. $\tan \frac{\pi}{12}$	π [÷] 12 [=] [tan]	0.26794919

■ *Try Exercise* **48**, *page 61.*

EXERCISE SET 2.2

In Exercises 1 to 12, find the six trigonometric function values with the given points on the terminal side of the angle in standard position.

1. $P(1, 3)$ **2.** $P(1, 7)$ **3.** $P(2, 2)$
4. $P(5, 2)$ **5.** $P(5, 5)$ **6.** $P(3, 5)$
7. $P(3, 2)$ **8.** $P(10, 6)$ **9.** $P(8, 4)$
10. $P(\sqrt{3}/2, 1)$ **11.** $P(\sqrt{2}, \sqrt{2})$ **12.** $P(2, \sqrt{3})$

In Exercises 13 to 15, if θ is an acute angle and $\sin \theta = 3/5$, use the definitions of the trigonometric functions to find the following:

13. $\tan \theta$ **14.** $\sec \theta$ **15.** $\cos \theta$

In Exercises 16 to 18, if θ is an acute angle and $\tan \theta = 4/3$, use the definitions of the trigonometric functions to find the following:

16. $\sin \theta$ **17.** $\cot \theta$ **18.** $\sec \theta$

In Exercises 19 to 21, if β is an acute angle and $\sec \beta = 13/12$, use the definitions of the trigonometric functions to find the following:

19. $\cos \beta$ **20.** $\cot \beta$ **21.** $\csc \beta$

In Exercises 22 to 24, if θ is an acute angle and $\cot \theta = 3$, use the definitions of the trigonometric functions to find the following:

22. $\sin \theta$ **23.** $\sec \theta$ **24.** $\tan \theta$

In Exercises 25 to 36, find the exact values of the trigonometric functions.

25. $\tan 45°$ **26.** $\sin 60°$ **27.** $\csc 30°$ **28.** $\cos 45°$
29. $\cot 30°$ **30.** $\sec 60°$ **31.** $\sin \pi/4$ **32.** $\cot \pi/6$
33. $\cos \pi/3$ **34.** $\sec \pi/6$ **35.** $\tan \pi/3$ **36.** $\csc \pi/4$

In Exercises 37 to 54, find the values of the trigonometric functions to 4 decimal places.

37. $\sin 12°$ **38.** $\cos 49°$ **39.** $\tan 32°$
40. $\sec 88°$ **41.** $\csc 63°20'$ **42.** $\cot 55°50'$
43. $\cos 34.7°$ **44.** $\tan 81.3°$ **45.** $\sec 5.9°$
46. $\sin \pi/5$ **47.** $\tan \pi/7$ **48.** $\sec 3\pi/8$
49. $\csc 1.2$ **50.** $\sin 0.45$ **51.** $\cos 1.25$
52. $\tan 3/4$ **53.** $\sec 5/8$ **54.** $\cot 3/5$

In Exercises 55 to 74, find the exact value of each expression.

55. $\sin 45° + \cos 45°$ **56.** $\csc 45° - \sec 45°$
57. $\sin 30° - \cos 60°$ **58.** $\tan 45° - \cot 45°$
59. $\sin 30° \cos 60° - \tan 45°$
60. $\csc 60° \sec 30° + \cot 45°$
61. $\cos 30° - \tan 30°$
62. $\sin 60° - \tan 60° \cos 0°$
63. $\sin 30° \cos 60° + \tan 45°$

68. $\csc \frac{\pi}{6} - \sec \frac{\pi}{3}$

70. $\sin \frac{\pi}{3} \cos \frac{\pi}{4} - \tan \frac{\pi}{4}$

71. ... $\frac{}{6}$

72. $\cos \frac{\pi}{4} \tan \frac{\pi}{6} + 2 \tan \frac{\pi}{3}$

73. $2 \csc \frac{\pi}{4} - \ldots \cos \frac{\pi}{6}$

74. $3 \tan \frac{\pi}{4} + \sec \frac{\pi}{6} \sin \frac{\pi}{3}$

Supplemental Exercises

In Exercises 75 to 86, find the value of the acute angle β in degrees and radians without using a calculator.

75. $\sin \beta = 1/2$
76. $\cos \beta = \sqrt{3}/2$
77. $\tan \beta = \sqrt{3}/3$
78. $\sec \beta = 2$
79. $\csc \beta = \sqrt{2}$
80. $\cot \beta = \sqrt{3}$
81. $\cos \beta = \sqrt{2}/2$
82. $\tan \beta = 1$
83. $\sec \beta = 2\sqrt{3}/3$
84. $\sin \beta = \sqrt{2}/2$
85. $\csc \beta = 2\sqrt{3}/3$
86. $\cot \beta = 1$

In Exercises 87 to 92, find the values of the trigonometric functions to 4 decimal places.

87. sin 36°23′4″
88. tan 67°38′26″
89. sec 5°45′34″
90. csc 34°49′17″
91. cos 50°45′9″
92. cot 55°22′12″

In Exercises 93 to 96, show that the equations are true.

93. $\sin(90° - 30°) = \cos 30°$
94. $\tan(90° - 60°) = \cot 60°$
95. $\cos(\pi/2 - \pi/3) = \sin \pi/3$
96. $\sec(\pi/2) - \pi/3) = \csc \pi/3$

In Exercises 97 to 100, use the definitions of the trigonometric functions to verify the equations.

97. $\sin^2\theta + \cos^2\theta = 1$
98. $1 + \tan^2\phi = \sec^2\phi$
99. $\tan \beta = \dfrac{\sin \beta}{\cos \beta}$
100. $\cot^2\phi + 1 = \csc^2\phi$

In Exercises 101 to 108, evaluate the given expressions.

101. $\sin^2 30° - 2\cos^2 45°$
102. $\cos 45° \tan^2 60° + 3 \cot^2 45°$
103. $\csc^2 \frac{\pi}{3} \sec^2 \frac{\pi}{6} + 2 \cot^2 \frac{\pi}{4}$
104. $\sin^3 \frac{\pi}{6} \csc^3 \frac{\pi}{6} + \cos^2 \frac{\pi}{4}$
105. $\sec 45° - \csc 60° \sin 60°$
106. $\cot 30° \tan 60° - \tan 60°$
107. $\sec \frac{\pi}{3} \sin \frac{\pi}{6} - \tan \frac{\pi}{4}$
108. $\sin^2 \frac{\pi}{6} - \tan^2 \frac{\pi}{3} \cos \frac{\pi}{6}$

In Exercises 109 to 112, determine which of the following equations are true.

109. $\tan \phi \cos \phi = \sin \phi$
110. $\cot \phi \sin \phi = \cos \phi$
111. $\sec \phi \sin \phi = \tan \phi$
112. $\csc \phi \cos \phi = \cot \phi$

2.3 Trigonometric Functions of Any Angle

The application of trigonometry would be quite limited if all angles had to be acute. Fortunately, this is not the case. In this section we extend the definition of a trigonometric function to include any angle.

We begin by considering an angle θ in standard position. Let $P(a, b)$ be any point along the terminal side of the angle θ. We then define the trigonometric functions according to the following definitions.

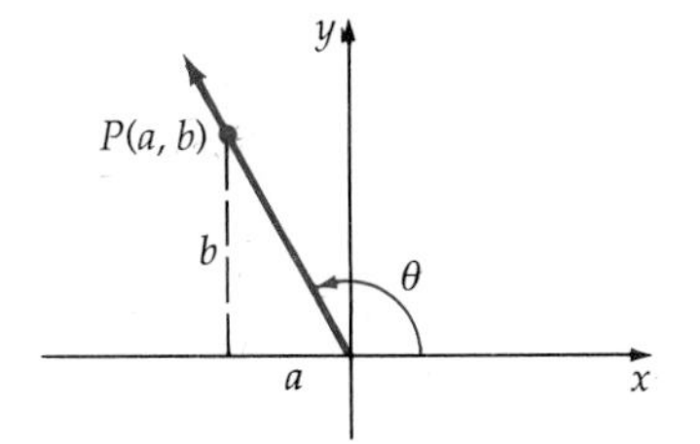

Figure 2.26

Definition of the Trigonometric Functions of Any Angl

Let $P(a, b)$ be any point, except the origin, on the terminal sid angle θ in standard position. Let $r = d(O, P)$, the distance from t origin to P. The six trigonometric functions of θ are

$$\sin\theta = \frac{b}{r} \qquad \cos\theta = \frac{a}{r} \qquad \tan\theta = \frac{b}{a}, \quad a \neq 0$$

$$\csc\theta = \frac{r}{b}, \quad b \neq 0 \qquad \sec\theta = \frac{r}{a}, \quad a \neq 0 \qquad \cot\theta = \frac{a}{b}, \quad b \neq 0$$

where $r = \sqrt{a^2 + b^2}$

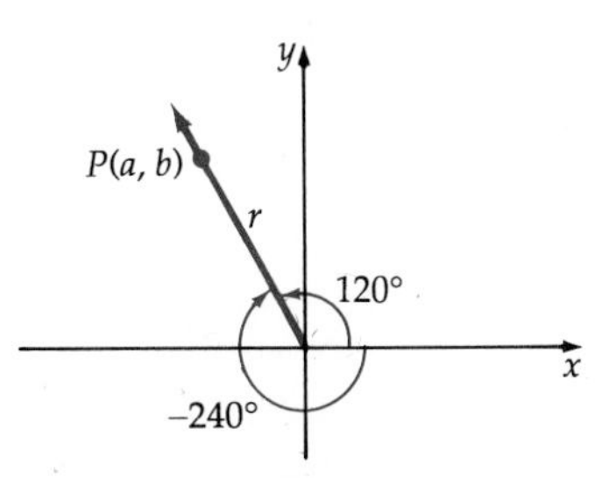

Figure 2.27

Remark Angle θ can be positive or negative. Note from Figure 2.27 that $\sin 120° = \sin(-240°)$, because point $P(a, b)$ is on the terminal side of the 120° angle and the −240° angle.

The advantage of our extended definition is that we can compute the trigonometric functions of any angle. The numbers a and b can be positive, negative, or zero, depending on where the terminal side of the angle is located.

Any point in a rectangular coordinate system can determine an angle in standard position. For example, the point $P(-4, 3)$ in the second quadrant (Figure 2.28) determines an angle θ in standard position with $r = \sqrt{(-4)^2 + 3^2} = 5$. The values of the trigonometric functions of θ are

$$\sin\theta = \frac{3}{5} \qquad \cos\theta = \frac{-4}{5} = -\frac{4}{5} \qquad \tan\theta = \frac{3}{-4} = -\frac{3}{4}$$

$$\csc\theta = \frac{5}{3} \qquad \sec\theta = \frac{5}{-4} = -\frac{5}{4} \qquad \cot\theta = \frac{-4}{3} = -\frac{4}{3}$$

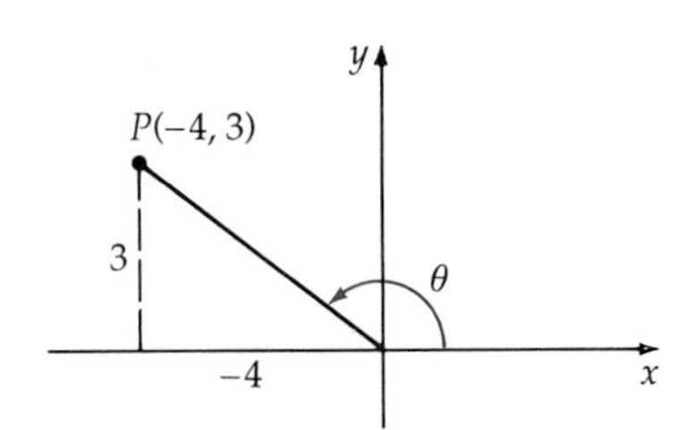

Figure 2.28

As this example shows, the sign of a trigonometric function depends on the quadrant in which the terminal side of the angle lies. For example, if θ is in the third quadrant and $P(a, b)$ is on the terminal side of angle θ, both a and b are negative. Thus only the tangent and cotangent functions are positive in the third quadrant. Since a and b are both negative, the quotient of b/a or a/b will be positive.

Table 2.3 lists the sign of the six trigonometric functions in each quadrant. Figure 2.29 indicates the quadrant in which the functions are positive.

y
sin, csc Positive
A11 Positive
x
tan, cot Positive
cos, sec Positive

Figure 2.29

Table 2.3

	θ in Quadrant			
	I	II	III	IV
$\sin\theta$ and $\csc\theta$	Positive	Positive	Negative	Negative
$\cos\theta$ and $\sec\theta$	Positive	Negative	Negative	Positive
$\tan\theta$ and $\cot\theta$	Positive	Negative	Positive	Negative

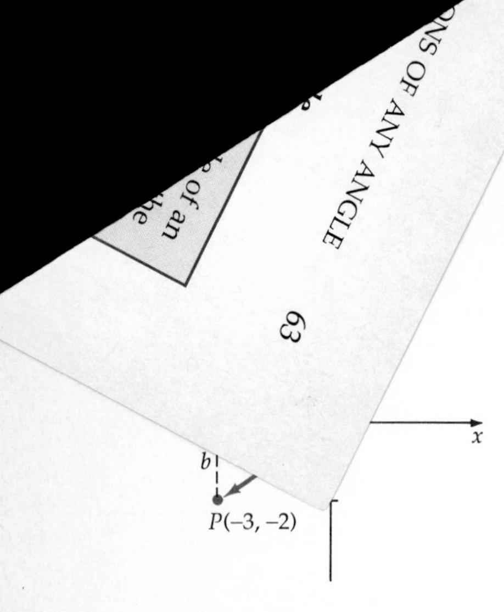

Figure 2.30

EXAMPLE 1 Evaluate the Trigonometric Functions of an Angle in the Third Quadrant

Find the six trigonometric functions of an angle whose terminal side contains the point $(-3, -2)$.

Solution Sketch the angle (Figure 6.30). Find r by using the Pythagorean Theorem.

$$r = \sqrt{(-3)^2 + (-2)^2} = \sqrt{9 + 4} = \sqrt{13} \quad a = -3,\ b = -2$$

$$\sin \phi = \frac{-2}{\sqrt{13}} = -\frac{2\sqrt{13}}{13} \qquad \cos \phi = \frac{-3}{\sqrt{13}} = -\frac{3\sqrt{13}}{13} \qquad \tan \phi = \frac{-2}{-3} = \frac{2}{3}$$

$$\csc \phi = \frac{\sqrt{13}}{-2} = -\frac{\sqrt{13}}{2} \qquad \sec \phi = \frac{\sqrt{13}}{-3} = -\frac{\sqrt{13}}{3} \qquad \cot \phi = \frac{-3}{-2} = \frac{3}{2}$$

■ *Try Exercise* **6**, *page 68.*

A **quadrantal angle** is an angle of 0° (0), 90° ($\pi/2$), 180° (π), or 270° ($3\pi/2$). The terminal side of a quadrantal angle coincides with the x- or y-axis. The trigonometric function value of a quadrantal angle can be found by choosing any point on the terminal side of the quandrantal angle and then applying the definition of the trigonometric function.

The terminal side of 0° coincides with the positive x-axes. Let $P(a, 0)$ be any point on the x-axis. Then $b = 0$ and $r = a$. The values of the six trigonometric functions at 0° are given by

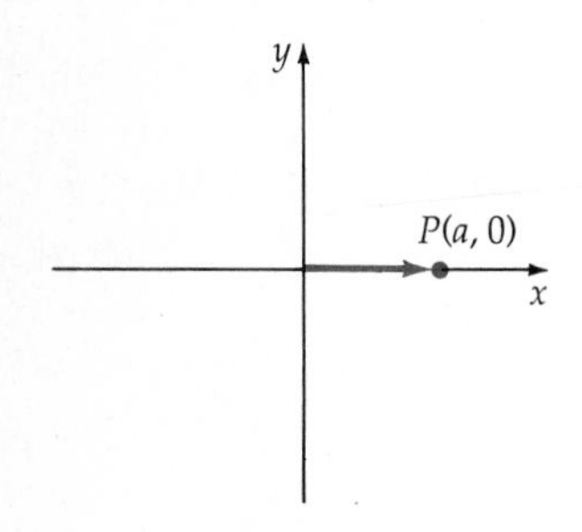

Figure 2.31

$$\sin 0° = \frac{0}{r} = 0 \qquad \cos 0° = \frac{a}{r} = \frac{a}{a} = 1 \qquad \tan 0° = \frac{0}{a} = 0$$

$$\csc 0° \text{ is undefined} \qquad \sec 0° = \frac{r}{a} = \frac{a}{a} = 1 \qquad \cot 0° \text{ is undefined}$$

Remark For the point $(a, 0)$, $\csc 0° = r/0$, which is undefined. A similar statement is true for $\cot 0°$.

In like manner, the trigonometric functions of 90°, 180°, and 270° can be found by using a point on the terminal side of each angle and using the definitions of the trigonometric functions. The results are shown in Table 2.4.

The values of the trigonometric functions in the table in Appendix III are for angles between 0° and 90°. To find the value of a trigonometric function for some other angle, a reference angle is used.

TABLE 2.4

θ	$\sin\theta$	$\cos\theta$	$\tan\theta$	$\csc\theta$	$\sec\theta$	$\cot\theta$
0°	0	1	0	Undefined	1	Undefined
90°	1	0	Undefined	1	Undefined	0
180°	0	−1	0	Undefined	−1	Undefined
270°	−1	0	Undefined	−1	Undefined	0

Reference Angle

> The **reference angle** for the angle α is the positive acute angle
> by the terminal side of α and the x-axis.

Figure 2.32 shows the reference angle β for an angle α whose terminal side lies in the second, third, or fourth quadrants. Quadrantal angles do not have a reference angle. The reference angle for an angle in the first quadrant is the given angle.

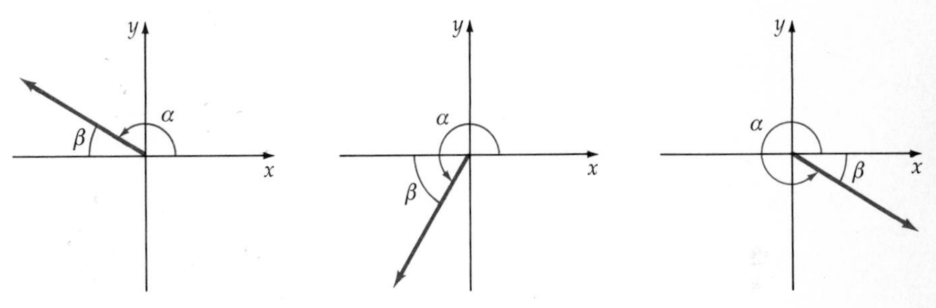

Figure 2.32

Figure 2.33 illustrates three examples of finding a reference angle:

1. The terminal side of a 224° angle lies in the third quadrant. The reference angle is the angle formed by the terminal side of the angle and the negative x-axis. The measurement of the reference angle is 44°.
2. The terminal side of a $-47°$ angle lies in the fourth quadrant. The reference angle is the angle formed by the terminal side of the angle and the positive x-axis. The measurement of the reference angle is 47°.
3. The terminal side of an angle with measure $-8\pi/3$ radians lies in the third quadrant. The reference angle is the angle formed by the terminal side of the angle and the negative x-axis. The measurement of the reference angle is $\pi/3$.

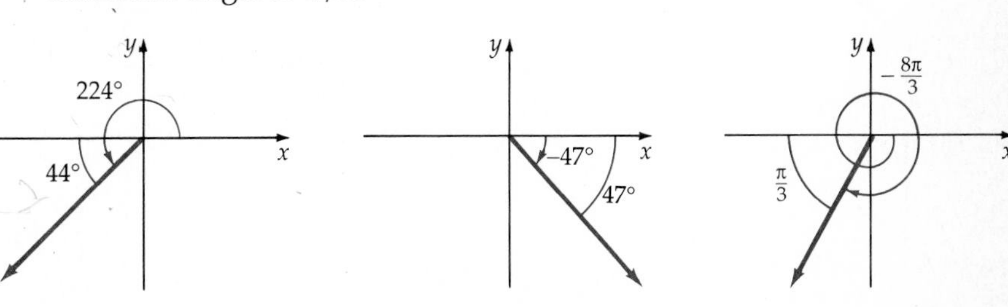

Figure 2.33

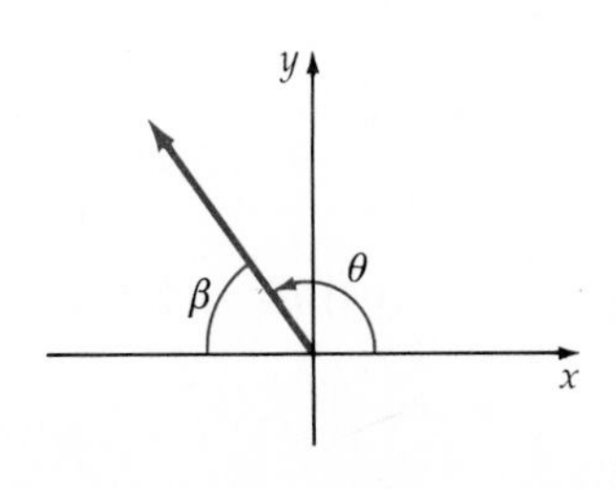

Figure 2.34

Value of a Trigonometric Function for Any Angle

> Let θ be any angle and β be the reference angle for θ. The trigonometric function of θ is the trigonometric function of β with the appropriate sign depending on the quadrant where the terminal side lies.

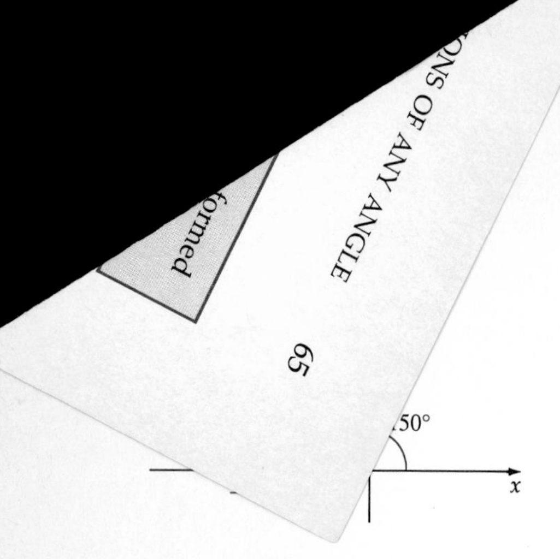

Figure 2.35

EXAMPLE 2 **Evaluate the Six Trigonometric Functions of an Angle By Using the Reference Angle**

Find the exact value of the six trigonometric functions of 150°.

Solution Sketch a 150° angle in standard position. The measure of the reference angle is 180° − 150° = 30°. Use the values of the trigonometric functions for a 30° angle and Table 2.3 to attach the appropriate sign.

$$\sin 150° = \frac{1}{2} \qquad \cos 150° = -\frac{\sqrt{3}}{2} \qquad \tan 150° = -\frac{\sqrt{3}}{3}$$

$$\csc 150° = 2 \qquad \sec 150° = -\frac{2\sqrt{3}}{3} \qquad \cot 150° = -\sqrt{3}$$

■ *Try Exercise* **10**, *page 68.*

EXAMPLE 3 **Evaluate Trigonometric Functions**

Given that $\cos \phi = \frac{-\sqrt{3}}{2}$ for an angle ϕ in the third quadrant. Find the exact values of the other five trigonomeric functions.

Solution Sketch a right triangle with the angle ϕ in standard position and b the unknown side.

$r^2 = a^2 + b^2$ — Use the Pythagorean Theorem to find b.

$2^2 = (-\sqrt{3})^2 + b^2$

$b = \pm 1$

$= -1$ — Because the terminal side of the angle is in the third quadrant, b is a negative number.

Use the definition of the trigonometric functions to find their exact values. We have $a = -\sqrt{3}$, $b = -1$, and $r = 2$ (Figure 2.36).

$$\sin \phi = \frac{b}{r} = -\frac{1}{2} \qquad \tan \phi = \frac{b}{a} = \frac{-1}{-\sqrt{3}} = \frac{\sqrt{3}}{3}$$

$$\csc \phi = \frac{r}{b} = \frac{2}{-1} = -2 \qquad \sec \phi = \frac{r}{a} = \frac{2}{-\sqrt{3}} = -\frac{2\sqrt{3}}{3}$$

$$\cot \phi = \frac{a}{b} = \frac{-\sqrt{3}}{-1} = \sqrt{3}$$

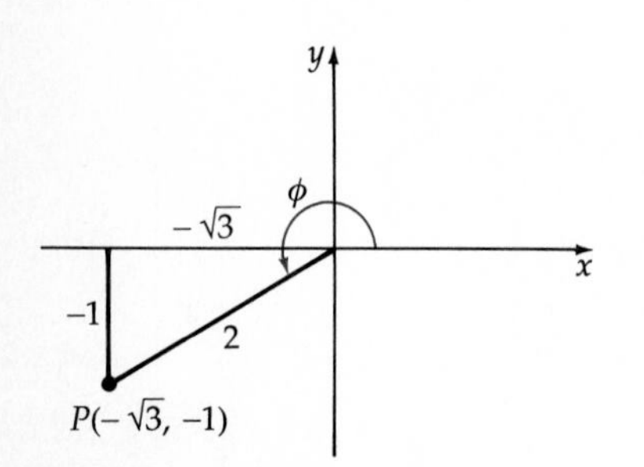

Figure 2.36

■ *Try Exercise* **40**, *page 69.*

EXAMPLE 4 **Evaluate a Trigonometric Expression of Special Angles**

Find the exact value of $\sin \frac{\pi}{2} \cos \frac{5\pi}{6} - \tan \frac{2\pi}{3}$.

Solution Sketch the angles in standard form and find the reference angles. Substitute the function values into the trigonometric expression and simplify.

$$\sin\frac{\pi}{2} = 1$$

$$\cos\frac{5\pi}{6} = -\cos\frac{\pi}{6} = -\frac{\sqrt{3}}{2}$$

$$\tan\frac{2\pi}{3} = -\tan\frac{\pi}{3} = -\sqrt{3}$$

$$\sin\frac{\pi}{2}\cos\frac{5\pi}{6} - \tan\frac{2\pi}{3} = (1)\left(-\frac{\sqrt{3}}{2}\right) - (-\sqrt{3}) = -\frac{\sqrt{3}}{2} + \sqrt{3} = \frac{\sqrt{3}}{2}$$

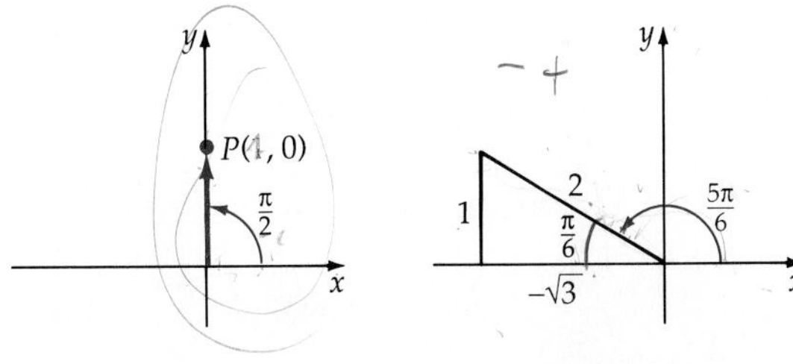

Figure 2.37

■ *Try Exercise* **26**, *page 69.*

When tables are used to find the value of a trigonometric function, the proper sign is attached to the value of the function depending on the quadrant in which the angle lies. A calculator, on the other hand, will correctly evaluate a trigonometric function, including the appropriate sign.

EXAMPLE 5 Evaluate a Trigonometric Expression By Using a Calculator

Find a. $\csc 322.3°$ b. $\tan\frac{7\pi}{12}$ c. $\sin 4.34$.

Solution Use a calculator and follow the key sequence shown.

Function	Key Sequence	Calculator Display
a. $\csc 322.3°$	322.3 [sin] [1/x]	-1.635250666
b. $\tan\frac{7\pi}{12}$	7 [×] π [÷] 12 [=] [tan]	-3.732050808
c. $\sin 4.34$	4.34 [sin]	-0.931460793

■ *Try Exercises* **70**, *page 69.*

Caution In b and c there are no units on the argument of the function. When this occurs, the argument of the function is in radians.

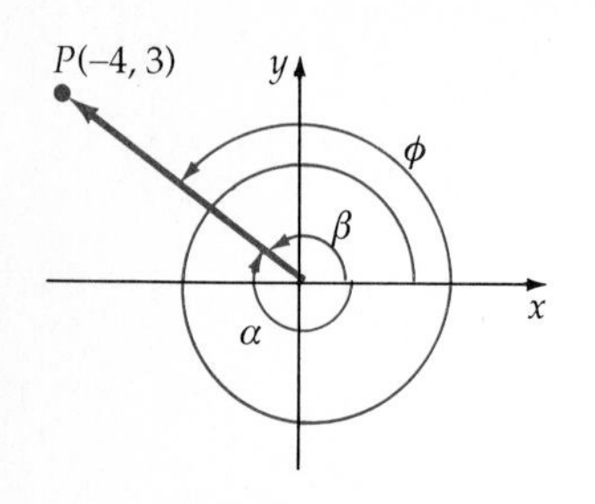

Figure 2.38

Figure 2.38 shows the point $P(-4, 3)$ on the terminal side of an angle β in standard position. The same point is also on the terminal side of the negative angle α. Note also that $\phi = \beta + 360$. Angles that have the same initial and terminal sides are called **coterminal** angles. The trigonometric function values of coterminal angles are equal. Therefore $\sin \beta = \sin \alpha$, and $\sin \beta = \sin \phi$.

To find the reference angle for an angle greater than 360°, first find the coterminal angle less than 360° and then find the reference angle. The terminal side of 660° lies in the fourth quadrant ($660° - 360° = 300°$). Thus the reference angle of 660° is the same as the reference angle of 300°. The reference angle is $360° - 300° = 60°$.

EXAMPLE 6 Evaluate Trigonometric Functions By Using a Coterminal Angle

Find the exact value of the six trigonometric functions of 585°.

Solution First we find the angle $0 \le \phi < 360°$ that is coterminal with 585°. Since 585° is less than two complete rotations, we subtract 360° from 585°: $585° - 360° = 225°$. Now we determine the reference angle of 225°. Since 225° is in the third quadrant, the reference angle is formed by the terminal side of ϕ and the negative x-axis. Thus the reference angle is $225° - 180° = 45°$.

Evaluate the trigonometric functions of 45° and attach the signs associated with an angle in the third quadrant.

$$\sin 585° = -\sin 45° = -\frac{\sqrt{2}}{2} \qquad \cos 585° = -\cos 45° = -\frac{\sqrt{2}}{2}$$

$$\tan 585° = \tan 45° = 1 \qquad \csc 585° = -\csc 45° = -\sqrt{2}$$

$$\sec 585° = -\sec 45° = -\sqrt{2} \qquad \cot 585° = \cot 45° = 1$$

■ *Try Exercise* **88**, *page 69.*

EXERCISE SET 2.3

In Exercises 1 to 8, find the six trigonometric function values with the given points on the terminal side of the angle.

1. $P(2, 3)$ **2.** $P(3, 7)$ **3.** $P(-2, 3)$ **4.** $P(-3, 5)$
5. $P(-8, -5)$ **6.** $P(-6, -9)$ **7.** $P(-5, 0)$ **8.** $P(0, 2)$

In Exercises 9 to 16, find the exact values of the six trigonometric functions of the given angles. Do not use a calculator.

9. 330° **10.** 225° **11.** 210° **12.** 315°
13. $\pi/3$ **14.** $7\pi/6$ **15.** $11\pi/6$ **16.** $3\pi/4$

In Exercises 17 to 28, find the exact value of each expression.

17. $\cos 0° - \sin 30° \tan 45°$

18. $\sin 90° \cos 60° + \cos 60°$

19. $\sin 210° - \cos 330° \tan 330°$

20. $\tan 225° + \sin 240° \cos 60°$

21. $\sin^2 30° + \cos^2 30°$

22. $\tan^2 60° - 2 \sin 30°$

23. $\cos \pi \sin(7\pi/4) - \tan(11\pi/6)$

24. $\cos(7\pi/6)\tan(3\pi/4) - \sin(\pi/6)$

25. $\sin(3\pi/2)\tan(\pi/4) - \cos(\pi/3)$

26. $\cos(7\pi/4)\tan(4\pi/3) + \cos(7\pi/6)$

27. $\sin^2(5\pi/4) + \cos^2(5\pi/4)$

28. $\tan^2(7\pi/4) - \sec^2(7\pi/4)$

In Exercises 29 to 34, let ϕ be an angle in standard position. State the quadrant in which the terminal side of ϕ lies.

29. $\sin\phi > 0$, $\cos\phi > 0$

30. $\tan\phi < 0$, $\sin\phi < 0$

31. $\cos\phi > 0$, $\tan\phi < 0$

32. $\sin\phi < 0$, $\cos\phi > 0$

33. $\sin\phi < 0$, $\cos\phi < 0$

34. $\tan\phi < 0$, $\cos\phi < 0$

In Exercises 35 to 48, find the value of each expression.

35. $\sin\phi = -1/2$; ϕ in quadrant III, find $\tan\phi$.

36. $\tan\phi = -\sqrt{3}$; ϕ in quadrant IV, find $\sin\phi$.

37. $\cos\phi = -\sqrt{3}/2$; ϕ in quadrant III, find $\sin\phi$.

38. $\cot\phi = -1$; ϕ in quadrant II, find $\cos\phi$.

39. $\csc\phi = \sqrt{2}$; ϕ in quadrant II, find $\cot\phi$.

40. $\sec\phi = 2\sqrt{3}/3$; ϕ in quadrant IV, find $\sin\phi$.

41. $\sin\phi = -1/2$ and $\cos\phi > 0$, find $\tan\phi$.

42. $\tan\phi = 1$ and $\sin < 0$, find $\cos\phi$.

43. $\cos\phi = 1/2$ and $\tan\phi = \sqrt{3}$, find $\csc\phi$.

44, $\tan\phi = 1$ and $\sin\phi = -\sqrt{2}/2$, find $\sec\phi$.

45. $\cos\phi = -1/2$ and $\sin\phi = \sqrt{3}/2$, find $\cot\phi$.

46. $\cot\phi = 1$ and $\csc\phi = -\sqrt{2}$, find $\cos\phi$.

47. $\sec\phi = 2\sqrt{3}/3$ and $\sin\phi = -1/2$, find $\cot\phi$.

48. $\sin\phi = -\sqrt{2}/2$ and $\sec\phi = -\sqrt{2}$, find $\cot\phi$.

Calculator Exercises

In Exercises 49 to 64, use a calculator to evaluate the following functions to 4 decimal places.

49. sin 127°	**50.** sin 257°	**51.** cos 116°	**52.** cos 355°
53. tan 548°	**54.** sin 398°	**55.** cos 578°	**56.** sin 740°
57. $\sin\pi/5$	**58.** $\cos 3\pi/7$	**59.** $\cos 9\pi/5$	**60.** $\tan 11\pi/8$
61. sin 4.12	**62.** sin 6.98	**63.** cos 4.45	**64.** cos 0.34

Supplemental Exercises

In Exercises 65 to 72, find the exact values of the angle ϕ, $0 \le \phi \le 360°$.

65. $\sin\phi = 1/2$

66. $\tan\phi = -\sqrt{3}$

67. $\cos\phi = -\sqrt{3}/2$

68. $\tan\phi = 1$

69. $\csc\phi = -\sqrt{2}$

70. $\cot\phi = -1$

71. $\csc\phi = -2\sqrt{3}/3$

72. $\sin\phi = \sqrt{3}/2$

In Exercises 73 to 80, find the exact values of the angle ϕ, $0 \le \phi < 2\pi$.

73. $\tan\phi = -1$

74. $\cos\phi = 1/2$

75. $\tan\phi = -\sqrt{3}/3$

76. $\sec\phi = -2\sqrt{3}/3$

77. $\sin\phi = \sqrt{3}/2$

78. $\cos\phi = -1/2$

79. $\cot\phi = \sqrt{3}$

80. $\sin\phi = \sqrt{2}/2$

In Exercises 81 to 94, verify the given equation.

81. $\sin^2 30° + \cos^2 30° = 1$

82. $1 + \tan^2 60° = \sec^2 60°$

83. $\sin(180° + 30°) = -\sin 30°$

84. $\sin(90° + 60°) = \cos 60°$

85. $\cos 90° = \cos 60° \cos 30° - \sin 60° \sin 30°$

86. $\tan(90° + 30°) = -\cot 30°$

87. $\sin 30° \tan 30° = \sec 30° - \cos 30°$

88. $\sin 120° \sec 120° = \tan 120°$

89. $\dfrac{\sin 45° + \cos 45°}{\cos 45°} = \tan 45° + 1$

90. $\tan^2 60° \cos^2 60° = \sin^2 60°$

91. $\cot^2 210° \sin^2 210° = \cos^2 210°$

92. $\dfrac{\sin 30° + \cos 30°}{\cos 30°} = \tan 30° + 1$

93. $1 + \tan^2 135° = \dfrac{1}{\cos^2 135°}$

94. $\dfrac{1}{\sin 60° \cos 60°} = \csc 60° \sec 60°$

In Exercises 95 to 98, use a calculator to find the trigonometric functions to 4 decimal places.

95. sin 34°23′12″

96. cos 123°14′56″

97. csc 235°48′5″

98. cot 145°34′51″

In Exercises 99 to 102, verify the equation by using the definitions of the trigonometric functions.

99. $\sin(-\phi) = -\sin\phi$

100. $\tan(-\phi) = -\tan\phi$

101. $\cos(-\phi) = \cos\phi$

102. $\sin(180° + \phi) = -\sin\phi$

103. The sine function can be approximated by the following polynomial function, where x is measured in radians.

$$\sin x \approx x - \frac{x^3}{6} + \frac{x^5}{120} - \frac{x^7}{5040}$$

a. Use the polynomial approximation of the sine function to evaluate sin 0.5. Compare with the calculator value of the function.

b. Use the polynomial approximation of the sine function to evaluate $\sin\pi/3$. Compare with the calculator value of the function.

2.4 Circular Functions

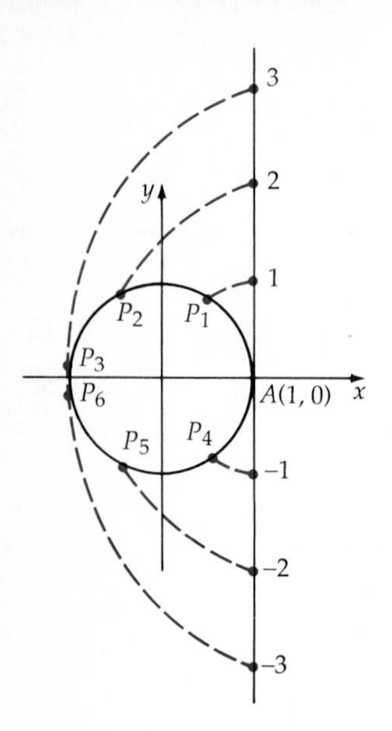

Figure 2.39

During the seventeenth century, applications of trigonometry were broadened to problems in physics and engineering. These kinds of problems required trigonometric functions whose domains of the functions were sets of real numbers rather than sets of angles. The definitions of trigonometric functions were extended by using a correspondence between a number and an angle.

One correspondence involves a circle with a radius of 1, called the **unit circle.** Consider a vertical number line tangent to a unit circle at the point $A(1, 0)$. Wrapping the line around the circle creates a one-to-one correspondence between a point t on the line and a point $P(x, y)$ on the unit circle. By wrapping the line *counterclockwise,* a positive real number is paired with a point on the unit circle. A *clockwise* wrapping of the line pairs a negative real number with a point on the unit circle.

Each real number t defines an arc $\widehat{AP}$ of measure t (see Figure 2.40). The arc $\widehat{AP}$ subtends an angle θ in standard position.

Recall that an arc length t on a circle of radius r subtends an angle θ such that $t = r\theta$ or $t = \theta$ for a unit circle, $r = 1$. Thus, on the unit circle, *the measure of a central angle and the length of an arc can be represented by the same real number* t.

As an example, consider a unit circle and the point $\pi/3$ on the number line tangent to the circle at $A(1, 0)$ paired with a point $P(x, y)$ on the circle. The length of the arc $t = \pi/3$. Since the measure of an angle and the length of the arc in a unit circle can be represented by the same number, $\theta = \pi/3$. From the definitions of the sine and cosine functions, we can find the coordinates of the point $P(x, y)$.

$$\cos\theta = \cos\frac{\pi}{3} = \frac{1}{2} = \frac{x}{r} = x \qquad r = 1$$

$$\sin\theta = \sin\frac{\pi}{3} = \frac{\sqrt{3}}{2} = \frac{y}{r} = y$$

Thus the coordinates of P are $P(\frac{1}{2}, \frac{\sqrt{3}}{2})$.

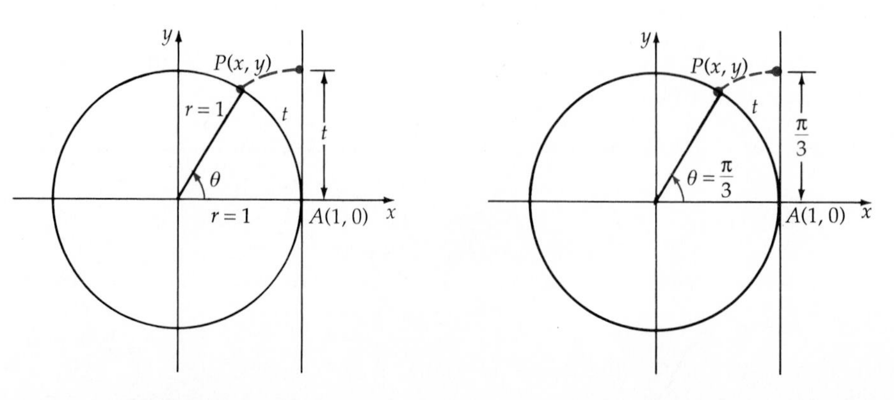

Figure 2.40

Figure 2.41

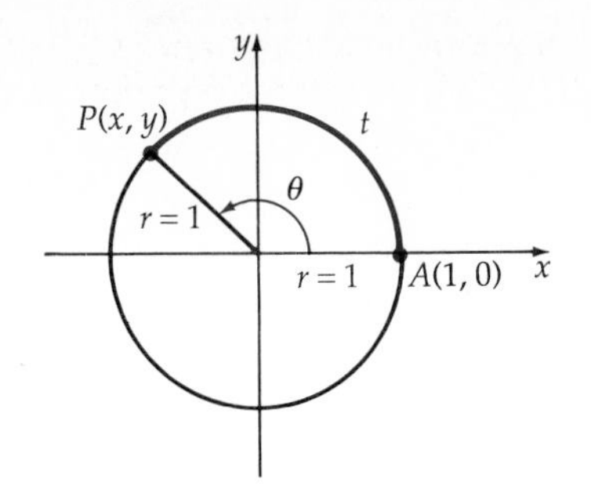

Figure 2.42

Remark The cosine of the real number $\pi/3$ is the x-coordinate and the sine of a real number is the y-coordinate.

To review, a point t on a line corresponds (by wrapping the line around the unit circle) to a point $P(x, y)$ on the unit circle, which in turn corresponds to an angle. With this correspondence, we can define the circular functions of a real number t.

Circular Functions

Let t be a real number and $P(x, y)$ be a point on a unit circle corresponding to t. Then

$\cos t = x$	$\sin t = y$	$\tan t = y/x,\ x \neq 0$
$\sec t = 1/x,\ x \neq 0$	$\csc t = 1/y,\ y \neq 0$	$\cot t = x/y,\ y \neq 0$

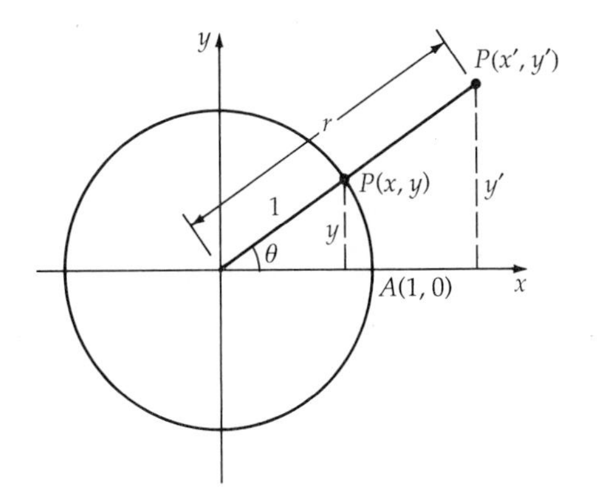

Figure 2.43

Remark The circular functions are defined for real numbers t, whereas the trigonometric functions are defined for angles. However, there is a relationship between the two.

Consider an angle θ (in radians) in standard position. Let $P(x, y)$ and $P'(x', y')$ be two points on the terminal side of the angle, with $x^2 + y^2 = 1$ and $(x')^2 + (y')^2 = r^2$. Let t be the length of the arc from $(1, 0)$ to (x, y). Then

$$\sin \theta = \frac{y}{1} = \frac{y'}{r} = \sin t\,.$$

Thus the value of the sine function of an angle measured in radians is equal to the value of the sine function of the real number t. This is true for each of the trigonometric functions.

The domain and range of the circular functions can be found from the definition of these functions. If t is any real number and $P(x, y)$ is the point on the unit circle corresponding to t, then by definition $\cos t = x$ and $\sin t = y$. Because t takes on all values of the real numbers, $\cos t$ and $\sin t$ are defined for all real numbers t. Thus the domain of the sine and cosine functions is the set of real numbers.

Because the radius of the unit circle is 1, we have $-1 \leq x \leq 1$ and $-1 \leq y \leq 1$. Therefore, with $x = \cos t$ and $y = \sin t$, we have

$$-1 \leq \cos t \leq 1 \quad \text{and} \quad -1 \leq \sin t \leq 1.$$

Thus, the range of the sine and cosine functions is $[-1, 1]$.

Domain and Range of the Sine and Cosine Functions

The domain of the sine and cosine functions is the set of real numbers. The range of the sine and cosine functions is the set of numbers $[-1, 1]$.

Using the definition of tangent and secant, we have

$$\tan t = \frac{y}{x} \quad \text{and} \quad \sec t = \frac{1}{x}.$$

Note the tangent and secant functions are undefined when $x = 0$. The domain of the tangent and secant functions is all real numbers t except those for which the x-coordinate of $P(x, y)$ is zero. The x-coordinate is zero when $t = \pm\frac{\pi}{2}, \pm\frac{3\pi}{2}, \pm\frac{5\pi}{2}$, and in general $\pm(2n + 1)\frac{\pi}{2}$, where n is a whole number. Thus the domain of the tangent and the secant function is all real numbers t except $t = \pm(2n + 1)\frac{\pi}{2}$, where n is a whole number.

Using the definition of cotangent and cosecant, we have

$$\cot t = \frac{x}{y} \quad \text{and} \quad \csc t = \frac{1}{y}.$$

The domain of the cotangent and cosecant functions is all real numbers t except those for which the y-coordinate of $P(x, y)$ is zero. The y-coordinate is zero when $t = 0, \pm\pi, \pm2\pi, \pm3\pi$, and in general $\pm n\pi$, where n is a whole number. Thus the domain of the cotangent and the cosecant function is all real numbers t except $t = \pm n\pi$, where n is a whole number.

We can use the same techniques we used before to evaluate a circular function of a real number t. Consider t an angle in standard position of radian measure t. Evaluate the trigonometric function of the angle by using a calculator (in radian mode) or the tables in Appendix III.

For example,

Function	Key Sequence	Calculator Display
sin(0.5)	0.5 [sin]	0.47942554
cot 20	20 [tan] [1/x]	0.44699511
sec(−2)	2 [+/−] [cos] [1/x]	−2.40299796

Period of the Sine and Cosine Functions

Because the circumference of the unit circle is 2π (Why?)[3], the point $P(x, y)$ on the circle that corresponds to, say, 2 on the number line will also correspond to $2 + 2\pi$, $2 + 4\pi$, $2 + 6\pi$, and in general $2 + n(2\pi)$, where n is a positive integer. Furthermore, it is also true that the point $P(x, y)$ that corresponds to 2 also corresponds to $2 - 2\pi$, $2 - 4\pi$, and in general $2 - n(2\pi)$, where n is a natural number.

For any real number t, the point $P(x, y)$ on the unit circle that corresponds to t also corresponds to $t + 2\pi$, $t + 4\pi$, $t + 6\pi$, and in general, $t + n(2\pi)$, where n is an integer. Thus $\cos t$ and $\cos(t + n \cdot 2\pi)$ correspond to the same x-coordinate. Similarly, $\sin t$ and $\sin(t + n \cdot 2\pi)$ correspond to the same y-coordinate. Therefore,

$$\cos t = \cos(t + n \cdot 2\pi) \quad \text{and} \quad \sin t = \sin(t + n \cdot 2\pi)$$

for all real numbers t and integers n. These equations state that the values of the cosine and sine functions are repeated in every interval of length 2π.

A function that repeats itself is said to be *periodic*.

[3] Since the radius is 1, $C = 2\pi r = 2\pi(1) = 2\pi$.

Period of a Function

> Let p be a constant. If $f(t) = f(t + p)$ for all t in the domain of f, then f is a **periodic function.** The **period** of f is the smallest positive value of p for which $f(t) = f(t + p)$.

The sine and cosine functions are periodic. The period of each of the functions is 2π.

EXAMPLE 1 Use the Periodic Property of the Sine Function

Use the periodic property of the sine function to evaluate $\sin \dfrac{20\pi}{3}$.

Solution

$$\sin \frac{20\pi}{3} = \sin\left(\frac{2\pi}{3} + \frac{18\pi}{3}\right) = \sin\left(\frac{2\pi}{3} + 6\pi\right)$$

$$= \sin\left(\frac{2\pi}{3} + 2(3\pi)\right)$$

$$= \sin \frac{2\pi}{3} \qquad \text{Sin has period } 2\pi.$$

$$= \frac{\sqrt{3}}{2}$$

■ *Try Exercise* **48,** *page 77*

EXAMPLE 2 Use the Unit Circle to Verify an Equation

Use the unit circle to show that $\sin(t + \pi) = -\sin t$.

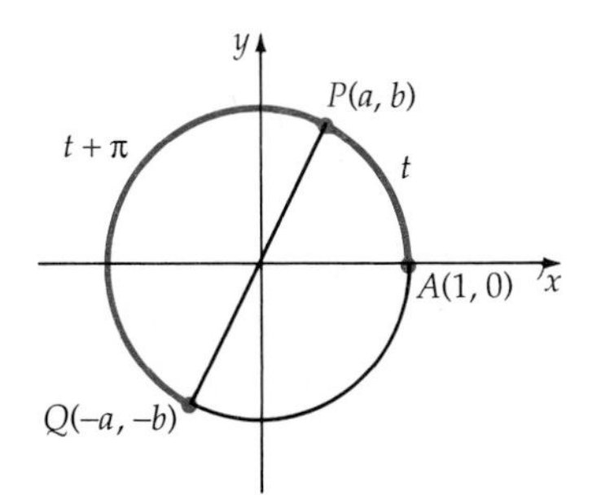

Figure 2.44

Solution Sketch a unit circle, and let P be the point on the unit circle corresponding to t. Draw a line from P through the origin. Label the point Q. Because PQ is a diameter, the length of PQ is π. Thus the arc is $t + \pi$. For any line through the origin, if (a, b) is a point on the line, then $(-a, -b)$ is also a point on the line. Thus if P has coordinates (a, b), then Q has coordinates $(-a, -b)$. From the definition of the sine function, we obtain

$$\sin t = a \qquad \text{and} \qquad \sin(t + \pi) = -a.$$

Therefore, $\sin(t + \pi) = -\sin t$

■ *Try Exercise* **58,** *page 77.*

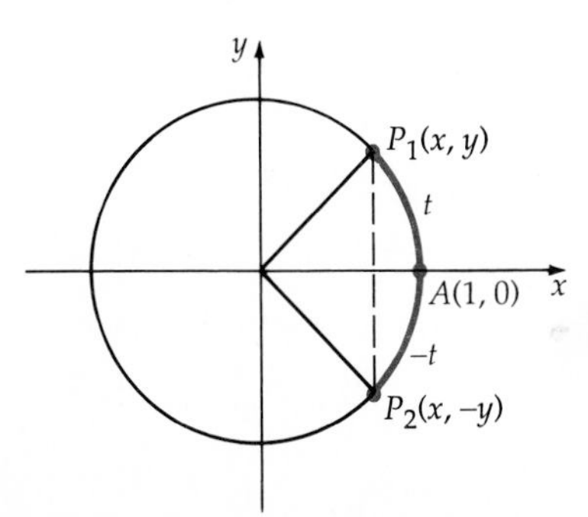

Figure 2.45

Certain important relationships exist among the trigonometric functions. Some of these, the reciprocal functions, we discussed previously. We now derive additional relationships.

Consider the two numbers t and $-t$ represented by the points P_1 and P_2, as shown on the unit circle in Figure 2.45. From the definition of the circular functions:

$$\sin t = y \text{ and } \sin(-t) = -y, \qquad \text{and} \qquad \cos t = x \text{ and } \cos(-t) = x.$$

Substituting $\sin t$ for y and $\cos t$ for x, we have

$$\sin(-t) = -\sin t \text{ and } \cos(-t) = \cos t.$$

Thus the sine function is an odd function and the cosine function is an even function.

Recall that an equation that is true for every number in the domain of the variable is an identity. The trigonometric reciprocal functions defined earlier are examples of trigonometric identities. The statement

$$\sin \theta = \frac{1}{\csc \theta} \qquad \csc \theta \neq 0$$

is a **trigonometric identity** because the two expressions produce the same result for all values of θ for which the functions are defined. By using the definitions of the trigonometric functions, we can prove other trigonometric identities.

The **ratio identities** are obtained by writing the tangent and cotangent functions in terms of the sine and cosine functions. Let $P(x, y)$ be the point on a unit circle corresponding to θ. Recall that $x = \cos \theta$, and $y = \sin \theta$. Substitute for x and y in the equations. $\tan \theta = y/x$ and $\cot \theta = x/y$. Substitute $x = \cos \theta$ and $y = \sin \theta$. The resulting ratio identities are

$$\tan \theta = \frac{y}{x} = \frac{\sin \theta}{\cos \theta}$$

and

$$\cot \theta = \frac{x}{y} = \frac{\cos \theta}{\sin \theta}.$$

The **Pythagorean identities** are based on the fact that $\cos \theta$ and $\sin \theta$ are, respectively, the x- and y-coordinates of a point on a unit circle. Thus

$$x^2 + y^2 = 1 \quad \text{Equation of a unit circle}$$

$$\cos^2\theta + \sin^2\theta = 1 \quad \text{Substitute } x = \cos \theta \text{ and } y = \sin \theta.$$

This is the first Pythagorean identity.

Dividing each term of the first Pythagorean identity by $\cos^2\theta$, we have

$$\frac{\cos^2\theta}{\cos^2\theta} + \frac{\sin^2\theta}{\cos^2\theta} = \frac{1}{\cos^2\theta} \qquad \cos \theta \neq 0$$

$$1 + \tan^2\theta = \sec^2\theta \qquad \frac{\sin \theta}{\cos \theta} = \tan \theta$$

This is the second Pythagorean identity.

Dividing each term of the first Pythagorean identity by $\sin^2\theta$, we have

$$\frac{\cos^2\theta}{\sin^2\theta} + \frac{\sin^2\theta}{\sin^2\theta} = \frac{1}{\sin^2\theta} \qquad \sin \theta \neq 0$$

$$\cot^2\theta + 1 = \csc^2\theta \qquad \frac{\sin \theta}{\cos \theta} = \tan \theta$$

This is the third Pythagorean identity.

The Fundamental Trigonometric Identities

The reciprocal, ratio, and Pythagorean identities are known as the eight **fundamental trigonometric identities** and are used to simplify and rewrite trigonometric expressions.

Reciprocal identities $\sin\theta = \dfrac{1}{\csc\theta}$ $\cos\theta = \dfrac{1}{\sec\theta}$ $\tan\theta = \dfrac{1}{\cot\theta}$

Ratio identities $\tan\theta = \dfrac{\sin\theta}{\cos\theta}$ $\cot\theta = \dfrac{\cos\theta}{\sin\theta}$

Pythagorean identities $\sin^2\theta + \cos^2\theta = 1$ $\tan^2\theta + 1 = \sec^2\theta$ $1 + \cot^2\theta = \csc^2\theta$

EXAMPLE 3 Evaluate Trigonometric Functions Using a Pythagorean Identity

Given that $\sin\theta = \dfrac{1}{2}$ for $\dfrac{\pi}{2} < \theta < \pi$, find $\cos\theta$.

Solution Use the Pythagorean identity $\sin^2\theta + \cos^2\theta = 1$.

$$\sin^2\theta + \cos^2\theta = 1$$

$$\left(\frac{1}{2}\right)^2 + \cos^2\theta = 1$$

$$\cos^2\theta = \frac{3}{4}$$

$$\cos\theta = -\frac{\sqrt{3}}{2} \qquad \text{• } \cos\theta \text{ is negative when } \pi/2 < \theta < \pi.$$

■ *Try Exercise* **96,** *page 77.*

EXAMPLE 4 Express a Trigonometric Expression in Terms of the Sine Function

Express the function $\cot^2 x + 1$ as an expression containing the sine function.

Solution

$$\cot^2 x + 1 = \frac{\cos^2 x}{\sin^2 x} + 1 = \frac{\cos^2 x}{\sin^2 x} + \frac{\sin^2 x}{\sin^2 x} \qquad \text{• } \cot x = \frac{\cos x}{\sin x}$$

$$= \frac{\cos^2 x + \sin^2 x}{\sin^2 x} = \frac{1}{\sin^2 x} \qquad \text{• } \sin^2 x + \cos^2 x = 1$$

■ *Try Exercise* **80,** *page 77.*

EXAMPLE 5 Express the Cotangent Function in Terms of the Cosine Function

Express $\cot \beta$ in terms of $\cos \beta$, where the terminal side of β is in the third quadrant.

Solution

$$\cot \beta = \frac{\cos \beta}{\sin \beta} \qquad \text{Ratio identity}$$

$$= -\frac{\cos \beta}{\sqrt{1 - \cos^2\beta}} \qquad \sin \beta = -\sqrt{1 - \cos^2\beta}\text{. Use the negative because } \sin \beta < 0 \text{ in quadrant III.}$$

■ *Try Exercise* **106,** *page 77.*

EXAMPLE 6 Evaluate a Trigonometric Function By Using Identities

Given that $\tan \beta = -\frac{1}{2}$, where β is an angle whose terminal side is in the fourth quadrant. Find $\sec \beta$ and $\cos \beta$.

Solution

$$1 + \tan^2\beta = \sec^2\beta \qquad \text{Pythagorean identity}$$

$$1 + \left(-\frac{1}{2}\right)^2 = \sec^2\beta \qquad \tan \beta = -1/2$$

$$\frac{5}{4} = \sec^2\beta$$

$$\frac{\sqrt{5}}{2} = \sec \beta \qquad \sec \beta > 0 \text{ in quadrant IV}$$

$$\cos \beta = \frac{1}{\sec \beta} = \frac{1}{\sqrt{5}/2} = \frac{2}{\sqrt{5}} = \frac{2\sqrt{5}}{5}$$

■ *Try Exercise* **98,** *page 77.*

EXERCISE SET 2.4

In Exercises 1 to 6, find the arc length (t) intercepted by the given angle (θ) in a unit circle.

1. $\theta = \pi/3$ radians
2. $\theta = 2\pi/3$ radians
3. $\theta = 7\pi/4$ radians
4. $\theta = 45°$
5. $\theta = 120°$
6. $\theta = 225°$

In Exercises 7 to 12, find the angle (in radians) subtended by the given arc (t) in a unit circle.

7. $t = \pi/4$
8. $t = 2\pi/3$
9. $t = 7\pi/4$
10. $t = 5\pi/4$
11. $t = 2.4$
12. $t = 3.4$

In Exercises 13 to 18, find the angle (in degrees) subtended by the given arc (t) in a unit circle.

13. $t = \pi/6$
14. $t = 5\pi/3$
15. $t = 11\pi/6$
16. $t = 5\pi/4$
17. $t = 2.2$
18. $t = 1.76$

In Exercises 19 to 34, find the coordinates of the point $P(x, y)$ on a unit circle that corresponds to the given real number t.

19. $t = \pi/3$
20. $t = \pi/4$
21. $t = 3\pi/4$
22. $t = 7\pi/4$
23. $t = 7\pi/6$
24. $t = 4\pi/3$
25. $t = 5\pi/3$
26. $t = \pi/6$
27. $t = 11\pi/6$
28. $t = 0$
29. $t = \pi$
30. $t = 9\pi/4$
31. $t = -\pi/3$
32. $t = -7\pi/4$
33. $t = -2\pi/3$
34. $t = -\pi$

In Exercises 35 to 46, evaluate the circular functions.

35. $\sin 1.22$
36. $\cos 4.22$
37. $\tan 5$
38. $\sec 3.5$
39. $\csc(-1.05)$
40. $\sin(-0.55)$
41. $\tan 11\pi/12$
42. $\cot 2\pi/5$
43. $\cos(-\pi/5)$
44. $\csc 8.2$
45. $\sec 1.55$
46. $\cot 2.11$

In Exercises 47 to 56, make use of the period to evaluate each function.

47. $\sin 17\pi/4$
48. $\sin 37\pi/4$
49. $\cos 13\pi/6$
50. $\cos 31\pi/6$
51. $\sin 29\pi/3$
52. $\sin 43\pi/6$
53. $\cos 43\pi/3$
54. $\cos 65\pi/6$
55. $\sin 41\pi/3$
56. $\cos 29\pi/6$

In Exercises 57 to 64, use the unit circle to show that the following expressions are true.

57. $\cos(-\phi) = \cos\phi$
58. $\tan(\phi - 180°) = \tan\phi$
59. $\cos\phi = -\cos(\phi + 180°)$
60. $\sin(-\phi) = -\sin\phi$
61. $\sin(\phi - 180°) = -\sin\phi$
62. $\sec(-\phi) = \sec\phi$
63. $\csc(\alpha) = -\csc\alpha$
64. $\tan(-\alpha) = -\tan\alpha$

In Exercises 65 to 85, use the fundamental identities to transform each expression in terms of the sine and cosine and simplify.

65. $\tan\phi\cos\phi$
66. $\cot\phi\sin\phi$
67. $\dfrac{\csc\phi}{\cot\phi}$
68. $\dfrac{\sec\phi}{\tan\phi}$
69. $1 + \tan^2\phi$
70. $1 + \cot^2\phi$
71. $\dfrac{\tan\phi}{\sec\phi}$
72. $\dfrac{\cot\phi}{\csc\phi}$
73. $\tan\phi + \cot\phi$
74. $\sec\phi + \csc\phi$
75. $1 - \sec^2\phi$
76. $1 - \csc^2\phi$
77. $\tan\phi - \dfrac{\sec^2\phi}{\tan\phi}$
78. $\dfrac{\csc^2\phi}{\cot\phi} - \cot\phi$
79. $\sec^2\phi + \csc^2\phi$
80. $\cos\phi\sec^2\phi - \dfrac{\cos\phi}{\cot^2\phi}$
81. $\dfrac{1 - \cos^2\phi}{\tan^2\phi}$
82. $\dfrac{1 - \sin^2\phi}{\cot^2\phi}$
83. $\sec\phi - \tan\phi\sin\phi$
84. $\dfrac{1}{1 - \cos\phi} + \dfrac{1}{1 + \cos\phi}$
85. $\dfrac{1}{1 - \sin\phi} + \dfrac{1}{1 + \sin\phi}$

In Exercises 86 to 95, simplify the first expression to the second expression.

86. $\dfrac{\sec^2\phi - \tan^2\phi}{\sec^2\phi}$; $\cos^2\phi$
87. $\dfrac{1 - \cos^2\phi}{\cos^2\phi}$; $\tan^2\phi$
88. $\dfrac{\tan\phi + \cot\phi}{\tan\phi}$; $\csc^2\phi$
89. $\dfrac{\csc\phi - \sin\phi}{\csc\phi}$; $\cos^2\phi$
90. $\dfrac{1 - \sin^2\phi}{1 - \cos^2\phi}$; $\cot^2\phi$
91. $\dfrac{1 + \tan^2\phi}{1 + \cot^2\phi}$; $\tan^2\phi$
92. $\sin^2\phi(1 + \cot^2\phi)$; 1
93. $\cos^2\phi(1 + \tan^2\phi)$; 1
94. $1 + \cot^2\phi$; $\csc^2\phi$
95. $\dfrac{1}{\tan\phi\cot\phi}$; 1

In Exercises 96 to 107, use the fundamental identities to find the other circular functions of ϕ.

96. $\csc\phi = \sqrt{2}$; ϕ in quadrant I, find $\sin\phi$.
97. $\cos\phi = 1/2$; ϕ in quadrant IV, find $\sin\phi$.
98. $\sin\phi = 1/2$; ϕ in quadrant II, find $\tan\phi$.
99. $\cot\phi = \sqrt{3}/3$; ϕ in quadrant III, find $\cos\phi$.
100. $\sec\phi = -2$; ϕ in quadrant III, find $\tan\phi$.
101. $\cot\phi = -1$; ϕ in quadrant IV, find $\tan\phi$.
102. Write $\sin\phi$ in terms of $\cos\phi$.
103. Write $\tan\phi$ in terms of $\sec\phi$.
104. Write $\csc\phi$ in terms of $\cot\phi$.
105. Write $\sec\phi$ in terms of $\tan\phi$.
106. Write $\tan\phi$ in terms of $\sin\phi$.
107. Write $\cot\phi$ in terms of $\csc\phi$.

Supplemental Exercises

108. Use the unit circle and the triangles in the figure to write each function in terms of the length of a line segment: **a.** $\sin\phi$; **b.** $\cos\phi$; **c.** $\tan\phi$.

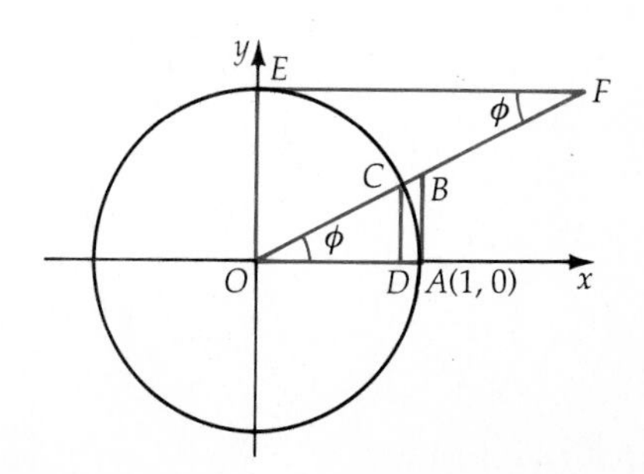

109. Use the unit circle and the triangles in the previous figure to write each function in terms of the length of a line segment: **a.** $\csc \phi$; **b.** $\sec \phi$; **c.** $\cot \phi$.

In Exercises 110 to 119, simplify the first expression to the second expression.

110. $\dfrac{\sin^2\phi + \cos^2\phi}{\sin^2\phi}$; $\csc^2\phi$

111. $\dfrac{\sin^2\phi + \cos^2\phi}{\cos^2\phi}$; $\sec^2\phi$

112. $(\cos \phi - 1)(\cos \phi + 1)$; $-\sin^2\phi$

113. $(\sec \phi - 1)(\sec \phi + 1)$; $\tan^2\phi$

114. $\sec^2\beta - 1$; $\tan^2\beta$

115. $1 - \sin^2\beta$; $\cos^2\beta$

116. $\dfrac{\sec \phi}{\cos \phi} + \dfrac{\csc \phi}{\sin \phi}$; $\sec^2\phi \csc^2\phi$

117. $\dfrac{1}{\sin \phi \cos \phi}$; $\tan \phi + \cot \phi$

118. $1 + \tan^2\phi$; $\dfrac{1}{\cos^2\phi}$

119. $1 + \cot^2\phi$; $\csc^2 \phi$

2.5 Graphs of the Sine and Cosine Functions

Graphing the Sine Function

The trigonometric functions of real numbers can be graphed on a rectangular coordinate system by plotting the points whose coordinates satisfy the function. We begin with the sine function.

Table 2.5 lists some of the ordered pairs (x, y), where $y = \sin x$ between 0 and 2π. Plot the ordered pairs and draw a smooth curve through the points to obtain the graph of $y = \sin x$.

TABLE 2.5

x	0	$\pi/4$	$\pi/2$	$3\pi/4$	π	$5\pi/4$	$3\pi/2$	$7\pi/4$	2π
$\sin x$	0	0.707	1	0.707	0	-0.707	-1	-0.707	0
(x, y)	$(0, 0)$	$\left(\frac{\pi}{4}, 0.707\right)$	$\left(\frac{\pi}{2}, 1\right)$	$\left(\frac{3\pi}{4}, 0.707\right)$	$(\pi, 0)$	$\left(\frac{5\pi}{4}, -0.707\right)$	$\left(\frac{3\pi}{2}, -1\right)$	$\left(\frac{7\pi}{4}, -0.707\right)$	$(2\pi, 0)$

Recall that the period of the sine function is 2π and the domain is the set of real numbers. Thus the graph of the sine function on the interval $[0, 2\pi]$ duplicates itself every 2π. Thus the graph in Figure 2.46 can be ex-

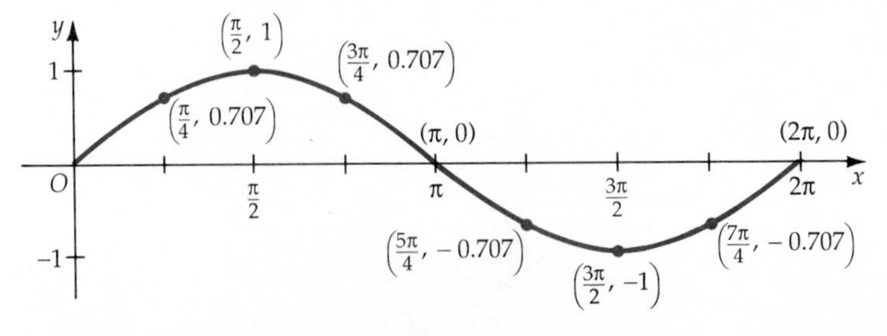

Figure 2.46
$y = \sin x,\ 0 \le x \le 2\pi$

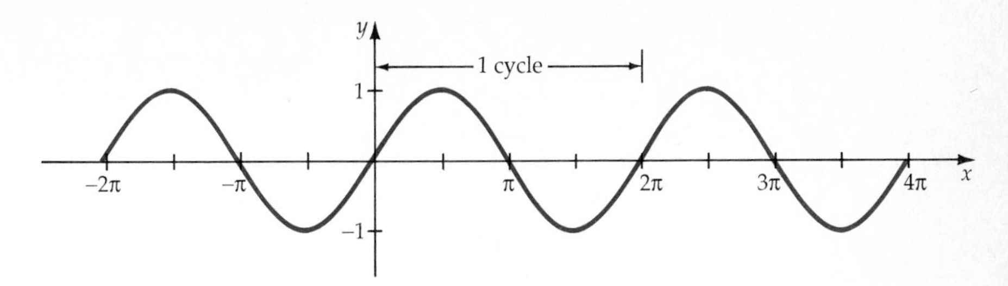

Figure 2.47
$y = \sin x, \ -2\pi \le x \le 4\pi$

tended indefinitely in both directions along the x-axes. The part of the graph corresponding to an interval of one period (2π) is referred to as one **cycle** of the graph.

Properties of the Sine Function

1. The domain of $\sin x$ is the set of real numbers.
2. The range of $\sin x$ is $[-1, 1]$.
3. The function $\sin x$ is periodic and the period is 2π.
4. The function $\sin x$ is an odd function. The graph of the function is symmetric with respect to the origin.

One period of the sine function can be graphed by using five key domain values; which include the maximum and minimum values of the sine as well as the zeros of the function (see Figure 2.48):

1. Beginning point of one cycle
2. Quarter point of one cycle
3. Middle point of one cycle
4. Three-quarter point of one cycle
5. Endpoint of one cycle

The maximum value attained by the function $f(x) = \sin x$ is 1 and the minimum value is -1. The **amplitude** of a sine or cosine function is defined as one-half the difference between the maximum and minimum val-

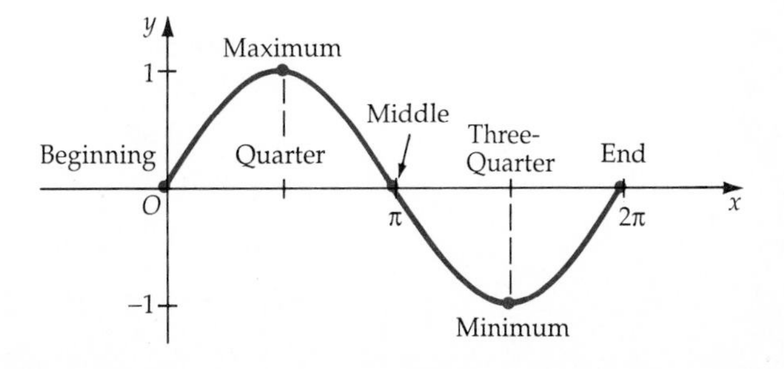

Figure 2.48

ues of the function. For example, if the maximum value of the function is M and the minimum value is m, then

$$\text{Amplitude} = \frac{1}{2}(M - m).$$

For $y = \sin x$, the amplitude is 1:

$$\text{Amplitude} = \frac{1}{2}(1 - (-1)) = 1.$$

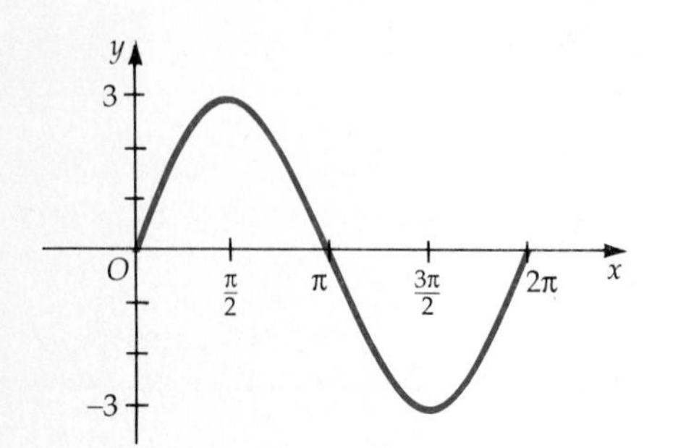

Figure 2.49

Figure 2.49 shows the graph of $f(x) = 3 \sin x$. The graph can be drawn by plotting key points.

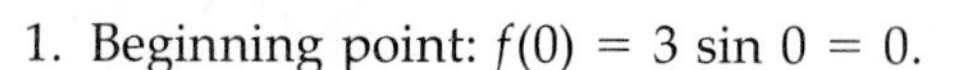

1. Beginning point: $f(0) = 3 \sin 0 = 0$.
2. Quarter point: $f(\pi/2) = 3 \sin \pi/2 = 3$ (maximum value).
3. Middle point $f(\pi) = 3 \sin \pi = 0$.
4. Three-quarter point: $f(3\pi/2) = 3 \sin 3\pi/2 = -3$ (minimum value).
5. Endpoint: $f(2\pi) = 3 \sin 2\pi = 0$.

x	0	$\pi/2$	π	$3\pi/2$	2π
$\sin x$	0	1	0	-1	0
$3 \sin x$	0	3	0	-3	0
(x, y)	$(0, 0)$	$\left(\frac{\pi}{2}, 3\right)$	$(\pi, 0)$	$\left(\frac{3\pi}{2}, -3\right)$	$(2\pi, 0)$

Remark Note that the value of the function $f(x) = 3 \sin x$ at the beginning, middle, and end of the period are zeros of the function. The y-value of the quarter point is the maximum of the function, and the y-value of the three-quarter point is the minimum of the function.

The amplitude of $f(x) = 3 \sin x$ is 3 because

$$\frac{1}{2}(M - m) = \frac{1}{2}(3 - (-3)) = 3.$$

From the graphs in Figures 2.48 and 2.49, the function $\sin x$ has an amplitude of 1, and the amplitude of $f(x) = 3 \sin x$ is 3. This suggests the following theorem.

Amplitude of $f(x) = a \sin x$

> Amplitude of function $f(x) = a \sin x$ is $|a|$.

EXAMPLE 1 **Graph a Sine Function With an Amplitude of 2**

Graph the sine function $f(x) = -2 \sin x$.

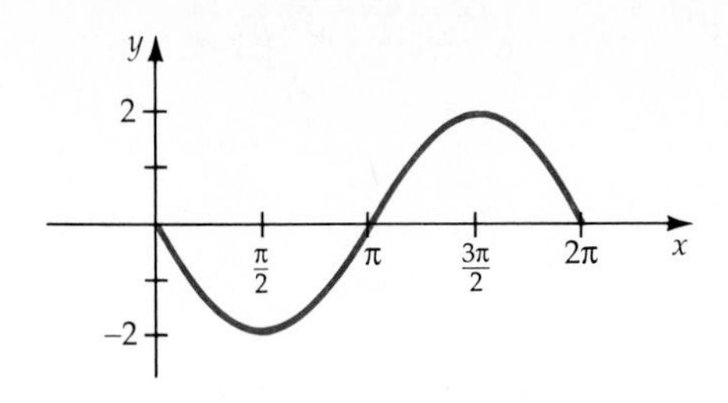

Figure 2.50
$f(x) = -2 \sin x$

Solution The amplitude of the function $f(x) = -2 \sin x$ is 2.

1. Beginning point: $f(0) = -2 \sin 0 = 0$.
2. Quarter point: $f(\pi/2) = -2 \sin \pi/2 = -2$ (minimum value)
3. Middle point: $f(\pi) = -2 \sin \pi = 0$.
4. Three-quarter point: $f(3\pi/2) = -2 \sin 3\pi/2 = 2$ (maximum value).
5. Endpoint: $f(2\pi) = -2 \sin 2\pi = 0$.

x	0	$\pi/2$	π	$3\pi/2$	2π
$\sin x$	0	1	0	-1	0
$-2 \sin x$	0	-2	0	2	0
(x, y)	$(0, 0)$	$\left(\frac{\pi}{2}, -2\right)$	$(\pi, 0)$	$\left(\frac{3\pi}{2}, 2\right)$	$(2\pi, 0)$

■ *Try Exercise* **20,** *page 87.*

Remark Note that the graph of $f(x) = -2 \sin x$ is a *reflection* across the x-axis of $2 \sin x$.

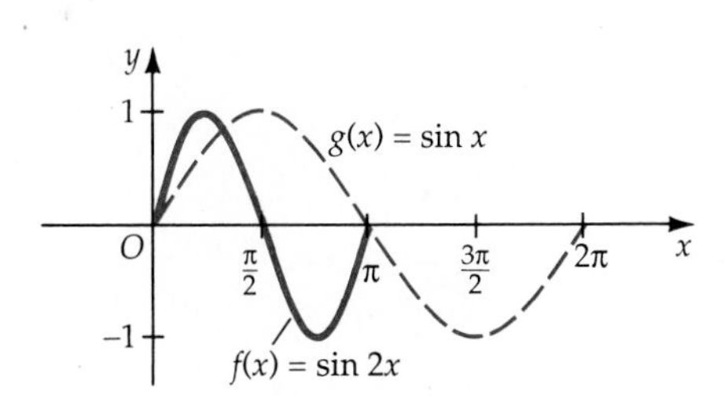

Figure 2.51

Figure 2.51 is the graph of the function $f(x) = \sin 2x$. The dashed graph is that of $g(x) = \sin x$. Because one cycle of the graph of $f(x) = \sin 2x$ is completed in an interval of length π, the period of f is π. Algebraically, one cycle of the function $f(x) = \sin 2x$ is completed as $2x$ varies from 0 to 2π. Thus

$0 \le 2x \le 2\pi$ or $0 \le x \le \pi$ The period of $f(x) = \sin 2x$ is π.

1. Beginning point: $f(0) = \sin 0 = 0$.
2. Quarter point: $f(\pi/4) = \sin \pi/2 = 1$.
3. Middle point: $f(\pi/2) = \sin \pi = 0$.
4. Three-quarter point: $f(3\pi/4) = \sin(3\pi/2) = -1$.
5. Endpoint: $f(\pi) = \sin 2\pi = 0$.

x	0	$\pi/4$	$\pi/2$	$3\pi/4$	π
$2x$	0	$\frac{\pi}{2}$	π	$\frac{3\pi}{2}$	2π
$\sin 2x$	0	1	0	-1	0
(x, y)	$(0, 0)$	$\left(\frac{\pi}{4}, 1\right)$	$\left(\frac{\pi}{2}, 0\right)$	$\left(\frac{3\pi}{4}, -1\right)$	$(\pi, 0)$

Figure 2.52 is the graph of the function $f(x) = \sin(x/2)$. Because one cycle of the graph of $f(x) = \sin(x/2)$ is completed in an interval of length 4π, the period of f is 4π. Algebraically, one cycle of the function $f(x) = \sin(x/2)$ is completed as $x/2$ varies from 0 to 2π. Thus

$0 \le x/2 \le 2\pi$ or $0 \le x \le 4\pi$ The period of $f(x) = \sin x/2$ is 4π.

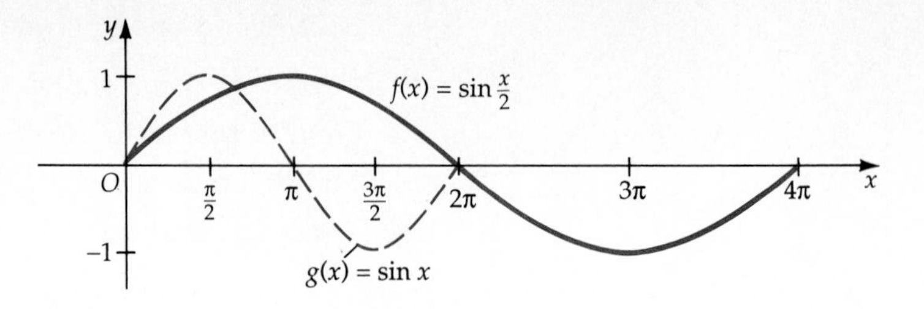

Figure 2.52

1. Beginning point: $f(0) = \sin 0 = 0$.
2. Quarter point: $f(\pi) = \sin \pi/2 = 1$.
3. Middle point: $f(2\pi) = \sin \pi = 0$.
4. Three-quarter point: $f(3\pi) = \sin(3\pi/2) = -1$.
5. Endpoint: $f(4\pi) = \sin 2\pi = 0$.

x	**0**	π	2π	3π	4π
$\frac{x}{2}$	0	$\frac{\pi}{2}$	π	$\frac{3\pi}{2}$	2π
$\sin \frac{x}{2}$	0	1	0	-1	0
(x, y)	$(0, 0)$	$(\pi, 1)$	$(2\pi, 0)$	$(3\pi, -1)$	$(4\pi, 0)$

The previous examples suggest that one cycle of the function $f(x) = \sin bx$, where $b > 0$, is completed as bx varies from 0 to 2π. Algebraically,

$$0 \leq bx \leq 2\pi$$

$$0 \leq x \leq \frac{2\pi}{b}. \quad \text{Divide by } b.$$

The length of the interval, $2\pi/b$, is the period of $f(x) = \sin bx$.

Remark If $b > 0$, then $f(x) = \sin(-bx) = -\sin bx$. Therefore, function with $b > 0$. The period is $2\pi/b$. Table 2.6 shows the amplitude and period for several sine functions.

TABLE 2.6

Function	Amplitude	Period
$f(x) = a \sin bx$	$\|a\|$	$\frac{2\pi}{b}$
$f(x) = 3 \sin(-2x)$	$\|3\| = 3$	$\frac{2\pi}{2} = \pi$
$f(x) = -\sin \frac{x}{3}$	$\|-1\| = 1$	$\frac{2\pi}{1/3} = 6\pi$
$f(x) = -2 \sin \frac{3x}{4}$	$\|-2\| = 2$	$\frac{2\pi}{3/4} = \frac{8\pi}{3}$

Here is a review of important properties used to graph the sine function $f(x) = a \sin bx$ where $0 \le x \le 2\pi/b$.

1. The amplitude of the function is $|a|$.
2. The period of the function is $2\pi/b$.
3. The zeros of the function are 0, π/b, and $2\pi/b$.

EXAMPLE 2 Graph a Sine Function

Graph one cycle of the function $f(x) = \dfrac{1}{2} \sin \dfrac{2\pi}{3} x$.

Solution Amplitude: $\left|\dfrac{1}{2}\right| = \dfrac{1}{2}$, Period: $\dfrac{2\pi}{2\pi/3} = 3$

Use the five key values to graph the function for values of x such that $0 \le x \le 3$.

1. Beginning point $f(0) = \frac{1}{2} \sin 0 = 0$.
2. Quarter point: $f(3/4) = \frac{1}{2} \sin \pi/2 = \frac{1}{2}$.
3. Middle point: $f(3/2) = \frac{1}{2} \sin \pi = 0$.
4. Three-quarter point: $f(9/4) = \frac{1}{2} \sin 3\pi/2 = -\frac{1}{2}$.
5. Endpoint: $f(3) = \frac{1}{2} \sin 2\pi = 0$.

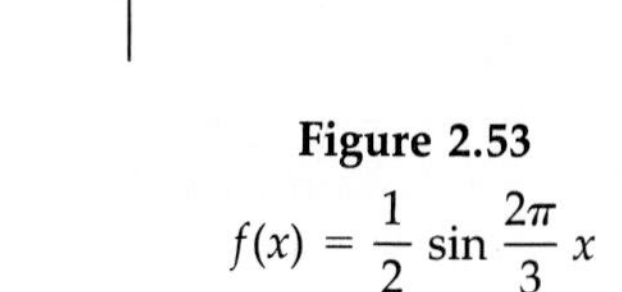

Figure 2.53
$f(x) = \dfrac{1}{2} \sin \dfrac{2\pi}{3} x$

x	0	3/4	3/2	9/4	3
$\dfrac{1}{2} \sin \dfrac{2\pi}{3} x$	0	$\dfrac{1}{2}$	0	$-\dfrac{1}{2}$	0
(x, y)	$(0, 0)$	$\left(\dfrac{3}{4}, \dfrac{1}{2}\right)$	$\left(\dfrac{3}{2}, 0\right)$	$\left(\dfrac{9}{4}, -\dfrac{1}{2}\right)$	$(3, 0)$

■ *Try Exercise* **32,** *page 87.*

The Cosine Function

Figure 2.54 shows the graph of the ordered pairs (x, y) where $y = \cos x$.

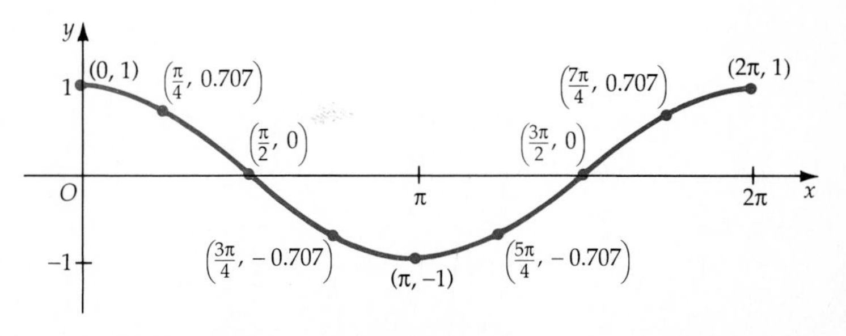

Figure 2.54
$y = \cos x,\ 0 \le x \le 2\pi$

Table 2.7 lists the ordered pairs (x, y), where $y = \cos x$ between 0 and 2π.

TABLE 2.7

x	0	$\pi/4$	$\pi/2$	$3\pi/4$	π	$5\pi/4$	$3\pi/2$	$7\pi/4$	2π
$\cos x$	1	0.707	0	-0.707	-1	-0.707	0	0.707	1
(x, y)	$(0, 1)$	$\left(\frac{\pi}{4}, 0.707\right)$	$\left(\frac{\pi}{2}, 0\right)$	$\left(\frac{3\pi}{4}, -0.707\right)$	$(\pi, -1)$	$\left(\frac{5\pi}{4}, -0.707\right)$	$\left(\frac{3\pi}{2}, 0\right)$	$\left(\frac{7\pi}{4}, 0.707\right)$	$(2\pi, 1)$

Recall that the cosine function is periodic with period 2π. Figure 2.55 shows the graph of the cosine function along the x-axis for $-2\pi \le x \le 4\pi$. The graph of one cycle is repeated every 2π units along the horizontal axes.

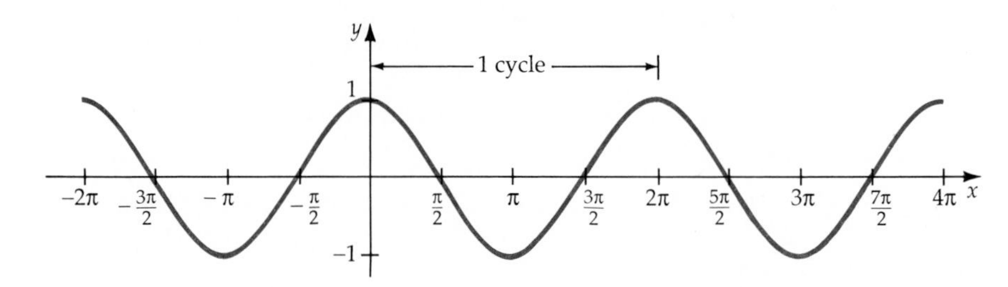

Figure 2.55
$f(x) = \cos x, \ -2\pi \le x \le 4\pi$

Properties of the Cosine Function

1. The domain of the function $\cos x$ is the set of real numbers.
2. The range of the function $\cos x$ is $[-1, 1]$.
3. The function $\cos x$ is periodic and the period is 2π.
4. The function $\cos x$ is an even function. The graph of $\cos x$ is symmetric to the y-axis.

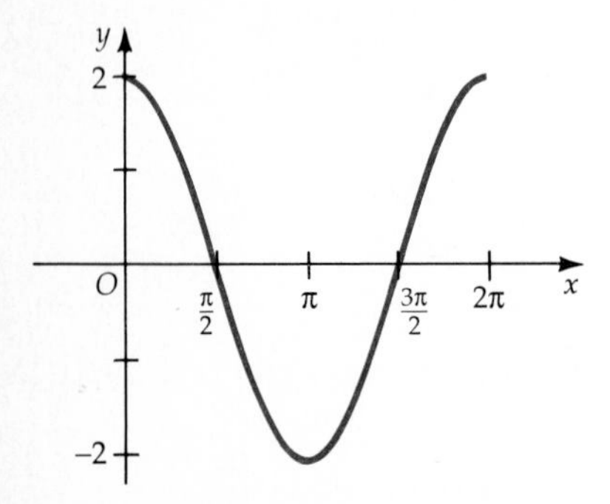

Figure 2.56
$f(x) = 2 \cos x$

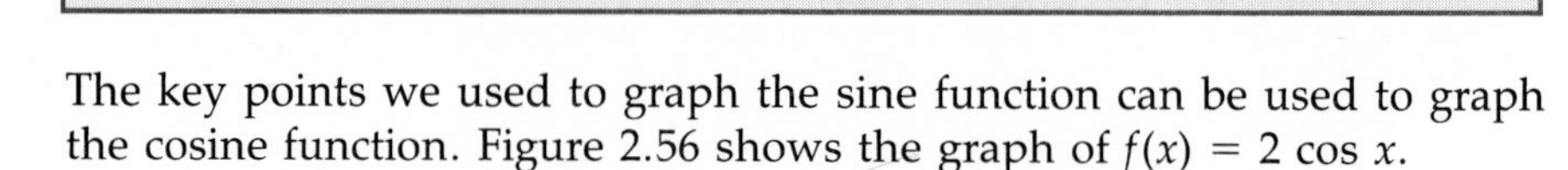

The key points we used to graph the sine function can be used to graph the cosine function. Figure 2.56 shows the graph of $f(x) = 2 \cos x$.

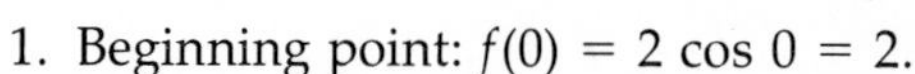

1. Beginning point: $f(0) = 2 \cos 0 = 2$.
2. Quarter point: $f(\pi/2) = 2 \cos \pi/2 = 0$.
3. Middle point: $f(\pi) = 2 \cos \pi = -2$.
4. Three-quarter point: $f(3\pi/2) = 2 \cos 3\pi/2 = 0$.
5. Endpoint: $f(2\pi) = 2 \cos 2\pi = 2$.

x	0	$\pi/2$	π	$3\pi/2$	2π
$\cos x$	1	0	-1	0	1
$2 \cos x$	2	0	-2	0	2
(x, y)	$(0, 2)$	$\left(\frac{\pi}{2}, 0\right)$	$(\pi, -2)$	$\left(\frac{3\pi}{2}, 0\right)$	$(2\pi, 2)$

It is clear from the previous graphs that the amplitude of $f(x) = \cos x$ is 1 and the amplitude of $f(x) = 2 \cos x$ is 2.

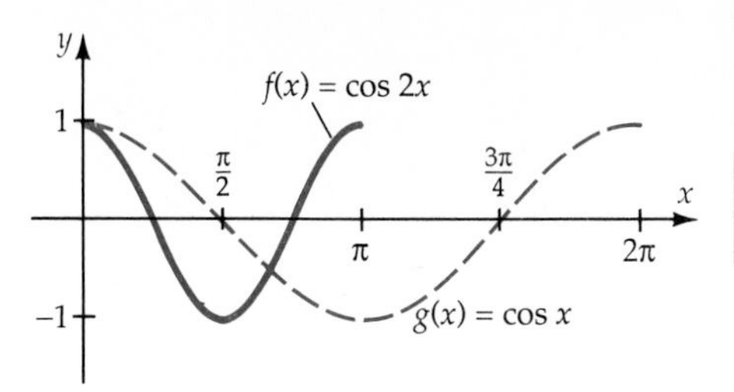

Figure 2.57

Amplitude of $f(x) = a \cos x$

> Amplitude of the function $f(x) = a \cos x$ is $|a|$.

The five key points have been used in Figure 6.57 to graph $f(x) = \cos 2x$. (The dashed graph is that of $g(x) = \cos x$.) The period of $f(x) = \cos 2x$ is π.

1. Beginning point: $f(0) = \cos 0 = 1$.
2. Quarter point: $f(\pi/4) = \cos \pi/2 = 0$.
3. Middle point: $f(\pi/2) = \cos \pi = -1$.
4. Three-quarter point: $f(3\pi/4) = \cos 3\pi/2 = 0$.
5. Endpoint: $f(\pi) = \cos 2\pi = 1$.

x	0	$\pi/4$	$\pi/2$	$3\pi/4$	π
$\cos 2x$	1	0	−1	0	1
(x, y)	$(0, 1)$	$\left(\frac{\pi}{4}, 0\right)$	$\left(\frac{\pi}{2}, -1\right)$	$\left(\frac{3\pi}{4}, 0\right)$	$(\pi, 1)$

As with the sine function, the period of $f(x) = \cos bx$, where $b > 0$, is $2\pi/b$.

Remark If $b > 0$, $f(x) = \cos(-bx) = \cos bx$. Therefore, f has period $2\pi/b$.

Table 2.8 shows the amplitude and period for several cosine functions.

TABLE 2.8

Function	Amplitude	Period
$a \cos bx$	$\lvert a\rvert$	$\frac{2\pi}{b}$
$2 \cos 3x$	$\lvert 2\rvert$	$\frac{2\pi}{3}$
$-3 \cos \frac{2x}{3}$	$\lvert -3\rvert = 3$	$\frac{2\pi}{2/3} = 3\pi$

Here is a review of important properties used to graph $f(x) = a \cos bx$ where $0 \le x \le 2\pi/b$.

1. The amplitude of $a \cos bx$ is $|a|$.
2. The period of $a \cos bx$ is $2\pi/b$.
3. The zeros of $a \cos bx$ are $\pi/2b$ and $3\pi/2b$.

EXAMPLE 3 Graph a Cosine Function

Graph one cycle of the function $f(x) = \frac{1}{2} \cos 3x$.

Solution Amplitude: $\left|\frac{1}{2}\right| = \frac{1}{2}$ Period: $\frac{2\pi}{3}$

Use the five key values to graph the function for values of x for $0 \le x \le 2\pi/3$.

1. Beginning point: $f(0) = 1/2 \cos 0 = 1/2$.
2. Quarter point: $f(\pi/6) = 1/2 \cos \pi/2 = 0$.
3. Middle point: $f(\pi/3) = 1/2 \cos \pi = -1/2$.
4. Three-quarter point: $f(\pi/2) = 1/2 \cos 3\pi/2 = 0$.
5. Endpoint: $f(2\pi/3) = 1/2 \cos 2\pi = 1/2$.

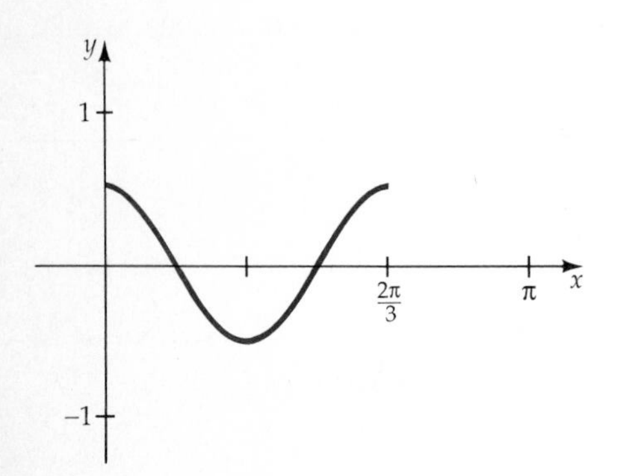

Figure 2.58
$f(x) = \frac{1}{2} \cos 3x$

x	0	$\pi/6$	$\pi/3$	$\pi/2$	$2\pi/3$
$\frac{1}{2} \cos 3x$	$1/2$	0	$-1/2$	0	$1/2$
(x, y)	$\left(0, \frac{1}{2}\right)$	$\left(\frac{\pi}{6}, 0\right)$	$\left(\frac{\pi}{3}, -\frac{1}{2}\right)$	$\left(\frac{\pi}{2}, 0\right)$	$\left(\frac{2\pi}{3}, \frac{1}{2}\right)$

■ *Try Exercise* **28,** *page 87.*

EXAMPLE 4 Graph a Cosine Function

Graph one cycle of the function $f(x) = -2 \cos \frac{\pi x}{2}$.

Solution Amplitude: $|-2| = 2,$ Period: $\frac{2\pi}{\pi/2} = 4$

Use the five key values to graph the function for the values of x for $0 \le x \le 4$.

1. Beginning point: $f(0) = -2 \cos 0 = -2$.
2. Quarter point: $f(1) = -2 \cos \pi/2 = 0$.
3. Middle point: $f(2) = -2 \cos \pi = 2$.
4. Three-quarter point: $f(3) = -2 \cos 3\pi/2 = 0$.
5. Endpoint: $f(4) = -2 \cos 2\pi = -2$.

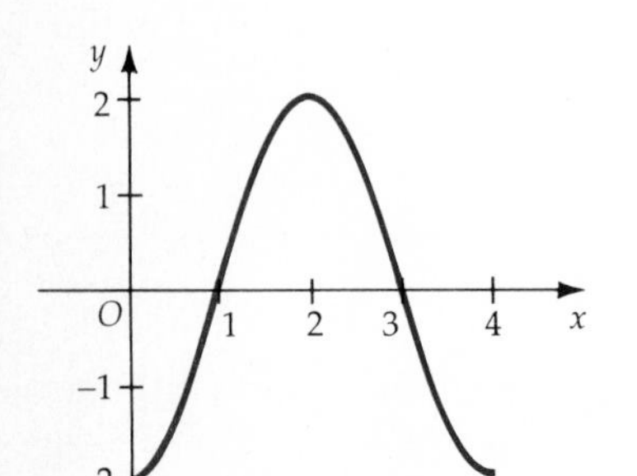

Figure 2.59
$f(x) = -2 \cos \frac{\pi x}{2}$

x	0	1	2	3	4
$-2 \cos \frac{\pi x}{2}$	-2	0	2	0	-2
(x, y)	$(0, -2)$	$(1, 0)$	$(2, 2)$	$(3, 0)$	$(4, -2)$

■ *Try Exercise* **34,** *page 87.*

EXAMPLE 5 **Graph the Absolute Value of the Cosine Function**

Graph the function $f(x) = |\cos x|$ from $0 \le x \le 2\pi$.

Solution Since $|\cos x| \ge 0$, the graph of $f(x) = |\cos x|$ can be drawn by reflecting the negative portions of the graph $y = \cos x$ across the x-axes.

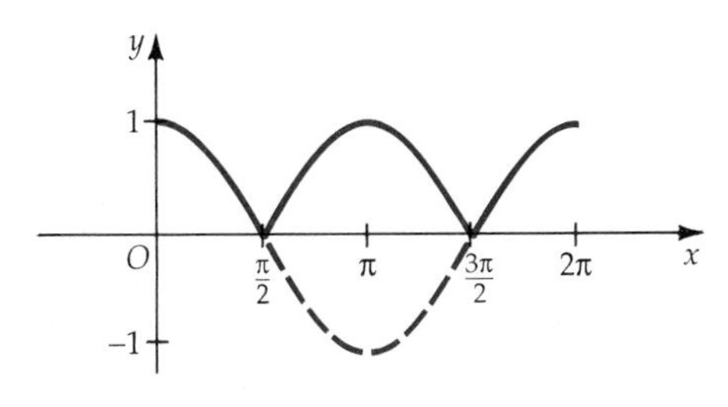

Figure 2.60
$f(x) = |\cos x|$

■ *Try Exercise* **40**, *page 87.*

EXERCISE SET 2.5

In Exercises 1 to 16, state the amplitude and period of the trigonometric function.

1. $f(x) = 2 \sin x$

2. $f(x) = -\dfrac{1}{2} \sin x$

3. $f(x) = \sin 2x$

4. $f(x) = \sin \dfrac{2x}{3}$

5. $f(x) = \dfrac{1}{2} \sin 2x$

6. $f(x) = 2 \sin \dfrac{x}{3}$

7. $f(x) = -2 \sin \dfrac{x}{2}$

8. $f(x) = -\dfrac{1}{2} \sin \dfrac{x}{2}$

9. $f(x) = \dfrac{1}{2} \cos x$

10. $f(x) = -3 \cos x$

11. $f(x) = \cos \dfrac{x}{4}$

12. $f(x) = \cos 3x$

13. $f(x) = 2 \cos \dfrac{x}{3}$

14. $f(x) = \dfrac{1}{2} \cos 2x$

15. $f(x) = -3 \cos \dfrac{2x}{3}$

16. $f(x) = \dfrac{3}{4} \cos 4x$

In Exercises 17 to 42, graph the trigonometric functions.

17. $f(x) = \dfrac{1}{2} \sin x$

18. $f(x) = \dfrac{3}{2} \cos x$

19. $f(x) = 3 \cos x$

20. $f(x) = \dfrac{3}{2} \sin x$

21. $f(x) = 4 \cos \dfrac{x}{2}$

22. $f(x) = 2 \cos \dfrac{3x}{4}$

23. $f(x) = -2 \cos \dfrac{x}{3}$

24. $f(x) = -\dfrac{4}{3} \cos 3x$

25. $f(x) = 2 \sin \pi x$

26. $f(x) = \dfrac{1}{2} \sin \dfrac{\pi x}{3}$

27. $f(x) = \dfrac{3}{2} \cos \dfrac{\pi x}{2}$

28. $f(x) = 3 \cos \dfrac{\pi x}{3}$

29. $f(x) = -4 \sin \dfrac{2\pi x}{3}$

30. $f(x) = 3 \cos \dfrac{3\pi x}{2}$

31. $f(x) = 2 \cos 2x$

32. $f(x) = \dfrac{1}{2} \sin 2.5x$

33. $f(x) = -2 \sin 1.5x$

34. $f(x) = -\dfrac{3}{4} \cos 5x$

35. $f(x) = \left|2 \sin \dfrac{x}{2}\right|$

36. $f(x) = \left|\dfrac{1}{2} \sin 3x\right|$

37. $f(x) = |-2 \cos 3x|$

38. $f(x) = \left|-\dfrac{1}{2} \cos \dfrac{x}{2}\right|$

39. $f(x) = -\left|2 \sin \dfrac{x}{3}\right|$

40. $f(x) = -\left|3 \sin \dfrac{2x}{3}\right|$

41. $f(x) = -|3 \cos \pi x|$

42. $f(x) = -\left|2 \cos \dfrac{\pi x}{2}\right|$

In Exercises 43 to 48, find an equation of each function shown in the accompanying graphs.

43.

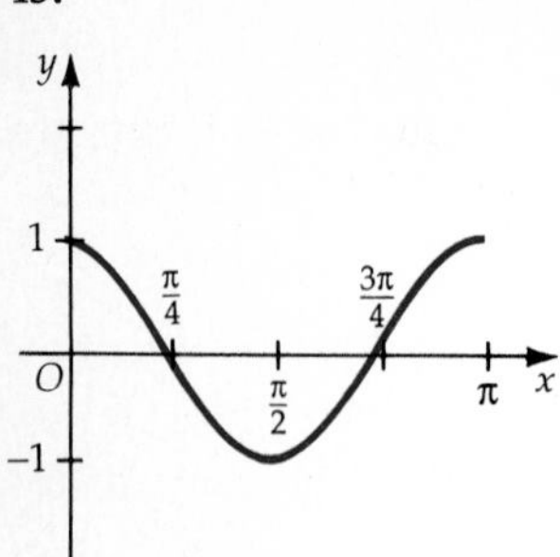

44.

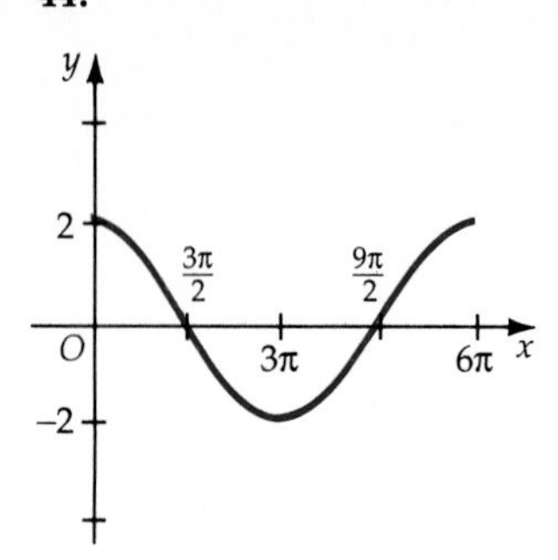

45.

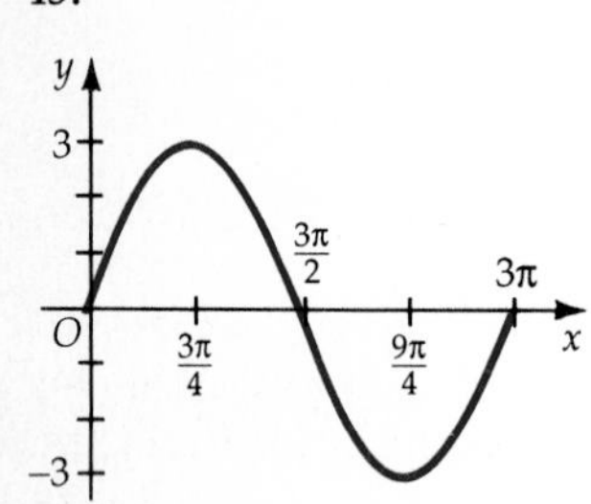

46.

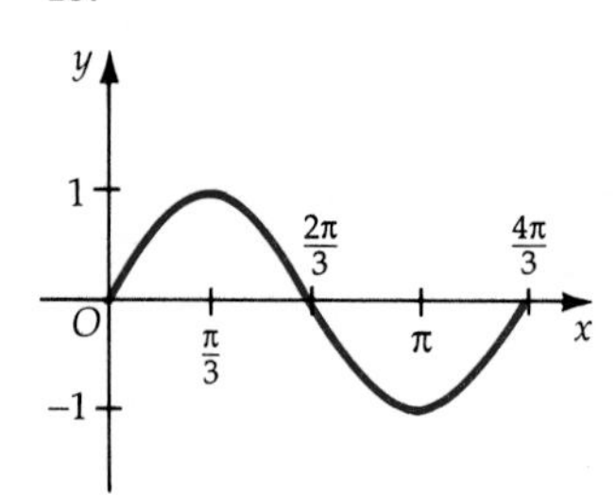

47.

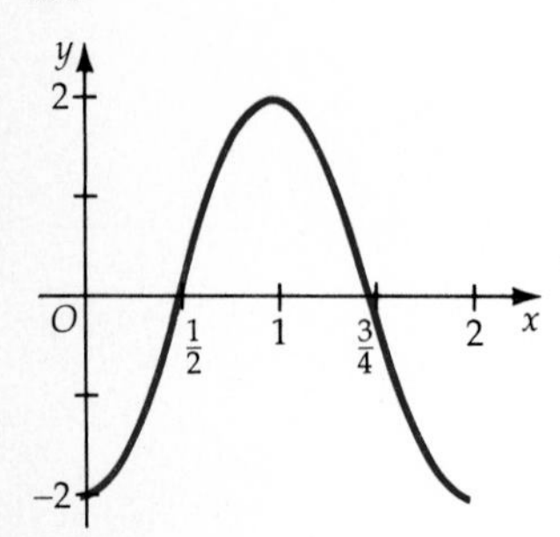

48.

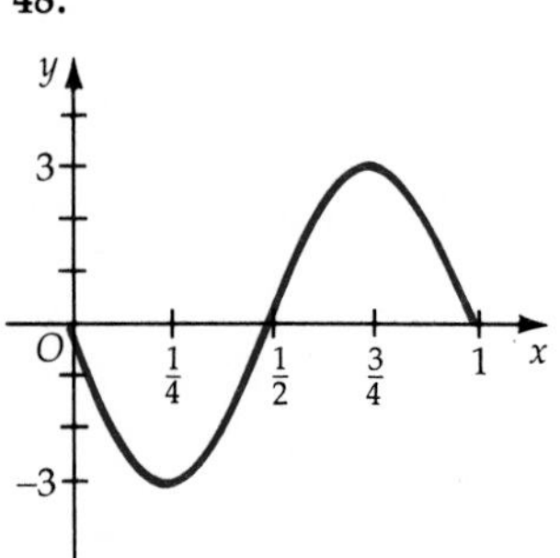

49. Sketch the graph of

$$f(x) = 2 \sin \frac{2x}{3} \text{ from } -3\pi \text{ to } 6\pi.$$

50. Sketch the graph of

$$f(x) = -3 \cos \frac{3x}{4} \text{ from } -2\pi \text{ to } 4\pi.$$

51. Sketch the graphs of

$$f(x) = 2 \cos \frac{x}{2} \quad \text{and} \quad g(x) = 2 \cos x$$

on the same set of axes from -2π to 4π.

52. Sketch the graphs of

$$f(x) = \sin \pi x \quad \text{and} \quad g(x) = \sin \frac{\pi x}{3}$$

on the same set of axes from -2 to 4.

Supplemental Exercises

53. Graph $f(x) = \sin^2 x$.

54. Graph $f(x) = 3^{\cos^2 x} \cdot 3^{\sin^2 x}$.

55. Graph $f(x) = \sin^2 x + \cos^2 x$.

56. Graph $f(x) = \cos |x|$.

In Exercises 57 and 58, discuss the symmetry of the trigonometric functions.

57. $f(x) = \sin x \cos x$

58. $f(x) = \dfrac{\sin x}{x}$

In Exercises 59 to 64, write an equation for a sine function with the given information.

59. Amplitude: 2, period: 3π

60. Amplitude: 5, period: $2\pi/3$

61. Amplitude: 0.5, period: $5\pi/4$

62. Amplitude: 1.5, period: 5π

63. Amplitude: 4, period: 2

64. Amplitude: 2.5, period: 3.2

In Exercises 65 to 70, write an equation for a cosine function with the given information.

65. Amplitude: 3, period: $\pi/2$

66. Amplitude: 0.8, period: 4π

67. Amplitude: 1.8, period: $3\pi/2$

68. Amplitude: 5, period: $7\pi/3$

69. Amplitude: 3, period: 2.5

70. Amplitude: 4.2, period: 1

71. A tidal wave that is caused by an earthquake under the sea is called a **tsunami wave**. These waves can be described by the formula $f(t) = A \cos Bt$. Find the equation of a tsunami wave that has a height of 60 feet, a period of 20 seconds, and travels at 120 feet per second. Graph two cycles of the equation and find the wavelength of one wave or the interval of one cycle.

2.6 Graphs of the Other Trigonometric Functions

The tangent function can be graphed by the same method of plotting points. Figure 2.61 shows the graph of the tangent function for the interval $-\pi/3 \le x \le \pi/3$.

TABLE 2.9

x	$-\pi/3$	$-\pi/4$	$-\pi/6$	0	$\pi/6$	$\pi/4$	$\pi/3$
tan x	-1.732	-1	-0.577	0	0.577	1	1.732
(x, y)	$\left(-\frac{\pi}{3}, -1.732\right)$	$\left(-\frac{\pi}{4}, -1\right)$	$\left(-\frac{\pi}{6}, -0.577\right)$	$(0, 0)$	$\left(\frac{\pi}{6}, 0.577\right)$	$\left(\frac{\pi}{4}, 1\right)$	$\left(\frac{\pi}{3}, 1.732\right)$

Because $\cos \pi/2 = 0$,

$$\tan x = \frac{\sin x}{\cos x}$$

is not defined at $\pi/2$. Table 2.10 shows that as x approaches $\pi/2$ from the left, tan x becomes larger and larger (increases without bound). We say that the vertical line $x = \pi/2$ is an asymptote of the tangent function. Vertical asymptotes of certain trigonometric functions will occur at the values of the domain for which the function is undefined. Figure 2.62 shows the graph of the tangent function for the interval $-\pi/2 < x < \pi/2$.

TABLE 2.10

x	1.5	1.52	1.55	1.57	1.5707
tan x	14.1	19.66	48.08	1255.8	10,381.3
(x, y)	(1.5, 14.1)	(1.52, 19.66)	(1.55, 48.08)	(1.57, 1255.8)	(1.5707, 10,381.3)

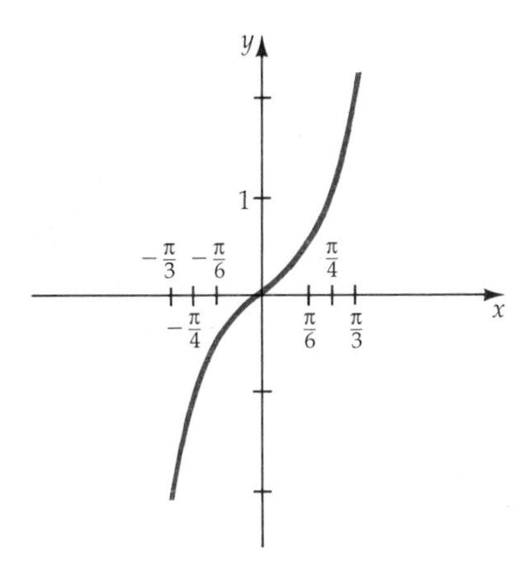

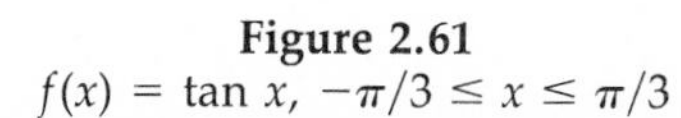
Figure 2.61
$f(x) = \tan x, -\pi/3 \le x \le \pi/3$

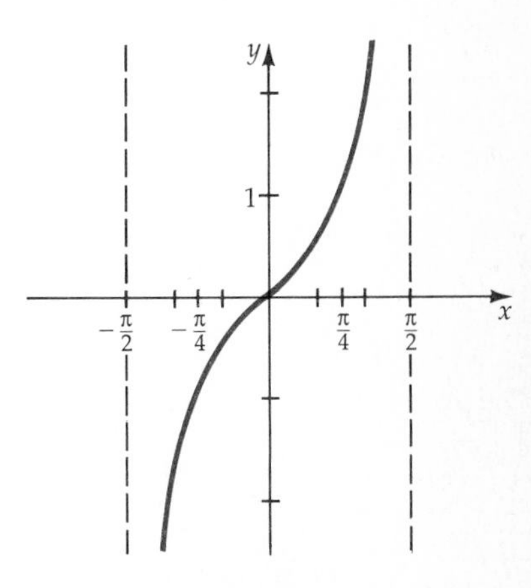

Figure 2.62
$f(x) = \tan x, -\pi/2 < x < \pi/2$

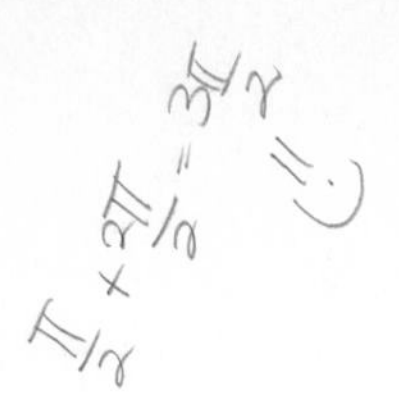

Remark The asymptotes of $f(x) = \tan x$ will occur at $x = \pi/2 + k\pi$, where k is an integer since

$$\tan x = \frac{\sin x}{\cos x}$$

asymptotes will occur where the tangent function is undefined. The tangent function is undefined where the cosine function is zero, or where $x = \pi/2 + k\pi$, k is an integer.

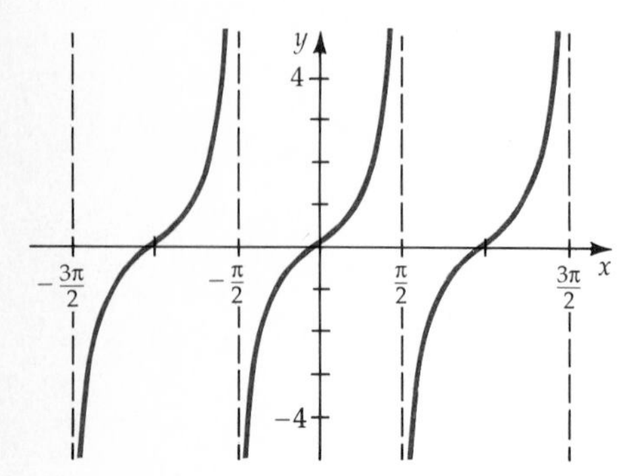

Figure 2.63
$f(x) = \tan x, -3\pi/2 < x < 3\pi/2$

Figure 2.63 shows the result of graphing the tangent function for values of x between $-3\pi/2$ and $3\pi/2$.

Important relationships of the function $f(x) = \tan x$ can be observed from the graph of the tangent function:

1. The domain of $f(x) = \tan x$ is the set of real number except the values $x = \pi/2 + k\pi$. Vertical asymptotes occur at the points $x = \pi/2 + k\pi$, where k is an integer.
2. The range of $f(x) = \tan x$ is the set of real numbers.
3. The function $f(x) = \tan x$ is periodic, and the period is π.
4. Since $\tan(-x) = -\tan x$, $\tan x$ is an odd function. Why?[4] The graph of the function is symmetric with respect to the origin.

Key values also can be used in graphing one period of the tangent function $f(x) = a \tan bx$. Since the tangent function is not bounded, the tangent function does not have an amplitude. The value a simply changes the rate of increase or decrease of the tangent function. The constant b determines the period of the function.

Since the period of $\tan x$ is π, the period of the tangent function $\tan bx$ is π/b. If $b < 0$, we use the identity $\tan(-bx) = -\tan bx$ to rewrite the function with $b > 0$. Normally we will show the graph of $f(x) = a \tan bx$ on the interval $-\pi/2b < x < \pi/2b$. Figure 2.64 shows the graph of the function $f(x) = 2 \tan 2x$.

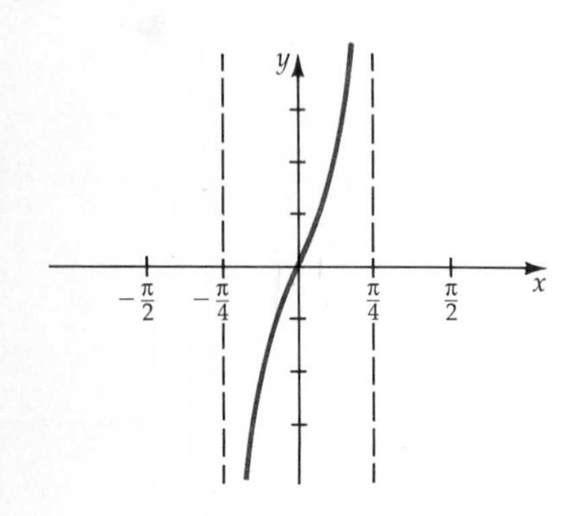

Figure 2.64
$f(x) = 2 \tan 2x$

$$\text{Period: } \frac{\pi}{2}$$

1. Beginning point: $f(-\pi/4) = 2\tan(-\pi/2)$ (undefined).
2. Quarter point: $f(-\pi/8) = 2\tan(-\pi/4) = -2$.
3. Middle point: $f(0) = 2\tan 0 = 0$.
4. Three-quarter point: $f(\pi/8) = 2\tan \pi/4 = 2$.
5. End point: $f(\pi/4) = 2\tan \pi/2$ (undefined).

x	$-\pi/4$	$-\pi/8$	0	$\pi/8$	$\pi/4$
$2x$	$-\frac{\pi}{2}$	$-\frac{\pi}{4}$	0	$\frac{\pi}{4}$	$\frac{\pi}{2}$
$2 \tan 2x$	Undefined	-2	0	2	Undefined
(x, y)	—	$\left(-\frac{\pi}{8}, -2\right)$	$(0, 0)$	$\left(\frac{\pi}{8}, 2\right)$	—

[4] Since $\tan(-x) = \frac{\sin(-x)}{\cos(-x)} = \frac{-\sin x}{\cos x} = -\tan x$, $\tan x$ is an odd function.

We can also use the five-point method to graph tangent functions. The vertical asymptotes of the tangent functions occur where the function is undefined.

EXAMPLE 1 Graph a Tangent Function

Graph the tangent function $f(x) = 2 \tan \frac{x}{2}$ from $-\pi$ to 3π.

Solution Period: $\frac{\pi}{1/2} = 2\pi$

Use the five-point method to graph one period of the function for values of x such that $-\pi < x < \pi$; then duplicate the graph on the interval $\pi < x < 3\pi$.

1. Beginning point: $f(-\pi) = 2 \tan(-\pi/2)$ (undefined).
2. Quarter point: $f(-\pi/2) = 2 \tan(-\pi/4) = -2$.
3. Middle point: $f(0) = 2 \tan 0 = 0$.
4. Three-quarter point: $f(\pi/2) = 2 \tan \pi/4 = 2$.
5. Endpoint: $f(\pi) = 2 \tan \pi/2$ (undefined).

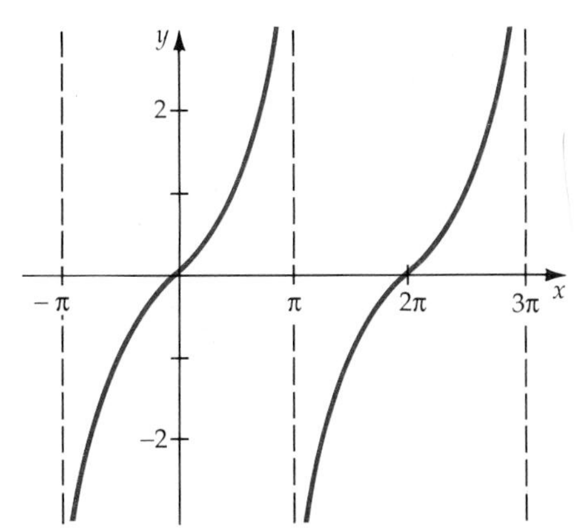

Figure 2.65
$f(x) = 2 \tan \frac{x}{2}$

x	$-\pi$	$-\pi/2$	0	$\pi/2$	π
$2 \tan \frac{x}{2}$	Undefined	-2	0	2	Undefined
(x, y)	—	$\left(-\frac{\pi}{2}, -2\right)$	$(0, 0)$	$\left(\frac{\pi}{2}, 2\right)$	—

■ *Try Exercise* **30,** *page 95.*

The Cotangent Function

Because the period of $\tan x$ is π and $\cot x = 1/\tan x$, the period of $\cot x$ is π. Since

$$\cot x = \frac{\cos x}{\sin x},$$

the cotangent function has vertical asymptotes where $\sin x = 0$; that is, when $x = k\pi$, where k is an integer.

Vertical asymptotes of $f(x) = \cot x$ will occur at $x = 0$ and $x = \pi$. Figure 2.66 shows the graph of one complete cycle of the cotangent function. The graph was achieved by using the five key values.

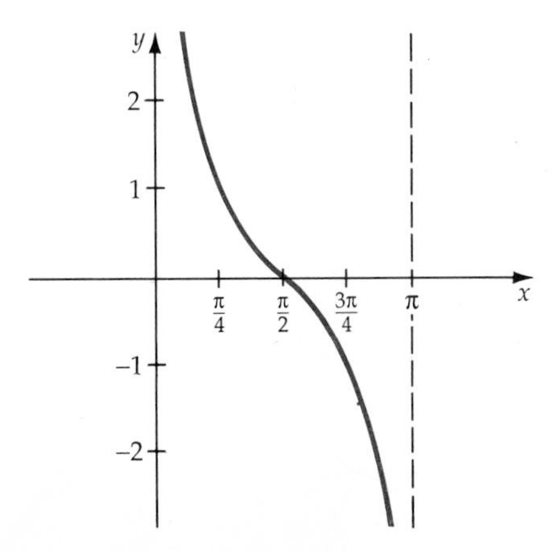

Figure 2.66
$f(x) = \cot x$

1. Beginning point: $f(0) = \cot 0$ (undefined).
2. Quarter point: $f(\pi/4) = \cot \pi/4 = 1$.
3. Middle point: $f(\pi/2) = \cot \pi/2 = 0$.
4. Three-quarter point: $f(3\pi/4) = \cot 3\pi/4 = -1$.
5. Endpoint: $f(x) = \cot \pi$ (undefined).

x	0	$\pi/4$	$\pi/2$	$3\pi/4$	π
$\cot x$	Undefined	1	0	-1	Undefined
(x, y)	—	$\left(\frac{\pi}{4}, 1\right)$	$\left(\frac{\pi}{2}, 0\right)$	$\left(\frac{3\pi}{4}, -1\right)$	—

Four important characteristics of the cotangent function $\cot x$ are now summarized:

1. The domain of $\cot x$ is the set of real numbers x except $x = k\pi$, k an integer. Vertical asymptotes occur, when $x = k\pi$, where k is an integer.
2. The range of $\cot x$ is the set of real numbers.
3. The function $\cot x$ is periodic, and the period is π.
4. The function $\cot x$ is an odd function. The function is symmetric to the origin.

Just as with the tangent function, the constant a in the function $f(x) = a \cot bx$ changes the rate of increase or decrease of the function. Since the period of $\cot x$ is π, the period of $a \cot bx$ is π/b. If $b > 0$, then $f(x) = \cot(-bx) = -\cot bx$. Therefore, the period of f is π/b. The interval of one period of the cotangent function is $0 \le x \le \pi/b$.

EXAMPLE 2 Graph a Cotangent Function

Graph one cycle of the function $f(x) = 3 \cot \frac{x}{2}$.

Solution Period: $\frac{\pi}{1/2} = 2\pi$

1. Beginning point: $f(0) = 3 \cot 0$ (undefined).
2. Quarter point: $f(\pi/2) = 3 \cot \pi/4 = 3$.
3. Middle point: $f(\pi) = 3 \cot \pi/2 = 0$.
4. Three-quarter point: $f(3\pi/2) = 3 \cot 3\pi/4 = -3$.
5. Endpoint: $f(2\pi) = 3 \cot \pi$ (undefined).

x	0	$\pi/2$	π	$3\pi/2$	2π
$3 \cot \frac{x}{2}$	Undefined	3	0	-3	Undefined
(x, y)	—	$\left(\frac{\pi}{2}, 3\right)$	$(\pi, 0)$	$\left(\frac{3\pi}{2}, -3\right)$	—

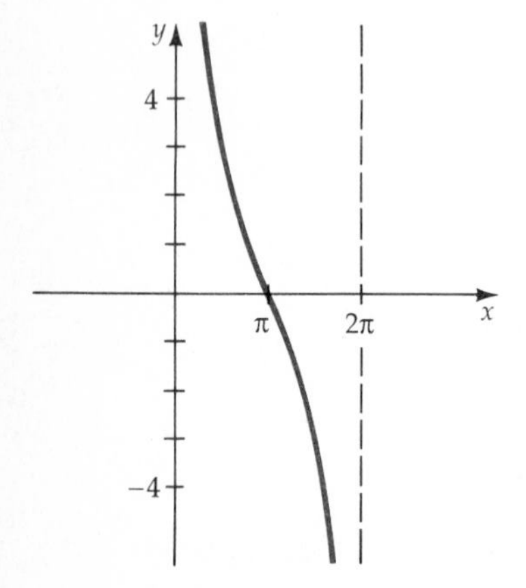

Figure 2.67
$f(x) = 3 \cot \frac{x}{2}$

■ *Try Exercise* **32**, *page 95.*

The Cosecant Function

The cosecant function $\csc x$ is the reciprocal of the sine function, that is $\csc x = 1/\sin x$. Therefore $\csc x$ is undefined at those values for which $\sin x = 0$. Vertical asymptotes of the graph of $y = \csc x$ will occur at those values. To find the value of $\csc x$, we take the reciprocal of the values of sin x. In Table 2.11 we have calculated csc x using this procedure for selected values of x.

TABLE 2.11

x	0	$\pi/4$	$\pi/2$	$3\pi/4$	π	$5\pi/4$	$3\pi/2$	$7\pi/4$	2π
sin x	0	0.707	1	0.707	0	-0.707	-1	-0.707	0
csc x	Undefined	1.414	1	1.414	Undefined	-1.414	-1	-1.414	Undefined
(x, y)	—	$\left(\frac{\pi}{4}, 1.414\right)$	$\left(\frac{\pi}{2}, 1\right)$	$\left(\frac{3\pi}{4}, 1.414\right)$	—	$\left(\frac{5\pi}{4}, -1.414\right)$	$\left(\frac{3\pi}{2}, -1\right)$	$\left(\frac{7\pi}{4}, -1.414\right)$	—

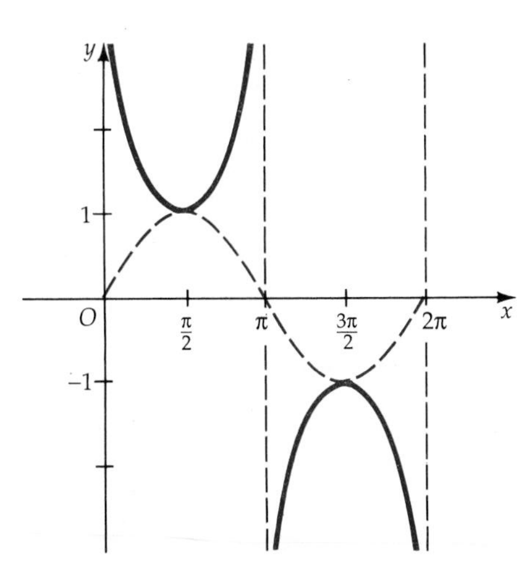

Figure 2.68
$f(x) = \csc x$

Remark The sine function has been sketched as a dashed line in Figure 2.68. Note the relationships among the zeros of the sine function and the asymptotes of the cosecant function. Note also that the turning points of the cosecant function will occur at the maximum and minimum points of the sine function.

Four important characteristics of the function $f(x) = \csc x$ are now summarized:

1. The domain of csc x is the set of real numbers, except at $x = k\pi$ where k is an integer. Vertical asymptotes occur at $x = k\pi$.
2. The range of csc x is the set of real numbers $\{y \mid y \geq 1, y \leq -1\}$.
3. The cosecant function has a period of 2π.
4. The function csc x is an odd function. The graph is symmetric with respect to the origin.

EXAMPLE 3 Graph a Cosecant Function

Graph one cycle of the function $f(x) = \csc 2x$.

Solution Period: $\frac{2\pi}{2} = \pi$.

Use the five key values to graph the function for $0 < x < \pi$. Note that the graph of $g(x) = \sin 2x$ is helpful in sketching the graph.

1. Beginning point: $f(0) = \csc 0$ (undefined).
2. Quarter point: $f(\pi/4) = \csc \pi/2 = 1$.
3. Middle point: $f(\pi/2) = \csc \pi$ (undefined).
4. Three-quarter point: $f(3\pi/4) = \csc 3\pi/2 = -1$.
5. Endpoint: $f(\pi) = \csc 2\pi$ (undefined).

x	0	$\pi/4$	$\pi/2$	$3\pi/4$	π
csc $2x$	Undefined	1	Undefined	-1	Undefined
(x, y)	—	$\left(\frac{\pi}{4}, 1\right)$	—	$\left(\frac{3\pi}{4}, -1\right)$	—

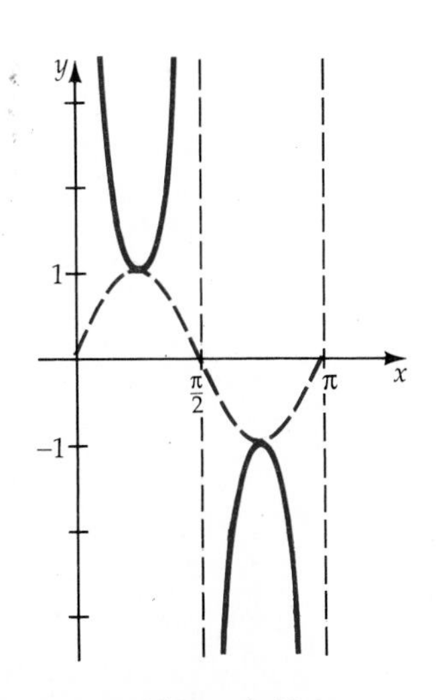

Figure 2.69
$f(x) = \csc 2x$

■ *Try Exercise* **38,** *page* 95.

The Secant Function

The secant function is the reciprocal of the cosine function: $\sec x = 1/\cos x$. The secant function is undefined at those points for which $\cos x = 0$. Vertical asymptotes of the secant will occur at those values. The values of the secant function are reciprocals of the values of the cosine function. Table 2.12 contains values of $\sec x$ for selected values of x.

TABLE 2.12

x	0	$\pi/4$	$\pi/2$	$3\pi/4$	π	$5\pi/4$	$3\pi/2$	$7\pi/4$	2π
$\cos x$	1	0.707	0	−0.707	−1	−0.707	0	0.707	1
$\sec x$	1	1.414	Undefined	−1.414	−1	−1.414	Undefined	1.414	1
(x, y)	$(0, 1)$	$\left(\frac{\pi}{4}, 1.414\right)$	—	$\left(\frac{3\pi}{4}, -1.414\right)$	$(\pi, -1)$	$\left(\frac{5\pi}{4}, -1.414\right)$	—	$\left(\frac{7\pi}{4}, 1.414\right)$	$(2\pi, 1)$

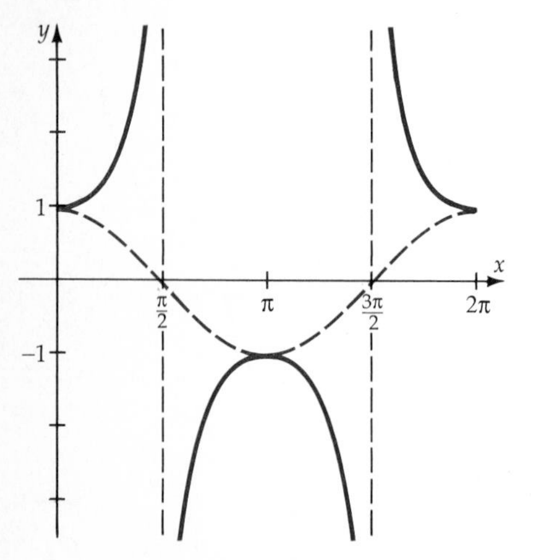

Figure 2.70
$f(x) = \sec x$

Four important relationships of the function $f(x) = \sec x$ are now summarized.

1. The domain of $\sec x$ is the set of real numbers x except $x = \pi/2 + k\pi$. Vertical asymptotes occur at $x = \pi/2 + k\pi$, where k is an integer.
2. The range of $\sec x$ is the set of real numbers $\{y \mid y \geq 1, y \leq -1\}$.
3. Since the cosine function has a period of 2π, the secant function has a period of 2π.
4. The function $\sec x$ is an even function. The graph of $y = \sec x$ is symmetric with respect to the y-axis.

EXAMPLE 4 **Graph a Secant Function**

Graph the function $f(x) = \sec 2x$ from 0 to 2π.

Solution Period: $\dfrac{2\pi}{2} = \pi$

Use the five key values to graph the function for values of x for $0 \leq x \leq \pi$, then duplicate the graph on the interval $\pi \leq x \leq 2\pi$. Note that the graph of the corresponding cosine curve $g(x) = \cos 2x$ is helpful in sketching the graph.

1. Beginning point: $f(0) = \sec 0 = 1$.
2. Quarter point: $f(\pi/4) = \sec \pi/2$ (undefined).
3. Middle point: $f(\pi/2) = \sec \pi = -1$.
4. Three-quarter point: $f(3\pi/4) = \sec(3\pi/2)$ (undefined).
5. Endpoint: $f(\pi) = \sec 2\pi = 1$.

x	0	$\pi/4$	$\pi/2$	$3\pi/4$	π
$\sec 2x$	1	Undefined	−1	Undefined	1
(x, y)	$(0, 1)$	—	$\left(\frac{\pi}{2}, -1\right)$	—	$(\pi, 1)$

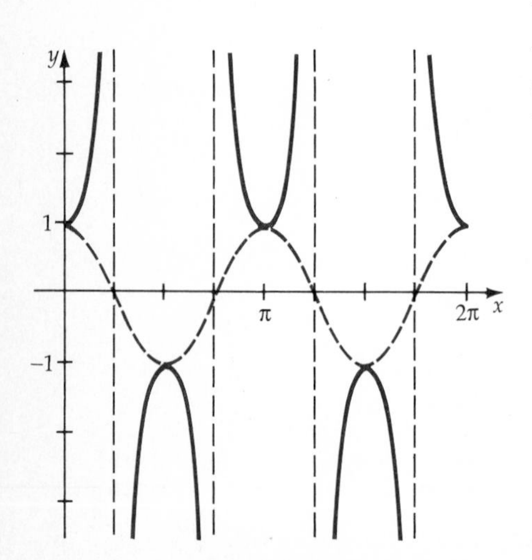

Figure 2.71
$f(x) = \sec 2x$

■ *Try Exercise* **42,** *page 95.*

EXERCISE SET 2.6

1. For what values of x is $y = \tan x$ undefined?
2. For what values of x is $y = \cot x$ undefined?
3. For what values of x is $y = \sec x$ undefined?
4. For what values of x is $y = \csc x$ undefined?

In Exercises 5 to 20, state the period of the given function.

5. $f(x) = \sec x$
6. $f(x) = \cot x$
7. $f(x) = \tan x$
8. $f(x) = \csc x$
9. $f(x) = 2 \tan \dfrac{x}{2}$
10. $f(x) = \dfrac{1}{2} \cot 2x$
11. $f(x) = \csc 3x$
12. $f(x) = \csc \dfrac{x}{2}$
13. $f(x) = -\tan 3x$
14. $f(x) = -3 \cot \dfrac{2x}{3}$
15. $f(x) = -3 \sec \dfrac{x}{4}$
16. $f(x) = -\dfrac{1}{2} \csc 2x$
17. $f(x) = \cot \pi x$
18. $f(x) = \cot \dfrac{\pi x}{3}$
19. $f(x) = 2 \csc \dfrac{\pi x}{2}$
20. $f(x) = -3 \cot \pi x$

In Exercises 21 to 40, sketch the graph of the given function.

21. $f(x) = 3 \tan x$
22. $f(x) = \dfrac{1}{3} \tan x$
23. $f(x) = \dfrac{3}{2} \cot x$
24. $f(x) = 4 \cot x$
25. $f(x) = 2 \sec x$
26. $f(x) = \dfrac{3}{4} \sec x$
27. $f(x) = \dfrac{1}{2} \csc x$
28. $f(x) = 2 \csc x$
29. $f(x) = 2 \tan \dfrac{x}{2}$
30. $f(x) = -3 \tan 3x$
31. $f(x) = -3 \cot \dfrac{x}{2}$
32. $f(x) = \dfrac{1}{2} \cot 2x$
33. $f(x) = -2 \csc \dfrac{x}{3}$
34. $f(x) = \dfrac{3}{2} \csc 3x$
35. $f(x) = \dfrac{1}{2} \sec 2x$
36. $f(x) = -3 \sec \dfrac{2x}{3}$
37. $f(x) = -2 \sec \pi x$
38. $f(x) = 3 \csc \dfrac{\pi x}{2}$
39. $f(x) = 3 \tan 2\pi x$
40. $f(x) = -\dfrac{1}{2} \cot \dfrac{\pi x}{2}$
41. Graph $f(x) = 2 \csc 3x$ from -2π to 2π
42. Graph $f(x) = \sec \dfrac{x}{2}$ from -4π to 4π
43. Graph $f(x) = 3 \sec \pi x$ from -2 to 4
44. Graph $f(x) = \csc \dfrac{\pi x}{2}$ from -4 to 4
45. Graph $f(x) = 2 \cot 2x$ from $-\pi$ to π
46. Graph $f(x) = \dfrac{1}{2} \tan \dfrac{x}{2}$ from -4π to 4π
47. Graph $f(x) = 3 \tan \pi x$ from -2 to 2
48. Graph $f(x) = \cot \dfrac{\pi x}{2}$ from -4 to 4

In Exercises 49 to 54, find an equation of each function.

49.

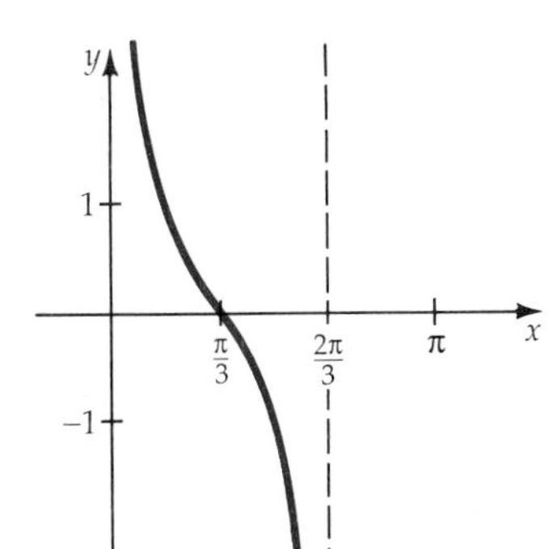

50.

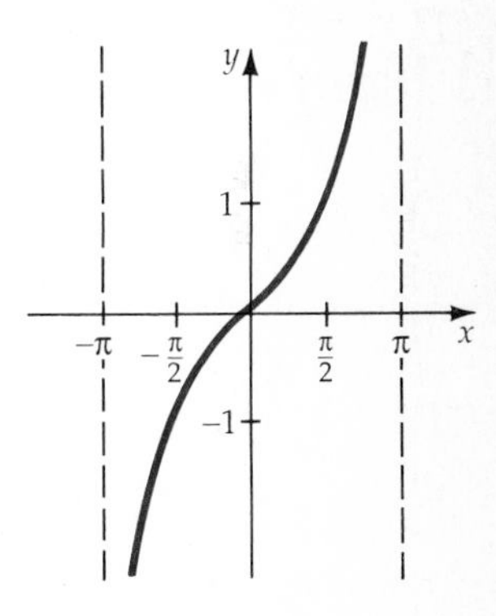

51.

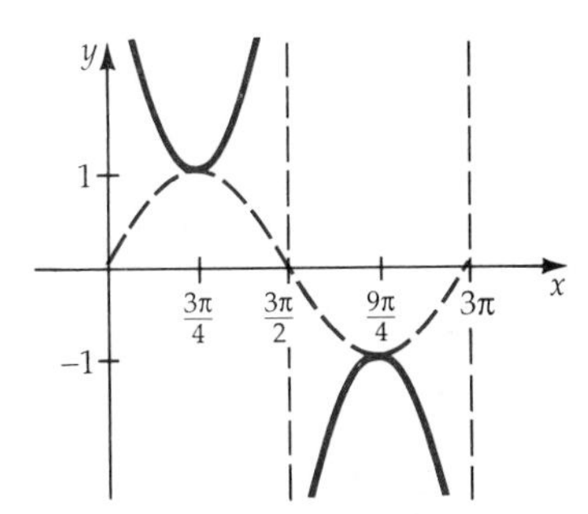

52.

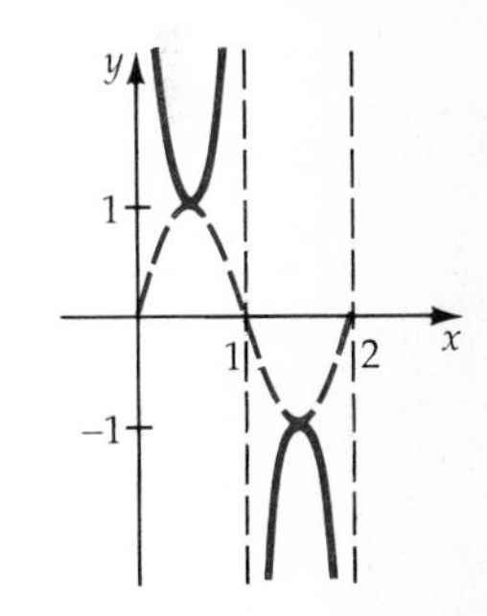

53.

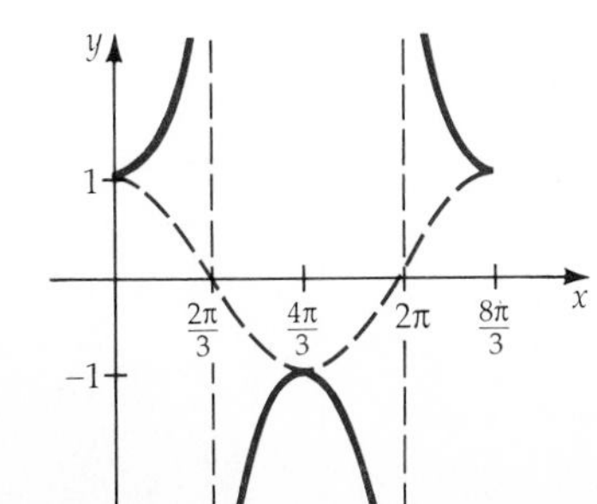

54.

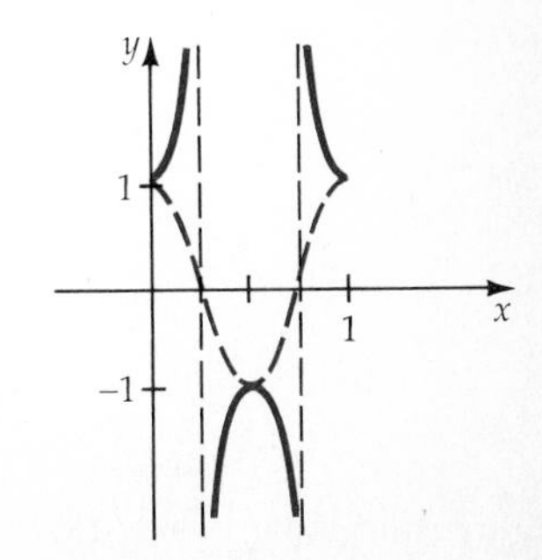

Supplemental Exercises

In Exercises 55 to 62, write an equation of the form $y = \tan bx$, $y = \cot bx$, $y = \sec bx$ or $y = \csc bx$ that satisifes the given conditions.

55. Tangent, period: $\pi/3$

56. Cotangent, period: $\pi/2$

57. Secant, period: $3\pi/4$

58. Cosecant, period: $5\pi/2$

59. Cotangent, period: 2

60. Tangent, period: 0.5

61. Cosecant, period: 1.5

62. Secant, period: 3

In Exercises 63 to 66, sketch the graph of the given function.

63. $f(x) = \tan |x|$

64. $f(x) = \sec |x|$

65. $f(x) = |\csc x|$

66. $f(x) = |\cot x|$

67. Graph $y = \tan x$ and $x = \tan y$ on the same coordinate axes.

68. Graph $y = \sin x$ and $x = \sin y$ on the same coordinate axes.

2.7 Phase Shift and Addition of Ordinates

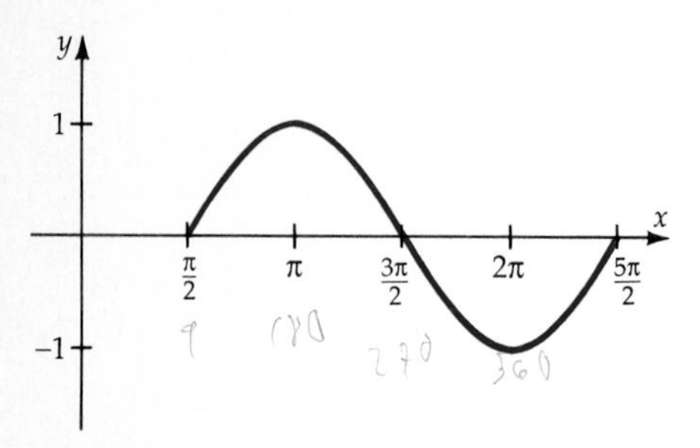

Figure 2.72
$f(x) = \sin(x - \pi/2)$

Recall that for $c > 0$, the graph of $y = f(x - c)$ is the graph of $y = f(x)$ shifted c units to the right on the x-axis and $y = f(x + c)$ is the graph of $y = f(x)$ shifted c units to the left on the x-axes. This property of functions that shifts a function horizontally is used to graph the functions of the form $f(x) = a \sin(bx \pm c)$ and $f(x) = a \cos(bx \pm c)$.

Figure 2.72 shows the graph of the function $f(x) = \sin(x - \pi/2)$. Each point on the curve $g(x) = \sin x$ is translated $\pi/2$ units to the right. This is called the **phase shift** of the graph.

We are now ready to graph sine or cosine functions of the form $y = a \cos(bx \pm c)$ or $y = a \sin(bx \pm c)$. The amplitude of each function is $|a|$ and the period is $2\pi/b$, $b > 0$. One cycle of the function can be determined by solving an inequality for x:

$$0 \le bx + c \le 2\pi$$

$$-\frac{c}{b} \le x \le -\frac{c}{b} + \frac{2\pi}{b}$$

The number $-\frac{c}{b}$ is the phase shift of the graph.

Properties of $y = a \sin(bx + c)$ and $y = a \cos(bx + c)$

The graphs of

$$y = a \sin(bx + c) \quad \text{and} \quad y = a \cos(bx + c)$$

have the following properties:

$$\text{amplitude} = |a|, \quad \text{period} = \frac{2\pi}{b}, \quad \text{phase shift} = -\frac{c}{b}.$$

The interval containing one cycle can be found by solving the inequality $0 \le bx + c \le 2\pi$.

EXAMPLE 1 Graph a Cosine Function with a Phase Shift

Graph one cycle of the function $f(x) = \cos\left(2x + \frac{\pi}{2}\right)$.

Solution To determine the interval of one cycle, we consider the inequality

$$0 \le 2x + \frac{\pi}{2} \le 2\pi$$
$$-\frac{\pi}{2} \le 2x \le \frac{3\pi}{2}$$
$$-\frac{\pi}{4} \le x \le \frac{3\pi}{4}$$

Amplitude $= |a| = 1$, Period $= \frac{2\pi}{2} = \pi$, Phase Shift $= \frac{-\pi/2}{2} = -\frac{\pi}{4}$.

The period of the function is equal to $\frac{2\pi}{b} = \frac{2\pi}{2} = \pi$.

Use the five key values starting at $-\frac{\pi}{4}$ and ending at $\frac{3\pi}{4}$.

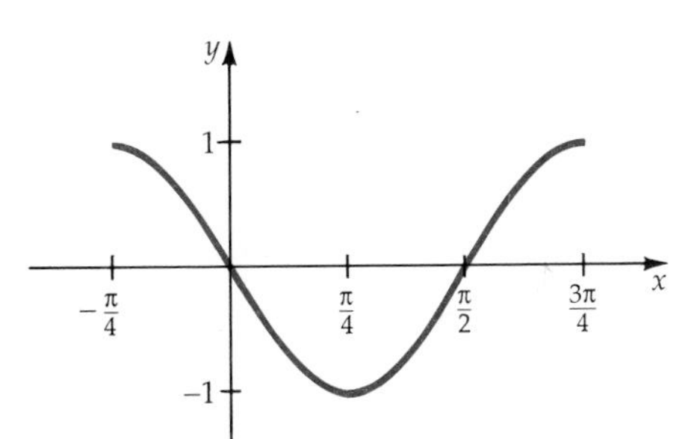

Figure 2.73
$f(x) = \cos\left(2x + \frac{\pi}{2}\right)$

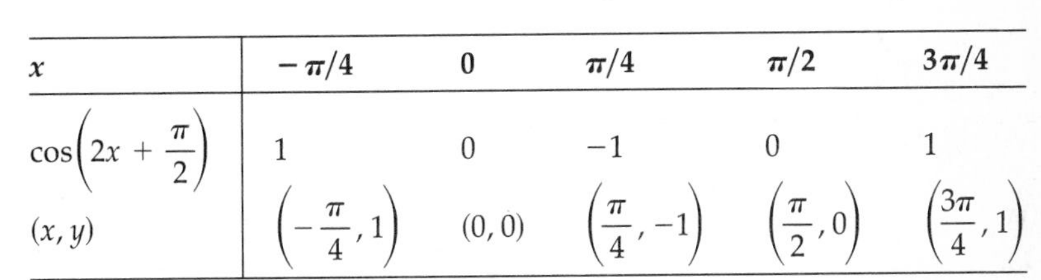

x	$-\pi/4$	0	$\pi/4$	$\pi/2$	$3\pi/4$
$\cos\left(2x + \frac{\pi}{2}\right)$	1	0	−1	0	1
(x, y)	$\left(-\frac{\pi}{4}, 1\right)$	$(0, 0)$	$\left(\frac{\pi}{4}, -1\right)$	$\left(\frac{\pi}{2}, 0\right)$	$\left(\frac{3\pi}{4}, 1\right)$

■ *Try Exercise* **20**, *page 101.*

The Phase Shift for the Tangent and Cotangent Functions

The phase shift of the graphs

$$f(x) = a\tan(bx \pm c) \text{ or } f(x) = a\cot(bx \pm c)$$

can be determined by using a technique similar to the one we used for $a\sin(bx + c)$.

Properties of $y = a\tan(bx + c)$ and $y = a\cot(bx + c)$

> The graphs of
>
> $$y = a\tan(bx + c) \qquad \text{and} \qquad y = a\cot(bx + c)$$
>
> have the following properties:
>
> $$\text{period} = \frac{\pi}{b}, \qquad \text{phase shift} = -\frac{c}{b}.$$
>
> Successive vertical asymptotes for $y = a\tan(bx + c)$ are the vertical lines $x = -c/b \pm \pi/2b$.
> Successive vertical asymptotes for $y = a\cot(bx + c)$ are the vertical lines $x = -c/b$ and $x = -c/b + \pi$.

Remark The interval containing one cycle of $y = a\tan(bx + c)$ can be found by solving the inequality $-\pi/2 < bx + c < \pi/2$. The corresponding interval for $y = a\cot(bx + c)$ can be found by solving the inequality $0 < bx + c < \pi$.

EXAMPLE 2 Graph a Tangent Function with a Phase Shift

Graph one cycle of the function $f(x) = 2\tan\left(x - \frac{\pi}{4}\right)$.

Solution To determine the interval of one cycle of the graph solve the inequality.

$$-\frac{\pi}{2} < x - \frac{\pi}{4} < \frac{\pi}{2}$$

$$-\frac{\pi}{4} < x < \frac{3\pi}{4}$$

$$\text{period} = \pi, \qquad \text{phase shift} = \frac{\pi}{4}.$$

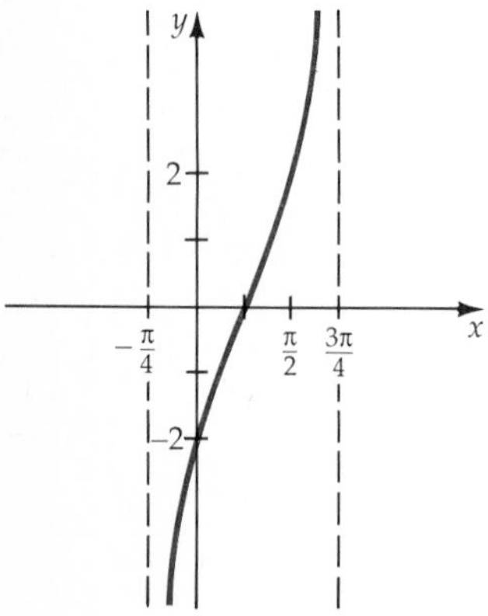

Figure 2.74
$f(x) = 2\tan\left(x - \frac{\pi}{4}\right)$

Use the five key values to graph one cycle of the function for values of x for $-\frac{\pi}{4} < x < \frac{3\pi}{4}$.

x	$-\pi/4$	0	$\pi/4$	$\pi/2$	$3\pi/4$
$2\tan\left(x - \frac{\pi}{4}\right)$	Undefined	-2	0	2	Undefined
(x, y)	—	$(0, -2)$	$\left(\frac{\pi}{4}, 0\right)$	$\left(\frac{\pi}{2}, 2\right)$	—

■ *Try Exercise* **22**, *page 101.*

The Graph of the Cosecant and Secant Functions

The graphs of the functions $y = a\csc(bx \pm c)$ or $y = a\sec(bx \pm c)$ can be found by finding the reciprocals of the corresponding sine or cosine functions and graphing point by point.

EXAMPLE 3 Graph a Cosecant Function

Graph one cycle of the function $f(x) = 2\csc(2x - \pi)$.

Solution First, sketch $y = 2\sin(2x - \pi)$ as a dashed graph.

$$\text{Amplitude} = 2, \qquad \text{Period} = \pi, \qquad \text{Phase Shift} = \frac{\pi}{2}.$$

Because $\csc x = \frac{1}{\sin x}$, we use the reciprocal of the y-coordinates of $y = 2\sin(2x - \pi)$ to produce the graph of $y = 2\csc(2x - \pi)$.

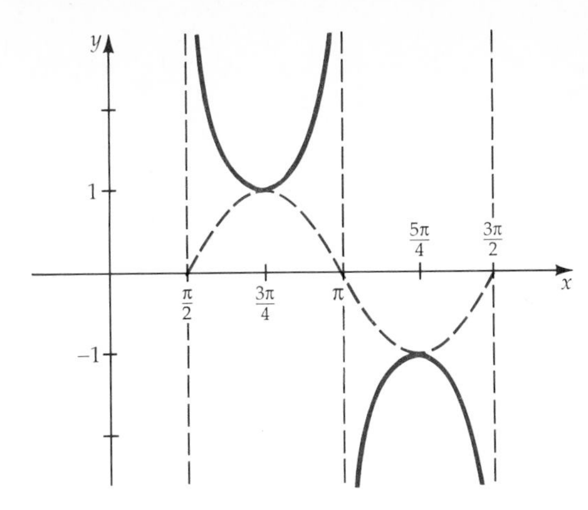

Figure 2.75
$f(x) = 2\csc(2x - \pi)$

■ *Try Exercise* **28,** *page 102.*

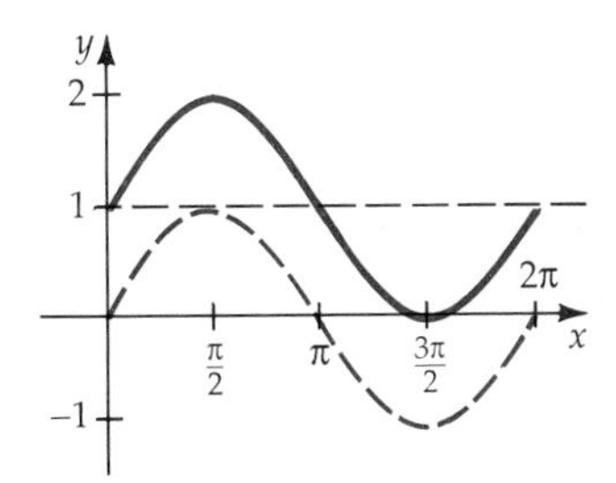

Figure 2.76

Vertical Translations of Circular Functions

Recall that for $c > 0$ the function $y = f(x) + c$ is the graph of $y = f(x)$ shifted vertically c units. For example, the graph of $f(x) = \sin x + 1$ is the graph of $g(x) = \sin x$ shifted upward one unit, as shown in Figure 2.76.

Remark Observe from the graph that a vertical shift has no effect on the amplitude, period, or phase shift.

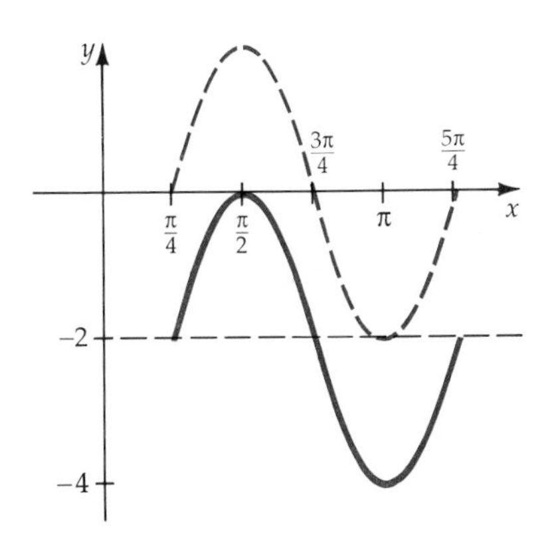

Figure 2.77
$f(x) = 2\sin\left(2x - \frac{\pi}{2}\right) - 2$

EXAMPLE 4 Graph a Sine Function with a Vertical Translation

Graph one cycle of the function $f(x) = 2\sin\left(2x - \frac{\pi}{2}\right) - 2$.

Solution First sketch $y = 2\sin\left(2x - \frac{\pi}{2}\right)$.

$$\text{Amplitude} = 2, \qquad \text{Period} = \pi, \qquad \text{Phase Shift} = \frac{\pi}{4}$$

The graph of $f(x) = 2\sin\left(2x - \frac{\pi}{2}\right) - 2$ is the graph of $y = 2\sin\left(2x - \frac{\pi}{2}\right)$ shifted 2 units downward.

■ *Try Exercise* **40,** *page 102.*

EXAMPLE 5 Graph a Cosecant Function with a Vertical Translation

Graph one cycle of the function $f(x) = \csc x + 1$.

Solution First sketch the graph of $y = \csc x$ as a dashed graph. (See Figure 2.78.) The graph of $f(x) = \csc x + 1$ is the graph of $y = \csc x$

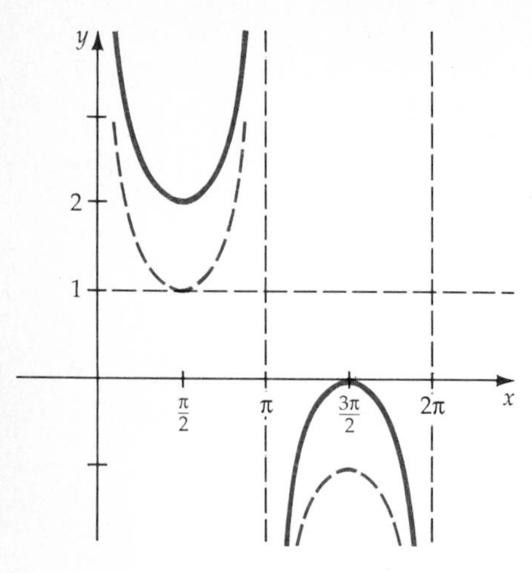

Figure 2.78
$f(x) = \csc x + 1$

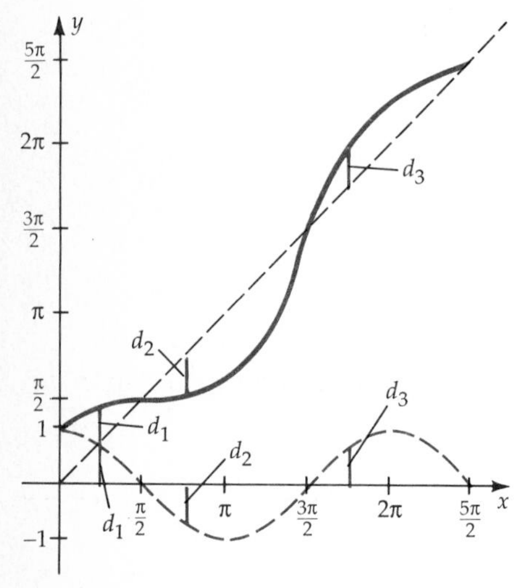

Figure 2.79
$f(x) = x + \cos x$

shifted one unit upward.

■ *Try Exercise* **44,** *page 102.*

Addition of Ordinates

Given two functions $g(x)$ and $h(x)$, the sum of the two functions is the function $f(x)$ defined by $f(x) = g(x) + h(x)$. The graph of the sum $f(x)$ can be obtained by graphing $g(x)$ and $h(x)$ separately and then geometrically adding the y coordinates of each function for a given x value. It is convenient when adding functions to pick zeros or maximum or minimum points of the function. Examples 6 and 7 illustrate this procedure.

EXAMPLE 6 Graph by Addition of Ordinates

Graph the function $f(x) = x + \cos x$ from 0 to 2π.

Solution Graph the functions $g(x) = x$ and $h(x) = \cos x$ on the same coordinate system. Then add the y-coordinates geometrically point by point. The graph in Figure 2.79 shows the results of adding, by using a ruler, the y-coordinates of the two functions for selected values of x.

■ *Try Exercise* **50,** *page 102.*

EXAMPLE 7 Graph by Addition of Ordinates

Graph the function $f(x) = \sin x - \cos x$ from 0 to 2π.

Solution The graph of $f(x) = \sin x - \cos x$ can be accomplished by sketching the graph of $y_1 = \sin x$ and the graph of $y_2 = -\cos x$ and adding the y-coordinates of selected values of x. A smooth curve is then drawn through the points.

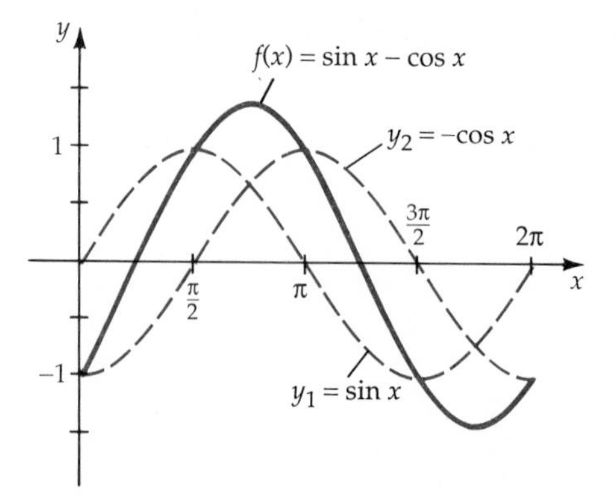

Figure 2.80
$f(x) = \sin x - \cos x$

■ *Try Exercise* **52,** *page 102.*

The points on the graph of the product of two functions can be found by multiplying the values of the y-coordinates at each given x-coordinate. It is convenient to choose values of x for which one of the functions is 0 or 1.

EXAMPLE 8 **Graph the Product of Functions**

Graph the function $f(x) = x \sin x$.

Solution This is the product of $y_1 = x$ and $y_2 = \sin x$. The graph in Figure 2.81 was obtained by selecting values for which y_1 or y_2 is 0 or 1.

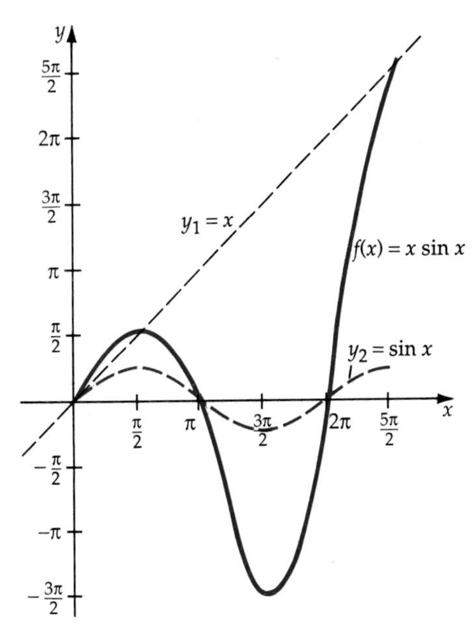

Figure 2.81
$f(x) = x \sin x$

■ *Try Exercise* **58**, *page 102.*

EXERCISE SET 2.7

In Exercises 1 to 8, find the amplitude, phase shift, and period for the given functions.

1. $f(x) = 2\sin\left(x - \frac{\pi}{2}\right)$

2. $f(x) = -3\sin(x + \pi)$

3. $f(x) = \cos\left(2x - \frac{\pi}{4}\right)$

4. $f(x) = \frac{3}{4}\cos\left(\frac{x}{2} + \frac{\pi}{3}\right)$

5. $f(x) = -4\sin\left(\frac{2x}{3} + \frac{\pi}{6}\right)$

6. $f(x) = \frac{3}{2}\sin\left(\frac{x}{4} - \frac{3\pi}{4}\right)$

7. $f(x) = \frac{5}{4}\cos(3x - 2\pi)$

8. $f(x) = 6\cos\left(\frac{x}{3} - \frac{\pi}{6}\right)$

In Exercises 9 to 16, find the phase shift and the period for the given functions.

9. $f(x) = 2\tan\left(2x - \frac{\pi}{4}\right)$

10. $f(x) = \frac{1}{2}\tan\left(\frac{x}{2} - \pi\right)$

11. $f(x) = -3\csc\left(\frac{x}{3} + \pi\right)$

12. $f(x) = -4\csc\left(3x - \frac{\pi}{6}\right)$

13. $f(x) = 2\sec\left(2x - \frac{\pi}{8}\right)$

14. $f(x) = 3\sec\left(\frac{x}{4} - \frac{\pi}{2}\right)$

15. $f(x) = -3\cot\left(\frac{x}{4} + 3\pi\right)$

16. $f(x) = \frac{3}{2}\cot\left(2x - \frac{\pi}{4}\right)$

In Exercises 17 to 32, graph one cycle of the given function.

17. $f(x) = \sin\left(x - \frac{\pi}{2}\right)$

18. $f(x) = \sin\left(x + \frac{\pi}{6}\right)$

19. $f(x) = \cos\left(\frac{x}{2} + \frac{\pi}{3}\right)$

20. $f(x) = \cos\left(2x - \frac{\pi}{3}\right)$

21. $f(x) = \tan\left(x + \frac{\pi}{4}\right)$

22. $f(x) = \tan(x - \pi)$

23. $f(x) = 2\cot\left(\frac{x}{2} - \frac{\pi}{8}\right)$

24. $f(x) = \frac{3}{2}\cot\left(3x + \frac{\pi}{4}\right)$

25. $f(x) = \sec\left(x + \frac{\pi}{4}\right)$

26. $f(x) = \csc(2x + \pi)$

27. $f(x) = \csc\left(\frac{x}{3} - \frac{\pi}{2}\right)$

28. $f(x) = \sec\left(2x + \frac{\pi}{6}\right)$

29. $f(x) = -2\sin\left(\frac{x}{3} - \frac{2\pi}{3}\right)$

30. $f(x) = -\frac{3}{2}\sin\left(2x + \frac{\pi}{4}\right)$

31. $f(x) = -3\cos\left(3x + \frac{\pi}{4}\right)$

32. $f(x) = -4\cos\left(\frac{3x}{2} + 2\pi\right)$

In Exercises 33 to 46, graph the functions by using vertical translations.

33. $f(x) = \sin x - 1$

34. $f(x) = -\sin x + 1$

35. $f(x) = -\cos x - 2$

36. $f(x) = 2\sin x + 3$

37. $f(x) = \sin 2x - 2$

38. $f(x) = -\cos\frac{x}{2} + 2$

39. $f(x) = \sin\left(x - \frac{\pi}{2}\right) - \frac{1}{2}$

40. $f(x) = -2\cos\left(x + \frac{\pi}{3}\right) + 3$

41. $f(x) = \tan\frac{x}{2} - 4$

42. $f(x) = \cot 2x + 3$

43. $f(x) = \sec 2x - 2$

44. $f(x) = \csc\frac{x}{3} + 4$

45. $f(x) = \csc\frac{x}{2} - 1$

46. $f(x) = \sec\left(x - \frac{\pi}{2}\right) + 1$

In Exercises 47 to 56, graph the given functions by using the addition of ordinates.

47. $f(x) = x - \sin x$

48. $f(x) = \frac{x}{2} + \cos x$

49. $f(x) = x + \sin 2x$

50. $f(x) = \frac{2x}{3} - \sin\frac{x}{2}$

51. $f(x) = \sin x + \cos x$

52. $f(x) = -\sin x + \cos x$

53. $f(x) = \sin x - \cos\frac{x}{2}$

54. $f(x) = 2\sin 2x - \cos x$

55. $f(x) = 2\cos x + \sin\frac{x}{2}$

56. $f(x) = -\frac{1}{2}\cos 2x + \sin\frac{x}{2}$

In Exercises 57 to 62, graph the following functions.

57. $f(x) = \frac{x}{2}\sin x$

58. $f(x) = x\cos x$

59. $f(x) = x\sin\frac{x}{2}$

60. $f(x) = \frac{x}{2}\cos\frac{x}{2}$

61. $f(x) = x\sin\left(x + \frac{\pi}{2}\right)$

62. $f(x) = x\cos\left(x - \frac{\pi}{2}\right)$

In Exercises 63 to 68, find an equation of the trigonometric function from the accompanying graph.

63.

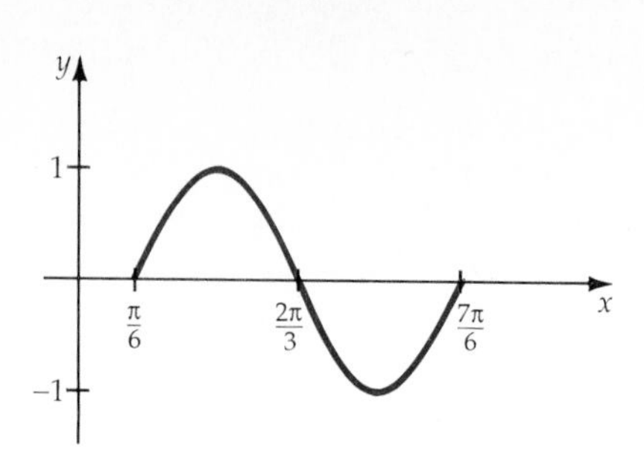

64.

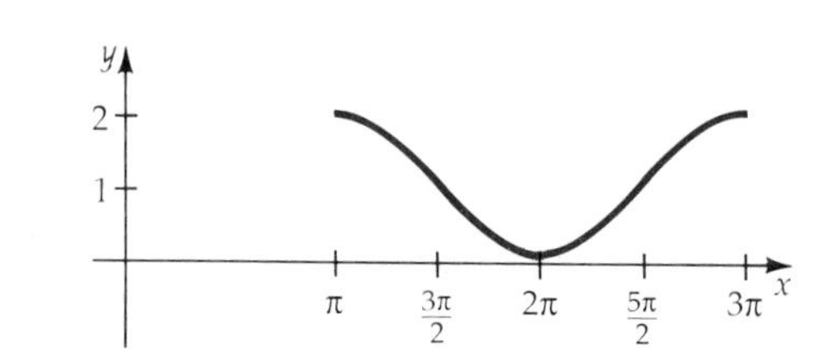

65. **66.**

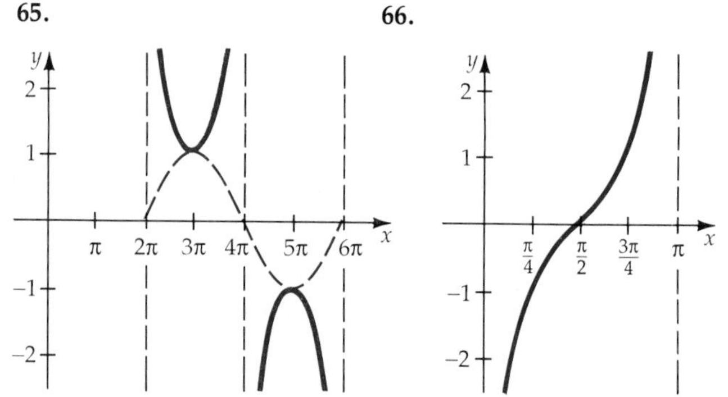

67.

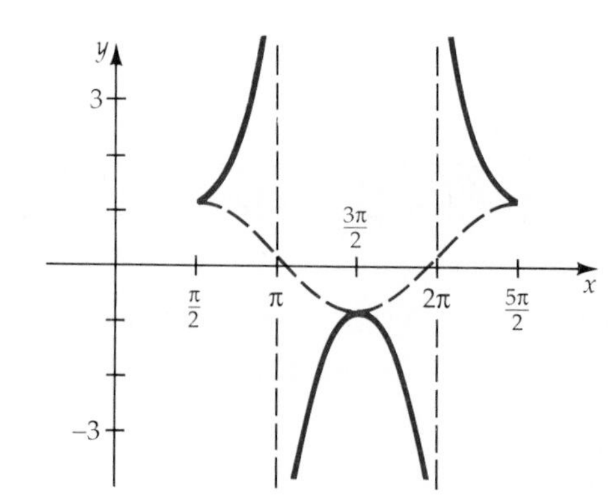

68.

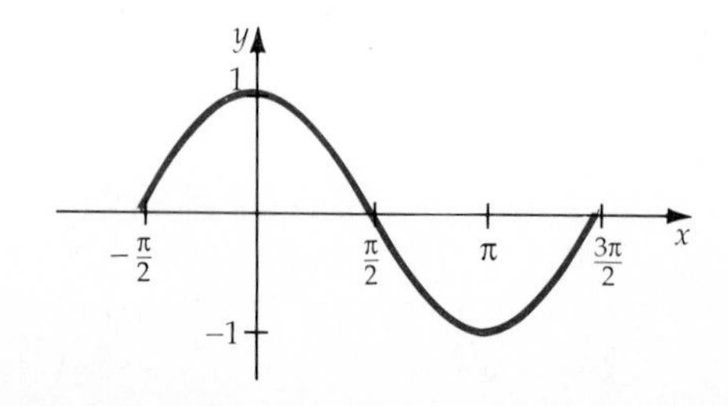

Supplemental Exercises

69. Find an equation of the sine function with amplitude 2, period π, and phase shift $\pi/3$.

70. Find an equation of the cosine function with amplitude 3, period 3π, and phase shift $-\pi/4$.

71. Find an equation of the tangent function with period 2π and phase shift $\pi/2$.

72. Find an equation of the cotangent function with period $\pi/2$ and phase shift $-\pi/4$.

73. Find an equation of the secant function with period 4π and phase shift $3\pi/4$.

74. Find an equation of the cosecant function with period $3\pi/2$ and phase shift $\pi/4$.

75. If $f(x) = \sin^2 x$ and $g(x) = \cos^2 x$, find $f(x) + g(x)$.

76. If $f(x) = 2 \sin x - 3$ and $g(x) = 4 \cos x + 2$, find the sum $f(x) + g(x)$.

77. If $f(x) = x^2 + 2$ and $g(x) = \cos x$, find $f[g(x)]$.

78. If $f(x) = \sin x$ and $g(x) = x^2 + 2x + 1$, find $g[f(x)]$.

In Exercises 79 to 82, sketch the graph of the given function.

79. $f(x) = \dfrac{\sin x}{x}$ **80.** $f(x) = 2 + \sec \dfrac{x}{2}$

81. $f(x) = |x| \sin x$ **82.** $f(x) = |x| \cos x$

83. The average depth at a pier in a small port is 9 feet. During a 24-hour day, the tides raise and lower the depth of water at the pier as shown in the figure. Write an equation in the form $f(t) = A \cos Bt + k$ and find the depth of the water at 6 P.M.

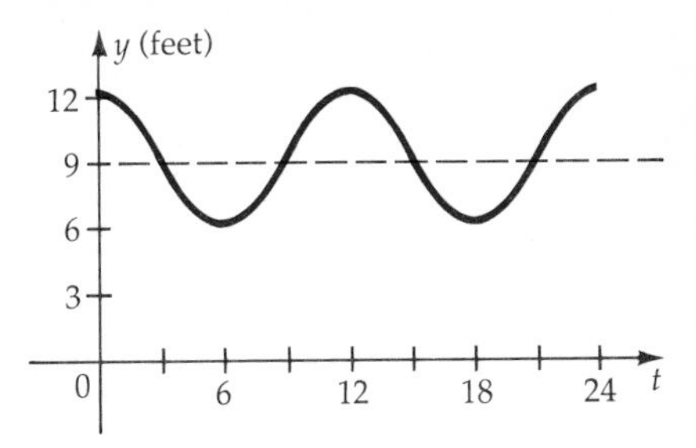

t is the number of hours from 6 AM

84. During a summer day, the ground temperature at a desert location was recorded and graphed as a function of time as shown in the figure. The graph can be approximated by the equation $f(x) = A \cos(bx + c) + k$. Find the equation and find the approximate temperature at 1:00 P.M.

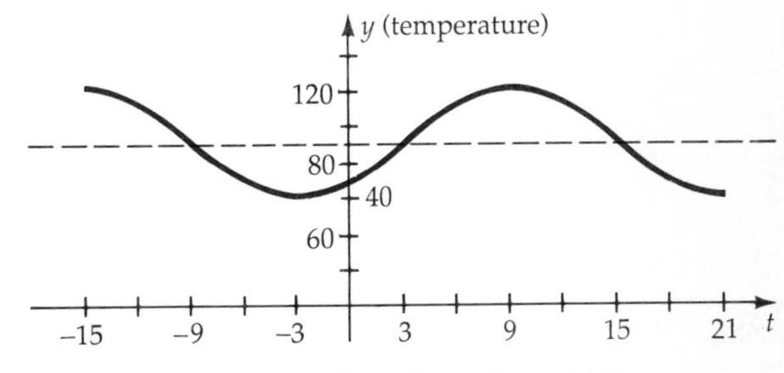

t is the number of hours from 6 AM

2.8 Simple Harmonic Motion—An Application of the Sine and Cosine Functions

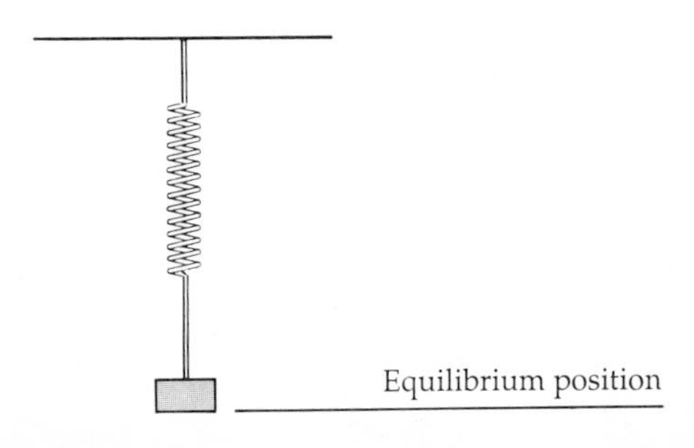

Figure 2.82

Many phenomena occur in nature that can be modeled by periodic functions, including vibrations in buildings, sound waves, electromagnetic waves, and vibrations of a swing or in a spring. These phenomena can be described by the *sinusoidal* functions, which are the sine and cosine functions or the sum of these two functions.

We will consider a mass on a spring to illustrate vibratory motion. Assume that we have placed a mass on a spring and allowed the spring to come to rest, as shown in Figure 2.82. The system is said to be in equilibrium when the mass is at rest. The point of rest is called the origin of the system. We consider the distance above the equilibrium point as positive and the distance below the equilibrium point as negative.

If the mass is now lifted a distance a and released, the mass will oscillate up and down in periodic motion, which means that the motion repeats itself in a certain period of time. The distance a is called the displacement from the origin. The number of times the mass oscillates in 1 second is called the frequency of the motion and the time for one oscillation is the period of the motion. For small oscillations, this period is a constant and the motion is referred to as simple harmonic motion. Figure 2.83 shows the position y of the mass for one oscillation for $t = 0$, $p/4$, $p/2$, $3p/4$, and p when the period is p.

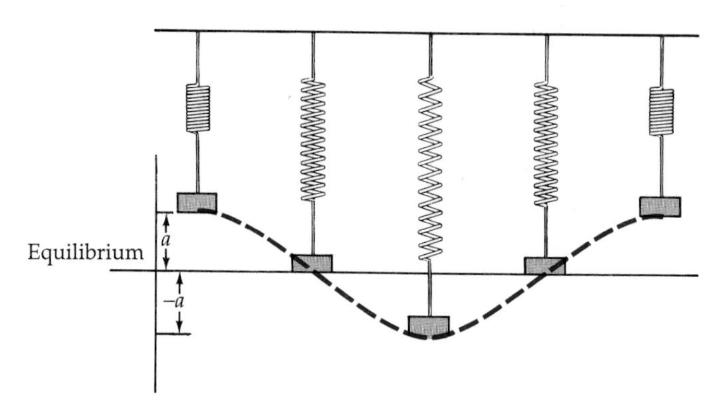

Figure 2.83

Remark Note that if we were to graph the displacement y as a function of t and draw a smooth line through the points, we would have a cosine curve.

There is a relationship between the frequency and the period. Assume that we have a mass that will make two oscillations (an oscillation is a back-and-forth motion) in 1 second. The time for one oscillation is $\frac{1}{2}$ second. Thus the period is $\frac{1}{2}$ second. The frequency and the period are related by the formula $f = 1/\text{period}$.

The maximum displacement from the equilibrium position is called the amplitude of the motion. Vibratory motion can be quite complicated. However, the motion that we have described with the mass on the spring is called simple harmonic motion and can be described by the following equation.

Definition of Simple Harmonic Motion

Simple harmonic motion is motion that can be modeled by one of the following equations:

$$y = a \cos 2\pi ft \qquad \text{or} \qquad y = a \sin 2\pi ft$$

where a is the amplitude (maximum displacement), f is the frequency, and $1/f$ is the period, y is the displacement and t is the time.

Remark We have been given two equations of simple harmonic motion. The cosine function is used if the displacement from the origin is at a maximum at time $t = 0$. The sine function is used if the displacement at time $t = 0$ is zero.

EXAMPLE 1 **Find the Equation of Motion of a Mass on a Spring**

A mass on a spring has been displaced 4 centimeters above the equilibrium point and released. The mass is vibrating with a frequency of $\frac{1}{2}$ cycles per second. Write the equation of motion and graph three cycles of the displacement as a function of time.

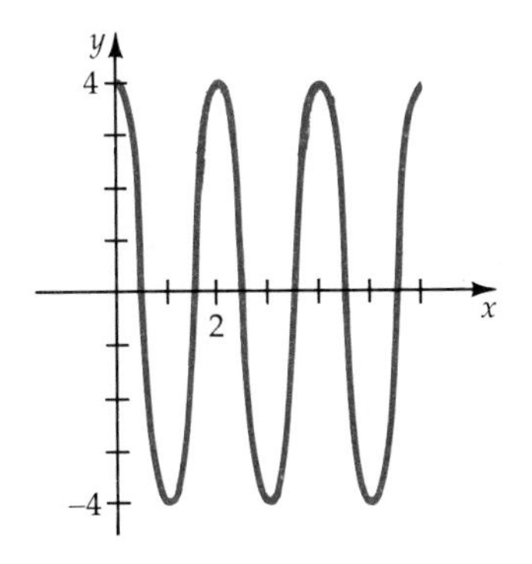

Figure 2.84
$y = 4 \cos \pi t$

Solution Since the maximum displacement is 4 centimeters when $t = 0$, use $y = a \cos 2\pi ft$.

$$y = a \cos 2\pi ft \qquad \text{Equation for simple harmonic motion}$$

$$= 4 \cos 2\pi\left(\frac{1}{2}\right)t \qquad a = 4,\ f = \tfrac{1}{2}.$$

$$= 4 \cos \pi t$$

■ *Try Exercise* **20**, *page 107.*

From physical laws determined by experiment, the frequency of oscillation of a mass on a spring is given by

$$f = \frac{1}{2\pi}\sqrt{\frac{k}{m}} \qquad k \text{ is a spring constant determined by experiment and } m \text{ is the mass.}$$

The motion of the mass on the spring can then be described by

$$y = a \cos 2\pi ft = a \cos 2\pi\left(\frac{1}{2\pi}\sqrt{\frac{k}{m}}\right)t$$

$$= a \cos \sqrt{\frac{k}{m}}\,t.$$

The equation of motion for zero displacement at $t = 0$ is

$$y = a \sin \sqrt{\frac{k}{m}}\,t.$$

EXAMPLE 2 **Find the Equation of Motion of a Mass on a Spring**

A mass of 2 units is in equilibrium suspended from a spring. The mass is pulled down 0.5 units and released. Find the period, frequency, and amplitude of the resulting motion. Write the equation of the motion if $k = 18$, and graph two cycles of the displacement as a function of time.

Solution At the start of the motion, the displacement is at a maximum but in the negative direction. The resulting motion is described by

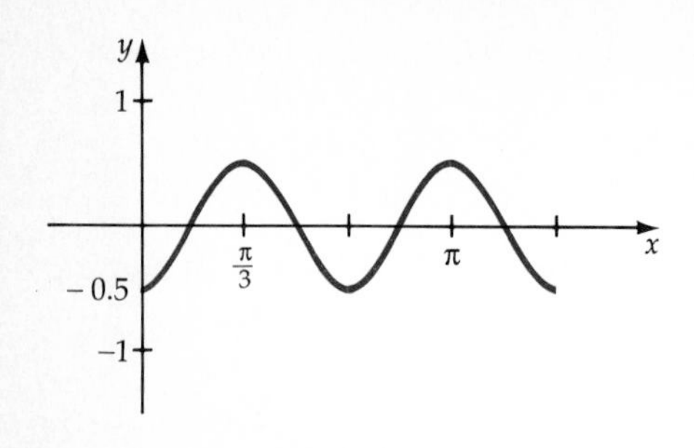

Figure 2.85
$y = -0.5 \cos 3t$

the cosine function. We know that $a = -0.5$, $k = 18$, $m = 2$.

$$y = a \cos \sqrt{\frac{k}{m}}\,t = -0.5 \cos \sqrt{\frac{18}{2}}\,t \quad \text{Substitute for } a, k, \text{ and } m.$$

$$= -0.5 \cos 3t \quad \text{Equation of motion}$$

$$\text{Period: } \frac{2\pi}{3}, \quad \text{Frequency: } \frac{3}{2\pi}, \quad \text{Frequency} = 1/\text{period}.$$

Amplitude: 0.5

■ *Try Exercise* **28**, *page 107.*

A simple pendulum is a system that exhibits approximate simple harmonic motion. A simple pendulum consists of a mass suspended from a string that is attached to a fixed point. If the mass is displaced through an angle θ and released, the pendulum will oscillate back and fourth in a plane. It can be shown from physical laws that if the angle θ is small, the equation of motion is approximated by the equation

$$y = a \cos 2\pi ft \qquad \text{or} \qquad y = a \sin 2\pi ft$$

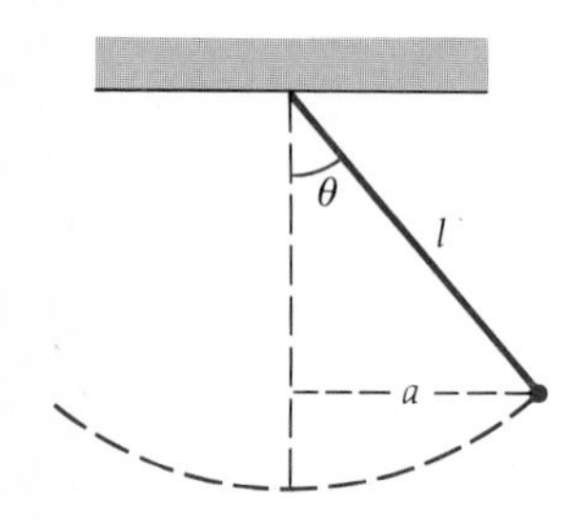

Figure 2.86

where y is the displacement at time t, a the maximum displacement, and f the frequency of motion. The period of the motion is $1/f$. For the pendulum, we will consider all displacements positive.

From measurements taken from experiments on a simple pendulum,

$$f = \frac{1}{2\pi}\sqrt{\frac{g}{l}}$$

where g is the gravitational constant 32 feet per second squared, and l is the length of the pendulum in feet.

Thus the equation for the motion of a pendulum is given by

$$y = a \cos \sqrt{\frac{g}{l}}\,t \qquad \text{or} \qquad y = a \sin \sqrt{\frac{g}{l}}\,t.$$

EXAMPLE 3 Find the Period and Frequency of the Motion of the Pendulum

Find the period and frequency of a pendulum with a length of 8 feet. Graph two cycles of the motion if the maximum displacement is 1.5 feet and the displacement is zero at $t = 0$.

Solution Use the formula to find the period and the frequency.

$$\text{Period: } \frac{2\pi}{\sqrt{g/l}} = \frac{2\pi}{\sqrt{32/8}} = \frac{2\pi}{2} = \pi, \qquad \text{Frequency: } 1/\text{period} = 1/\pi$$

To graph the motion, find the equation of motion. We use the equation $y = a \sin \sqrt{\frac{g}{l}}\,t$ because the displacement is zero at $t = 0$.

$$y = 1.5 \sin 2t \quad a = 1.5, \text{ period} = \pi.$$

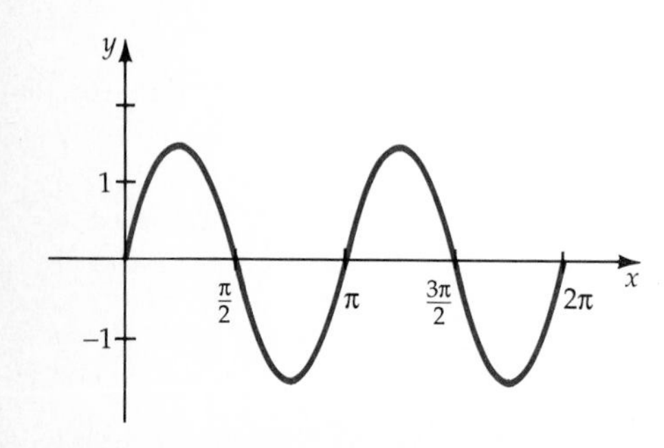

Figure 2.87
$y = 1.5 \sin 2t$

■ *Try Exercise* **30**, *page 107.*

EXERCISE SET 2.8

In Exercises 1 to 8, find the amplitude, period, and frequency of the harmonic motion.

1. $y = 2 \sin 2t$

2. $y = \dfrac{2}{3} \cos \dfrac{t}{3}$

3. $y = 3 \cos \dfrac{2t}{3}$

4. $y = 4 \sin 3t$

5. $y = 4 \cos \pi t$

6. $y = 2 \sin \dfrac{\pi t}{3}$

7. $y = \dfrac{3}{4} \sin \dfrac{\pi t}{2}$

8. $y = 5 \cos 2\pi t$

In Exercises 9 to 12, write the equation of motion and graph the amplitude as a function of time for the following given values. Assume that the motion is harmonic motion and maximum displacement occurs at $t = 0$.

9. frequency = 1.5 cycles per second, a = 4 inches

10. frequency = 0.8 cycle per second, a = 4 centimeters

11. period = 1.5 seconds, $a = \frac{3}{2}$ feet

12. period = 0.6 second, a = 1 meter

In Exercises 13 to 18, find the equation of simple harmonic motion with the given conditions. Assume zero displacement at $t = 0$.

13. Amplitude 2 centimeters, period π seconds

14. Amplitude 4 inches, period $\pi/2$ seconds

15. Amplitude 1 inch, period 2 seconds

16. Amplitude 3 centimeters, period 1 second

17. Amplitude 2 centimeters, frequency 1 second

18. Amplitude 4 inches, frequency 4 seconds

In Exercises 19 to 26, write an equation for simple harmonic motion. Assume that the maximum displacement occurs when $t = 0$.

19. Amplitude $\frac{1}{2}$ centimeters, frequency $2/\pi$ cycles per second

20. Amplitude 3 inches, frequency $1/\pi$ cycles per second

21. Amplitude 2.5 inches, frequency 0.5 cycles per second

22. Amplitude 5 inches, frequency $\frac{1}{8}$ cycles per second

23. Amplitude $\frac{1}{2}$ inch, period 3 seconds

24. Amplitude 5 centimeters, period 5 seconds

25. Amplitude 4 inches, period $\pi/2$ seconds

26. Amplitude 2 centimeters, period π seconds

27. A mass of 32 units is in equilibrium suspended from a spring. The mass is pulled down 2 feet and released. Find the period, frequency, and amplitude of the resulting motion. Write the equation of motion. Let $k = 8$.

28. A mass of 27 units is in equilibrium suspended from a spring. The mass is pulled down 1.5 feet and released. Find the period, frequency, and amplitude of the resulting motion. Write the equation of motion. Let $k = 3$.

29. A pendulum 6 feet long is displaced a distance of 1 foot and released. Write the equation for the displacement as a function of time. Find the period and frequency of the motion.

30. A pendulum 20 feet long is displaced a distance of 4 feet and released. Write the equation for the displacement as a function of time. Find the period and frequency of the motion.

31. A mass of 5 units is suspended from a spring. The spring is compressed 6 inches and released. Find the period, frequency, and amplitude of the resulting motion. The constant $k = 2$. Write the equation of motion.

32. A pendulum of 3 feet long is displaced a distance of 6 inches and released. Find the period and frequency of the motion and write the equation of motion.

Supplemental Exercises

33. A pendulum with a length of 4 feet is released with an initial displacement of 6 inches. Write the equation for the displacement and find the frequency of the motion.

34. A mass of 0.5 unit is suspended from a spring with a constant of 32. The mass is displaced a distance of 10 inches and released. Write the equation for the motion and find the frequency of the motion.

35. A weight on a spring is displaced 6 inches from its equilibrium position and then released. The weight oscillated with a frequency of 1.5 seconds. Find the period and the equation of the motion.

36. A weight on a spring is displaced 9 inches from its equilibrium position and then released. The weight oscillated with a frequency of 2 seconds. Find the period and the equation of the motion.

37. What effect does doubling the length of a pendulum have on the period of the pendulum?

38. The length of a pendulum is changed until the period is cut in half. Find the length of the pendulum as compared to the original length.

Chapter 2 Review

2.1 Measuring Angles

An angle is formed by rotating a ray about a point 0.

An angle in standard position has the initial side along the positive x-axis and the vertex at the origin of the coordinate axes.

Coterminal angles have the same initial and terminal sides.

Degrees and radians are used for the measure of angles.

Arc length of a circle is the product of the radius r and central angle θ in radians is $s = r\theta$.

The area of a sector of radius r and central angle θ is given by the formula $A = \frac{1}{2}r^2\theta$.

Angular velocity is given by the formula: $\omega = \theta/t$.

The velocity of a point on a rotating body is given by the formula $V = r\omega$.

2.2 Trigonometric Functions of Acute Angles

The six trigonometric functions of an angle θ in standard position are defined by

$$\sin\theta = \frac{b}{r} \qquad \cos\theta = \frac{a}{r} \qquad \tan\theta = \frac{b}{a},\ a \neq 0$$

$$\csc\theta = \frac{r}{b},\ b \neq 0 \qquad \sec\theta = \frac{r}{a},\ a \neq 0 \qquad \cot\theta = \frac{a}{b},\ b \neq 0$$

2.3 Trigonometric Functions of Any Angle

Quadrantal angles are angles in which the terminal side coincides with the x- or y-axis.

A reference angle is the positive acute angle made by the terminal side of an angle in standard position and the x-axis.

2.4 Circular Functions

A circle of radius one is a unit circle.

The circular functions are defined in terms of the unit circle and the domain of the circular functions is the set of real numbers. If $P(x, y)$ is a point on a unit circle corresponding to the real number t, then

$$\cos t = x \qquad \sin t = y \qquad \tan t = \frac{y}{x},\ x \neq 0$$

$$\sec t = \frac{1}{x},\ x \neq 0 \qquad \csc t = \frac{1}{y},\ y \neq 0 \qquad \cot t = \frac{x}{y},\ y \neq 0$$

The reciprocal identities are:

$$\sin \theta = \frac{1}{\csc \theta} \qquad \cos \theta = \frac{1}{\sec \theta} \qquad \tan \theta = \frac{1}{\cot \theta}$$

The ratio identities are

$$\tan \theta = \frac{\sin \theta}{\cos \theta} \qquad \cos \theta = \frac{\cos \theta}{\sin \theta}$$

The Pythagorean identities are

$$\cos^2\theta + \sin^2\theta = 1 \qquad 1 + \tan^2\theta = \sec^2\theta \qquad 1 + \cot^2\theta = \csc^2\theta$$

2.5 Graphs of the Sine and Cosine Functions

The amplitude of $y = a \sin bx$ and $y = a \cos bx$ is $|a|$. The period of each function is given by $2\pi/b$ $(b > 0)$.

2.6 Graphs of the Other Trigonometric Functions

The period of $y = a \tan bx$ and $y = a \cot bx$ is π/b. The graphs of $y = a \sec bx$ and $y = a \csc bx$ are obtained by using the reciprocals of the y-coordinates of the graphs of $y = a \cos bx$ and $y = a \sin bx$, respectively.

2.7 Phase Shift and Addition of Ordinates

The phase shift of $y = a \sin(bx + c)$ and $y = a \cos(bx + c)$, where $b > 0$, is $-c/b$.

The phase shift of $y = a \tan(bx + c)$ and $y = \cot(bx + c)$ is $-c/b$.

The graph of the sum of two trigonometric functions can be obtained by the method of addition of ordinates.

2.8 Simple Harmonic Motion

The equations of simple harmonic motion are $y = a \cos 2\pi ft$ or $y = a \sin 2\pi ft$, where a is the amplitude and f is the frequency.

CHALLENGE EXERCISES

In Exercises 1 to 14, answer true or false. If the statement is false, give an example.

1. An angle is in standard position when the vertex is at the origin of a coordinate system.
2. The angle θ is in radians in standard position with the terminal side in the second quadrant. The reference angle of θ is $\pi - \theta$.
3. In the formula $s = r\theta$, the angle θ must be measured in radians.
4. If $\tan \theta < 0$ and $\cos \theta > 0$, then the terminal side θ is in the third quadrant.
5. $\sec^2\theta + \tan^2\theta = 1$ is an identity.
6. The amplitude of the function $f(x) = 2 \tan x$ is 2.
7. The period of the function $\cos \theta$ is π.

8. The graph of the function $\sin\theta$ is symmetric to the origin.
9. For any acute angle θ, $\sin\theta + \cos(90° - \theta) = 1$.
10. $\sin(x + y) = \sin x + \sin y$.
11. $\sin^2 x = \sin x^2$.
12. The phase shift of the function $f(x) = 2\sin\left(2x - \frac{\pi}{3}\right)$ is $\frac{\pi}{3}$.
13. One radian has approximately the same measure as one degree.
14. The measure of one radian differs depending on the radius of the circle used.

REVIEW EXERCISES

1. Sketch the angles in standard position: a. 120°; b. −135°.
2. Convert the angular measure to decimal form: a. 37° 34′; b. −142° 46′ 8″.
3. Convert the angular measurement to DMS: a. 114.8°; b. −38.38°.
4. Convert the radian measure to degree measure: a. $7\pi/4$ rad; b. 2 rad.
5. Convert the angle measurement to radian measure: a. 315°; b. 97.4°.
6. Find the arc length subtended by an angle with a measure of 75° in a circle with a 3-meter radius.
7. Find the arc length subtended by an angle with a measure of 42° in a circle with a 5-inch radius.
8. Find the radian measure of a central angle subtended by an arc length of 12 centimeters in a circle with a 40-centimeter radius.
9. Find the degree measure of a central angle subtended by an arc length of 5 inches in a circle with a 20-inch radius.
10. Find the area of a sector of a circle with a 16-inch radius. The central angle of the sector is 40°.
11. Find the area of a sector of a circle with a 4-meter radius. The central angle of the sector is 25°.
12. A car with a 16-inch wheel is moving with a velocity of 50 miles per hour. Find the angular velocity of the wheel in radians per second.
13. A truck with an 18-inch wheel is moving with a velocity of 55 miles per hour. Find the angular velocity of the wheel in radians per second.
14. Find the six trigonometric function values of an angle in standard position with the point $P(-2, 5)$ on the terminal side of the angle.
15. Find the six trigonometric function values of an angle in standard position with the point $P(1, -3)$ on the terminal side of the angle.

In Exercises 16 to 19, $\csc\theta = 3/2$, in the first quadrant. Evaluate the following trigonometric functions.

16. $\cos\theta$ 17. $\cot\theta$ 18. $\sin\theta$ 19. $\sec\theta$

20. Find the exact values of: **a.** sec 150°; **b.** $\tan(-3\pi/4)$.
21. Find the exact values of: **a.** cot(−225°); **b.** $\cos 2\pi/3$.
22. Find the value of: **a.** cos 123°; **b.** cot 4.22.
23. Find the value of: **a.** sec 612°; **b.** $\tan 7\pi$.
24. $\cos\phi = -\sqrt{3}/2$, ϕ in quadrant three, find the exact values of: **a.** $\sin\phi$; **b.** $\tan\phi$.
25. $\tan\phi = -\sqrt{3}/3$, ϕ in quadrant two, find the exact values of: **a.** $\sec\phi$; **b.** $\cot\phi$.
26. $\tan\phi = 1$, ϕ in quadrant three, find the exact values of: **a.** $\csc\phi$; **b.** $\cot\phi$.
27. $\sin\phi = -\sqrt{2}/2$, ϕ in quadrant four, find the exact values of: **a.** $\sec\phi$; **b.** $\cot\phi$.

In Exercises 28 to 31, find the value of each expression.

28. $\sin 120° \tan 45° - \cos 315°$.
29. $\tan\frac{3\pi}{4} - \sin\frac{\pi}{3}\cos\frac{3\pi}{4}$
30. $\tan 225° - \cot 30° \sin 150°$.
31. $\cos\frac{5\pi}{3}\cot\frac{\pi}{6} + \csc\frac{7\pi}{4}$

In Exercises 32 to 35, find the exact value of the indicated function.

32. $\sec\phi = -\sqrt{2}$, ϕ in quadrant II, find $\tan\phi$.
33. $\cot\phi = \sqrt{3}$, ϕ in quadrant III, find $\sin\phi$.
34. $\sin\phi = -1/2$, ϕ in quadrant IV, find $\cot\phi$.
35. $\tan\phi = 1$, ϕ in quadrant III, find $\cos\phi$.

In Exercises 36 to 39, find the arc length intercepted by the given angle in a unit circle.

36. $2\pi/5$ radians 37. 112°
38. 0.6 radians 39. 2°

In Exercises 40 to 43, find the angle (in radians) subtended by the given arc in a unit circle.

40. $3\pi/4$ 41. 2.6 42. 2 43. 0.85

44. Use the unit circle to show that the following expressions are true:

a. $\cos(180° + \phi) = -\cos\phi$; **b.** $\tan(-\phi) = -\tan\phi$.

In Exercises 45 to 50, simplify the first expression to show that it is equivalent to the second expression.

45. $1 + \dfrac{\sin^2\phi}{\cos^2\phi}$: $\sec^2\phi$

46. $\dfrac{\tan\phi + 1}{\cot\phi + 1}$: $\tan\phi$

47. $\dfrac{\cos^2\phi + \sin^2\phi}{\csc\phi}$: $\sin\phi$

48. $\sin^2\phi(\tan^2\phi + 1)$: $\tan^2\phi$

49. $1 + \dfrac{1}{\tan^2\phi}$: $\csc^2\phi$

50. $\dfrac{\cos^2\phi}{1 - \sin\phi}$: $1 + \sin\phi$

In Exercises 51 to 64, graph the given functions.

51. $f(x) = 2\cos\pi x$

52. $f(x) = -\sin\dfrac{2x}{3}$

53. $f(x) = 2\sin\dfrac{3x}{2}$

54. $f(x) = \cos\left(x - \dfrac{\pi}{2}\right)$

55. $f(x) = \dfrac{1}{2}\sin\left(2x + \dfrac{\pi}{4}\right)$

56. $f(x) = 3\cos 3(x - \pi)$

57. $f(x) = -\tan\dfrac{x}{2}$

58. $f(x) = 2\cot 2x$

59. $f(x) = \tan\left(x - \dfrac{\pi}{2}\right)$

60. $f(x) = -\cot\left(2x + \dfrac{\pi}{4}\right)$

61. $f(x) = -2\csc\left(2x - \dfrac{\pi}{3}\right)$

62. $f(x) = 3\sec\left(x + \dfrac{\pi}{4}\right)$

63. $f(x) = 2 - \sin 2x$

64. $f(x) = \sin x - \sqrt{3}\cos x$

65. Find the amplitude, period, and frequency of the harmonic motion given by the equation $y = 2.5\sin 50t$.

66. A pendulum 5 feet long is displaced a distance of 9 inches and released. Write the equation for the displacement as a function of time. Find the period and frequency of the motion.

67. A mass of 5 kilograms is in equilibrium suspended from a spring. The mass is pulled down 0.5 feet and released. Find the period, frequency, and amplitude of the resulting motion. Write the equation of motion. (Let $k = 20$.)

3 Trigonometric Identities and Equations

In most sciences, one generation tears down what another has built, and what one has established, the next undoes. In mathematics alone, each generation builds a new story to the old structure.

HERMANN HANKEL (1839–1873)

RESONANCE PHENOMENA IN A PHYSICAL SYSTEM

Many vibrating systems can be described by simple harmonic motion. Simple harmonic motion can be described by equations of the form $y = a \cos 2\pi ft$. In pure simple harmonic motion, the amplitude of the vibrations remain constant. In an actual physical system, energy will be lost, and the amplitude of the vibrations will gradually decrease. When energy is put into a vibrating system, the amplitude of the vibrations may increase. For example, the amplitude of a swing will increase if impulses of energy are applied to the system at the proper times. *Resonance* is the building up of larger amplitudes of a vibrating system when small impulses are applied at the proper time.

An equation for a motion with resonance is $y = at \cos 2\pi ft$. The graph shows how the vibrations in a resonance system would increase with time.

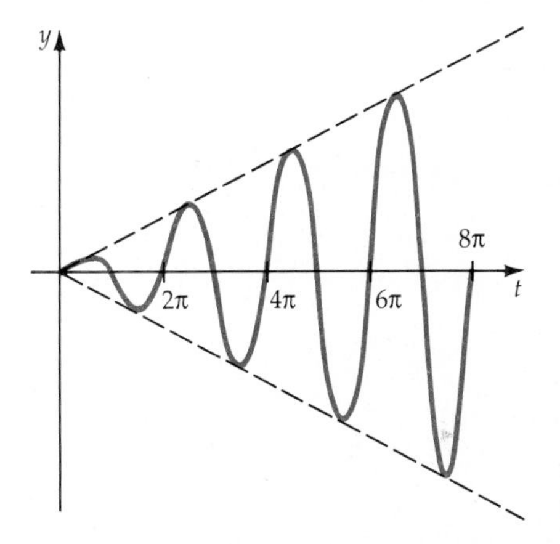

Resonance phenomena can occur in almost any physical system. It is possible for soldiers marching across a bridge to cause large vibrations, if the frequency of march is the same as the natural frequency of the bridge. Another example of resonance is the Tacoma Narrows bridge collapse, which occurred on July 2, 1940. A wind coming down a canyon created a

The Tacoma Narrows Bridge at Puget Sound, Washington. Wide World.

vibration in the bridge that continued to increase as the day passed. Vibrations in the concrete and steel structure reached an amplitude of 8 feet and the bridge collapsed.

3.1 The Fundamental Trigonometric Identities

An equation that is true for all replacements of the variables for which all terms of the equation are defined is called an **identity**. Trigonometric identities are used to simplify a trigonometric expression or to write a trigonometric expression in an equivalent form.

As for algebraic equations, the domain of a trigonometric identity will be all values of the variable for which the expression is meaningful. For example, the identity

$$\frac{\sin x \cos x}{\sin x} = \cos x$$

is true for all real numbers x such that $\sin x \neq 0$. Because $\sin x = 0$ when $x = n\pi$ and n is an integer, the domain of the equation must exclude all integral multiples of π. To verify an identity means to show that each side of the equation represents the same expression. For example, to verify the identity above, simplify the left side of the equation. We now have $\cos x = \cos x$. The left and right sides of the equation are the same expression.

There is no one method that can be used to verify identities. Generally we work only on one side of the equation. The list below provides some suggestions for verifying trigonometric identities.

1. If one side of the identity is more complex than the other, simplify the more complex side.
2. Perform algebraic operations in the identity such as
   a. Squaring
   b. Factoring
   c. Adding fractions
   d. Multiplying the numerator and denominator by a nonzero factor
3. Rewrite the identity in terms of sine and cosine functions.
4. Rewrite one side of the identity in terms of a single function.

TABLE 3.1 Fundamental Trigonometric Identities

Reciprocal identities	$\sin x = \dfrac{1}{\csc x}$	$\cos x = \dfrac{1}{\sec x}$	$\tan x = \dfrac{1}{\cot x}$
Ratio identities	$\tan x = \dfrac{\sin x}{\cos x}$	$\cot x = \dfrac{\cos x}{\sin x}$	
Pythagorean identities	$\sin^2 x + \cos^2 x = 1$	$\tan^2 x + 1 = \sec^2 x$	$1 + \cot^2 x = \csc^2 x$

Table 3.1 lists the eight fundamental identities established in the previous chapter. These identities will be valuable when we verify other trigonometric identities.

EXAMPLE 1 Verify a Trigonometric Identity by Converting to Sines and Cosines

Verify the identity $\sec x - \cos x = \sin x \tan x$.

Solution Simplify the left side of the equation.

$$\sec x - \cos x = \frac{1}{\cos x} - \cos x = \frac{1 - \cos^2 x}{\cos x}$$ Write as a single fraction with a common denominator

$$= \frac{\sin^2 x}{\cos x} = \sin x \cdot \frac{\sin x}{\cos x}$$ $1 - \cos^2 x = \sin^2 x$

$$= \sin x \tan x$$ $\frac{\sin x}{\cos x} = \tan x$

Since the left side of the identity has been rewritten to be the right, we have verified the identity.

■ *Try Exercise* **30**, *page 117.*

EXAMPLE 2 Verify a Trigonometric Identity by Using a Pythagorean Identity

Verify the identity $1 - 2\sin^2 x = 2\cos^2 x - 1$.

Solution The Pythagorean identity $\sin^2 x + \cos^2 x = 1$ implies that $\cos^2 x = 1 - \sin^2 x$. Use this identity; rewrite the right side of the equation.

$$2\cos^2 x - 1 = 2(1 - \sin^2 x) - 1$$ $\cos^2 x = 1 - \sin^2 x$

$$= 2 - 2\sin^2 x - 1$$

$$= 1 - 2\sin^2 x$$

■ *Try Exercise* **40**, *page 117.*

EXAMPLE 3 Verify a Trigonometric Identity by Factoring

Verify the identity $\csc^2 x - \cos^2 x \csc^2 x = 1$.

Solution

$$\csc^2 x - \cos^2 x \csc^2 x = \csc^2 x(1 - \cos^2 x)$$ Factor out $\csc^2 x$.

$$= \csc^2 x \sin^2 x$$

$$= \frac{1}{\sin^2 x} \cdot \sin^2 x = 1$$ $\csc^2 x = \frac{1}{\sin^2 x}$

■ *Try Exercise* **52**, *page 117.*

EXAMPLE 4 Verify a Trigonometric Identity by Using a Conjugate

Verify the identity $\dfrac{\sin x}{1 + \cos x} = \dfrac{1 - \cos x}{\sin x}$.

Solution Multiply the numerator and denominator of the left side of the identity by the conjugate of $1 + \cos x$.

$$\frac{\sin x}{1 + \cos x} = \frac{\sin x}{1 + \cos x} \cdot \frac{1 - \cos x}{1 - \cos x} = \frac{\sin x(1 - \cos x)}{1 - \cos^2 x}$$

$$= \frac{\sin x(1 - \cos x)}{\sin^2 x} = \frac{1 - \cos x}{\sin x}$$

■ *Try Exercise* **62**, *page 117.*

EXAMPLE 5 Verify a Trigonometric Identity

Verify the identity $\dfrac{\sin x + \tan x}{1 + \cos x} = \tan x$.

Solution Rewrite the left side of the identity in terms of sines and cosines.

$$\frac{\sin x + \tan x}{1 + \cos x} = \frac{\sin x + \dfrac{\sin x}{\cos x}}{1 + \cos x} \qquad \tan x = \frac{\sin x}{\cos x}$$

$$= \frac{\dfrac{\sin x \cos x + \sin x}{\cos x}}{1 + \cos x} \qquad \text{Write the numerator with a common denominator.}$$

$$= \frac{\sin x \cos x + \sin x}{\cos x(1 + \cos x)} \qquad \text{Simplify.}$$

$$= \frac{\sin x\,(1 + \cos x)}{\cos x\,(1 + \cos x)}$$

$$= \tan x$$

■ *Try Exercise* **72**, *page 118.*

EXERCISE SET 3.1

In Exercises 1 to 10, find the product.

1. $(\cos x - 2)(\cos x - 3)$

2. $(\sin x - 2)(\sin x + 7)$

3. $(\sin^2 x - 5)(\sin^2 x - 12)$

4. $(\tan^2 x + 5)(\tan^2 x + 3)$

5. $(3 \sin x - 2)(4 \sin x + 5)$

6. $(2 \cos x + 5)(4 \cos x + 1)$

7. $(\cos x - 2)(\sin x + 7)$

8. $(\tan x + 5)(\cot x - 3)$

9. $(\sin x - 2 \cos x)(3 \sin x + 4 \cos x)$

10. $(2 \tan x - 3 \cot x)(5 \tan x - \cot x)$

In Exercises 11 to 20, factor completely.

11. $\sin^2 x - 2 \sin x + 1$

12. $\tan^2 x + 2 \tan x + 1$

13. $8 \tan^2 x - 26x + 15$

14. $6 \sin^2 x + \sin x - 15$

15. $4 \sin^2 x - 1$

16. $\tan^2 x - 64$

17. $\sin^3 x - 27$

18. $1 + \tan^2 x$

19. $\sin x \cos x + \sin x + \cos x + 1$

20. $\tan x \cot x + \tan x - 2 \cot x - 2$

In Exercises 21 to 28, simplify by performing the indicated operation.

21. $\dfrac{1}{\sin x} + \dfrac{3}{\cos x}$

22. $\dfrac{2}{\cos x} - \dfrac{3}{\sin x}$

23. $\dfrac{2 \sin x}{\cos x} - \dfrac{\cos x}{\sin x}$

24. $\dfrac{4 \tan x}{\cot x} + \dfrac{5}{\tan x}$

25. $\dfrac{3}{3 \cos x + 1} + \dfrac{2}{\cos x}$

26. $\dfrac{2}{\sin x - 2} - \dfrac{4}{\sin x}$

27. $\dfrac{1 + \dfrac{1}{\sin x}}{1 - \dfrac{1}{\sin x}}$

28. $\dfrac{\sin x - 2 + \dfrac{1}{\sin x}}{\sin x - \dfrac{1}{\sin x}}$

In Exercises 29 to 82, verify the identities.

29. $\tan x \csc x \cos x = 1$

30. $\sin x \cot x \sec x = 1$

31. $\dfrac{4 \sin^2 x - 1}{2 \sin x + 1} = 2 \sin x - 1$

32. $\dfrac{\sin^2 x - 2 \sin x + 1}{\sin x - 1} = \sin x - 1$

33. $(\sin x - \cos x)(\sin x + \cos x) = 1 - 2 \cos^2 x$

34. $\tan x(1 - \cot x) = \tan x - 1$

35. $\dfrac{1}{\sin x} - \dfrac{1}{\cos x} = \dfrac{\cos x - \sin x}{\sin x \cos x}$

36. $\dfrac{1}{\sin x} + \dfrac{3}{\cos x} = \dfrac{\cos x + 3 \sin x}{\sin x \cos x}$

37. $\dfrac{\cos x}{1 - \sin x} = \sec x + \tan x$

38. $\dfrac{\sin x}{1 - \cos x} = \csc x + \cot x$

39. $\dfrac{1 - \tan^4 x}{\sec^2 x} = 1 - \tan^2 x$

40. $\sin^4 x - \cos^4 x = \sin^2 x - \cos^2 x$

41. $\dfrac{1 + \tan^3 x}{1 + \tan x} = 1 - \tan x + \tan^2 x$

42. $\dfrac{\cos x \tan x - \sin x}{\cot x} = 0$

43. $\dfrac{\sin x - 2 + \dfrac{1}{\sin x}}{\sin x - \dfrac{1}{\sin x}} = \dfrac{\sin x - 1}{\sin x + 1}$

44. $\dfrac{\sin x}{1 - \cos x} - \dfrac{\sin x}{1 + \cos x} = 2 \cot x$

45. $(\sin x + \cos x)^2 = 1 + 2 \sin x \cos x$

46. $(\tan x + 1)^2 = \sec^2 x + 2 \tan x$

47. $\dfrac{\cos x}{1 + \sin x} = \sec x - \tan x$

48. $\dfrac{\sin x}{1 + \cos x} = \csc x - \cot x$

49. $\csc x = \dfrac{\cot x + \tan x}{\sec x}$

50. $\sec x = \dfrac{\cot x + \tan x}{\csc x}$

51. $\dfrac{\cos x \tan x + 2 \cos x - \tan x - 2}{\tan x + 2} = \cos x - 1$

52. $\dfrac{2 \sin x \cot x + \sin x - 4 \cot x - 2}{2 \cot x + 1} = \sin x - 2$

53. $\sec x - \tan x = \dfrac{1 - \sin x}{\cos x}$

54. $\cot x - \csc x = \dfrac{\cos x - 1}{\sin x}$

55. $\sin^2 x - \cos^2 x = 2 \sin^2 x - 1$

56. $\sin^2 x - \cos^2 x = 1 - 2 \cos^2 x$

57. $\dfrac{1}{\sin^2 x} + \dfrac{1}{\cos^2 x} = \csc^2 x \sec^2 x$

58. $\dfrac{1}{\tan^2 x} - \dfrac{1}{\cot^2 x} = \csc^2 x - \sec^2 x$

59. $\sec x - \cos x = \sin x \tan x$

60. $\tan x + \cot x = \sec x \csc x$

61. $\dfrac{\dfrac{1}{\sin x} + 1}{\dfrac{1}{\sin x} - 1} = \tan^2 x + 2 \tan x \sec x + \sec^2 x$

62. $\dfrac{\dfrac{1}{\sin x} + \dfrac{1}{\cos x}}{\dfrac{1}{\sin x} - \dfrac{1}{\cos x}} = \dfrac{\cos^2 x - \sin^2 x}{1 - 2 \cos x \sin x}$

63. $\sin^4 x - \cos^4 x = 2\sin^2 x - 1$

64. $\sin^6 x + \cos^6 x = \sin^4 x - \sin^2 x \cos^2 x + \cos^4 x$

65. $\dfrac{1}{1 - \cos x} = \dfrac{1 + \cos x}{\sin^2 x}$

66. $1 + \sin x = \dfrac{\cos^2 x}{1 - \sin x}$

67. $\dfrac{\sin x}{1 - \sin x} - \dfrac{\cos x}{1 - \sin x} = \dfrac{1 - \cot x}{\csc x - 1}$

68. $\dfrac{\tan x}{1 + \tan x} - \dfrac{\cot x}{1 + \tan x} = 1 - \cot x$

69. $\dfrac{1}{1 + \cos x} - \dfrac{1}{1 - \cos x} = -2\cot x \csc x$

70. $\dfrac{1}{1 - \sin x} - \dfrac{1}{1 + \sin x} = 2\tan x \sec x$

71. $\dfrac{\dfrac{1}{\sin x} + \csc x}{\dfrac{1}{\sin x} - \sin x} = \dfrac{2}{\cos^2 x}$

72. $\dfrac{\dfrac{1}{\tan x} + \cot x}{\dfrac{1}{\tan x} + \tan x} = \dfrac{2}{\sec^2 x}$

73. $\sqrt{\dfrac{1 + \sin x}{1 - \sin x}} = \dfrac{1 + \sin x}{\cos x}, \quad \cos x > 0$

74. $\dfrac{\cos x + \cot x \sin x}{\cot x} = 2\sin x$

75. $\dfrac{\sin^3 x + \cos^3 x}{\sin x + \cos x} = 1 - \sin x \cos x$

76. $\dfrac{1 - \sin x}{1 + \sin x} - \dfrac{1 + \sin x}{1 - \sin x} = -4\sec x \tan x$

77. $\dfrac{\sec x - 1}{\sec x + 1} - \dfrac{\sec x + 1}{\sec x - 1} = -4\csc x \cot x$

78. $\dfrac{1}{1 - \cos x} - \dfrac{\cos x}{1 + \cos x} = 2\csc^2 x - 1$

79. $\dfrac{1 + \sin x}{\cos x} - \dfrac{\cos x}{1 - \sin x} = 0$

80. $(\sin x + \cos x + 1)^2 = 2(\sin x + 1)(\cos x + 1)$

81. $\dfrac{\sec x + \tan x}{\sec x - \tan x} = \dfrac{(\sin x + 1)^2}{\cos^2 x}$

82. $\dfrac{\sin^3 x - \cos^3 x}{\sin x + \cos x} = \dfrac{\csc^2 x - \cot x - 2\cos^2 x}{1 - \cot^2 x}$

Supplemental Exercises

83. Express $\cos x$ in terms of $\sin x$.

84. Express $\tan x$ in terms of $\cos x$.

85. Express $\sec x$ in terms of $\sin x$.

86. Express $\csc x$ in terms of $\sec x$.

In Exercises 87 to 92, verify the identity.

87. $\dfrac{1 - \sin x + \cos x}{1 + \sin x + \cos x} = \dfrac{\cos x}{\sin x + 1}$

88. $\dfrac{1 - \tan x + \sec x}{1 + \tan x - \sec x} = \dfrac{1 + \sec x}{\tan x}$

89. $\dfrac{2\sin^4 x + 2\sin^2 x \cos^2 x - 3\sin^2 x - 3\cos^2 x}{2\sin^2 x} = 1 - \dfrac{3}{2}\csc^2 x$

90. $\dfrac{4\tan x \sec^2 x - 4\tan x - \sec^2 x + 1}{4\tan^3 x - \tan^2 x} = 1$

91. $\dfrac{\sin x(\tan x + 1) - 2\tan x \cos x}{\sin x - \cos x} = \tan x$

92. $\dfrac{\sin^2 x \cos x + \cos^3 x - \sin^3 x \cos x - \sin x \cos^3 x}{1 - \sin^2 x} = \dfrac{\cos x}{1 + \sin x}$

93. Verify the identity $\sin^4 x + \cos^4 x = 1 - 2\sin^2 x \cos^2 x$ by completing the square of the left side of the identity.

94. Verify the identity $\tan^4 x + \sec^4 x = 1 + 2\tan^2 x \sec^2 x$ by completing the square of the left side of the identity.

3.2 Sum and Difference Identities

There are several useful identities relating the sum and difference of two angles $(\alpha \pm \beta)$. We begin by finding an identity for $\cos(\alpha - \beta)$.

In Figure 3.1, angles α and β are drawn in standard position, with OA and OB as the terminal sides of α and β, respectively. The coordinates of A are $(\cos \alpha, \sin \alpha)$, and the coordinates of B are $(\cos \beta, \sin \beta)$. The angle $(\alpha - \beta)$ is formed by the terminal sides of the angles α and β (angle AOB).

An angle equal in measure to angle $(\alpha - \beta)$ is placed in standard position in the same figure. The chords AB and CB are equal, because there is a theorem from geometry that states that if two central angles of a circle have the same measure, then the measure of their chords are also equal. Using the distance formula, we can calculate the lengths of the chords AB and CD.

$$d(A, B) = \sqrt{(\sin \alpha - \sin \beta)^2 + (\cos \alpha - \cos \beta)^2}$$
$$d(C, D) = \sqrt{[\cos(\alpha - \beta) - 1]^2 + [\sin(\alpha - \beta) - 0]^2}$$

Since $d(A, B) = d(C, D)$, we have

$$\sqrt{(\sin \alpha - \sin \beta)^2 + (\cos \alpha - \cos \beta)^2} = \sqrt{[\cos(\alpha - \beta) - 1]^2 + [\sin(\alpha - \beta) - 0]^2}$$

Squaring each side of the equation and simplifying, we obtain

$$(\sin \alpha - \sin \beta)^2 + (\cos \alpha - \cos \beta)^2 = [\cos(\alpha - \beta) - 1]^2 + [\sin(\alpha - \beta) - 0]^2$$

$$\sin^2\alpha - 2 \sin \alpha \sin \beta + \sin^2\beta + \cos^2\alpha - 2 \cos \alpha \cos \beta + \cos^2\beta = \cos^2(\alpha - \beta) - 2 \cos(\alpha - \beta) + 1 + \sin^2(\alpha - \beta)$$

$$\sin^2\alpha + \cos^2\alpha + \sin^2\beta + \cos^2\beta - 2 \sin \alpha \sin \beta - 2 \cos \alpha \cos \beta = \sin^2(\alpha - \beta) + \cos^2(\alpha - \beta) + 1 - 2 \cos(\alpha - \beta)$$

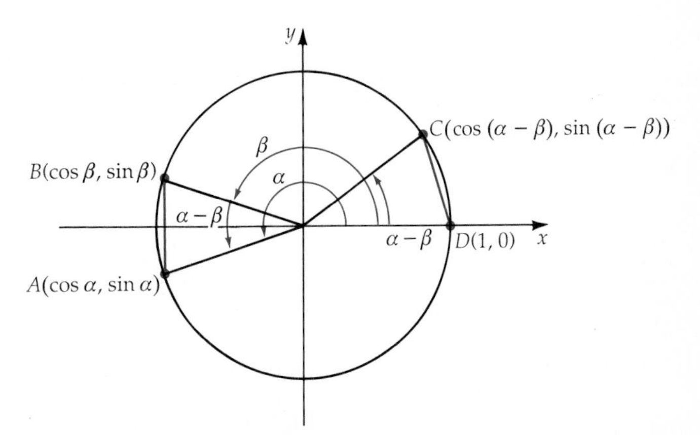

Figure 3.1

Simplifying by using the Pythagorean identity $\sin^2\theta + \cos^2\theta = 1$, we have

$$2 - 2\sin\alpha\sin\beta - 2\cos\alpha\cos\beta = 2 - 2\cos(\alpha - \beta).$$

Solving for $\cos(\alpha - \beta)$ gives us

$$\cos(\alpha - \beta) = \cos\alpha\cos\beta + \sin\alpha\sin\beta.$$

To derive the identity for the cosine of the sum of two angles, substitute $-\beta$ for β in $\cos(\alpha - \beta)$.

$$\cos(\alpha - \beta) = \cos[\alpha - (-\beta)] = \cos\alpha\cos(-\beta) + \sin\alpha\sin(-\beta)$$

Recall that $\cos(-\beta) = \cos\beta$ and $\sin(-\beta) = -\sin\beta$. Substituting into the previous equation, we obtain the identity

$$\cos(\alpha + \beta) = \cos\alpha\cos\beta - \sin\alpha\sin\beta.$$

EXAMPLE 1 Evaluate a Trigonometric Function

Evaluate $\cos\frac{5\pi}{12}$.

Solution Since $\frac{5\pi}{12} = \frac{\pi}{4} + \frac{\pi}{6}$, we can substitute $\alpha = \frac{\pi}{4}$ and $\beta = \frac{\pi}{6}$ in the identity for the cosine of the sum of two angles.

$$\cos\frac{5\pi}{12} = \cos\left(\frac{\pi}{4} + \frac{\pi}{6}\right) = \cos\frac{\pi}{4}\cos\frac{\pi}{6} - \sin\frac{\pi}{4}\sin\frac{\pi}{6}$$

$$= \frac{\sqrt{2}}{2}\cdot\frac{\sqrt{3}}{2} - \frac{\sqrt{2}}{2}\cdot\frac{1}{2} = \frac{\sqrt{6}}{4} - \frac{\sqrt{2}}{4} = \frac{\sqrt{6} - \sqrt{2}}{4}$$

■ *Try Exercise* **22,** *page 124.*

EXAMPLE 2 Verify an Identity

Verify the identity $\cos(\pi - \theta) = -\cos\theta$.

Solution Use the identity for the cosine of the difference of two angles. Recall that $\cos\pi = -1$ and $\sin\pi = 0$.

$$\cos(\pi - \theta) = \cos\pi\cos\theta + \sin\pi\sin\theta = -1\cdot\cos\theta + 0\cdot\sin\theta = -\cos\theta$$

■ *Try Exercise* **70,** *page 125.*

EXAMPLE 3 Evaluate a Trigonometric Function

Given $\sin\alpha = \frac{1}{2}$ for α in quadrant II and $\cos\beta = \frac{\sqrt{3}}{2}$ for β in quadrant IV, find $\cos(\alpha + \beta)$.

Solution We sketch the diagrams of angles α and β with the lengths

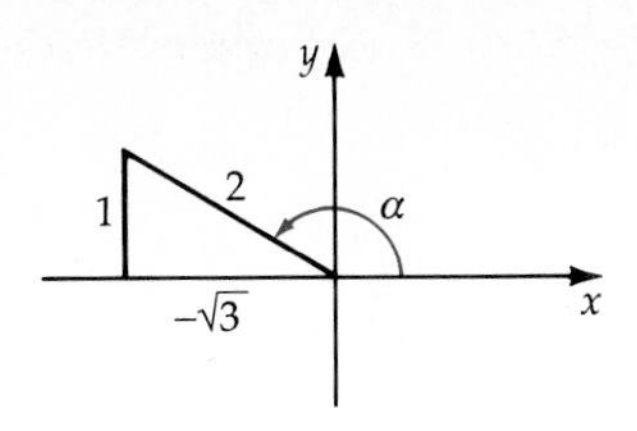

of the sides indicated. Use the Pythagorean Theorem to evaluate $\cos\alpha$ and $\sin\beta$. Use the identity of the cosine of the sum of two angles.

$$\begin{aligned}\cos(\alpha+\beta) &= \cos\alpha\cos\beta - \sin\alpha\sin\beta \\ &= \frac{-\sqrt{3}}{2}\cdot\frac{\sqrt{3}}{2} - \frac{1}{2}\cdot\frac{-1}{2} = \frac{-3}{4} + \frac{1}{4} = -\frac{1}{2}\end{aligned}$$

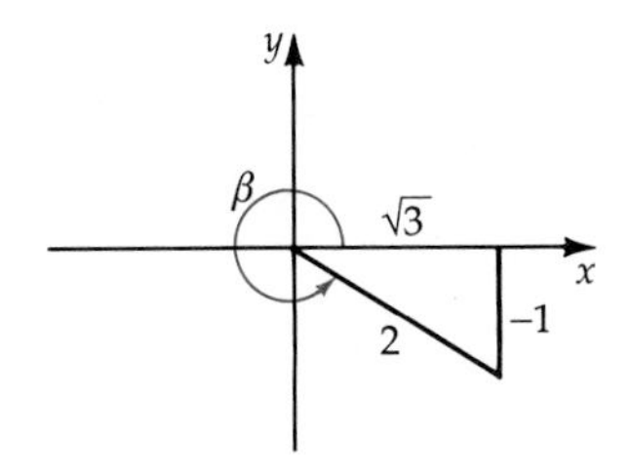

Figure 3.2

EXAMPLE 4 **Verify an Identity**

Verify the identity $\dfrac{\cos 4\theta}{\sin\theta} - \dfrac{\sin 4\theta}{\cos\theta} = \dfrac{\cos 5\theta}{\sin\theta\cos\theta}$.

Solution Subtract the fractions on the left side of the equation.

$$\begin{aligned}\frac{\cos 4\theta}{\sin\theta} - \frac{\sin 4\theta}{\cos\theta} &= \frac{\cos 4\theta\cos\theta - \sin 4\theta\sin\theta}{\sin\theta\cos\theta} \\ &= \frac{\cos(4\theta+\theta)}{\sin\theta\cos\theta} = \frac{\cos 5\theta}{\sin\theta\cos\theta}\end{aligned}$$

Use the identity for $\cos(\alpha+\beta)$.

■ *Try Exercise* **82,** *page 125.*

If we apply the difference identity for the cosine function to the expression $\cos(90^\circ - \alpha)$, we obtain the following result:

$$\cos(90^\circ-\alpha) = \cos 90^\circ\cos\alpha + \sin 90^\circ\sin\alpha = 0\cdot\cos\alpha + 1\cdot\sin\alpha$$

$$\cos(90^\circ-\alpha) = \sin\alpha$$

Thus the sine of an angle is equal to the cosine of its complement. Similarly, we have

$$\begin{aligned}\cos\alpha &= \cos[90^\circ - (90^\circ-\alpha)] \\ &= \cos 90^\circ\cos(90^\circ-\alpha) + \sin 90^\circ\sin(90^\circ-\alpha) \\ &= 0\cdot\cos(90^\circ-\alpha) + 1\cdot\sin(90^\circ-\alpha)\end{aligned}$$

which gives us

$$\cos\alpha = \sin(90^\circ-\alpha)$$

Thus the cosine of an angle is equal to the sine of its complement. Any pair of functions with this property are said to be **cofunctions**.

We can use the ratio identity to show that the tangent and cotangent functions are cofunctions. (The secant and cosecant are also cofunctions.)

$$\begin{aligned}\tan(90^\circ-\theta) &= \frac{\sin(90^\circ-\theta)}{\cos(90^\circ-\theta)} \\ &= \frac{\cos\theta}{\sin\theta} = \cot\theta\end{aligned} \qquad \begin{aligned}\cot(90^\circ-\theta) &= \frac{\cos(90^\circ-\theta)}{\sin(90^\circ-\theta)} \\ &= \frac{\sin\theta}{\cos\theta} = \tan\theta\end{aligned}$$

To derive the identity for the sine of the sum of two angles, substitute $\theta = \alpha + \beta$ in the cofunction identity $\sin \theta = \cos(90° - \theta)$.

$$\begin{aligned}
\sin \theta &= \cos(90° - \theta) \\
\sin(\alpha + \beta) &= \cos[90° - (\alpha + \beta)] \\
&= \cos[(90° - \alpha) - \beta] \quad \text{Rewrite as the difference of two angles.} \\
&= \cos(90° - \alpha) \cos \beta + \sin(90° - \alpha) \sin \beta
\end{aligned}$$

$$\sin(\alpha + \beta) = \sin \alpha \cos \beta + \cos \alpha \sin \beta$$

We also can derive the difference identity for the sine by substituting $-\beta$ for β in the sum identity for the $\sin(\alpha + \beta)$.

$$\begin{aligned}
\sin(\alpha - \beta) &= \sin[\alpha + (-\beta)] \\
&= \sin \alpha \cos(-\beta) + \cos \alpha \sin(-\beta) \quad \cos(-\beta) = \cos \beta,\ \sin(-\beta) = -\sin \beta.
\end{aligned}$$

$$\sin(\alpha - \beta) = \sin \alpha \cos \beta - \cos \alpha \sin \beta$$

EXAMPLE 5 Evaluate a Trigonometric Function

Find the exact value of $\sin 105°$.

Solution Since 105° is the sum of two of the special angles, 60° and 45°, use the sum identity for the sine.

$$\begin{aligned}
\sin 105° &= \sin(60° + 45°) = \sin 60° \cos 45° + \cos 60° \sin 45° \\
&= \frac{\sqrt{3}}{2} \cdot \frac{\sqrt{2}}{2} + \frac{1}{2} \cdot \frac{\sqrt{2}}{2} = \frac{\sqrt{6} + \sqrt{2}}{4}
\end{aligned}$$

■ *Try Exercise* **20,** *page 124.*

EXAMPLE 6 Verify an Identity

Verify the identity $\sin(\alpha + \beta) \cdot \sin(\alpha - \beta) = \sin^2\alpha - \sin^2\beta$.

Solution Work on the left side of the identity.

$$\begin{aligned}
&\sin(\alpha + \beta) \cdot \sin(\alpha - \beta) \\
&\quad = (\sin \alpha \cos \beta + \cos \alpha \sin \beta)(\sin \alpha \cos \beta - \cos \alpha \sin \beta) \\
&\quad = \sin^2\alpha \cos^2\beta - \cos^2\alpha \sin^2\beta \\
&\quad = \sin^2\alpha(1 - \sin^2\beta) - (1 - \sin^2\alpha)(\sin^2\beta) \\
&\quad = \sin^2\alpha - \sin^2\alpha \sin^2\beta - \sin^2\beta + \sin^2\alpha \sin^2\beta \\
&\quad = \sin^2\alpha - \sin^2\beta
\end{aligned}$$

■ *Try Exercise* **86,** *page 125.*

The sum identity for the tangent function is a direct result of the ratio identity $\tan\theta = \sin\theta/\cos\theta$ and the sum identity of the sine and cosine:

$$\tan(\alpha+\beta) = \frac{\sin(\alpha+\beta)}{\cos(\alpha+\beta)} = \frac{\sin\alpha\cos\beta + \cos\alpha\sin\beta}{\cos\alpha\cos\beta - \sin\alpha\sin\beta}$$

$$= \frac{\dfrac{\sin\alpha\cos\beta}{\cos\alpha\cos\beta} + \dfrac{\cos\alpha\sin\beta}{\cos\alpha\cos\beta}}{\dfrac{\cos\alpha\cos\beta}{\cos\alpha\cos\beta} - \dfrac{\sin\alpha\sin\beta}{\cos\alpha\cos\beta}}$$

Divide each term by $\cos\alpha\cos\beta$ and simplify.

$$\tan(\alpha+\beta) = \frac{\tan\alpha + \tan\beta}{1 - \tan\alpha\tan\beta}$$

The tangent function is an odd function; thus $\tan(-\theta) = -\tan\theta$. Substituting $-\beta$ for β in $\tan(\alpha+\beta)$, we derive the difference identity for $\tan(\alpha-\beta)$.

$$\tan[\alpha + (-\beta)] = \frac{\tan\alpha + \tan(-\beta)}{1 - \tan\alpha\tan(-\beta)}$$

$$\tan(\alpha-\beta) = \frac{\tan\alpha - \tan\beta}{1 + \tan\alpha\tan\beta}$$

EXAMPLE 7 Evaluate a Trigonometric Function

Find the exact value of $\tan 75°$.

Solution Since $75° = 45° + 30°$, use the identity for the tangent of the sum of two angles.

$$\tan 75° = \tan(45° + 30°) = \frac{\tan 45° + \tan 30°}{1 - \tan 45°\tan 30°}$$

$$= \frac{1 + \dfrac{\sqrt{3}}{3}}{1 - (1)\dfrac{\sqrt{3}}{3}} = \frac{3+\sqrt{3}}{3-\sqrt{3}}$$

$$= \frac{(3+\sqrt{3})(3+\sqrt{3})}{(3-\sqrt{3})(3+\sqrt{3})}$$

$$= \frac{9 + 6\sqrt{3} + 3}{9 - 3} = 2 + \sqrt{3}$$

■ *Try Exercise* **24,** *page 124.*

The identities we developed in this section are used frequently in problems involving trigonometric functions. Following is a list of the identities for convenient reference.

Sum or Difference of Two Angle Identities

$$\cos(\alpha - \beta) = \cos\alpha\cos\beta + \sin\alpha\sin\beta$$
$$\cos(\alpha + \beta) = \cos\alpha\cos\beta - \sin\alpha\sin\beta$$
$$\sin(\alpha - \beta) = \sin\alpha\cos\beta - \cos\alpha\sin\beta$$
$$\sin(\alpha + \beta) = \sin\alpha\cos\beta + \cos\alpha\sin\beta$$
$$\tan(\alpha + \beta) = \frac{\tan\alpha + \tan\beta}{1 - \tan\alpha\tan\beta}$$
$$\tan(\alpha - \beta) = \frac{\tan\alpha - \tan\beta}{1 + \tan\alpha\tan\beta}$$

Cofunction Identities

$\sin(90° - \theta) = \cos\theta$ $\quad\cos(90° - \theta) = \sin\theta$ $\quad\tan(90° - \theta) = \cot\theta$

$\csc(90° - \theta) = \sec\theta$ $\quad\sec(90° - \theta) = \csc\theta$ $\quad\cot(90° - \theta) = \tan\theta$

The cofunction identities are also true when 90° is replaced by $\pi/2$ radians.

EXERCISE SET 3.2

In Exercises 1 to 34, find the exact value of the given function or expression.

1. $\sin(45° + 30°)$
2. $\sin(330° + 45°)$
3. $\cos(45° - 30°)$
4. $\cos(120° - 45°)$
5. $\tan(45° + 135°)$
6. $\tan(240° - 45°)$
7. $\sin\left(\frac{5\pi}{4} - \frac{\pi}{6}\right)$
8. $\sin\left(\frac{4\pi}{3} + \frac{\pi}{4}\right)$
9. $\cos\left(\frac{3\pi}{4} + \frac{\pi}{6}\right)$
10. $\cos\left(\frac{\pi}{4} - \frac{\pi}{3}\right)$
11. $\tan\left(\frac{\pi}{6} + \frac{\pi}{4}\right)$
12. $\tan\left(\frac{11\pi}{6} - \frac{\pi}{4}\right)$

13. $\sin 15°$
14. $\sin 285°$
15. $\cos 195°$
16. $\cos 165°$
17. $\tan 75°$
18. $\tan 285°$
19. $\sin\frac{5\pi}{12}$
20. $\sin\frac{11\pi}{12}$
21. $\cos\frac{13\pi}{12}$
22. $\cos\frac{\pi}{12}$
23. $\tan\frac{19\pi}{12}$
24. $\tan\frac{7\pi}{12}$

25. $\cos 212° \cos 122° + \sin 212° \sin 122°$
26. $\cos 82° \cos 37° + \sin 82° \sin 37°$
27. $\sin 167° \cos 107° - \cos 167° \sin 107°$
28. $\sin 178° \cos 58° - \cos 178° \sin 58°$
29. $\sin\frac{5\pi}{12}\cos\frac{\pi}{4} - \cos\frac{5\pi}{12}\sin\frac{\pi}{4}$
30. $\cos\frac{7\pi}{12}\cos\frac{\pi}{4} + \sin\frac{7\pi}{12}\sin\frac{\pi}{4}$
31. $\cos\frac{\pi}{12}\cos\frac{\pi}{4} - \sin\frac{\pi}{12}\sin\frac{\pi}{4}$
32. $\sin\frac{\pi}{3}\cos\frac{\pi}{6} + \cos\frac{\pi}{3}\sin\frac{\pi}{6}$
33. $\dfrac{\tan 7\pi/12 - \tan \pi/4}{1 + \tan 7\pi/12 \tan \pi/4}$
34. $\dfrac{\tan \pi/6 + \tan \pi/3}{1 - \tan \pi/6 \tan \pi/3}$

In Exercises 35 to 44, evaluate each expression.

35. $\cos 30° \cos 40° + \sin 30° \sin 40°$

36. $\cos 125° \cos 65° - \sin 125° \sin 65°$

37. $\sin 8° \cos 48° - \cos 8° \sin 48°$

38. $\sin 230° \cos 88° + \cos 230° \sin 88°$

39. $\sin \dfrac{\pi}{12} \cos \dfrac{\pi}{3} + \cos \dfrac{\pi}{12} \sin \dfrac{\pi}{3}$

40. $\sin \dfrac{11\pi}{6} \cos \dfrac{\pi}{4} - \cos \dfrac{11\pi}{6} \sin \dfrac{\pi}{4}$

41. $\cos \dfrac{\pi}{4} \cos \dfrac{\pi}{6} - \sin \dfrac{\pi}{4} \sin \dfrac{\pi}{6}$

42. $\cos \dfrac{\pi}{3} \cos \dfrac{\pi}{4} + \sin \dfrac{\pi}{3} \sin \dfrac{\pi}{4}$

43. $\dfrac{\tan \pi/3 + \tan \pi/4}{1 - \tan \pi/3 \tan \pi/4}$

44. $\dfrac{\tan 11\pi/6 - \tan \pi/4}{1 + \tan 11\pi/6 \tan \pi/4}$

In Exercises 45 to 56, write each expression in terms of a single trigonometric function.

45. $\sin 7x \cos 2x - \cos 7x \sin 2x$

46. $\sin x \cos 3x + \cos 3x \sin x$

47. $\cos x \cos 2x + \sin x \sin 2x$

48. $\cos 4x \cos 2x - \sin 4x \sin 2x$

49. $\sin 7x \cos 3x - \cos 7x \sin 3x$

50. $\cos x \cos 5x - \sin x \sin 5x$

51. $\cos 4x \cos(-2x) - \sin 4x \sin(-2x)$

52. $\sin(-x) \cos 3x - \cos(-x) \sin 3x$

53. $\sin \dfrac{x}{3} \cos \dfrac{2x}{3} + \cos \dfrac{2x}{3} \sin \dfrac{x}{3}$

54. $\cos \dfrac{3x}{4} \cos \dfrac{x}{4} + \sin \dfrac{3x}{4} \sin \dfrac{x}{4}$

55. $\dfrac{\tan 3x + \tan 4x}{1 - \tan 3x \tan 4x}$

56. $\dfrac{\tan 2x - \tan 3x}{1 + \tan 2x \tan 3x}$

In Exercises 57 to 60, evaluate each expression.

57. $\sin 2 \cos 3 - \cos 2 \sin 3$

58. $\cos 4 \cos 2 - \sin 4 \sin 2$

59. $\cos 5 \cos 8 + \sin 5 \sin 8$

60. $\sin 4 \cos 6 + \cos 4 \sin 6$

In Exercises 61 to 68, find the exact value of the given functions.

61. Given $\sin \alpha = \frac{3}{5}$ in quadrant I and $\cos \beta = -\frac{5}{13}$ in quadrant II, find **a.** $\sin(\alpha - \beta)$ and **b.** $\cos(\alpha + \beta)$.

62. Given $\sin \alpha = \frac{24}{25}$ in quadrant II and $\cos \beta = -\frac{4}{5}$ in quadrant III, find **a.** $\cos(\beta - \alpha)$ and **b.** $\sin(\alpha + \beta)$.

63. Given $\sin \alpha = -\frac{4}{5}$ in quadrant III and $\cos \beta = -\frac{12}{13}$ in quadrant II, find **a.** $\sin(\alpha - \beta)$, **b.** $\cos(\alpha + \beta)$, and **c.** $\tan(\alpha + \beta)$.

64. Given $\sin \alpha = -\frac{7}{25}$ in quadrant IV and $\cos \beta = \frac{8}{17}$ in quadrant IV, find **a.** $\sin(\alpha + \beta)$, **b.** $\cos(\alpha - \beta)$, and **c.** $\tan(\alpha + \beta)$.

65. Given $\cos \alpha = \frac{15}{17}$ in quadrant I and $\sin \beta = -\frac{3}{5}$ in quadrant III, find **a.** $\sin(\alpha + \beta)$, **b.** $\cos(\alpha - \beta)$, and **c.** $\tan(\alpha - \beta)$.

66. Given $\cos \alpha = -\frac{7}{25}$ in quadrant II and $\sin \beta = -\frac{12}{13}$ in quadrant IV, find **a.** $\sin(\alpha + \beta)$, **b.** $\cos(\alpha + \beta)$, and **c.** $\tan(\alpha - \beta)$.

67. Given $\cos \alpha = -\frac{3}{5}$ in quadrant III and $\sin \beta = \frac{5}{13}$ in quadrant I, find **a.** $\sin(\alpha - \beta)$, **b.** $\cos(\alpha + \beta)$, and **c.** $\tan(\alpha + \beta)$.

68. Given $\cos \alpha = \frac{8}{17}$ in quadrant IV and $\sin \beta = -\frac{24}{25}$ in quadrant III, find **a.** $\sin(\alpha - \beta)$, **b.** $\cos(\alpha + \beta)$, and **c.** $\tan(\alpha + \beta)$.

In Exercises 69 to 92, verify the identities.

69. $\cos\left(\dfrac{\pi}{2} - \theta\right) = \sin \theta$

70. $\cos(\theta + \pi) = -\cos \theta$

71. $\sin\left(\theta + \dfrac{\pi}{2}\right) = \cos \theta$

72. $\sin(\theta + \pi) = -\sin \theta$

73. $\tan\left(\theta + \dfrac{\pi}{4}\right) = \dfrac{\tan \theta + 1}{1 - \tan \theta}$

74. $\tan 2\theta = \dfrac{2 \tan \theta}{1 - \tan^2\theta}$

75. $\cos\left(\dfrac{3\pi}{2} - \theta\right) = -\sin \theta$

76. $\sin\left(\dfrac{3\pi}{2} + \theta\right) = -\cos \theta$

77. $\cot\left(\dfrac{\pi}{2} - \theta\right) = \tan \theta$

78. $\cot(\pi + \theta) = \cot \theta$

79. $\csc(\pi - \theta) = \csc \theta$

80. $\sec\left(\dfrac{\pi}{2} - \theta\right) = \csc \theta$

81. $\sin 6x \cos 2x - \cos 6x \sin 2x = 2 \sin 2x \cos 2x$

82. $\cos 5x \cos 3x + \sin 5x \sin 3x = \cos^2 x - \sin^2 x$

83. $\cos(\alpha + \beta) + \cos(\alpha - \beta) = 2 \cos \alpha \cos \beta$

84. $\cos(\alpha - \beta) - \cos(\alpha + \beta) = 2 \sin \alpha \sin \beta$

85. $\sin(\alpha + \beta) + \sin(\alpha - \beta) = 2 \sin \alpha \cos \beta$

86. $\sin(\alpha - \beta) - \sin(\alpha + \beta) = -2 \cos \alpha \sin \beta$

87. $\dfrac{\cos(\alpha - \beta)}{\sin(\alpha + \beta)} = \dfrac{\cot \alpha + \tan \beta}{1 + \cot \alpha \tan \beta}$

88. $\dfrac{\sin(\alpha + \beta)}{\sin(\alpha - \beta)} = \dfrac{1 + \cot \alpha \tan \beta}{1 - \cot \alpha \tan \beta}$

89. $\sin(\pi/2 + \alpha - \beta) = \cos \alpha \cos \beta + \sin \alpha \sin \beta$

90. $\cos(\pi/2 + \alpha + \beta) = -(\sin \alpha \cos \beta + \cos \alpha \sin \beta)$

91. $\sin 3x = 3 \sin x - 4 \sin^3 x$

92. $\cos 3x = 4 \cos^3 x - 3 \cos x$

Supplemental Exercises

In Exercises 93 to 99, verify the identities.

93. $\sin(x - y) \cdot \sin(x + y) = \sin^2 x \cos^2 y - \cos^2 x \sin^2 y$

94. $\sin(x + y + z) = \sin x \cos y \cos z + \cos x \sin y \cos z + \cos x \cos y \sin z - \sin x \sin y \sin z$

95. $\cos(x + y + z) = \cos x \cos y \cos z - \sin x \sin y \cos z - \sin x \cos y \sin z - \cos x \sin y \sin z$

96. $\dfrac{\sin(x + y)}{\sin x \sin y} = \cot x + \cot y$

97. $\dfrac{\cos(x - y)}{\cos x \sin y} = \cot y + \tan x$

98. $\dfrac{\sin(x + h) - \sin x}{h} = \cos x \dfrac{\sin h}{h} + \sin x \dfrac{\cos h - 1}{h}$

99. $\dfrac{\cos(x + h) - \cos x}{h} = \cos x \dfrac{\cos h - 1}{h} - \sin x \dfrac{\sin h}{h}$

3.3 Double- and Half-Angle Identities

By using the sum identities, we can derive identities for twice the argument of a trigonometic function. These are called the *double-angle* identities. To find the sine of a double angle, substitute α for β in the identity for $\sin(\alpha + \beta)$ to obtain

$$\sin(\alpha + \beta) = \sin \alpha \cos \beta + \cos \alpha \sin \beta$$

$$\sin(\alpha + \alpha) = \sin \alpha \cos \alpha + \cos \alpha \sin \alpha \quad \text{Let } \alpha = \beta.$$

$$\sin 2\alpha = 2 \sin \alpha \cos \alpha .$$

The double-angle identity for cosine is derived in a similar manner.

$$\cos(\alpha + \beta) = \cos \alpha \cos \beta - \sin \alpha \sin \beta$$

$$\cos(\alpha + \alpha) = \cos \alpha \cos \alpha - \sin \alpha \sin \alpha \quad \text{Let } \alpha = \beta .$$

$$\cos 2\alpha = \cos^2\alpha - \sin^2\alpha$$

There are two alternative forms of the double-angle identity for the cosine. Using a Pythagorean identity, we can rewrite the identity for $\cos 2\alpha$ as follows:

$$\cos 2\alpha = \cos^2\alpha - \sin^2\alpha$$

$$\cos 2\alpha = (1 - \sin^2\alpha) - \sin^2\alpha \quad \cos^2\alpha = 1 - \sin^2\alpha$$

$$\cos 2\alpha = 1 - 2 \sin^2\alpha$$

In addition, we can rewrite $\cos 2\alpha$ as

$$\cos 2\alpha = \cos^2\alpha - \sin^2\alpha$$

$$\cos 2\alpha = \cos^2\alpha - (1 - \cos^2\alpha) \qquad \sin^2\alpha = 1 - \cos^2\alpha$$

$$\cos 2\alpha = 2\cos^2\alpha - 1$$

The double-angle identity for the tangent function is derived from the identity for the tangent of the sum of two angles.

$$\tan(\alpha + \beta) = \frac{\tan\alpha + \tan\beta}{1 - \tan\alpha\tan\beta}$$

$$\tan(\alpha + \alpha) = \frac{\tan\alpha + \tan\alpha}{1 - \tan\alpha\tan\alpha} \qquad \text{Let } \alpha = \beta.$$

$$\tan 2\alpha = \frac{2\tan\alpha}{1 - \tan^2\alpha}$$

EXAMPLE 1 Evaluate a Trigonometric Function

For an angle α in quadrant I, $\sin\alpha = \dfrac{4}{5}$. Find $\sin 2\alpha$.

Solution We will use $\sin 2\alpha = 2\sin\alpha\cos\alpha$. Find $\cos\alpha$ by substituting for $\sin\alpha$ in $\sin^2\alpha + \cos^2\alpha = 1$ and solve for $\cos\alpha$.

$$\cos\alpha = \sqrt{1 - \sin^2\alpha} = \sqrt{1 - \left(\frac{4}{5}\right)^2} = \frac{3}{5} \qquad \cos\alpha > 0 \text{ in quadrant I.}$$

Substitute the values of $\sin\alpha$ and $\cos\alpha$ in the double angle-formula for the sine function.

$$\sin 2\alpha = 2\sin\alpha\cos\alpha = 2\left(\frac{4}{5}\right)\left(\frac{3}{5}\right) = \frac{24}{25}$$

■ *Try Exercise* **26**, *page 131.*

EXAMPLE 2 Verify a Double-Angle Identity

Verify the identity $\csc 2\alpha = \dfrac{1}{2}(\tan\alpha + \cot\alpha)$.

Solution Work on the right-hand side of the equation.

$$\frac{1}{2}(\tan\alpha + \cot\alpha) = \frac{1}{2}\left(\frac{\sin\alpha}{\cos\alpha} + \frac{\cos\alpha}{\sin\alpha}\right) = \frac{1}{2}\left(\frac{\sin^2\alpha + \cos^2\alpha}{\cos\alpha\sin\alpha}\right)$$

$$= \frac{1}{2\cos\alpha\sin\alpha} = \frac{1}{\sin 2\alpha} = \csc 2\alpha$$

■ *Try Exercise* **54**, *page 132.*

EXAMPLE 3 Verify a Double-Angle Identity

Verify the identity $\sin^2 x = \frac{1}{2}(1 - \cos 2x)$.

Solution Work on the right side of the equation.

$$\frac{1}{2}(1 - \cos 2x) = \frac{1}{2}[1 - (1 - 2\sin^2 x)] = \frac{1}{2}(1 - 1 + 2\sin^2 x) = \sin^2 x$$

■ *Try Exercise* **62,** *page 132.*

EXAMPLE 4 Verify a Double-Angle Identity

Verify the identity $\tan 2x = \dfrac{2}{\cot x - \tan x}$.

Solution Work on the right side of the equation.

$$\frac{2}{\cot x - \tan x} = \frac{2}{\dfrac{1}{\tan x} - \tan x} = \frac{2}{\dfrac{1}{\tan x} - \tan x} \cdot \frac{\tan x}{\tan x}$$

$$= \frac{2\tan x}{1 - \tan^2 x} = \tan 2x$$

■ *Try Exercise* **58,** *page 132.*

EXAMPLE 5 Verify an Identity

Verify the identity $\sin 3\theta = 3\sin\theta - 4\sin^3\theta$.

Solution Work on the left side of the identity.

$$\begin{aligned}
\sin 3\theta &= \sin(2\theta + \theta) = \sin 2\theta\cos\theta + \cos 2\theta\sin\theta \\
&= (2\sin\theta\cos\theta)\cos\theta + (1 - 2\sin^2\theta)\sin\theta \\
&= 2\sin\theta\cos^2\theta + \sin\theta - 2\sin^3\theta \\
&= 2\sin\theta(1 - \sin^2\theta) + \sin\theta - 2\sin^3\theta \\
&= 2\sin\theta - 2\sin^3\theta + \sin\theta - 2\sin^3\theta \\
&= 3\sin\theta - 4\sin^3\theta
\end{aligned}$$

■ *Try Exercise* **68,** *page 132.*

The double-angle identity for the cosine function can be used to derive the *half-angle* identities. To derive an identity for $\sin \alpha/2$, we solve for $\sin^2\theta$ in the double-angle identity for $\cos\theta$.

$$\cos 2\theta = 1 - 2\sin^2\theta$$

$$\sin^2\theta = \frac{1 - \cos 2\theta}{2}$$

Substitute $\alpha/2$ for θ and take square root of both sides of the equation

$$\sin^2 \frac{\alpha}{2} = \frac{1 - \cos 2(\alpha/2)}{2}$$

$$\sin \frac{\alpha}{2} = \pm \sqrt{\frac{1 - \cos \alpha}{2}}$$

The sign of the radical is determined by the quadrant in which the terminal side of angle $\alpha/2$ lies.

In a similar manner, we derive an identity for $\cos \alpha/2$.

$$\cos 2\theta = 2 \cos^2\theta - 1$$

$$\cos^2\theta = \frac{1 + \cos 2\theta}{2}$$

Substitute $\alpha/2$ for θ and take the square root of both sides of the equation.

$$\cos^2 \frac{\alpha}{2} = \frac{1 + \cos 2(\alpha/2)}{2}$$

$$\cos \frac{\alpha}{2} = \pm \sqrt{\frac{1 + \cos \alpha}{2}}$$

A half-angle identity for the tangent is derived from the ratio identity for the tangent function. Two different forms of the half-angle identity for the tangent function are possible.

$$\tan \alpha/2 = \frac{\sin \alpha/2}{\cos \alpha/2} = \frac{\sin \alpha/2}{\cos \alpha/2} \cdot \frac{2 \cos \alpha/2}{2 \cos \alpha/2}$$

$$= \frac{2 \sin \alpha/2 \cos \alpha/2}{2 \cos^2\alpha/2}$$

$$= \frac{\sin 2(\alpha/2)}{1 + \cos 2(\alpha/2)} = \frac{\sin \alpha}{1 + \cos \alpha}$$

$$\tan \frac{\alpha}{2} = \frac{\sin \alpha}{1 + \cos \alpha}.$$

To obtain an equivalent identity for $\tan \alpha/2$, multiply by the conjugate of the denominator.

$$\tan \frac{\alpha}{2} = \frac{\sin \alpha}{1 + \cos \alpha} \cdot \frac{1 - \cos \alpha}{1 - \cos \alpha}$$

$$= \frac{\sin \alpha(1 - \cos \alpha)}{1 - \cos^2\alpha}$$

$$= \frac{\sin \alpha(1 - \cos \alpha)}{\sin^2\alpha}$$

$$\tan \frac{\alpha}{2} = \frac{1 - \cos \alpha}{\sin \alpha}$$

EXAMPLE 6 **Evaluate a Trigonometric Function**

Find sin 15° by using the half-angle identity.

Solution Use the half-angle for the sine function in the first quadrant. Since $15° = \dfrac{30°}{2}$, let $\alpha = 30°$.

$$\sin\frac{\alpha}{2} = \sqrt{\frac{1-\cos\alpha}{2}}$$

$$\sin\frac{30°}{2} = \sqrt{\frac{1-\cos 30°}{2}}$$

$$\sin 15° = \sqrt{\frac{1-\sqrt{3}/2}{2}} = \sqrt{\frac{2-\sqrt{3}}{4}} = \frac{\sqrt{2-\sqrt{3}}}{2}$$

■ *Try Exercise* **38,** *page 132.*

EXAMPLE 7 **Verify a Half-Angle Identity**

Verify the identity $2\csc x\cos^2\dfrac{x}{2} = \dfrac{\sin x}{1-\cos x}$.

Solution Work on the left side of the identity.

$$2\csc x\cos^2\frac{x}{2} = 2\csc x\frac{1+\cos x}{2} \qquad \cos^2\frac{x}{2} = \frac{1+\cos x}{2}$$

$$= \frac{1+\cos x}{\sin x} \qquad \csc x = \frac{1}{\sin x}$$

$$= \frac{1-\cos x}{1-\cos x}\cdot\frac{1+\cos x}{\sin x}$$

Multiply the numerator and denominator by the conjugate of the numerator.

$$= \frac{1-\cos^2 x}{(1-\cos x)\sin x}$$

$$= \frac{\sin^2 x}{(1-\cos x)(\sin x)} \qquad 1-\cos^2 x = \sin^2 x$$

$$= \frac{\sin x}{1-\cos x}$$

■ *Try Exercise* **72,** *page 132.*

EXAMPLE 8 **Verify a Half-Angle Identity**

Verify the identity $\tan\dfrac{\alpha}{2} = \sin\alpha + \cos\alpha\cot\alpha - \cot\alpha$.

Solution Work on the left side of the identity.

$$\tan\frac{\alpha}{2} = \frac{1-\cos\alpha}{\sin\alpha} = \frac{\sin^2\alpha + \cos^2\alpha - \cos\alpha}{\sin\alpha} \qquad 1 = \sin^2\alpha + \cos^2\alpha$$

$$= \frac{\sin^2\alpha}{\sin\alpha} + \frac{\cos^2\alpha}{\sin\alpha} - \frac{\cos\alpha}{\sin\alpha} \qquad \text{Divide each term by } \sin\alpha.$$

$$= \sin\alpha + \cos\alpha\cot\alpha - \cot\alpha$$

■ *Try Exercise* **78**, *page 132.*

Here is a summary of the double- and half-angle identities:

Double-Angle Identities

$$\sin 2\alpha = 2\sin\alpha\cos\alpha$$

$$\cos 2\alpha = \cos^2\alpha - \sin^2\alpha = 1 - 2\sin^2\alpha = 2\cos^2\alpha - 1$$

$$\tan 2\alpha = \frac{2\tan\alpha}{1-\tan^2\alpha}$$

Half-Angle Identities

$$\sin\frac{\alpha}{2} = \pm\sqrt{\frac{1-\cos\alpha}{2}} \qquad \tan\frac{\alpha}{2} = \frac{\sin\alpha}{1+\cos\alpha} = \frac{1-\cos\alpha}{\sin\alpha}$$

$$\cos\frac{\alpha}{2} = \pm\sqrt{\frac{1+\cos\alpha}{2}}$$

EXERCISE SET 3.3

In Exercises 1 to 8, write the trigonometric expressions in terms of a single trigonometric function.

1. $2\sin 2\alpha\cos 2\alpha$ **2.** $2\sin 3\theta\cos 3\theta$

3. $1 - 2\sin^2 5\beta$ **4.** $2\cos^2 2\beta - 1$

5. $\cos^2 3\alpha - \sin^2 3\alpha$ **6.** $\cos^2 6\alpha - \sin^2 6\alpha$

7. $\dfrac{2\tan 3\alpha}{1-\tan^2 3\alpha}$ **8.** $\dfrac{2\tan 4\theta}{1-\tan^2 4\theta}$

In Exercises 9 to 24, use the half-angle identities to evaluate the trigonometric expressions.

9. $\sin 75°$ **10.** $\cos 105°$ **11.** $\tan 67.5°$

12. $\tan 165°$ **13.** $\cos 157.5°$ **14.** $\sin 112.5°$

15. $\sin 22.5°$ **16.** $\cos 67.5°$ **17.** $\sin\dfrac{7\pi}{8}$

18. $\cos\dfrac{5\pi}{8}$ **19.** $\cos\dfrac{5\pi}{12}$ **20.** $\sin\dfrac{3\pi}{8}$

21. $\tan\dfrac{7\pi}{12}$ **22.** $\tan\dfrac{3\pi}{8}$

23. $\cos\dfrac{\pi}{12}$ **24.** $\sin\dfrac{\pi}{8}$

In Exercises 25 to 36, find the exact value of $\sin 2\theta$, $\cos 2\theta$ and $\tan 2\theta$ given the following information.

25. $\cos\theta = -\dfrac{4}{5}$, θ is in quadrant II.

26. $\cos\theta = \dfrac{24}{25}$, θ is in quadrant IV.

27. $\sin \theta = \frac{8}{17}$, θ is in quadrant II.

28. $\sin \theta = -\frac{9}{41}$, θ is in quadrant III.

29. $\tan \theta = -\frac{24}{7}$, θ is in quadrant IV.

30. $\tan \theta = \frac{4}{3}$, θ is in quadrant I.

31. $\sin \theta = \frac{15}{17}$, θ is in quadrant I.

32. $\sin \theta = -\frac{3}{5}$, θ is in quadrant III.

33. $\cos \theta = \frac{40}{41}$, θ is in quadrant IV.

34. $\cos \theta = \frac{4}{5}$, θ is in quadrant IV.

35. $\tan \theta = \frac{15}{8}$, θ is in quadrant III.

36. $\tan \theta = -\frac{40}{9}$, θ is in quadrant II.

In Exercises 37 to 48, find the value of the sine, cosine, and tangent of $\alpha/2$ given the following information.

37. $\sin \alpha = \frac{5}{13}$, α is in quadrant II.

38. $\sin \alpha = -\frac{7}{25}$, α is in quadrant III.

39. $\cos \alpha = -\frac{8}{17}$, α is in quadrant III.

40. $\cos \alpha = \frac{12}{13}$, α is in quadrant I.

41. $\tan \alpha = \frac{4}{3}$, α is in quadrant I.

42. $\tan \alpha = -\frac{8}{15}$, α is in quadrant II.

43. $\cos \alpha = \frac{24}{25}$, α is in quadrant IV.

44. $\sin \alpha = -\frac{9}{41}$, α is in quadrant IV.

45. $\sec \alpha = \frac{17}{15}$, α is in quadrant I.

46. $\csc \alpha = -\frac{5}{3}$, α is in quadrant IV.

47. $\cot \alpha = \frac{8}{15}$, α is in quadrant III.

48. $\sec \alpha = -\frac{13}{5}$, α is in quadrant II.

In Exercises 49 to 94, verify the given identity.

49. $\sin 3x \cos 3x = \frac{1}{2} \sin 6x$

50. $\cos 8x = \cos^2 4x - \sin^2 4x$

51. $\sin^2 x + \cos 2x = \cos^2 x$

52. $\frac{\cos 2x}{\sin^2 x} = \cot^2 x - 1$

53. $\frac{1 + \cos 2x}{\sin 2x} = \cot x$

54. $\frac{1}{1 - \cos 2x} = \frac{1}{2} \csc^2 x$

55. $\frac{\sin 2x}{1 - \sin^2 x} = 2 \tan x$

56. $\frac{\cos^2 x - \sin^2 x}{2 \sin x \cos x} = \cot 2x$

57. $1 - \tan^2 x = \frac{\cos 2x}{\cos^2 x}$

58. $\tan 2x = \frac{2 \sin x \cos x}{\cos^2 x - \sin^2 x}$

59. $\sin 2x - \tan x = \tan x \cos 2x$

60. $\sin 2x - \cot x = -\cot x \cos 2x$

61. $\cos^4 x - \sin^4 x = \cos 2x$

62. $\sin 4x = 4 \sin x \cos^3 x - 4 \cos x \sin^3 x$

63. $\cos^2 x - 2 \sin^2 x \cos^2 x - \sin^2 x + 2 \sin^4 x = \cos^2 2x$

64. $2 \cos^4 x - \cos^2 x - 2 \sin^2 x \cos^2 x + \sin^2 x = \cos^2 2x$

65. $\cos 4x = 1 - 8 \cos^2 x + 8 \cos^4 x$

66. $\sin 4x = 4 \sin x \cos x - 8 \cos x \sin^3 x$

67. $\cos 3x - \cos x = 4 \cos^3 x - 4 \cos x$

68. $\sin 3x + \sin x = 4 \sin x - 4 \sin^3 x$

69. $\sin^3 x + \cos^3 x = (\sin x + \cos x)\left(1 - \frac{1}{2} \sin 2x\right)$

70. $\cos^3 x - \sin^3 x = (\cos x - \sin x)\left(1 + \frac{1}{2} \sin 2x\right)$

71. $\sin^2 \frac{x}{2} = \frac{\sec x - 1}{2 \sec x}$

72. $\cos^2 \frac{x}{2} = \frac{\sec x + 1}{2 \sec x}$

73. $\tan \frac{x}{2} = \csc x - \cot x$

74. $\tan \frac{x}{2} = \frac{\tan x}{\sec x + 1}$

75. $2 \sin \frac{x}{2} \cos \frac{x}{2} = \sin x$

76. $\cos^2 \frac{x}{2} - \sin^2 \frac{x}{2} = \cos x$

77. $\left(\cos \frac{x}{2} + \sin \frac{x}{2}\right)^2 = 1 + \sin x$

78. $\tan^2 \frac{x}{2} = \frac{\sec x - 1}{\sec x + 1}$

79. $\sin^2 \frac{x}{2} \sec x = \frac{1}{2}(\sec x - 1)$

80. $\cos^2 \frac{x}{2} \sec x = \frac{1}{2}(\sec x + 1)$

81. $\cos^2 \frac{x}{2} - \cos x = \sin^2 \frac{x}{2}$

82. $\sin^2 \frac{x}{2} + \cos x = \cos^2 \frac{x}{2}$

83. $\sin^2 \frac{x}{2} - \cos^2 \frac{x}{2} = -\cos x$

84. $\cos^2 \frac{x}{2} - \sin^2 \frac{x}{2} = \frac{1}{2} \csc x \sin 2x$

85. $\sin 2x - \cos x = \cos x(2 \sin x - 1)$

86. $\frac{\cos 2x}{\sin^2 x} = \csc^2 x - 2$

87. $\tan 2x = \frac{2}{\cot x - \tan x}$

88. $\frac{2 \cos 2x}{\sin 2x} = \cot x - \tan x$

89. $2 \tan \frac{x}{2} = \frac{\sin^2 x + 1 - \cos^2 x}{\sin x(1 + \cos x)}$

90. $\frac{1}{2} \csc^2 \frac{x}{2} = \csc^2 x + \cot x \csc x$

91. $\csc 2x = \frac{1}{2} \csc x \sec x$

92. $\sec 2x = \frac{\sec^2 x}{2 - \sec^2 x}$

93. $\cos \frac{x}{5} = 1 - 2 \sin^2 \frac{x}{10}$

94. $\sec^2 \frac{x}{2} = \frac{2}{1 + \cos x}$

Supplemental Exercises

In Exercises 95 to 98, verify the identities.

95. $\frac{\sin^3 x + \cos^3 x}{\sin x + \cos x} = 1 - \frac{1}{2} \sin 2x$

96. $\cos^4 x = \frac{1}{8} \cos 4x + \frac{1}{2} \cos 2x + \frac{3}{8}$

97. $\sin \frac{x}{2} - \cos \frac{x}{2} = \sqrt{1 - \sin x}, \ 0 \le x \le 90°$

98. $\frac{\sin x - \sin 2x}{\cos x + \cos 2x} = -\tan \frac{x}{2}$

99. If $x + y = 90°$; verify $\sin(x - y) = -\cos 2x$.

100. If $x + y = 90°$; verify $\sin(x - y) = \cos 2y$.

101. If $x + y = 180°$; verify $\sin(x - y) = -\sin 2x$.

102. If $x + y = 180°$; verify $\cos(x - y) = -\cos 2x$.

3.4 Identities Involving the Sum of Trigonometric Functions

Some applications require that a product of trigonometric functions be written as a sum or difference of these functions. Other applications require that the sum or difference of trigonometric functions be represented as a product of these functions. The product-to-sum identities are particularly useful to these types of problems.

The Product-to-Sum Identities

The product-to-sum identities can be derived by using the sum and difference identities. Adding the identities for $\sin(\alpha + \beta)$ and $\sin(\alpha - \beta)$, we have

$$\sin(\alpha + \beta) = \sin \alpha \cos \beta + \cos \alpha \sin \beta$$

$$\underline{\sin(\alpha - \beta) = \sin \alpha \cos \beta - \cos \alpha \sin \beta}$$

$$\sin(\alpha + \beta) + \sin(\alpha - \beta) = 2 \sin \alpha \cos \beta$$

Solving for $\sin \alpha \cos \beta$, we obtain the first product-to-sum identity:

$$\sin \alpha \cos \beta = \frac{1}{2}[\sin(\alpha + \beta) + \sin(\alpha - \beta)].$$

The identity for $\cos \alpha \sin \beta$ is obtained when $\sin(\alpha - \beta)$ is subtracted from $\sin(\alpha + \beta)$. The result is

$$\cos \alpha \sin \beta = \frac{1}{2}[\sin(\alpha + \beta) - \sin(\alpha - \beta)].$$

In like manner, the identities for $\cos(\alpha + \beta)$ and $\cos(\alpha - \beta)$ are used to derive the identities for $\cos \alpha \cos \beta$ and $\sin \alpha \sin \beta$.

$$\cos \alpha \cos \beta = \frac{1}{2}[\cos(\alpha + \beta) + \cos(\alpha - \beta)]$$

$$\sin \alpha \sin \beta = \frac{1}{2}[\cos(\alpha - \beta) - \cos(\alpha + \beta)]$$

EXAMPLE 1 Use the Product-to-Sum Identity to Evaluate an Expression

Use a product-to-sum identity to evaluate $\sin 75° \cos 15°$.

Solution

$$\sin 75° \cos 15° = \frac{1}{2}(\sin[75° + 15°] + \sin[75° - 15°])$$

$$= \frac{1}{2}(\sin 90° + \sin 60°)$$

$$= \frac{1}{2}\left(1 + \frac{\sqrt{3}}{2}\right) = \frac{1}{2} + \frac{\sqrt{3}}{4} = \frac{2 + \sqrt{3}}{4}$$

■ *Try Exercise* **12,** *page 139.*

EXAMPLE 2 Verify an Identity By Using the Product-to-Sum Identity

Verify the identity $\cos 2x \sin 5x = \frac{1}{2}(\sin 7x + \sin 3x)$.

Solution

$$\cos 2x \sin 5x = \frac{1}{2}[\sin(2x + 5x) - \sin(2x - 5x)]$$ Use the product-to-sum identity: $\cos \alpha \sin \beta$.

$$= \frac{1}{2}[\sin 7x - \sin(-3x)]$$

$$= \frac{1}{2}(\sin 7x + \sin 3x)$$ $\sin(-3x) = -\sin 3x$

■ *Try Exercise* **36,** *page 139.*

The Sum-to-Product Identities

The sum-to-product identities can be derived from the product-to-sum identities. To derive the sum-to-product identity for $\sin x + \sin y$ first we let $x = \alpha + \beta$ and $y = \alpha - \beta$. Then

$$x + y = \alpha + \beta + \alpha - \beta \quad \text{and} \quad x - y = \alpha + \beta - (\alpha - \beta)$$

$$x + y = 2\alpha \qquad x - y = 2\beta$$

$$\alpha = \frac{x + y}{2} \qquad \beta = \frac{x - y}{2} \tag{1}$$

Substituting α and β in the product-to-sum identity

$$\frac{1}{2}[\sin(\alpha + \beta) + \sin(\alpha - \beta)] = \sin \alpha \cos \beta,$$

we have

$$\sin\left(\frac{x + y}{2} + \frac{x - y}{2}\right) + \sin\left(\frac{x + y}{2} - \frac{x - y}{2}\right) = 2 \sin \frac{x + y}{2} \cos \frac{x - y}{2}.$$

Simplifying the left side, we have a sum-to-product identity.

$$\sin x + \sin y = 2 \sin \frac{x + y}{2} \cos \frac{x - y}{2}$$

To derive the sum to product identity for $\cos x - \cos y$, substitute the expressions for α and β from equation (1) in the product identity $\frac{1}{2}[\cos(\alpha - \beta) - \cos(\alpha + \beta)] = \sin \alpha \sin \beta$ to get

$$\cos\left(\frac{x + y}{2} - \frac{x - y}{2}\right) - \cos\left(\frac{x + y}{2} + \frac{x - y}{2}\right) = 2 \sin \frac{x + y}{2} \sin \frac{x - y}{2}.$$

Simplifying the left side, we have a sum-to-product identity.

$$\cos y - \cos x = 2 \sin \frac{x + y}{2} \sin \frac{x - y}{2}$$

or, multiplying by -1,

$$\cos x - \cos y = -2 \sin \frac{x + y}{2} \sin \frac{x - y}{2}$$

In like manner, two other sum-to-product identities can be derived from the other product-to-sum identities. The proof of the identities are left as exercises.

$$\sin x - \sin y = 2 \cos \frac{x + y}{2} \sin \frac{x - y}{2}$$

$$\cos x + \cos y = 2 \cos \frac{x + y}{2} \cos \frac{x - y}{2}$$

EXAMPLE 3 **Write the Difference of Trigonometric Expressions As a Product**

Write $\sin 142° - \sin 80°$ as the product of two functions.

Solution

$$\sin 142° - \sin 80° = 2 \cos \frac{142° + 80°}{2} \sin \frac{142° - 80°}{2} = 2 \cos 111° \sin 31°$$

■ *Try Exercise* **22,** *page 139.*

EXAMPLE 4 **Verify a Sum-to-Product Identity**

Verify the identity $\dfrac{\sin 6x + \sin 2x}{\sin 6x - \sin 2x} = \tan 4x \cot 2x$.

Solution

$$\frac{\sin 6x + \sin 2x}{\sin 6x - \sin 2x} = \frac{2 \sin \dfrac{6x + 2x}{2} \cos \dfrac{6x - 2x}{2}}{2 \cos \dfrac{6x + 2x}{2} \sin \dfrac{6x - 2x}{2}} = \frac{\sin 4x \cos 2x}{\cos 4x \sin 2x}$$

$$= \tan 4x \cot 2x$$

■ *Try Exercise* **44,** *page 139.*

Functions of the Form $f(x) = a \sin x + b \cos x$

The function $f(x) = a \sin x + b \cos x$ can be written as $f(x) = k \sin(x + \alpha)$. This form of the function is useful in graphing and engineering applications because the amplitude, period, and phase shift can be readily calculated.

Let $P(a, b)$ be a point on a coordinate plane and let α represent an angle in standard position, as shown in Figure 3.3. To rewrite $y = a \sin x + b \cos x$, multiply and divide the expression $a \sin x + b \cos x$ by $\sqrt{a^2 + b^2}$.

$$a \sin x + b \cos x = \frac{\sqrt{a^2 + b^2}}{\sqrt{a^2 + b^2}}(a \sin x + b \cos x)$$

$$= \sqrt{(a^2 + b^2)}\left(\frac{a}{\sqrt{a^2 + b^2}} \sin x + \frac{b}{\sqrt{a^2 + b^2}} \cos x\right) \quad (1)$$

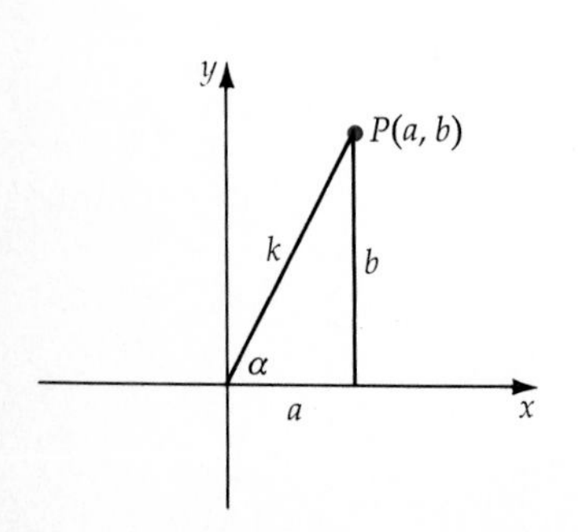

Figure 3.3

From the definition of the sine and cosine of an angle in standard position, let

$$k = \sqrt{a^2 + b^2}, \quad \cos \alpha = \frac{a}{\sqrt{a^2 + b^2}}, \quad \text{and} \quad \sin \alpha = \frac{b}{\sqrt{a^2 + b^2}}.$$

Substitute these expressions into Equation (1). We have

$$a \sin x + b \cos x = k(\cos \alpha \sin x + \sin \alpha \cos x).$$

Now using the identity for the sum of two angles, we have

$$a \sin x + b \cos x = k \sin(x + \alpha).$$

Thus $a \sin x + b \cos x = k \sin(x + \alpha)$, where $k = \sqrt{a^2 + b^2}$ and α is the angle for which $\sin \alpha = b/\sqrt{a^2 + b^2}$ and $\cos \alpha = a/\sqrt{a^2 + b^2}$.

EXAMPLE 5 Write the Sum of the Sine and Cosine in Terms of the Sine

Write $f(x) = \sin x + \cos x$ in terms of $f(x) = k \sin(x + \alpha)$.

Solution Since $a = 1$, $b = 1$, then we have $k = \sqrt{1^2 + 1^2} = \sqrt{2}$, $\sin \alpha = \dfrac{1}{\sqrt{2}}$, and $\cos \alpha = \dfrac{1}{\sqrt{2}}$. Thus, $\alpha = 45°$.

$$\sin x + \cos x = k \sin(x + \alpha) = \sqrt{2} \sin(x + 45°)$$

■ *Try Exercise* **62,** *page 139.*

The expression $y = a \sin x + b \cos x$ can be graphed by the addition of ordinates. However, it is easier to graph the function as a sine function with a phase shift.

EXAMPLE 6 Use an Identity to Graph a Trigonometric Function

Graph the function $f(x) = -\sin x + \sqrt{3} \cos x$.

Solution First, we write $f(x)$ as $k \sin(x + \alpha)$. Let $a = -1$, $b = \sqrt{3}$, then $k = \sqrt{(-1)^2 + (\sqrt{3})^2} = 2$. The point $P(-1, \sqrt{3})$ is the second quadrant. Find the reference angle β.

$$\sin \beta = \frac{\sqrt{3}}{2} \qquad \beta \text{ is the reference angle of } \alpha.$$

$$\beta = \frac{\pi}{3}$$

$$\alpha = \pi - \frac{\pi}{3} = \frac{2\pi}{3}$$

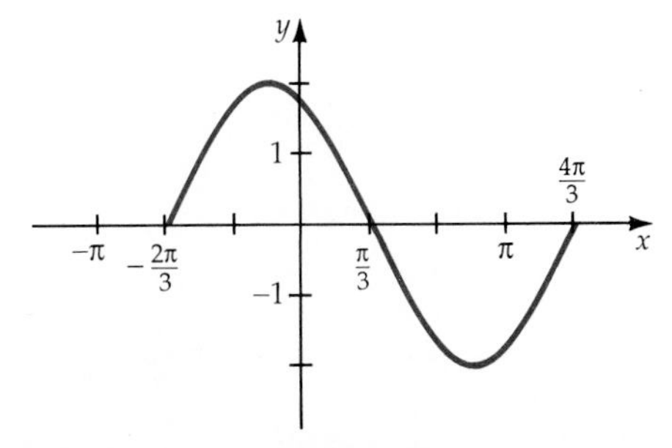

Figure 3.4
$f(x) = -\sin x + \sqrt{3} \cos x$

Substituting the values of k and α in the formula $y = k \sin(x + \alpha)$, we have

$$y = 2 \sin\left(x + \frac{2\pi}{3}\right).$$

Graph by using the key values:

$$\text{Amplitude} = 2 \qquad \text{Period} = 2\pi \qquad \text{Phase Shift} = -\frac{2\pi}{3}$$

1. Beginning point: $f(-2\pi/3) = 2 \sin 0 = 0$.
2. Quarter point: $f(-\pi/6) = 2 \sin \pi/2 = 2$.
3. Middle point: $f(\pi/3) = 2 \sin \pi = 0$.

4. Three-quarter point: $f(5\pi/6) = 2 \sin 3\pi/2 = -2$.
5. End point: $f(4/3) = 2 \sin 2\pi = 0$.

x	$-2\pi/3$	$-\pi/6$	$\pi/3$	$5\pi/6$	$4\pi/3$
$2 \sin\left(x + \frac{2\pi}{3}\right)$	0	2	0	-2	0
(x, y)	$\left(-\frac{2\pi}{3}, 0\right)$	$\left(-\frac{\pi}{6}, 2\right)$	$\left(\frac{\pi}{3}, 0\right)$	$\left(\frac{5\pi}{6}, -2\right)$	$\left(\frac{4\pi}{3}, 0\right)$

■ *Try Exercise* **70**, *page 139.*

We now list all the product-to-sum and sum-to-product identities.

Product-to-Sum Identities

$$\sin \alpha \cos \beta = \frac{1}{2}[\sin(\alpha + \beta) + \sin(\alpha - \beta)]$$

$$\cos \alpha \sin \beta = \frac{1}{2}[\sin(\alpha + \beta) - \sin(\alpha - \beta)]$$

$$\cos \alpha \cos \beta = \frac{1}{2}[\cos(\alpha + \beta) + \cos(\alpha - \beta)]$$

$$\sin \alpha \sin \beta = \frac{1}{2}[\cos(\alpha - \beta) - \cos(\alpha + \beta)]$$

Sum-to-Product Identities

$$\sin x + \sin y = 2 \sin \frac{x + y}{2} \cos \frac{x - y}{2}$$

$$\cos x - \cos y = -2 \sin \frac{x + y}{2} \sin \frac{x - y}{2}$$

$$\sin x - \sin y = 2 \cos \frac{x + y}{2} \sin \frac{x - y}{2}$$

$$\cos x + \cos y = 2 \cos \frac{x + y}{2} \cos \frac{x - y}{2}$$

Sums of the Form $a \sin x + b \cos x$

$$a \sin x + b \cos x = k \sin(x + \alpha),$$

where $k = \sqrt{a^2 + b^2}$, $\sin \alpha = \dfrac{b}{\sqrt{a^2 + b^2}}$, and $\cos \alpha = \dfrac{a}{\sqrt{a^2 + b^2}}$.

EXERCISE SET 3.4

In Exercises 1 to 8, write each expression as the sum or difference of two functions.

1. $2 \sin x \cos 2x$
2. $2 \sin 4x \sin 2x$
3. $\cos 6x \sin 2x$
4. $\cos 3x \cos 5x$
5. $2 \sin 5x \cos 3x$
6. $2 \sin 2x \cos 6x$
7. $\sin x \sin 5x$
8. $\cos 3x \sin x$

In Exercises 9 to 16, evaluate each of the expressions. Do not use a calculator or tables.

9. $\cos 75° \cos 15°$
10. $\sin 105° \cos 15°$
11. $\cos 157.5° \sin 22.5°$
12. $\sin 195° \cos 15°$
13. $\sin \frac{13\pi}{12} \cos \frac{\pi}{12}$
14. $\sin \frac{11\pi}{12} \sin \frac{7\pi}{12}$
15. $\sin \frac{\pi}{12} \cos \frac{7\pi}{12}$
16. $\cos \frac{17\pi}{12} \sin \frac{7\pi}{12}$

In Exercises 17 to 24, write each expression as the product of two functions.

17. $\sin 4\theta + \sin 2\theta$
18. $\cos 5\theta - \cos 3\theta$
19. $\cos 3\theta + \cos \theta$
20. $\sin 7\theta - \sin 3\theta$
21. $\cos 6\theta - \cos 2\theta$
22. $\cos 3\theta + \cos 5\theta$
23. $\cos \theta + \cos 3\theta$
24. $\sin 3\theta + \sin 7\theta$

In Exercises 25 to 32, evaluate each of the expressions. Do not use a calculator or tables.

25. $\sin 75° + \sin 15°$
26. $\cos 105° + \cos 165°$
27. $\cos 105° - \cos 15°$
28. $\sin 255° - \sin 165°$
29. $\sin \frac{13\pi}{12} + \sin \frac{5\pi}{12}$
30. $\sin \frac{11\pi}{12} - \sin \frac{5\pi}{12}$
31. $\cos \frac{\pi}{12} + \cos \frac{5\pi}{12}$
32. $\cos \frac{7\pi}{8} + \cos \frac{\pi}{8}$

In Exercises 33 to 48, verify the identities.

33. $2 \cos \alpha \cos \beta = \cos(\alpha + \beta) + \cos(\alpha - \beta)$
34. $2 \sin \alpha \sin \beta = \cos(\alpha - \beta) - \cos(\alpha + \beta)$
35. $2 \cos 3x \sin x = 2 \sin x \cos x - 8 \cos x \sin^3 x$
36. $\sin 5x \cos 3x = \sin 4x \cos 4x + \sin x \cos x$
37. $2 \cos 5x \cos 7x = \cos^2 6x - \sin^2 6x + 2 \cos^2 x - 1$
38. $\sin 3x \cos x = \sin x \cos x(3 - 4 \sin^2 x)$
39. $\sin 3x - \sin x = 2 \sin x - 4 \sin^3 x$
40. $\cos 5x - \cos 3x = -8 \sin^2 x(2 \cos^3 x - \cos x)$
41. $\sin 2x + \sin 4x = 2 \sin x \cos x(4 \cos^2 x - 1)$
42. $\cos 3x + \cos x = 4 \cos^3 x - 2 \cos x$
43. $\dfrac{\sin 3x - \sin x}{\cos 3x - \cos x} = -\cot 2x$
44. $\dfrac{\cos 5x - \cos 3x}{\sin 5x + \sin 3x} = -\tan x$
45. $\dfrac{\sin 5x + \sin 3x}{4 \sin x \cos^3 x - 4 \sin^3 x \cos x} = 2 \cos x$
46. $\dfrac{\cos 4x - \cos 2x}{\sin 2x - \sin 4x} = \tan 3x$
47. $\sin(x + y) \cos(x - y) = \sin x \cos x + \sin y \cos y$
48. $\sin(x + y) \sin(x - y) = \sin^2 x - \sin^2 y$

In Exercises 49 to 58, write the given equation in the form $y = k \sin(x + \alpha)$, where the measure of α is in degrees.

49. $y = -\sin x - \cos x$
50. $y = \sqrt{3} \sin x - \cos x$
51. $y = \frac{1}{2} \sin x - \frac{\sqrt{3}}{2} \cos x$
52. $y = \frac{\sqrt{3}}{2} \sin x - \frac{1}{2} \cos x$
53. $y = \frac{1}{2} \sin x - \frac{1}{2} \cos x$
54. $y = -\frac{\sqrt{3}}{2} \sin x - \frac{1}{2} \cos x$
55. $y = 8 \sin x + 15 \cos x$
56. $y = -7 \sin x + 24 \cos x$
57. $y = 8 \sin x - 3 \cos x$
58. $y = -4 \sin x + 7 \cos x$

In Exercises 59 to 66, write the given equations in the form $k \sin(x + \alpha)$ where the measure of α is in radians.

59. $y = -\sin x + \cos x$
60. $y = -\sqrt{3} \sin x - \cos x$
61. $y = \frac{\sqrt{3}}{2} \sin x + \frac{1}{2} \cos x$
62. $y = \sin x + \sqrt{3} \cos x$
63. $y = -4 \sin x + 9 \cos x$
64. $y = -3 \sin x + 5 \cos x$
65. $y = -5 \sin x + 5 \cos x$
66. $y = 3 \sin x - 3 \cos x$

In Exercises 67 to 76, graph one cycle of the following equations.

67. $y = -\sin x - \sqrt{3} \cos x$
68. $y = -\sqrt{3} \sin x + \cos x$
69. $y = 2 \sin x + 2 \cos x$
70. $y = \sin x + \sqrt{3} \cos x$
71. $y = -\sqrt{3} \sin x - \cos x$
72. $y = -\sin x + \cos x$
73. $y = 2 \sin x - 5 \cos x$
74. $y = -6 \sin x - 10 \cos x$
75. $y = 3 \sin x - 4 \cos x$
76. $y = -5 \sin x + 9 \cos x$

Supplemental Exercises

77. Derive the sum-to-product identity:

$$\cos x + \cos y = 2 \cos \frac{x+y}{2} \cos \frac{x-y}{2}$$

78. Derive the product-to-sum identity:

$$\sin x \sin y = \frac{1}{2}[\cos(x-y) - \cos(x+y)]$$

79. If $x + y = 180°$; show that $\sin x + \sin y = 2 \sin x$.

80. If $x + y = 360°$; show that $\cos x + \cos y = 2 \cos x$.

In Exercises 81 to 86, verify the identities.

81. $\sin 2x + \sin 4x + \sin 6x = 4 \sin 3x \cos 2x \cos x$

82. $\sin 4x - \sin 2x + \sin 6x = 4 \cos 3x \sin 2x \cos x$

83. $\dfrac{\cos 10x + \cos 8x}{\sin 10x - \sin 8x} = \cot x$

84. $\dfrac{\sin 10x + \sin 2x}{\cos 10x + \cos 2x} = \dfrac{2 \tan 3x}{1 - \tan^2 3x}$

85. $\dfrac{\sin 2x + \sin 4x + \sin 6x}{\cos 2x + \cos 4x + \cos 6x} = \tan 4x$

86. $\dfrac{\sin 2x + \sin 6x}{\cos 6x - \cos 2x} = -\cot 2x$

87. Verify $\cos^2 x - \sin^2 x = \cos 2x$ by using a product-to-sum identity.

88. Verify $2 \sin x \cos x = \sin 2x$ by using a product-to-sum identity.

89. Verify that $a \sin x + b \cos x = k \cos(x - \alpha)$, where $k = \sqrt{a^2 + b^2}$ and $\tan \alpha = a/b$.

90. Verify that $a \sin cx + b \cos cx = k \sin(cx + \alpha)$, where $k = \sqrt{a^2 + b^2}$ and $\tan \alpha = b/a$.

In Exercises 91 to 96, find the amplitude, phase shift, and period and then graph the function.

91. $y = \sin \dfrac{x}{2} - \cos \dfrac{x}{2}$

92. $y = -\sqrt{3} \sin \dfrac{x}{2} + \cos \dfrac{x}{2}$

93. $y = \sqrt{3} \sin 2x - \cos 2x$

94. $y = -\sin 2x + \cos 2x$

95. $y = \sin \pi x + \sqrt{3} \cos \pi x$

96. $y = 3 \sin 2\pi x - 4 \cos 2\pi x$

97. Two nonvertical lines intersect in a plane. The slope of l_1 is m_1 and the slope of l_2 is m_2. Show that the tangent of the angle θ formed by the two lines is given by the expression:

$$\tan \theta = \frac{m_1 - m_2}{1 + m_1 m_2}.$$

98. Use the equation from Exercise 97 to find the angle from the line $y = x + 5$ to the line $y = 3x - 4$.

99. Use the equation from Exercise 97 to find the angle from the line $y = -3x/2 - 4$ to the line $y = 2x/3 + 3$.

3.5

Inverse Trigonometric Functions

Because the sine function is not a one-to-one function it does not have an inverse function. Figure 3.5 shows the graphs of the function $f(x) = \sin x$ and the equation $x = \sin y$. The sine function is one-to-one on the interval $-\pi/2 \le x \le \pi/2$. Thus the sine function, with domain restricted to that interval, does have an inverse function. Using this restricted domain, the graphs of the restricted sine function and the inverse sine function are shown in Figure 3.6. The range values of the inverse sine function are called the **principal values** of the function.

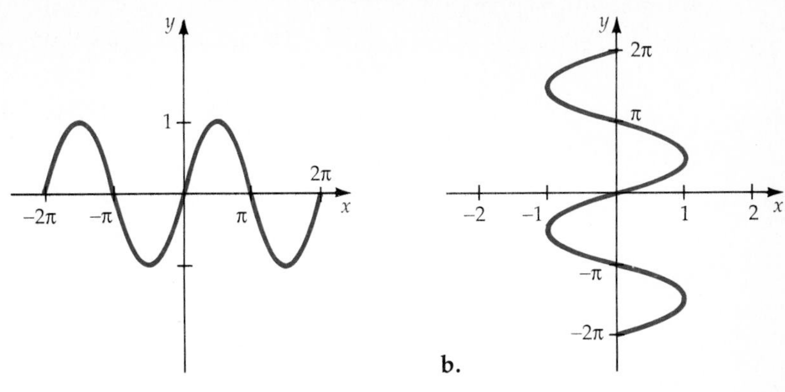

Figure 3.5
a. $y = \sin x$; b. $x = \sin y$

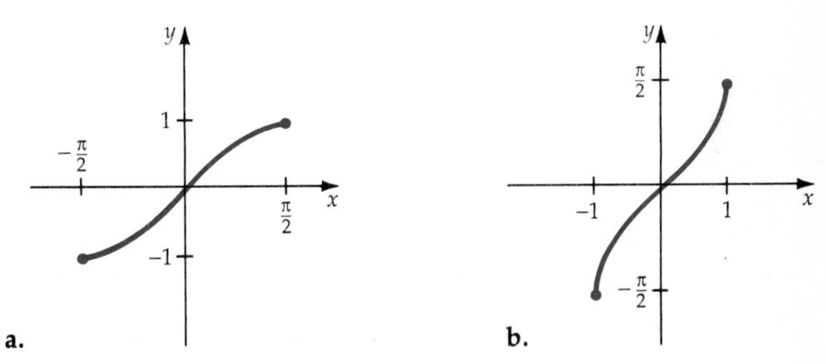

Figure 3.6
a. $y = \sin x$: $-\pi/2 \leq x \leq \pi/2$;
b. $x = \sin y$: $-\pi/2 \leq y \leq \pi/2$.

The graph shows that the domain of the sine function $y = \sin x$ with $-\pi/2 \leq x \leq \pi/2$ is the range of the inverse sine function $x = \sin y$. We can attempt to find an equation for the inverse sine function via the method we used for linear functions. We begin with

$$y = \sin x$$

Interchange the x and y. $\qquad x = \sin y$

Unfortunately, there is no algebraic solution for y. Thus we establish new notation and write

$$y = \sin^{-1} x$$

which is read "y is the inverse sine of x." Some textbooks use the notation $y = \arcsin x$ for the inverse sine function.

Caution The -1 in $\sin^{-1}x$ is not an exponent. The -1 is used to denote the inverse function. To use -1 as an exponent for the sine function, enclose the function with parentheses.

$$(\sin x)^{-1} = \frac{1}{\sin x} \qquad \sin^{-1} x \neq \frac{1}{\sin x}$$

This convention is used for other inverse trigonometric functions presented in this text.

From our discussion, we can write the following definition of the inverse sine function:

Definition of the Inverse Sine Function

$$y = \sin^{-1} x \text{ if and only if } x = \sin y$$

The domain of $y = \sin^{-1} x$ is $\{x \mid -1 \le x \le 1\}$. The range is $\{y \mid -\pi/2 \le y \le \pi/2\}$.

Recall that a function f and its inverse function f^{-1} have the property that $f[f^{-1}(x)] = x$ and $f^{-1}[f(x)] = x$. Thus, if $f(x) = \sin x$ and $f^{-1}(x) = \sin^{-1} x$, then

$$f[f^{-1}(x)] = \sin(\sin^{-1} x) = x \quad \text{where } -1 \le x \le 1,$$

$$f^{-1}[f(x)] = \sin^{-1}(\sin x) = x \quad \text{where } -\frac{\pi}{2} \le x \le \frac{\pi}{2}$$

It is convenient to think of the value of an inverse trigonometric function as an angle. For example, $\sin^{-1}\frac{1}{2}$ is an angle whose sine is $\frac{1}{2}$. There is an infinite number of angles whose sign is $\frac{1}{2}$. (Why?[1]) However, the function values of the inverse sine function are restricted to angles whose measure y is $-\pi/2 \le y \le \pi/2$. Thus $\sin^{-1}\frac{1}{2} = \pi/6$.

EXAMPLE 1 Evaluate the Inverse Sine Function

Find the exact value of $y = \sin^{-1}\left(-\dfrac{\sqrt{3}}{2}\right)$, where y is a real number.

Solution The meaning of $\sin^{-1}\left(-\dfrac{\sqrt{3}}{2}\right)$ is the angle whose sine is $-\dfrac{\sqrt{3}}{2}$. Since the range of the inverse sine function is $-\dfrac{\pi}{2} \le y \le \dfrac{\pi}{2}$, the angle is in the fourth quadrant because the inverse sine function is negative.

$$y = \sin^{-1}\left(-\frac{\sqrt{3}}{2}\right)$$

$$\sin y = -\frac{\sqrt{3}}{2}$$ y is the angle whose sine is $-\sqrt{3}/2$, and $-\pi/2 \le y \le \pi/2$.

$$y = -\frac{\pi}{3}$$

■ *Try Exercise* **2**, *page 151.*

[1] The sine function is a periodic function. Thus, $\sin\frac{1}{2} = \sin(\frac{1}{2} + 2k\pi)$, where k is an integer. There is an infinite number of values because the set of integers is infinite.

EXAMPLE 2 Evaluate the Inverse Sine Function

Evaluate $\sin^{-1}\left(\sin \frac{2\pi}{3}\right)$.

Solution Recall that $\sin^{-1}(\sin x) = x$ when $-\frac{\pi}{2} \le x \le \frac{\pi}{2}$. Because $\frac{2\pi}{3}$ is not in this interval, we must first evaluate $\sin \frac{2\pi}{3}$.

$$y = \sin^{-1}\left(\sin \frac{2\pi}{3}\right) = \sin^{-1}\left(\frac{\sqrt{3}}{2}\right) \qquad \sin 2\pi/3 = \sqrt{3}/2.$$

$$\sin y = \frac{\sqrt{3}}{2} \qquad -\frac{\pi}{2} \le y \le \frac{\pi}{2} \qquad \text{By the definition of } y = \sin^{-1} x$$

$$y = \frac{\pi}{3}$$

■ *Try Exercise* **54,** *page 152.*

Caution In this case, $\sin^{-1}(\sin x) \ne x$. The range of the inverse sine function is $\{y \mid -\pi/2 \le y \le \pi/2\}$ and $2\pi/3$ is not in the range of the function.

EXAMPLE 3 Evaluate Inverse Sine Functions with a Calculator

Use a calculator to evaluate the following expressions:

a. $\sin^{-1} 0.8$ b. $\sin^{-1}(\sin 120°)$ c. $\sin^{-1}(\sin 1.2)$ d. $\sin(\sin^{-1} 2)$

Solution

Function	Key Sequence	Calculator Display
a. $\sin^{-1} 0.8$	0.8 INV sin	0.92729522
b. $\sin^{-1}(\sin 120°)$	120 sin INV sin	60
c. $\sin^{-1}(\sin 1.2)$	1.2 sin INV sin	1.2
d. $\sin(\sin^{-1} 2)$	2 INV sin	Error

■ *Try Exercise* **30,** *page 151.*

Remark In part d, $\sin^{-1} 2$ is undefined. (Why?[2]) Therefore, $\sin(\sin^{-1} 2)$ results in an error message.

The cosine function is not a one-to-one function so it does not have an inverse function. Figure 3.7 shows the graphs of the function $y = \cos x$ and the equation $x = \cos y$.

[2] If $x = \sin^{-1} 2$, then $\sin x = 2$. However, the range of $\sin x$ is $-1 \le \sin x \le 1$. Therefore, there is no value of x for $\sin x = 2$.

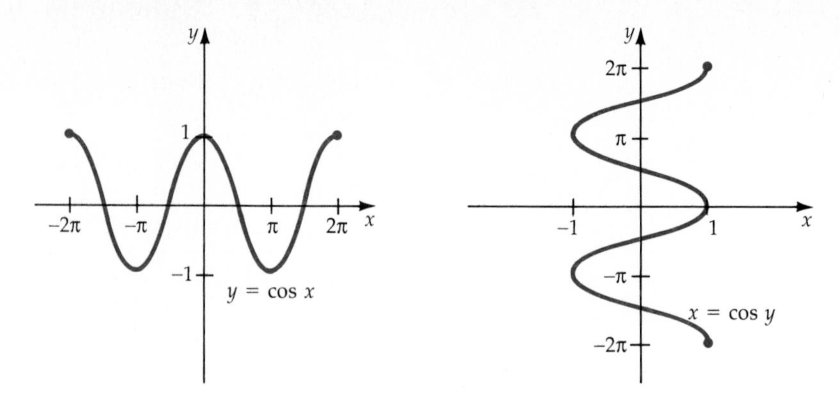

Figure 3.7

The cosine function is one-to-one in the interval $0 \le x \le \pi$. Thus the cosine function, with the domain restricted to that interval, has an inverse function. Using this restricted domain, the graphs of the restricted cosine function and the inverse cosine function are shown in Figure 3.8.

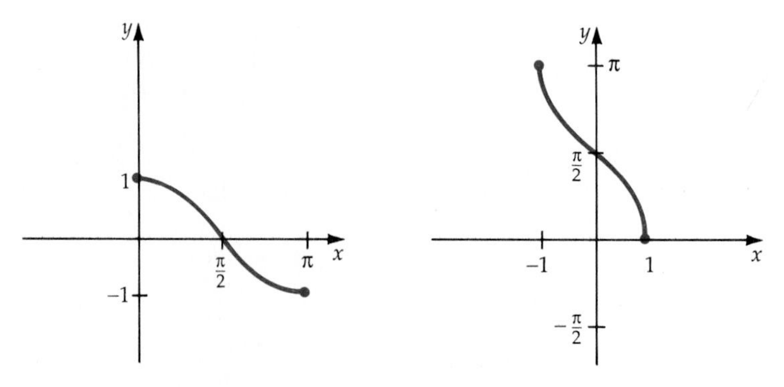

Figure 3.8
$y = \cos x$: $0 \le x \le \pi$; $x = \cos y$: $0 \le y \le \pi$.

Definition of the Inverse Cosine Function

$$y = \cos^{-1} x \text{ if and only if } x = \cos y$$

The domain of $y = \cos^{-1} x$ is $\{x \mid -1 \le x \le 1\}$. The range is $\{y \mid 0 \le y \le \pi\}$

A function f and its inverse f^{-1} have the property that $f[f^{-1}(x)] = x$ and $f^{-1}[f(x)] = x$. Thus, if $f(x) = \cos x$ and $f^{-1}(x) = \cos^{-1} x$, then

$$f[f^{-1}(x)] = \cos(\cos^{-1} x) = x \quad \text{where } -1 \le x \le 1$$

$$f^{-1}[f(x)] = \cos^{-1}(\cos x) = x \quad \text{where } 0 \le x \le \pi$$

EXAMPLE 4 Evaluate the Inverse Cosine Function with a Calculator

Use a calculator to evaluate the following expressions:

a. $\cos^{-1}(-0.2388)$ b. $\cos(\cos^{-1} 2)$ c. $\cos^{-1}(\cos 5)$

Solution

Function	Key Sequence	Calculator Display
a. $\cos^{-1}(-0.2388)$	−0.2388 [INV] [cos]	1.811926238
b. $\cos(\cos^{-1} 2)$	2 [INV] [cos]	Error
c. $\cos^{-1}(\cos 5)$	5 [cos] [INV] [cos]	1.283185307

■ *Try Exercise* **58**, *page 152.*

The domains of the other trigonometric functions can be restricted so that each has an inverse function. Table 3.2 shows the restricted function and the inverse function for tan x, csc x, sec x, and cot x. Pay close attention to the asymptotes of the function and the inverse function.

TABLE 3.2

	$y = \tan x$	$y = \tan^{-1} x$	$y = \csc x$	$y = \csc^{-1} x$
Domain	$-\frac{\pi}{2} < x < \frac{\pi}{2}$	$-\infty < x < \infty$	$-\frac{\pi}{2} \le x \le \frac{\pi}{2}$, $x \ne 0$	$x \le -1$ or $x \ge 1$
Range	$-\infty < y < \infty$	$-\frac{\pi}{2} < y < \frac{\pi}{2}$	$x \le -1, x \ge 1$	$-\frac{\pi}{2} \le y \le \frac{\pi}{2}$, $y \ne 0$
Asymptotes	$x = -\frac{\pi}{2}, x = \frac{\pi}{2}$ 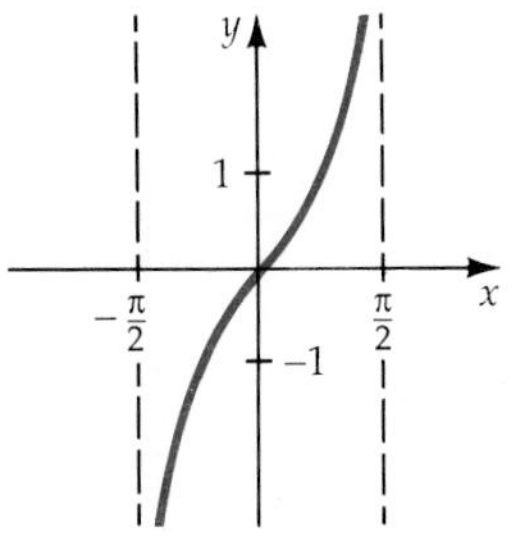	$y = -\frac{\pi}{2}, y = \frac{\pi}{2}$ 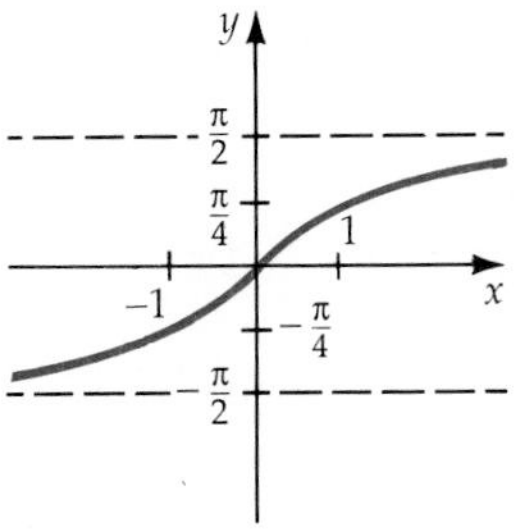	$x = 0$ 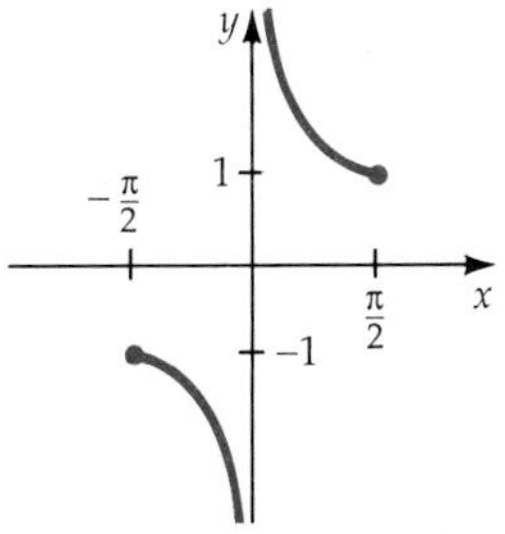	$y = 0$ 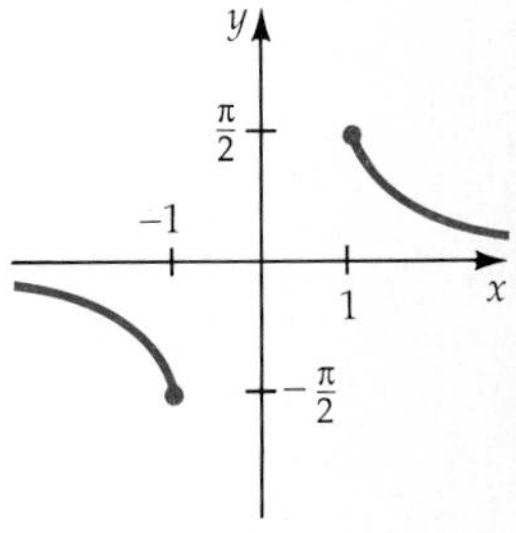

(Continued)

TABLE 3.2 *(continued)*

	$y = \sec x$	$y = \sec^{-1} x$	$y = \cot x$	$y = \cot^{-1} x$
Domain	$0 \le x \le \pi,$ $x \ne \frac{\pi}{2}$	$x \le -1$ or $x \ge 1$	$0 < x < \pi$	$-\infty < x < \infty$
Range	$y \le -1$ or $y \ge 1$	$0 \le y \le \pi,$ $y \ne \frac{\pi}{2}$	$-\infty < y < \infty$	$0 < y < \pi$
Asymptotes	$x = \frac{\pi}{2}$	$y = \frac{\pi}{2}$	$x = 0,\ x = \pi$	$y = 0,\ y = \pi$

EXAMPLE 5 Evaluate the Inverse Tangent Function

Find the exact value of $y = \tan^{-1}\left(-\frac{\sqrt{3}}{3}\right)$.

Solution We know that y is the real number whose tangent is $-\frac{\sqrt{3}}{3}$. The range of the inverse tangent function is $-\pi/2 < y < \pi/2$. Thus y must be in the interval $-\pi/2 < y < 0$ because it is a negative number.

$$y = \tan^{-1}\left(-\frac{\sqrt{3}}{3}\right)$$

$$\tan y = -\frac{\sqrt{3}}{3}$$

$$y = -\frac{\pi}{6}$$

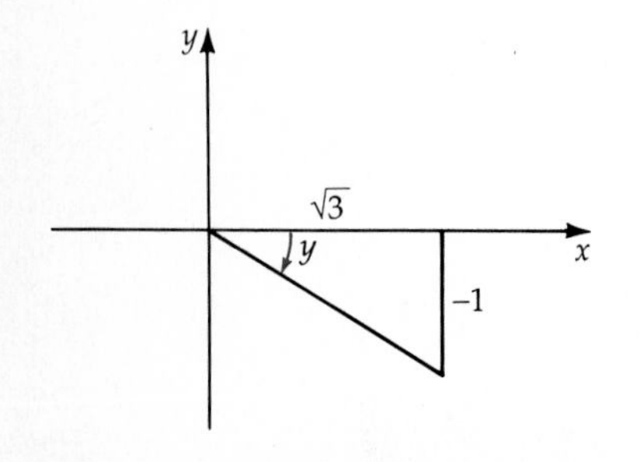

Figure 3.9

■ *Try Exercise* **18**, *page 151.*

A calculator may not have keys for the inverse secant, cosecant, and cotangent functions. The following procedure shows an identity for the inverse cosecant function in terms of the inverse sine function.

If we need to determine y, which is the angle (or number) whose cosecant is x, we can rewrite $y = \csc^{-1} x$ as follows.

$y = \csc^{-1} x$	Domain: $x \le -1$ or $x \ge 1$ Range: $-\pi/2 \le y \le \pi/2,\ y \ne 0$.
$\csc y = x$	Definition of inverse function
$\dfrac{1}{\sin y} = x$	Substitute $1/\sin y$ for $\csc y$.
$\sin y = \dfrac{1}{x}$	Solve for $\sin y$.
$y = \sin^{-1} \dfrac{1}{x}$	
$\csc^{-1} x = \sin^{-1} \dfrac{1}{x}$	

Thus the inverse cosecant of x is the same as the inverse sine of $1/x$.

There are similar expressions for the inverse secant and inverse cotangent.

$$\sec^{-1} x = \cos^{-1} 1/x \qquad \cot^{-1} x = \tan^{-1} 1/x$$

EXAMPLE 6 Find the Inverse Cosecant Function with a Calculator

Find $\csc^{-1} 4.9873$ using a calculator.

Solution The inverse cosecant of an argument is the inverse sine of the reciprocal of the argument. Be sure your calculator is in the radian mode.

Function	Key Sequence	Calculator Display
$\csc^{-1} 4.9873$	4.9873 [1/x] [INV] [sin]	0.20187774

■ *Try Exercise* **40**, *page 151.*

EXAMPLE 7 Evaluate the Inverse Secant Function

Find the exact value of $y = \sec^{-1} 2$.

Solution Use the identity $\sec^{-1} x = \cos^{-1} \dfrac{1}{x}$.

$$\sec^{-1} 2 = \cos^{-1} \frac{1}{2} = \frac{\pi}{3}$$

■ *Try Exercise* **10**, *page 151.*

EXAMPLE 8 **Evaluate an Inverse Trigonometric Expression**

Evaluate $\sin\left[\cos^{-1}\left(-\frac{3}{5}\right)\right]$.

Solution Let $y = \cos^{-1}\left(-\frac{3}{5}\right)$, then $\cos y = -\frac{3}{5}$.

$$\sin^2 y + \cos^2 y = 1$$

$$\sin^2 y = 1 - \cos^2 y = 1 - \left(-\frac{3}{5}\right)^2 = \frac{16}{25}$$

Because $\cos y = -\frac{3}{5}$, we know $\frac{\pi}{2} < y < \pi$ and thus $\sin y > 0$. Therefore,

$$\sin y = \frac{4}{5}$$

The same problem worked with a calculator is as follows:

Function	Key Sequence	Calculator Display
$\sin\left(\cos^{-1}\left(-\frac{3}{5}\right)\right)$	3/5 [+/−] [INV] [cos] [sin]	[0.8]

■ *Try Exercise* **70**, *page 152.*

EXAMPLE 9 **Evaluate an Inverse Trigonometric Expression**

Evaluate $\sin\left(\sin^{-1}\frac{3}{5} + \cos^{-1}\frac{5}{13}\right)$.

Solution Let $x = \sin^{-1}\frac{3}{5}$ and let $y = \cos^{-1}\frac{5}{13}$, then $\sin x = \frac{3}{5}$ and $\cos y = \frac{5}{13}$ with $0 < x < \frac{\pi}{2}$ and $0 < y < \frac{\pi}{2}$. Therefore,

$$\cos x = \sqrt{1 - \sin^2 x} = \sqrt{1 - \left(\frac{3}{5}\right)^2} = \frac{4}{5}$$

$$\sin y = \sqrt{1 - \cos^2 y} = \sqrt{1 - \left(\frac{5}{13}\right)^2} = \frac{12}{13}$$

$$\sin\left(\sin^{-1}\frac{3}{5} + \cos^{-1}\frac{5}{13}\right) = \sin(x + y) = \sin x \cos y + \cos x \sin y$$

$$= \frac{3}{5}\cdot\frac{5}{13} + \frac{4}{5}\cdot\frac{12}{13}$$

$$= \frac{15}{65} + \frac{48}{65} = \frac{63}{65}$$

■ *Try Exercise* **84**, *page 152.*

EXAMPLE 10 Solve an Inverse Trigonometric Equation

Solve the inverse trigonometric equation $\sin^{-1} x + \cos^{-1} \dfrac{3}{5} = \pi$.

Solution Solve for $\sin^{-1} x$, then take the sine of both sides of the equation.

$$\sin^{-1} x + \cos^{-1} \frac{3}{5} = \pi$$

$$\sin^{-1} x = \pi - \cos^{-1} \frac{3}{5}$$

$$\sin(\sin^{-1} x) = \sin\left(\pi - \cos^{-1} \frac{3}{5}\right)$$

$$x = \sin(\pi - \alpha) \qquad \text{Let } \alpha = \cos^{-1} 3/5.$$

$$= \sin \pi \cos \alpha - \cos \pi \sin \alpha$$

$$= (0) \cos \alpha - (-1) \sin \alpha$$

$$= \sin \alpha \qquad \sin \alpha = 4/5 \text{ (see Figure 3.10)}$$

$$x = \frac{4}{5}$$

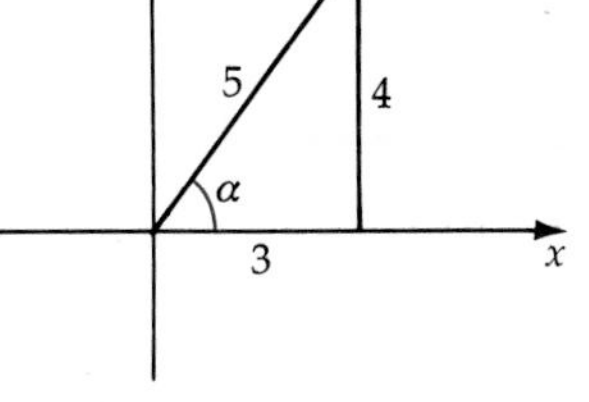

Figure 3.10

■ *Try Exercise* **94,** *page 152.*

EXAMPLE 11 Verify an Inverse Trigonometric Identity

Verify the identity $\sin^{-1} x + \cos^{-1} x = \dfrac{\pi}{2}$.

Solution Let $\alpha = \sin^{-1} x$ and $\beta = \cos^{-1} x$ which implies that $\sin \alpha = x$ and $\cos \beta = x$. Then $\cos \alpha = \sqrt{1 - \sin^2\alpha} = \sqrt{1 - x^2}$ and in $\beta = \sqrt{1 - \cos^2\beta} = \sqrt{1 - x^2}$. Working with the left side of the identity produces

$$\sin^{-1} x + \cos^{-1} x = \alpha + \beta$$

$$= \sin^{-1}[\sin(\alpha + \beta)]$$

$$= \sin^{-1}[\sin \alpha \cos \beta + \cos \alpha \sin \beta]$$

$$= \sin^{-1}[x \cdot x + \sqrt{1 - x^2}\sqrt{1 - x^2}]$$

$$= \sin^{-1}[x^2 + 1 - x^2]$$

$$= \sin^{-1} 1$$

$$= \frac{\pi}{2}$$

■ *Try Exercise* **102,** *page 152.*

The inverse trigonometric functions can be graphed by finding the upper and lower values of the range and then finding as many other points as needed. To find points on the graph, it is sometimes easier to solve for the independent variable x and assigning values to the dependent variable y. This procedure is shown by graphing the function $y = 2 \sin^{-1} 2x$.

$$y = 2 \sin^{-1} 2x$$

$$y/2 = \sin^{-1} 2x \quad \text{By dividing each side by 2}$$

We know that the range of the inverse sine function is from $-\pi/2$ to $\pi/2$. Thus we have

$$-\frac{\pi}{2} \le \sin^{-1} 2x \le \frac{\pi}{2},$$

and we can substitute $y/2$ for the inverse sine

$$-\frac{\pi}{2} \le \frac{y}{2} \le \frac{\pi}{2}$$

$$-\pi \le y \le \pi \quad \text{Multiplying by 2}$$

The range of $y = 2 \sin^{-1} 2x$ is $-\pi \le y \le \pi$. Thus the upper value of the range is π and the lower value is $-\pi$.

Now solve for x.

$$y = 2 \sin^{-1} 2x$$

$$\frac{y}{2} = \sin^{-1} 2x$$

$$\sin \frac{y}{2} = 2x$$

$$x = \frac{1}{2} \sin \frac{y}{2}$$

We have found the range of the function and solved the equation for x. Choose values from the range for y and calculate x to find points on the graph of the function. Assign values to y between $-\pi$ and π.

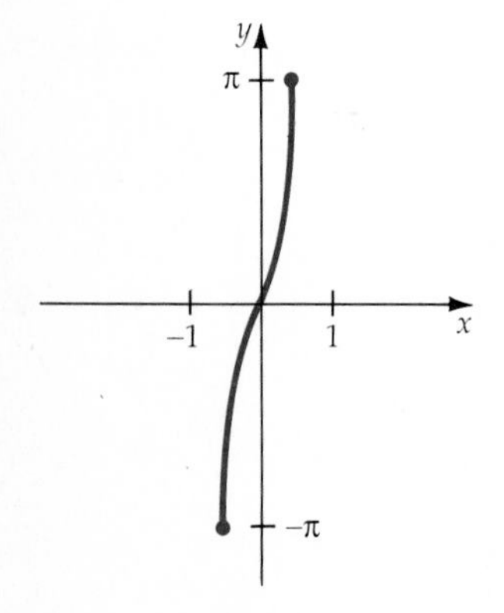

Figure 3.11
$y = 2 \sin^{-1} 2x$

	y	$\sin \frac{y}{2}$	$x = \frac{1}{2} \sin \frac{y}{2}$	(x, y)
Lower value	$-\pi$	-1	$-\frac{1}{2}$	$\left(-\frac{1}{2}, -\pi\right)$
Middle value	0	0	0	$(0, 0)$
Upper value	π	1	$\frac{1}{2}$	$\left(\frac{1}{2}, \pi\right)$

EXAMPLE 12 **Graph the Inverse Cosine Function**

Graph the function $y = 0.5\cos^{-1}(x + 2)$.

Solution The graph of $y = 0.5\cos^{-1}(x + 2)$ is the graph of $y = \cos^{-1}x$ shifted left horizontally 2 units and shrunk towards the x-axis, so that each y-coordinate is $\frac{1}{2}$ of its previous value. See Figure 3.12.

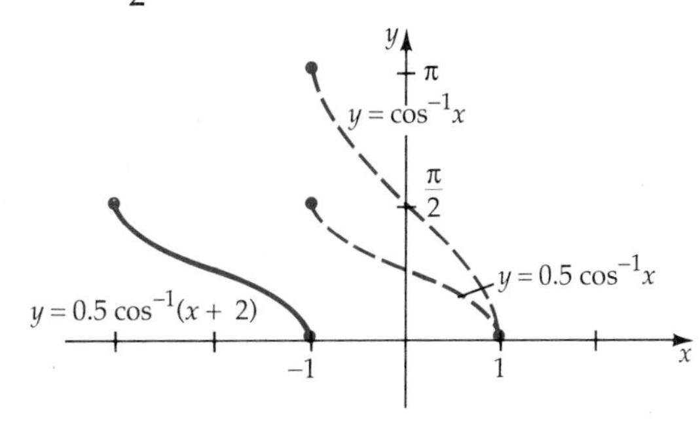

Figure 3.12
$y = 0.5\cos^{-1}(x + 2)$

■ *Try Exercise* **106**, *page 152.*

EXERCISE SET 3.5

In Exercises 1 to 12, find the exact radian value for the given inverse functions.

1. $\sin^{-1}1$ **2.** $\sin^{-1}\frac{\sqrt{2}}{2}$ **3.** $\cos^{-1}\left(-\frac{\sqrt{3}}{2}\right)$

4. $\cos^{-1}\left(-\frac{1}{2}\right)$ **5.** $\tan^{-1}(-1)$ **6.** $\tan^{-1}\sqrt{3}$

7. $\cot^{-1}\frac{\sqrt{3}}{3}$ **8.** $\cot^{-1}1$ **9.** $\sec^{-1}2$

10. $\sec^{-1}\frac{2\sqrt{3}}{3}$ **11.** $\csc^{-1}(-\sqrt{2})$ **12.** $\csc^{-1}(-2)$

In Exercises 13 to 20, find the exact degree value of the given inverse functions.

13. $\sin^{-1}\left(-\frac{\sqrt{3}}{2}\right)$ **14.** $\sin^{-1}\frac{1}{2}$ **15.** $\cos^{-1}\left(-\frac{1}{2}\right)$

16. $\cos^{-1}\frac{\sqrt{3}}{2}$ **17.** $\tan^{-1}\frac{\sqrt{3}}{3}$ **18.** $\tan^{-1}1$

19. $\cot^{-1}\sqrt{3}$ **20.** $\cot^{-1}(-1)$

In Exercises 21 to 28, find the approximate radian value for the given functions to 4 significant digits.

21. $\sin^{-1}0.4555$ **22.** $\sin^{-1}0.8700$

23. $\cos^{-1}(-0.2357)$ **24.** $\cos^{-1}(-0.1298)$

25. $\tan^{-1}(-1.4344)$ **26.** $\tan^{-1}(-5.2691)$

27. $\cot^{-1}0.9823$ **28.** $\cot^{-1}4.2317$

In Exercises 29 to 40, find the approximate degree value for the given inverse functions to the nearest tenth of a degree.

29. $\sin^{-1}(-0.2781)$ **30.** $\sin^{-1}(-0.9650)$

31. $\cos^{-1}0.5555$ **32.** $\cos^{-1}0.1598$

33. $\tan^{-1}(-2.0440)$ **34.** $\tan^{-1}10.0050$

35. $\cot^{-1}(-0.9752)$ **36.** $\cot^{-1}1.0578$

37. $\sec^{-1}(-3.4785)$ **38.** $\sec^{-1}9.4455$

39. $\csc^{-1}1.0056$ **40.** $\csc^{-1}(-10.9856)$

In Exercises 41 to 60, evaluate the given expression.

41. $\cos\left(\cos^{-1}\frac{1}{2}\right)$ **42.** $\cos(\cos^{-1}2)$

43. $\tan(\tan^{-1}2)$ **44.** $\tan\left(\tan^{-1}\frac{1}{2}\right)$

45. $\sin\left(\tan^{-1}\frac{3}{4}\right)$ **46.** $\cos\left(\sin^{-1}\frac{5}{13}\right)$

47. $\tan\left(\sin^{-1}\frac{\sqrt{2}}{2}\right)$ **48.** $\sin\left[\cos^{-1}\left(-\frac{\sqrt{3}}{2}\right)\right]$

49. $\cos[\sec^{-1}(2)]$

50. $\sin^{-1}(\sin 2)$

51. $\sin^{-1}\left(\sin \frac{\pi}{6}\right)$

52. $\sin^{-1}\left(\sin \frac{5\pi}{6}\right)$

53. $\cos^{-1}\left(\sin \frac{\pi}{4}\right)$

54. $\sin^{-1}\left(\cos \frac{7\pi}{6}\right)$

55. $\sin^{-1}\left(\tan \frac{\pi}{3}\right)$

56. $\cos^{-1}\left(\tan \frac{2\pi}{3}\right)$

57. $\tan^{-1}\left(\sin \frac{\pi}{6}\right)$

58. $\cot^{-1}\left(\cos \frac{2\pi}{3}\right)$

59. $\sin^{-1}\left[\cos\left(-\frac{2\pi}{3}\right)\right]$

60. $\cos^{-1}\left[\tan\left(-\frac{\pi}{3}\right)\right]$

In Exercises 61 to 100, solve the following inverse trigonometric equations.

61. $y = \sin^{-1}\left(-\frac{\sqrt{3}}{2}\right)$

62. $y = \tan^{-1}\frac{\sqrt{3}}{3}$

63. $y = \cos^{-1}(-0.5669)$

64. $y = \csc^{-1}(2.3033)$

65. $y = \cot^{-1}(-1.0886)$

66. $y = \sec^{-1}(-2.9071)$

67. $y = \cos^{-1}\frac{\pi}{4}$

68. $y = \tan^{-1}\left(-\frac{\pi}{3}\right)$

69. $y = \cos\left(\sin^{-1}\frac{7}{25}\right)$

70. $y = \tan\left(\cos^{-1}\frac{3}{5}\right)$

71. $y = \sec\left(\tan^{-1}\frac{12}{5}\right)$

72. $y = \csc\left(\sin^{-1}\frac{12}{13}\right)$

73. $y = \sin^{-1}\left(\sin \frac{2\pi}{3}\right)$

74. $y = \tan^{-1}\left(\tan \frac{5\pi}{4}\right)$

75. $y = \cos^{-1}\left[\cos\left(-\frac{\pi}{6}\right)\right]$

76. $y = \sin^{-1}\left(\sin \frac{5\pi}{3}\right)$

77. $y = \tan^{-1}\left(\tan \frac{3\pi}{4}\right)$

78. $y = \cos^{-1}\left(\cos \frac{5\pi}{6}\right)$

79. $y = \cos\left(2 \sin^{-1}\frac{\sqrt{2}}{2}\right)$

80. $y = \tan\left(2 \sin^{-1}\frac{\sqrt{3}}{2}\right)$

81. $y = \sin\left(2 \sin^{-1}\frac{4}{5}\right)$

82. $y = \cos(2 \tan^{-1}1)$

83. $\sin^{-1}x = \cos^{-1}\frac{5}{13}$

84. $\tan^{-1}x = \sin^{-1}\frac{24}{25}$

85. $\sin^{-1}\left(y - 1\right) = \frac{\pi}{2}$

86. $\cos^{-1}\left(y - \frac{1}{2}\right) = \frac{\pi}{3}$

87. $\tan^{-1}\left(y + \frac{\sqrt{2}}{2}\right) = \frac{\pi}{4}$

88. $\sin^{-1}\left(y - 2\right) = -\frac{\pi}{6}$

89. $y = \sin\left(\sin^{-1}\frac{2}{3} + \cos^{-1}\frac{1}{2}\right)$

90. $y = \cos\left(\sin^{-1}\frac{3}{4} + \cos^{-1}\frac{5}{13}\right)$

91. $y = \tan\left(\cos^{-1}\frac{1}{2} - \sin^{-1}\frac{3}{4}\right)$

92. $y = \sec\left(\cos^{-1}\frac{2}{3} + \sin^{-1}\frac{2}{3}\right)$

93. $\sin^{-1}\frac{3}{5} + \cos^{-1}x = \frac{\pi}{4}$

94. $\sin^{-1}x + \cos^{-1}\frac{4}{5} = \frac{\pi}{6}$

95. $\sin^{-1}x - \cos^{-1}\frac{\sqrt{2}}{2} = \frac{2\pi}{3}$

96. $\cos^{-1}x + \sin^{-1}\frac{\sqrt{3}}{2} = \frac{\pi}{4}$

In Exercises 97 to 100, evaluate each expression.

97. $y = \cos(\sin^{-1}x)$

98. $y = \tan(\cos^{-1}x)$

99. $y = \sin(\sec^{-1}x)$

100. $y = \sec(\sin^{-1}x)$

In Exercises 101 to 104, verify the identity.

101. $\sin^{-1}x + \sin^{-1}(-x) = 0$

102. $\cos^{-1}x + \cos^{-1}(-x) = \pi$

103. $\tan^{-1}x + \tan^{-1}\frac{1}{x} = \frac{\pi}{2}$

104. $\sec^{-1}\frac{1}{x} + \csc^{-1}\frac{1}{x} = \frac{\pi}{2}$

In Exercises 105 to 116, graph the given inverse functions.

105. $y = 2 \sin^{-1}x$

106. $y = \frac{1}{2} \sin^{-1}x$

107. $y = 2 \cos^{-1}\frac{x}{2}$

108. $y = \frac{1}{2} \cos^{-1}2x$

109. $y = 2 \sin^{-1}(x - 2)$

110. $y = 3 \sin^{-1}(x + 1)$

111. $y = 2 \cos^{-1}(x + 3)$

112. $y = \frac{1}{3} \cos^{-1}(x - 2)$

113. $y = \csc^{-1}2x$

114. $y = 0.5 \sec^{-1}\frac{x}{2}$

115. $y = \sec^{-1}(x - 1)$

116. $y = \sec^{-1}(x + \pi)$

Supplemental Exercises

In Exercises 117 to 120, show that the identities are true.

117. $\cos(\sin^{-1}x) = \sqrt{1 - x^2}$

118. $\sec(\sin^{-1}x) = \frac{\sqrt{1 - x^2}}{1 - x^2}$

119. $\tan(\csc^{-1}x) = \frac{\sqrt{x^2 - 1}}{x^2 - 1}$

120. $\sin(\cot^{-1}x) = \frac{\sqrt{x^2 - 1}}{x^2 + 1}$

In Exercises 121 to 124, solve for y in terms of x.

121. $5x = \tan^{-1}3y$

122. $2x = \frac{1}{2}\sin^{-1}2y$

123. $x - \frac{\pi}{3} = \cos^{-1}(y - 3)$

124. $x + \frac{\pi}{2} = \tan^{-1}(2y - 1)$

In Exercises 125 to 128, graph the given inverse function.

125. $y = 2\tan^{-1} 2x$

126. $y = \tan^{-1}(x - 1)$

127. $y = \cot^{-1} \frac{x}{3}$

128. $y = 2\cot^{-1}(x - 1)$

In Exercises 129 and 130, use the following formula. In dot-matrix printing, the *blank-area factor* is the ratio of the blank area (unprinted area) to the total area of the line. If circular dots are used to print, then the blank-area factor is given by

$$\frac{A}{SD} = 1 - \frac{1}{2}\left[1 - \left(\frac{S}{D}\right)^2 + \frac{D}{S}\sin^{-1}\left(\frac{S}{D}\right)\right]$$

where $A = A_1 + A_2$, with A_1 and A_2 shown in the figure, S equals the distance between centers of overlapping dots, and D is the diameter of a dot.

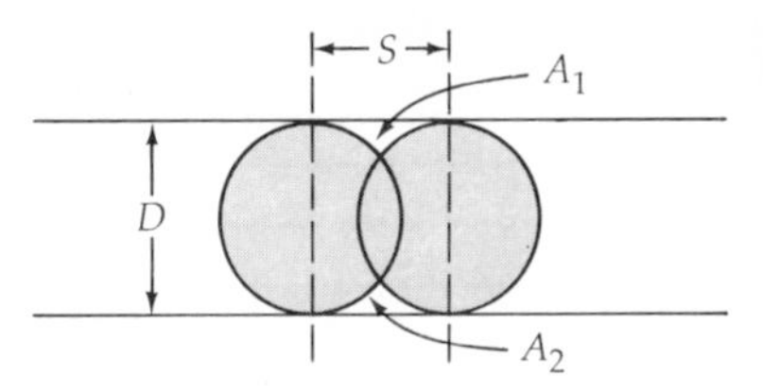

129. Calculate the blank-area factor where $D = 0.2$ millimeters and $S = 0.1$ millimeters.

130. Calculate the blank-area factor where $D = 0.16$ millimeters and $S = 0.1$ millimeters.

3.6 Trigonometric Equations

Consider the equation $\sin x = 1/2$. The graph of $y = \sin x$ along with the line $y = 1/2$ is shown in Figure 3.13. The intersections of the two graphs are the solutions of $\sin x = 1/2$. The solutions in the interval $0 \le x < 2\pi$ are $x = \pi/6$ and $5\pi/6$.

If we remove the restriction $0 \le x < 2\pi$, there are many more possible solutions. Because the sine function is periodic with a period of 2π, other solutions are obtained by adding $2k\pi$, k an integer, to either of the above solutions. Thus, the solutions of $\sin x = 1/2$ are

$$x = \frac{\pi}{6} + 2k\pi, \quad k \text{ is an integer,}$$

$$x = \frac{5\pi}{6} + 2k\pi, \quad k \text{ is an integer.}$$

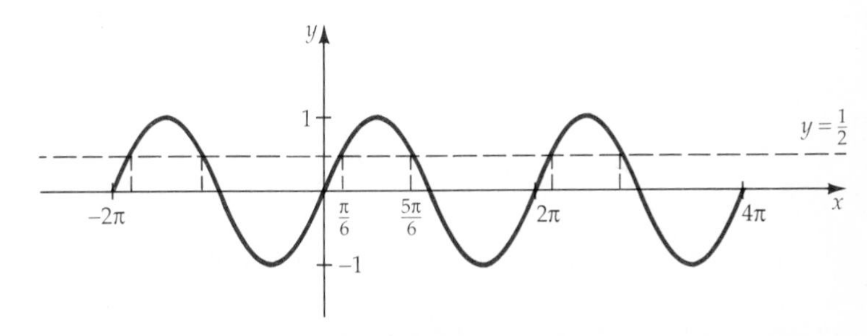

Figure 3.13

The identity for the sine of the sum of two angles can be used to check the above solutions.

$$\sin\left(\frac{\pi}{6} + 2k\pi\right) = \sin\frac{\pi}{6}\cos 2k\pi + \cos\frac{\pi}{6}\sin 2k\pi$$

$$= \sin\frac{\pi}{6}(1) + \cos\frac{\pi}{6}(0) = \sin\frac{\pi}{6} = \frac{1}{2}$$

The same check will show that $x = 5\pi/6 + 2k\pi$ are also solutions.

Algebraic methods and trigonometric identities are used frequently to find the solutions of trigonometric equations. Algebraic methods frequently used are solving by factoring, solving by using the quadratic formula, and squaring each side of the equation.

Remark Squaring both sides of an equation may not produce an equivalent equation. Thus, when this method is used, the proposed solutions must be checked to eliminate the extraneous solutions.

EXAMPLE 1 Solve a Trigonometric Equation by Factoring

Solve the equation $2\sin^2 x\cos x - \cos x = 0$, $0 \le x < 2\pi$.

Solution Factor the left side of the equation and set each factor equal to zero.

$$2\sin^2 x\cos x - \cos x = 0$$

$$\cos x(2\sin^2 x - 1) = 0$$

$$\cos x = 0 \quad \text{or} \quad 2\sin^2 x - 1 = 0$$

$$x = \frac{\pi}{2}, \frac{3\pi}{2} \qquad \sin^2 x = \frac{1}{2}$$

$$\sin x = \pm\frac{\sqrt{2}}{2}$$

$$x = \frac{\pi}{4}, \frac{3\pi}{4}, \frac{5\pi}{4}, \frac{7\pi}{4}$$

The solutions in the interval $0 \le x < 2\pi$ are $\frac{\pi}{4}, \frac{\pi}{2}, \frac{3\pi}{4}, \frac{5\pi}{4}, \frac{3\pi}{2}$, and $\frac{7\pi}{4}$.

■ *Try Exercise* **44,** *page 158.*

EXAMPLE 2 Solve a Trigonometric Equation

Solve the equation $\sin x + \cos x = 1$, $0 \le x < 2\pi$.

Solution Solve for $\sin x$, then square each side of the equation.

$$\sin x + \cos x = 1$$
$$\sin x = 1 - \cos x$$
$$\sin^2 x = (1 - \cos x)^2 = 1 - 2\cos x + \cos^2 x$$
$$1 - \cos^2 x = 1 - 2\cos x + \cos^2 x$$
$$2\cos^2 x - 2\cos x = 0$$

Factor the left side and set each factor equal to zero.

$$2\cos x(\cos x - 1) = 0$$
$$2\cos x = 0 \quad \text{or} \quad \cos x = 1$$
$$x = \frac{\pi}{2}, \frac{3\pi}{2} \qquad x = 0$$

Squaring each side of an equation may introduce an extraneous root. Therefore, we must check the solution. A check will show that 0 and $\frac{\pi}{2}$ are solutions, but $\frac{3\pi}{2}$ is not a solution.

■ *Try Exercise* **52,** *page 158.*

EXAMPLE 3 Solve a Trigonometric Equation by Using the Quadratic Formula

Solve the equation $3\cos^2 x - 5\cos x - 4 = 0$, $0° \le x < 360°$.

Solution This equation is quadratic in form and cannot be factored. However, we can use the quadratic formula to solve for $\cos x$.

$$3\cos^2 x - 5\cos x - 4 = 0 \qquad a = 3,\ b = -5,\ c = -4.$$
$$\cos x = \frac{-(-5) \pm \sqrt{(-5)^2 - 4(3)(-4)}}{(2)(3)}$$
$$= \frac{5 \pm \sqrt{73}}{6}$$

The equation $\cos x = \frac{5 - \sqrt{73}}{6}$ has no solution. Why?[3] Thus

$$\cos x = \frac{5 + \sqrt{73}}{6}$$
$$x \approx 126.2° \text{ or } 233.8°$$

■ *Try Exercise* **56,** *page 158.*

Caution The condition $0° \le x < 360°$ requires that you list both 126.2° and 233.8° as solutions of the equation.

[3] The range of the cosine function is $-1 \le \cos x \le 1$. The value $(5 + \sqrt{73})/6$ is not in the range of the function.

When solving equations containing multiple angles, use care to make sure all the solutions of the equation are found for the given interval. Consider the equation $\sin 2x = \frac{1}{2}$. We first solve for $2x$.

$$\sin 2x = \frac{1}{2}$$

$$2x = \frac{\pi}{6} + 2k\pi \quad \text{or} \quad 2x = \frac{5\pi}{6} + 2k\pi \quad k \text{ is an integer.}$$

Solving for x, we have $x = \pi/12 + k\pi$ or $x = 5\pi/12 + k\pi$. Substituting integers for k, we obtain

$$k = 0 \qquad x = \frac{\pi}{12} \quad \text{or} \quad x = \frac{5\pi}{12}$$

$$k = 1 \qquad x = \frac{13\pi}{12} \quad \text{or} \quad x = \frac{17\pi}{12}$$

$$k = 2 \qquad x = \frac{25\pi}{12} \quad \text{or} \quad x = \frac{29\pi}{12}$$

Note that for $k > 1$, $x > 2\pi$ and the solutions to $\sin 2x = \frac{1}{2}$ are not in the interval $0 \le x < 2\pi$. Thus for $0 \le x < 2\pi$ the solutions are $\pi/12$, $5\pi/12$, $13\pi/12$, and $17\pi/12$.

EXAMPLE 4 Solve a Trigonometric Equation

Solve the equation $\sin 3x = 1$.

Solution The equation

$$\sin 3x = 1$$

implies

$$3x = \frac{\pi}{2} + 2k\pi, \quad k \text{ an integer}$$

$$x = \frac{\pi}{6} + \frac{2k\pi}{3}, \quad k \text{ an integer}$$

■ *Try Exercise* **66,** *page 158.*

EXAMPLE 5 Solve a Trigonometric Equation

Solve the equation $\sin^2 2x - \frac{\sqrt{3}}{2} \sin 2x + \sin 2x - \frac{\sqrt{3}}{2} = 0$, $0° \le x < 360°$.

Solution The required solutions for x are in the interval $0° \le x < 360°$. The interval for $2x$ is two times as long.

$$0° \le x < 360°$$

$$0° \le 2x < 720°$$

Now factor the left side of the equation by grouping and then set each factor equal to zero.

$$\sin^2 2x - \frac{\sqrt{3}}{2}\sin 2x + \sin 2x - \frac{\sqrt{3}}{2} = 0$$

$$\sin 2x\left(\sin 2x - \frac{\sqrt{3}}{2}\right) + \left(\sin 2x - \frac{\sqrt{3}}{2}\right) = 0$$

$$(\sin 2x + 1)\left(\sin 2x - \frac{\sqrt{3}}{2}\right) = 0$$

$$\sin 2x + 1 = 0 \quad \text{or} \quad \sin 2x - \frac{\sqrt{3}}{2} = 0$$

$$\sin 2x = -1 \qquad\qquad \sin 2x = \frac{\sqrt{3}}{2}$$

The equation $\sin 2x = -1$ implies that $2x = 270° + 360° \cdot k$, k an integer. Thus $x = 135° + 180° \cdot k$. The solutions of this equation with $0 \leq x < 360°$ are 135° and 315°. Similarly, the solutions of the equation $\sin 2x = \frac{\sqrt{3}}{2}$ with $0 \leq x < 360°$ are

$$2x = 60°, 120°, 420°, 480°$$

$$x = 30°, 60°, 210°, 240°$$

The solutions are 30°, 60°, 135°, 210°, 240°, 315°.

■ *Try Exercise* **84**, *page 158.*

EXERCISE SET 3.6

In Exercises 1 to 22, solve the equation for all values in the interval $0 \leq x < 2\pi$.

1. $\sec x - \sqrt{2} = 0$
2. $2 \sin x = \sqrt{3}$
3. $\tan x - \sqrt{3} = 0$
4. $\cos x - 1 = 0$
5. $2 \sin x \cos x = \sqrt{2} \cos x$
6. $2 \sin x \cos x = \sqrt{3} \sin x$
7. $\sin^2 x - 1 = 0$
8. $\cos^2 x - 1 = 0$
9. $4 \sin x \cos x - 2\sqrt{3} \sin x - 2\sqrt{2} \cos x + \sqrt{6} = 0$
10. $\sec^2 x + \sqrt{3} \sec x - \sqrt{2} \sec x - \sqrt{6} = 0$
11. $\csc x - \sqrt{2} = 0$
12. $3 \cot x + \sqrt{3} = 0$
13. $2 \sin^2 x + 1 = 3 \sin x$
14. $2 \cos^2 x + 1 = -3 \cos x$
15. $4 \cos^2 x - 3 = 0$
16. $2 \sin^2 x - 1 = 0$
17. $2 \sin^3 x = \sin x$
18. $4 \cos^3 x = 3 \cos x$
19. $4 \sin^2 x + 2\sqrt{3} \sin x - \sqrt{3} = 2 \sin x$
20. $\tan^2 x + \tan x - \sqrt{3} = \sqrt{3} \tan x$
21. $\sin^4 x = \sin^2 x$
22. $\cos^4 x = \cos^2 x$

In Exercises 23 to 60, solve the following equations, where $0 \leq x < 360°$. Round to the nearest tenth of a degree.

23. $\cos x - 0.75 = 0$
24. $\sin x + 0.432 = 0$
25. $3 \sin x - 5 = 0$
26. $4 \cos x - 1 = 0$
27. $3 \sec x - 8 = 0$
28. $4 \csc x + 9 = 0$
29. $\cos x + 3 = 0$
30. $\sin x - 4 = 0$
31. $3 - 5 \sin x = 4 \sin x + 1$
32. $4 \cos x - 5 = \cos x - 3$
33. $\frac{1}{2} \sin x + \frac{2}{3} = \frac{3}{4} \sin x + \frac{3}{5}$
34. $\frac{2}{5} \cos x - \frac{1}{2} = \frac{1}{3} - \frac{1}{2} \cos x$

35. $3 \tan^2 x - 2 \tan x = 0$

36. $4 \cot^2 x + 3 \cot x = 0$

37. $3 \cos x + \sec x = 0$

38. $5 \sin x - \csc x = 0$

39. $\tan^2 x = 3 \sec^2 x - 2$

40. $\csc^2 x - 1 = 3 \cot^2 x + 2$

41. $2 \sin^2 x = 1 - \cos x$

42. $\cos^2 x + 4 = 2 \sin x - 3$

43. $3 \cos^2 x + 5 \cos x - 2 = 0$

44. $2 \sin^2 x + 5 \sin x + 3 = 0$

45. $2 \tan^2 x - \tan x - 10 = 0$

46. $2 \cot^2 x - 7 \cot x + 3 = 0$

47. $3 \sin x \cos x - \cos x = 0$

48. $\tan x \sin x - \sin x = 0$

49. $2 \sin x \cos x - \sin x - 2 \cos x + 1 = 0$

50. $6 \cos x \sin x - 3 \cos x - 4 \sin x + 2 = 0$

51. $2 \sin x - \cos x = 1$

52. $\sin x + 2 \cos x = 1$

53. $2 \sin x - 3 \cos x = 1$

54. $\sqrt{3} \sin x + \cos x = 1$

55. $3 \sin^2 x - \sin x - 1 = 0$

56. $2 \cos^2 x - 5 \cos x - 5 = 0$

57. $2 \cos x - 1 + 3 \sec x = 0$

58. $3 \sin x - 5 + \csc x = 0$

59. $\cos^2 x - 3 \sin x + 2 \sin^2 x = 0$

60. $\sin^2 x = 2 \cos x + 3 \cos^2 x$

In Exercises 61 to 70, solve the trigonometric equations.

61. $\tan 2x - 1 = 0$

62. $\sec 3x - \dfrac{2\sqrt{3}}{3} = 0$

63. $\sin 5x = 1$

64. $\cos 4x = -\dfrac{\sqrt{2}}{2}$

65. $\sin 2x - \sin x = 0$

66. $\cos 2x = -\dfrac{\sqrt{3}}{2}$

67. $\sin\left(2x + \dfrac{\pi}{6}\right) = -\dfrac{1}{2}$

68. $\cos\left(2x - \dfrac{\pi}{4}\right) = -\dfrac{\sqrt{2}}{2}$

69. $\sin^2 \dfrac{x}{2} + \cos x = 1$

70. $\cos^2 \dfrac{x}{2} - \cos x = 1$

In Exercises 71 to 84, solve each equation where $0 \le x < 2\pi$.

71. $\cos 2x = 1 - 3 \sin x$

72. $\cos 2x = 2 \cos x - 1$

73. $\sin 4x - \sin 2x = 0$

74. $\sin 4x - \cos 2x = 0$

75. $\tan \dfrac{x}{2} = \sin x$

76. $\tan \dfrac{x}{2} = 1 - \cos x$

77. $\sin 2x \cos x + \cos 2x \sin x = 0$

78. $\cos 2x \cos x - \sin 2x \sin x = 0$

79. $\sin x \cos 2x - \cos x \sin 2x = \dfrac{\sqrt{3}}{2}$

80. $\cos 2x \cos x + \sin 2x \sin x = -1$

81. $\sin 3x - \sin x = 0$

82. $\cos 3x + \cos x = 0$

83. $2 \sin x \cos x + 2 \sin x - \cos x - 1 = 0$

84. $2 \sin x \cos x - 2\sqrt{2} \sin x - \sqrt{3} \cos x + \sqrt{6} = 0$

Supplemental Exercises

In Exercises 85 to 94, solve the trigonometric equations for $0 \le x < 2\pi$.

85. $\sqrt{3} \sin x + \cos x = \sqrt{3}/2$

86. $\sin x - \cos x = 1$

87. $-\sin x + \sqrt{3} \cos x = \sqrt{3}$

88. $-\sqrt{3} \sin x - \cos x = 1$

89. $\cos 5x - \cos 3x = 0$

90. $\cos 5x - \cos x - \sin 3x = 0$

91. $\sin 3x + \sin x = 0$

92. $\sin 3x + \sin x - \sin 2x = 0$

93. $\cos 4x + \cos 2x = 0$

94. $\cos 4x + \cos 2x - \cos 3x = 0$

95. Find the area of the sector in terms of r and θ in the accompanying figure.

96. Find the area of triangle OAB in terms of r and θ in the figure.

97. Find the area of the shaded part of the figure. Find that area in terms of r and θ.

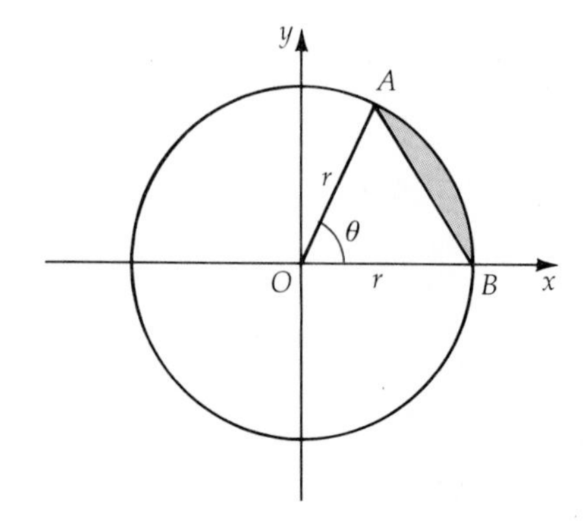

undamental Trigonometric Identities

metric identities are verified by using algebraic methods and pre-proven identities.

$$\sin x = \frac{1}{\csc x} \qquad \cos x = \frac{1}{\sec x} \qquad \tan x = \frac{1}{\cot x}$$

$$\tan x = \frac{\sin x}{\cos x} \qquad \cot x = \frac{\cos x}{\sin x}$$

$$x + \cos^2 x = 1 \qquad \tan^2 x + 1 = \sec^2 x \qquad 1 + \cot^2 x = \csc^2 x$$

n and Difference Identities

and difference identities for the cosine function are

$$\cos(\alpha - \beta) = \cos\alpha\cos\beta + \sin\alpha\sin\beta$$

$$\cos(\alpha + \beta) = \cos\alpha\cos\beta - \sin\alpha\sin\beta$$

Sum and difference identities for the sine function are

$$\sin(\alpha - \beta) = \sin\alpha\cos\beta - \cos\alpha\sin\beta$$

$$\sin(\alpha + \beta) = \sin\alpha\cos\beta + \cos\alpha\sin\beta$$

Sum and difference identities for the tangent function are

$$\tan(\alpha + \beta) = \frac{\tan\alpha + \tan\beta}{1 - \tan\alpha\tan\beta}$$

$$\tan(\alpha - \beta) = \frac{\tan\alpha - \tan\beta}{1 + \tan\alpha\tan\beta}$$

3.3 Double- and Half-Angle Identities

The double-angle identities are

$$\sin 2\alpha = 2\sin\alpha\cos\alpha$$

$$\cos 2\alpha = \cos^2\alpha - \sin^2\alpha$$

$$= 1 - 2\sin^2\alpha$$

$$= 2\cos^2\alpha - 1$$

$$\tan 2\alpha = \frac{2\tan\alpha}{1 - \tan^2\alpha}$$

The half-angle identities are

$$\sin\frac{\alpha}{2} = \pm\sqrt{\frac{1-\cos\alpha}{2}} \qquad \tan\frac{\alpha}{2} = \frac{\sin\alpha}{1+\cos\alpha}$$

$$\cos\frac{\alpha}{2} = \pm\sqrt{\frac{1+\cos\alpha}{2}} \qquad = \frac{1-\cos\alpha}{\sin\alpha}$$

3.4 Identities Involving the Sums of Trigonometric Functions

The product-to-sum identities are

$$\sin\alpha\cos\beta = \frac{1}{2}[\sin(\alpha+\beta) + \sin(\alpha-\beta)]$$

$$\cos\alpha\sin\beta = \frac{1}{2}[\sin(\alpha+\beta) - \sin(\alpha-\beta)]$$

$$\cos\alpha\cos\beta = \frac{1}{2}[\cos(\alpha+\beta) + \cos(\alpha-\beta)]$$

$$\sin\alpha\sin\beta = \frac{1}{2}[\cos(\alpha-\beta) - \cos(\alpha+\beta)]$$

The sum-to-product identities are

$$\sin x + \sin y = 2\sin\frac{x+y}{2}\cos\frac{x-y}{2}$$

$$\cos x - \cos y = -2\sin\frac{x+y}{2}\sin\frac{x-y}{2}$$

$$\sin x - \sin y = 2\cos\frac{x+y}{2}\sin\frac{x-y}{2}$$

$$\cos x + \cos y = 2\cos\frac{x+y}{2}\cos\frac{x-y}{2}$$

3.5 Inverse Trigonometric Functions

The inverse of $y = \sin t$ is $y = \sin^{-1}t$, with $-1 \le t \le 1$ and $-\pi/2 \le y \le \pi/2$.

The inverse of $y = \cos t$ is $y = \cos^{-1}t$, with $-1 \le t \le 1$ and $0 \le y \le \pi$.

The inverse of $y = \tan t$ is $y = \tan^{-1}t$, with $-\infty < t < \infty$ and $-\pi/2 < y < \pi/2$.

The inverse of $y = \cot t$ is $y = \cot^{-1}t$, with $-\infty < t < \infty$ and $0 < y < \pi$.

The inverse of $y = \csc t$ is $y = \csc^{-1}t$, with $t \le -1$ or $t \ge 1$ and $-\pi/2 \le y \le \pi/2$, $y \ne 0$.

The inverse of $y = \sec t$ is $y = \sec^{-1}t$, with $t \le -1$ or $t \ge 1$ and $0 \le y \le \pi$, $y \ne \pi/2$.

3.6 Trigonometric Equations

Algebraic methods and the use of identities are used to solve trigonometric equations. Since the trigonometric functions are periodic, there may be an infinite number of solutions.

CHALLENGE EXERCISES

In Exercises 1 to 12, answer true or false. If the statement is false, give an example.

1. $\dfrac{\tan \alpha}{\tan \beta} = \dfrac{\alpha}{\beta}$

2. $\dfrac{\sin x}{\cos y} = \tan \dfrac{x}{y}$

3. $\sin^{-1} x = \csc x$

4. $\sin 2\alpha = 2 \sin \alpha$ for all α

5. $\sin(\alpha + \beta) = \sin \alpha + \sin \beta$

6. An equation that has an infinite number of solutions is an identity.

7. If $\tan \alpha = \tan \beta$, then $\alpha = \beta$.

8. $\cos^{-1}(\cos x) = x$

9. $\cos(\cos^{-1} x) = x$

10. $\csc^{-1} \dfrac{1}{\alpha} = \dfrac{1}{\csc \alpha}$

11. If $0° \le \theta \le 90°$, then $\cos \theta = \sin(180° - \theta)$

12. $\sin^2 \theta = \sin \theta^2$

REVIEW EXERCISES

In Exercises 1 to 6, find the exact value of the given function.

1. $\cos(45° + 30°)$

2. $\tan(210° - 45°)$

3. $\sin\left(\dfrac{2\pi}{3} + \dfrac{\pi}{4}\right)$

4. $\sec\left(\dfrac{4\pi}{3} - \dfrac{\pi}{4}\right)$

5. $\sin(60° - 135°)$

6. $\cos\left(\dfrac{5\pi}{3} - \dfrac{7\pi}{4}\right)$

In Exercises 7 to 10, evaluate the functions by using the half-angle identities.

7. $\sin\left(22\dfrac{1}{2}\right)^{\circ}$

8. $\cos 105°$

9. $\tan\left(67\dfrac{1}{2}\right)^{\circ}$

10. $\sin 112.5°$

In Exercises 11 to 14, find the exact value of the given functions.

11. Given $\sin \alpha = \frac{1}{2}$, α in quadrant I and $\cos \beta = \frac{1}{2}$, β in quadrant IV, find a. $\cos(\alpha - \beta)$, b. $\tan 2\alpha$, c. $\sin \beta/2$.

12. Given $\sin \alpha = \frac{\sqrt{3}}{2}$, α in quadrant II and $\cos \beta = -\frac{1}{2}$, β in quadrant III, find a. $\sin(\alpha + \beta)$, b. $\sec 2\beta$ c. $\cos \frac{\alpha}{2}$.

13. Given $\sin \alpha = -\frac{1}{2}$, α in quadrant IV and $\cos \beta = -\frac{\sqrt{3}}{2}$, β in quadrant III, find a. $\sin(\alpha - \beta)$, b. $\tan 2\alpha$, c. $\cos \beta/2$.

14. Given $\sin \alpha = \frac{\sqrt{2}}{2}$, α in quadrant I and $\cos \beta = \frac{\sqrt{3}}{2}$, β in quadrant IV, find a. $\cos(\alpha - \beta)$, b. $\tan 2\beta$, c. $\sin 2\alpha$.

In Exercises 15 to 20, write the given expression as a single trigonometric function.

15. $2 \sin 3x \cos 3x$

16. $\dfrac{\tan 2x + \tan x}{1 - \tan 2x \tan x}$

17. $\sin 4x \cos x - \cos 4x \sin x$

18. $\cos^2 2\theta - \sin^2 2\theta$

19. $1 - 2 \sin^2 \dfrac{\beta}{2}$

20. $\pm\sqrt{\dfrac{1 - \cos 4\theta}{2}}$

In Exercises 21 to 24, evaluate each expression.

21. $\sin 47° \sin 22°$

22. $\cos 14° \cos 92°$

23. $2 \sin \dfrac{\pi}{3} \cos \dfrac{2\pi}{3}$

24. $2 \cos \dfrac{\pi}{4} \cos \dfrac{3\pi}{2}$

In Exercises 25 to 28, write each expression as the product of two functions.

25. $\cos 2\theta - \cos 4\theta$

26. $\sin 3\theta - \sin 5\theta$

27. $\sin 6\theta + \sin 2\theta$

28. $\sin 5\theta - \sin \theta$

In Exercises 29 to 46, verify the identities.

29. $\dfrac{1}{\sin x - 1} + \dfrac{1}{\sin x + 1} = -2 \tan x \sec x$

30. $\dfrac{\sin x}{1 - \cos x} = \csc x + \cot x, \quad 0 < x < \dfrac{\pi}{2}$

31. $\dfrac{1 + \sin x}{\cos^2 x} = \tan^2 x + 1 + \tan x \sec x$

32. $\cos^2 x - \sin^2 x - \sin 2x = \dfrac{\cos^2 2x - \sin^2 2x}{\cos 2x + \sin 2x}$

33. $\dfrac{1}{\cos x} - \cos x = \tan x \sec x$

34. $\sin(270° - \theta) - \cos(270° - \theta) = \sin \theta - \cos \theta$

35. $\sin\left(\dfrac{\pi}{4} - \alpha\right) = \dfrac{\sqrt{2}}{2}(\cos \alpha - \sin \alpha)$

36. $\sin(180° - \alpha + \beta) = \sin \alpha \cos \beta - \cos \alpha \sin \beta$

37. $\dfrac{\sin 4x - \sin 2x}{\cos 4x - \cos 2x} = -\cot 3x$

38. $2 \sin x \sin 3x = (1 - \cos 2x)(1 + 2 \cos 2x)$

39. $\sin x - \cos 2x = (2 \sin x - 1)(\sin x + 1)$

40. $\cos 4x = 1 - 8 \sin^2 x + 8 \sin^4 x$

41. $\tan 4x = \dfrac{4 \tan x - 4 \tan^3 x}{1 - 6 \tan^2 x + \tan^4 x}$

42. $\dfrac{\sin 2x - \sin x}{\cos 2x + \cos x} = \dfrac{1 - \cos x}{\sin x}$

43. $2 \cos 4x \sin 2x = 2 \sin 3x \cos 3x - 2 \sin x \cos x$

44. $2 \sin x \sin 2x = 4 \cos x \sin^2 x$

45. $\cos(x + y) \cos(x - y) = \cos^2 x + \cos^2 y - 1$

46. $\cos(x + y) \sin(x - y) = \sin x \cos x - \sin y \cos y$

In Exercises 47 to 52, solve the equation.

47. $y = \sec\left(\sin^{-1} \dfrac{12}{13}\right)$

48. $y = \cos\left(\sin^{-1} \dfrac{3}{5}\right)$

49. $2 \sin^{-1}(x - 1) = \dfrac{\pi}{3}$

50. $y = \cos\left(\sin^{-1}\left(-\dfrac{3}{5}\right) + \cos^{-1} \dfrac{5}{13}\right)$

51. $\sin^{-1} x - \cos^{-1} \dfrac{4}{5} = \dfrac{\pi}{2}$

52. $y = \cos\left[2 \sin^{-1}\left(\dfrac{3}{5}\right)\right]$

In Exercises 53 to 58, solve the equations for $0° \le x < 360°$.

53. $4 \sin^2 x + 2\sqrt{3} \sin x - 2 \sin x - \sqrt{3} = 0$

54. $2 \sin x \cos x - \sqrt{2} \cos x - 2 \sin x + \sqrt{2} = 0$

In Exercises 55 and 56, solve the trigonometric equation.

55. $3 \cos^2 x + \sin x = 1$

56. $\tan^2 x - 2 \tan x - 3 = 0$

In Exercises 57 and 58, solve the equation for $0 \le x < 2\pi$.

57. $\sin 3x \cos x - \cos 3x \sin x = \dfrac{1}{2}$

58. $\cos\left(2x - \dfrac{\pi}{3}\right) = -\dfrac{\sqrt{3}}{2}$

In Exercises 59 to 62, find the amplitude and phase shift of each function. Graph each function.

59. $f(x) = \sqrt{3} \sin x + \cos x$

60. $f(x) = -2 \sin x - 2 \cos x$

61. $f(x) = -\sin x - \sqrt{3} \cos x$

62. $f(x) = \dfrac{\sqrt{3}}{2} \sin x - \dfrac{1}{2} \cos x$

In Exercises 63 to 66, graph each function.

63. $f(x) = 2 \cos^{-1} x$

64. $f(x) = \sin^{-1}(x - 1)$

65. $f(x) = \sin^{-1} \dfrac{x}{2}$

66. $f(x) = \sec^{-1} 2x$

4 Applications of Trigonometry

When you can measure what you are talking about and express it in numbers, you know something about it.

LORD KELVIN (1824–1907)

THE ELECTROMAGNETIC SPECTRUM

Electromagnetic waves are generated by oscillations of an electrically charged particle. All electromagnetic waves travel through space at approximately 186,000 miles per second or 300,000 kilometers per second. This speed is frequently referred to as the speed of light.

Visible light however, comprises only a small portion of what is called the *electromagnetic spectrum.* The electromagnetic spectrum consists of bands of electromagnetic waves of different wavelengths. The major types of waves are gamma rays, x-rays, ultraviolet light, visible light, infrared rays, microwaves, and radio waves. Gamma rays have the shortest wavelength and radio waves have the longest wavelength.

The electromagnetic spectrum is shown below. The lengths of the waves are given in meters, and the frequency of the wave is given in terms of hertz. One hertz is 1 wave cycle per second. No sharp line can be drawn between various portions of the spectrum. The only difference between the waves is the wavelength or frequency of the wave.

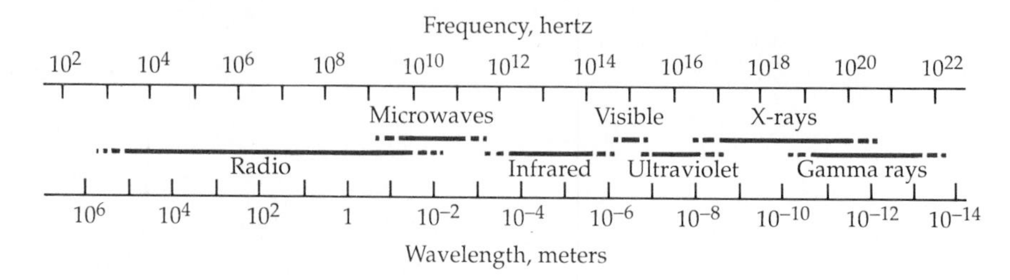

The Electromagnetic Spectrum

Notice that the wavelength of visible light is somewhere between 10^{-7} and 10^{-6} meters. The spectrum of visible light ranges over wavelengths of 4×10^{-7} to 7×10^{-7} meters.

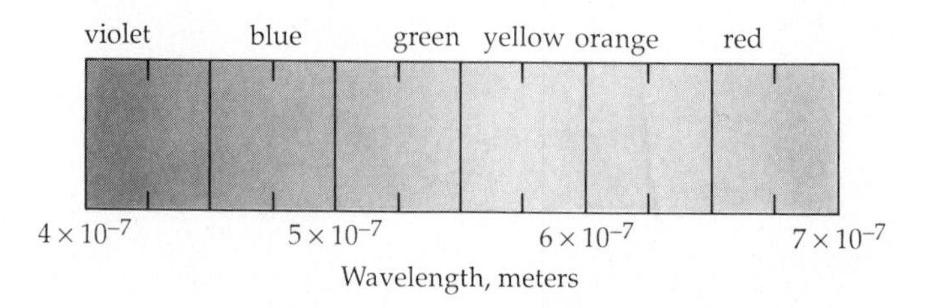

The Visible Light Spectrum

The frequencies of the waves that are broadcast on the AM (Amplitude Modulated) band of a radio range approximately from 530,000 hertz to 1,600,000 hertz. Setting the radio dial to 760 means that you are tuning your radio to receive radio waves that are broadcast at 760,000 hertz. The frequencies in the FM (Frequency Modulated) band range from 88,000,000 hertz to 108,000,000 hertz.

4.1 Right Triangles

A right triangle is a triangle with a right angle and two acute angles that are complementary. The sides opposite the acute angles are called the legs and the side opposite the right angle is called the hypotenuse. The angles in a right triangle are usually labeled with the capital letters A, B, and C with C being reserved for the right angle. The side opposite angle A is a, b is opposite angle B, and c is the hypotenuse.

Solving a triangle involves finding the lengths of all sides and the measures of all angles in a triangle. A right triangle can be solved when an acute angle and one side of the triangle are known, or when two sides are known. The Pythagorean Theorem and the trigonometric functions are used to solve a right triangle.

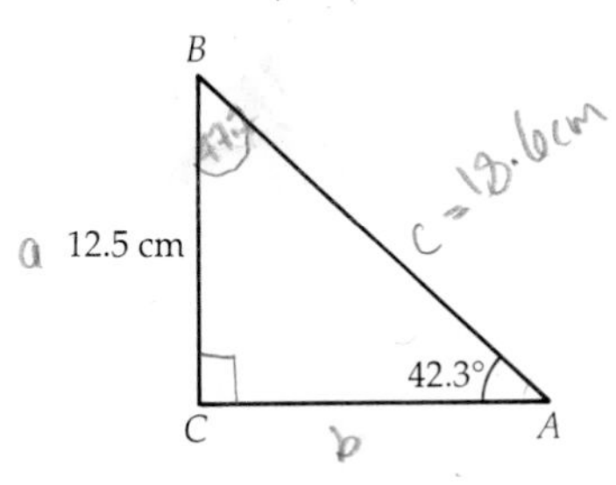

Figure 4.1

The right triangle in Figure 4.1 has an acute angle of 42.3° and the side opposite the 42.3° angle is 12.5 centimeters. To solve the triangle, find the unknown parts: angle B, the hypotenuse (c) and the unknown leg (b). Angle A and angle B are complementary, that is their sum is 90°. Thus, we have

$$42.3^\circ + B = 90^\circ$$

$$B = 47.7^\circ$$

Since angle A and the side opposite A are known, we can use the sine function to find the length of the hypotenuse:

$$\sin 42.3^\circ = \frac{12.5}{c}$$

$$c = \frac{12.5}{\sin 42.3^\circ} \approx 18.6$$

To find the length of side b, we use the tangent function.

$$\tan 42.3^\circ = \frac{12.5}{b}$$

$$b = \frac{12.5}{\tan 42.3^\circ} \approx 13.7$$

Angle $B = 47.7^\circ$, hypotenuse $c \approx 18.6$ cm, and side $b \approx 13.7$ cm.

If a calculator were used for the calculations above, it would show that the calculator displays more digits than the problem warrants. The precision of the calculator is misleading given that the original values were measured in tenths of a centimeter and tenths of a degree. The following rounding convention will be used when solving triangles.

Angle Measure to the Nearest	Significant Digits of the Lengths
Degree	Two
Tenths of a degree	Three
Hundredths of a degree	Four

For example, because the angle of 42.3° in Figure 4.1 is measured to the nearest tenth of a degree, c and b are rounded to three significant digits. To minimize the error in rounding, the intermediate steps in a problem will not be rounded. Only round the final result using the rounding convention.

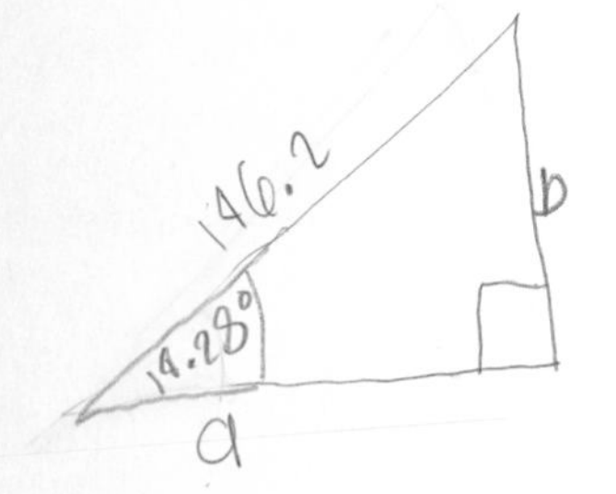

EXAMPLE 1 Solve a Right Triangle

Solve the right triangle with an acute angle of 14.28° and a hypotenuse of 146.2 feet.

Solution Since A and B are complementary angles,

$$A = 90.00° - 14.28° = 75.72°$$

$$\sin 14.28° = \frac{b}{146.2} \quad \text{The sine function is used to find side } b.$$

$$b = 146.2(\sin 14.28°)$$

$$\approx 36.06 \text{ ft}$$

Now find a.

$$\cos 14.28° = \frac{a}{146.2}$$

$$a = 146.2(\cos 14.28°) \approx 141.7 \text{ ft}$$

Angle $B = 75.72°$, $b \approx 36.06$ ft, and $a \approx 141.7$ ft.

■ *Try Exercise* **14**, *page 168.*

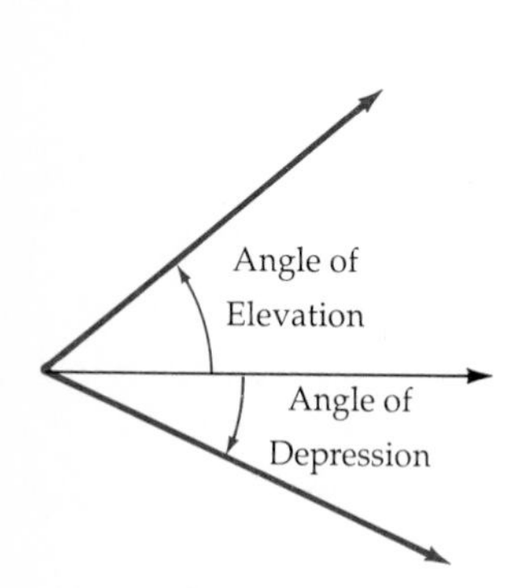

Figure 4.2

In some application problems, a horizontal line is used as a reference. An angle measured above the line to a line of sight is called an **angle of elevation,** and the angle measured below the line to a line of sight is called an **angle of depression.** See Figure 4.2.

EXAMPLE 2 Solve an Application Involving Angle of Elevation

The measure of the angle of elevation from a position 62 feet from the base of the flagpole to the top of the flagpole is 34°. Find the height of the flagpole.

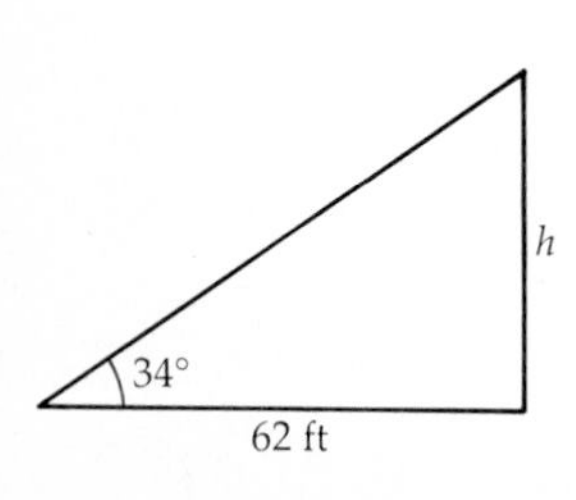

Figure 4.3

Solution Sketch the figure representing the given conditions. Because the length of the side opposite the angle is unknown and the measure of the adjacent angle is known, the tangent function is used.

$$\tan 34° = \frac{h}{62}$$

$$h = 62 \tan 34° \approx 42 \text{ ft.}$$

■ *Try Exercise* **28**, *page 168.*

Two terms are used in navigation and surveying to represent direction: heading and bearing. **Heading** is the measure of an angle clockwise from north. Figure 4.4 shows a heading of 45° and a heading of 225°. **Bearing** is the measure of the acute angle formed by a north-south line and the line of direction. Figure 4.5 shows a bearing of N38°W and a bearing of S15°E.

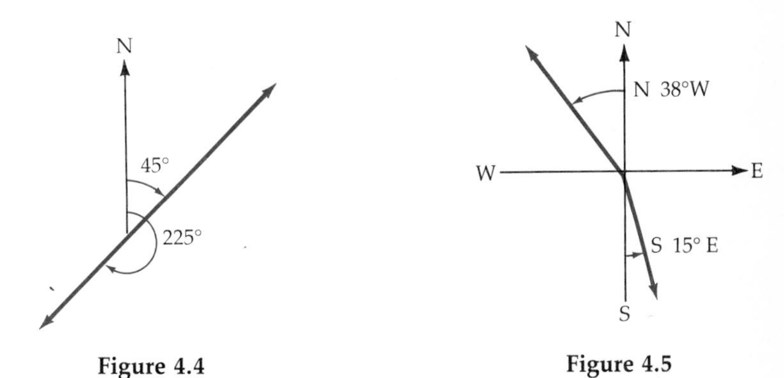

Figure 4.4 **Figure 4.5**

EXAMPLE 3 Solve an Application Involving Bearing

A motorist drove 40 miles at a bearing of S25°E. The motorist then drove 32 miles at a bearing S65°W. Find the distance of the motorist from the starting point.

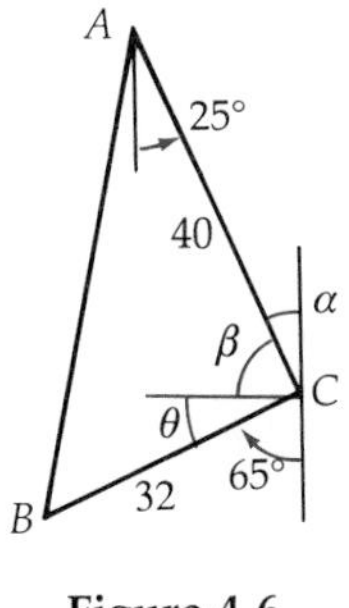

Solution Sketch a diagram showing the information given. From the figure, we see that angle α is 25° because alternate interior angles of parallel lines are equal. The angles α and β are complementary; thus β is 65°. Angle θ is complementary to 65°. Thus $\theta = 25°$.

$\beta + \theta = 65° + 25° = 90°$. Thus, angle ACB is a right angle. Use the Pythagorean Theorem to solve the right triangle for the hypotenuse.

$$c = \sqrt{a^2 + b^2} = \sqrt{32^2 + 40^2} \qquad a = 32,\ b = 40$$

$$= \sqrt{1024 + 1600}$$

$$= \sqrt{2624} \approx 51 \text{ mi} \qquad \text{Round to two significant digits.}$$

Figure 4.6

■ *Try Exercise* **32,** *page 168.*

EXAMPLE 4 Solve an Application Involving Heading

A car travels at 60 mph for 1 hour at a heading of 315°. The car then travels at 45 mph for 2 hours at a heading of 225°. At the end of 3 hours, how far is the car from its starting point?

Solution Sketch a diagram. Triangle ABC can be shown to be a right triangle. Angle $\theta = 45°$ by subtracting 315° from 360°. Angle $\alpha = 45°$ by alternate interior angles of parallel lines. Angle $\beta = 45°$ by subtracting 180°

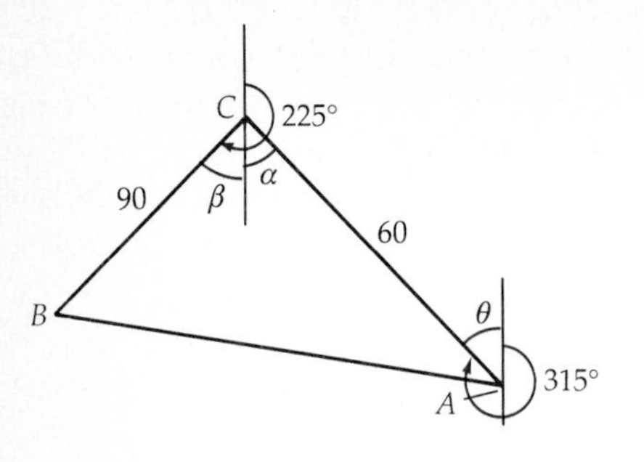

Figure 4.7

from 225°. Therefore angle C is 90° and triangle ABC is a right triangle.

$b = 60$ Traveling 60 mph for 1 hour

$a = 90$ Traveling 45 mph for 2 hours

Use the Pythagorean Theorem to find c.

$$c = \sqrt{60^2 + 90^2} = \sqrt{3600 + 8100}$$

$$= \sqrt{11,700} \approx 110 \text{ mi}$$ Round to two significant digits.

■ *Try Exercise* **34**, *page 168.*

EXERCISE SET 4.1

In Exercises 1 to 8, find the indicated part of the right triangles.

1. $a = 12$, $b = 15$; find angle B

2. $a = 40$, $b = 15$; find angle A

3. $a = 12$, $c = 30$; find angle A

4. $b = 25$, $c = 40$; find angle A

5. $b = 623$, $B = 23.4°$; find c

6. $a = 185$, $B = 72.7°$; find b

7. $a = 4.08$, $B = 40.4°$; find b

8. $b = 10.2$, $A = 50.5°$; find c

In Exercises 9 to 16, solve the given right triangle.

9. $A = 56°$, $a = 120$

10. $a = 340$, $b = 520$

11. $A = 23.8°$, $b = 125$

12. $B = 78.4°$, $a = 12.8$

13. $B = 23.47°$, $c = 1248$

14. $A = 45.89°$, $b = 1.228$

15. $a = 32.35$, $b = 12.67$

16. $a = 12.38$, $b = 4.367$

17. Find the leg opposite the 60° angle in a 30°-60° right triangle with a hypotenuse of 16 meters.

18. Find the leg of a 45°-45° right triangle with a hypotenuse of 60 centimeters.

19. The legs of a right triangle are $a = 4.0$ and $b = 8.0$. Find the six trigonometric functions of angle A.

20. The hypotenuse of a right triangle is 80, and one leg is 50. Find the six trigonometric functions of the angle opposite the given leg.

21. The sine of angle A of a right triangle is $\frac{3}{5}$. Find the cosine of angle B.

22. The sine of angle A of a right triangle is $\frac{2}{3}$. Find the cosecant of angle A.

23. The cotangent of angle B of a right triangle is $\frac{5}{8}$. Find the tangent of angle A.

24. The cosecant of angle A of a right triangle is 2. Find the sine of angle A.

25. An angle of a right triangle is 42° and the length of the side opposite the angle is 10 inches. Solve the triangle.

26. The length of the two sides of a right triangle are 8.0 inches and 12 inches. Solve the triangle.

27. The length of the hypotenuse of a right triangle is 6.0 feet, and one leg is 4.5 feet. Solve the triangle.

28. A telephone pole casts a shadow of 5.2 meters. Find the height of the telephone pole if the angle of elevation from the tip of the shadow to the top of the pole is 70°.

29. A flagpole casts a shadow of 18 feet. Find the height of the flagpole if the angle of elevation from the tip of the shadow to the top of the pole is 44°.

30. The angle of depression from a hotel room 82 feet from the ground to the base of a nearby building is 17°. Find the distance between the two buildings.

31. The angle of depression of one side of a lake measured from a balloon 2500 feet high is 43°. The angle of depression to the opposite side of the lake is 27°. Find the width of the lake.

32. A ship is 453 yards from a lighthouse and has a bearing of S33.8°W. A second ship is 1520 yards from the lighthouse at a bearing of S56.2°E. Find the distance between the two ships.

33. Two lookout stations are located 5.25 miles apart on an east-west line. A fire was observed from one lookout at a heading of 63.4° and from the other lookout at a heading of 296.6°. Find the distance of the fire from each lookout station.

34. A plane leaves an airport traveling at 215 mph at a heading of 65.4° at the same time another plane leaves the airport traveling at 480 mph at a heading of 335.4°. Find the distance between the two planes at the end of 2 hours.

35. A car goes 34 miles at a bearing of S70°E and then turns and travels 18 miles at a bearing of S20°W. Find the distance from the starting point.

Supplemental Exercises

36. Two buildings are 240 feet apart. The angle of elevation from the top of the shorter building to the top of the other building is 22°. If the shorter building is 80 feet high, how high is the taller building?

37. A circle is inscribed in a regular hexagon with a side of 6.0 meters. Find the radius of the circle.

38. The angle of elevation to the top of a radio antenna on the top of a building is 53.4°. After moving 200 feet closer to the building, the angle of elevation is 64.3°. Find the height of the building if the height of the antenna is 180 feet.

39. The angle of elevation to the top of a building is 32°. After moving 50 feet closer to the building, the angle of elevation is 53°. Find the height of the building.

40. An airplane is flying at an altitude of 30,000 feet. The pilot sees one side of a canyon at an angle of depression of 33.5°. The other side of the canyon is at an angle of depression of 27.4°. Find the width of the canyon accurate to three significant digits.

41. An airplane traveling 240 mph descends 42.0 feet in 1 second. Find the angle of descent.

42. An airplane flies directly overhead at 32,000 feet. Twenty seconds later the plane is observed at an angle of elevation of 67°. Find the velocity of the plane in miles per hour.

43. A submarine traveling at 9.0 mph is diving at an angle of depression of 5°. How long does it take the submarine to reach a depth of 80 feet?

44. Let $y = mx + b$, $m \neq 0$, be the equation of a line. Show that $\tan \alpha = m$, where m is the slope of the straight line and α is the angle made by the line and the positive x-axis.

4.2 The Law of Sines

An oblique triangle is a triangle that does not have a right angle. The *Law of Sines* can be used to solve oblique triangles when the following information is given:

1. Two angles and a side
2. Two sides and an angle opposite one of the given sides

In Figure 4.8, the altitude CD is drawn from C perpendicular to the opposite side. The length of the altitude is h. Triangles ACD and BCD are right triangles.

The sines of the angles A and B are

$$\sin A = \frac{h}{b} \qquad \sin B = \frac{h}{a}$$

$$h = b \sin A \qquad h = a \sin B.$$

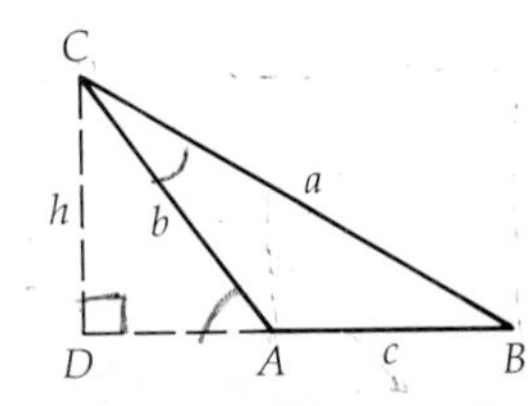

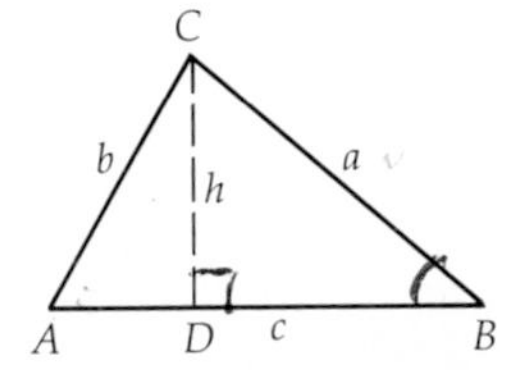

Figure 4.8

Thus $$b \sin A = a \sin B,$$

and by dividing each side of the equation by $\sin A \sin B$, we obtain

$$\frac{b}{\sin B} = \frac{a}{\sin A}.$$

Similarly, when the perpendicular is drawn to another side, the following formulas result:

$$\frac{c}{\sin C} = \frac{b}{\sin B} \quad \text{and} \quad \frac{c}{\sin C} = \frac{a}{\sin A}.$$

The Law of Sines

> If A, B, and C are the measures of the angles of a triangle and a, b, and c are the lengths of the sides opposite these angles, then
>
> $$\frac{a}{\sin A} = \frac{b}{\sin B} = \frac{c}{\sin C}.$$

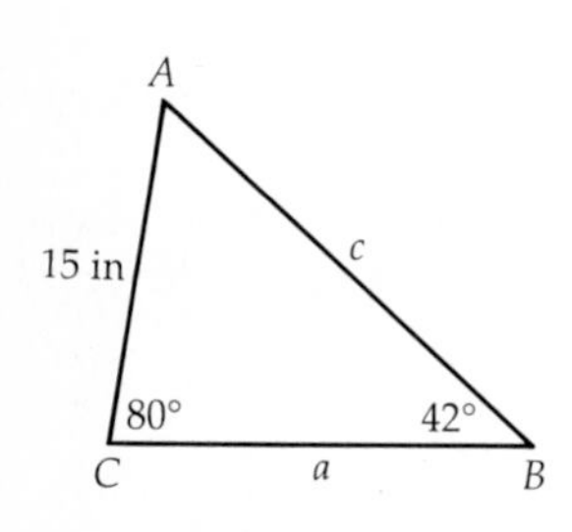

Figure 4.9

EXAMPLE 1 Solve a Triangle Using the Law of Sines

Solve triangle ABC if $B = 42°$, $C = 80°$, and $b = 15$ inches.

Solution The sum of the angles in a triangle equals 180°.

$$A + B + C = 180°$$

$$A + 80° + 42° = 180°$$

$$A = 58°$$

Since we know A, B, and b, we can use the Law of Sines to find side a.

$$\frac{b}{\sin B} = \frac{a}{\sin A}$$

$$\frac{15}{\sin 42°} = \frac{a}{\sin 58°}$$

$$a = \frac{15 \sin 58°}{\sin 42°} \approx 19 \text{ in.}$$

Now we use the Law of Sines to find side c.

$$\frac{b}{\sin B} = \frac{c}{\sin C}$$

$$\frac{15}{\sin 42°} = \frac{c}{\sin 80°}$$

$$c = \frac{15 \sin 80°}{\sin 42°} \approx 22 \text{ in.}$$

$A = 58°$, $a \approx 19$ in., $c \approx 22$ in.

■ *Try Exercise* **4**, *page 173.*

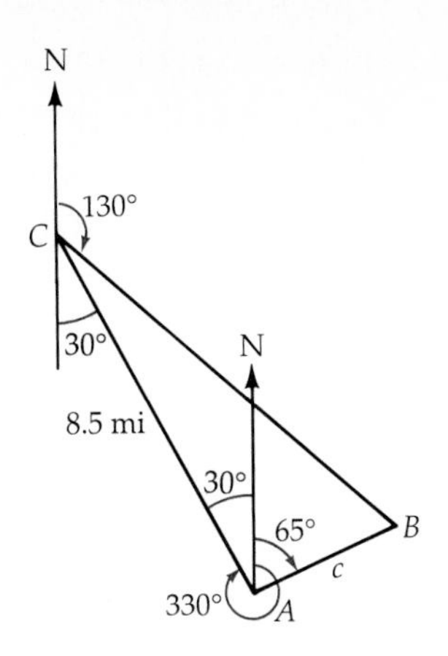

Figure 4.10

EXAMPLE 2 **Solve an Application Using the Law of Sines**

A ship with a heading of 330° first sighted a lighthouse at a heading of 65°. After traveling 8.5 miles, the ship observed the lighthouse at a heading of 130°. Find the distance to the lighthouse from the ship when the first sighting was made.

Solution From Figure 8.10 we see that angle $A = 65° + 30° = 95°$; angle $C = 180° - 30° - 130° = 20°$; angle $B = 180° - 95° - 20° = 65°$. Use the Law of Sines to find c.

$$\frac{b}{\sin B} = \frac{c}{\sin C}$$

$$\frac{8.5}{\sin 65°} = \frac{c}{\sin 20°}$$

$$c = \frac{8.5 \sin 20°}{\sin 65°} \approx 3.2$$

The lighthouse was 3.2 miles, to the nearest tenth of a mile, from the ship when the first sighting was made.

■ *Try Exercise* **20,** *page 174.*

Not all triangles with two sides and an angle opposite one of the given sides are unique. Some information may result in two triangles, and some may result in no triangle at all. The case of two sides and an angle opposite one of them is called the ambiguous case of the Law of Sines.

Suppose we are given sides a and c and the nonincluded angle A and asked to solve triangle ABC. First we determine h, the height of the triangle, which is found by dropping a perpendicular from B to side b. Note that since $\sin A = h/c$, we have $h = c \sin A$. Now we examine what happens for various values of a. We consider two cases:

Case 1 A is an acute angle. In Figure 4.11a, $a < h$, no triangle. In 4.11b, $a = h$; one triangle. In 4.11c, $h \leq a \leq c$; two triangles. In 4.11c, $a > c$; one triangle.

Case 2 A is an obtuse angle. In Figure 4.11e, $a < c$; no triangle. In 4.11f, $a > c$; one triangle.

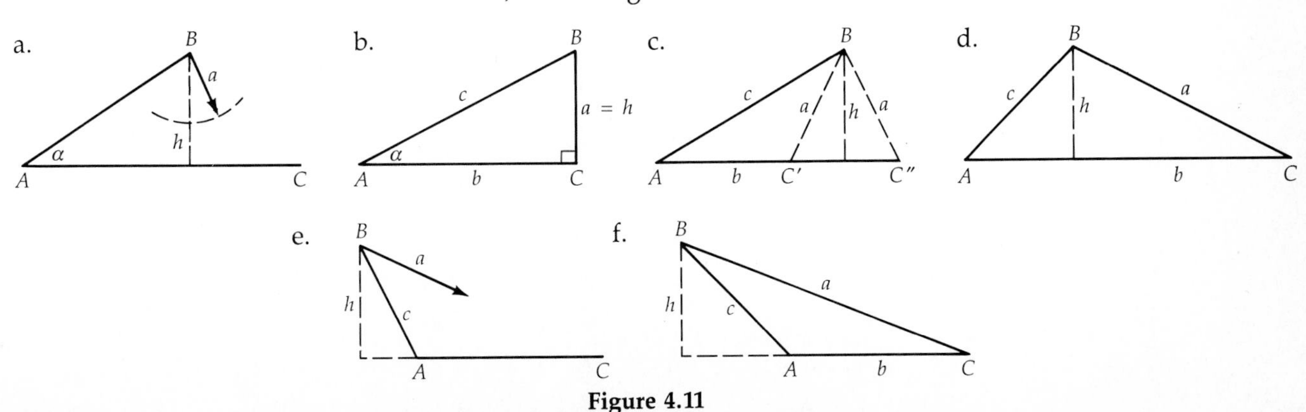

Figure 4.11

Remark It is not necessary to remember all these cases. In solving such problems, the values in the problem will dictate the outcome that is possible in each case.

To find the possible solutions of the triangle in Figure 4.12 where the measure of $A = 57°$ and sides are $a = 15$ feet and $c = 20$ feet, we must first find h.

$$h = 20 \sin 57° \approx 17$$

Since $a < h$, no triangle is formed, and thus there is no solution.

Note the result when the Law of Sines is used to find angle C.

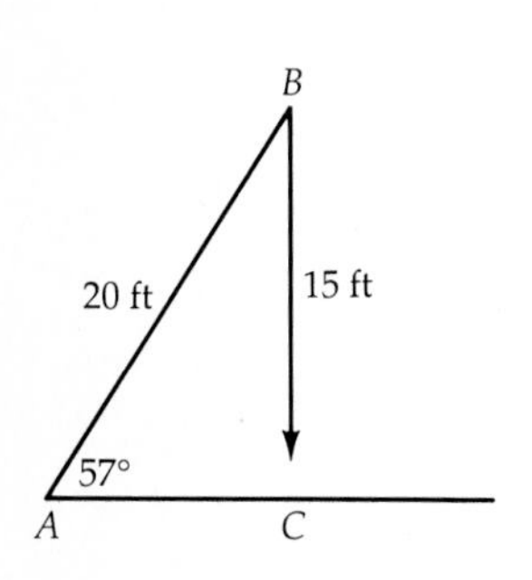

Figure 4.12

$$\frac{a}{\sin A} = \frac{c}{\sin C}$$

$$\frac{15}{\sin 57°} = \frac{20}{\sin C}$$

$$\sin C = \frac{20 \sin 57°}{15}$$

$$\approx 1.1182$$

Because 1.1182 is not in the domain of the sine function, there is no triangle for these values of A, a, and c.

EXAMPLE 3 **Solve the Ambiguous Case of the Law of Sines**

Given the triangle ABC with $B = 32°$, $a = 42$, and $b = 30$, find angle A.

Solution Find h to determine if this is the ambiguous case.

$$h = 42 \sin 32° = 22$$

We have $h < b < a$; thus we know there are two solutions. Use the Law of Sines to find the two possible values of angle A.

$$\frac{b}{\sin B} = \frac{a}{\sin A}$$

$$\frac{30}{\sin 32°} = \frac{42}{\sin A}$$

$$\sin A = \frac{42 \sin 32°}{30°}$$

$$\approx 0.7419$$

$$A \approx 48°$$

$$A \approx 132°$$

The two angles that have a sine of 0.719 are approximately 48° and 132°.

Angle $A \approx 48°$ or $A \approx 132°$.

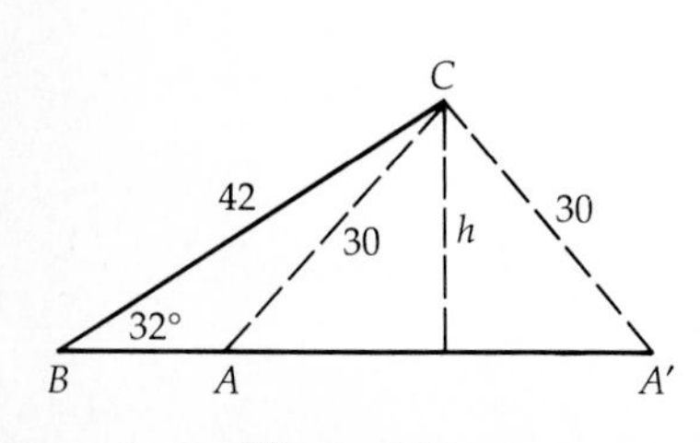

Figure 4.13

■ *Try Exercise* **14**, *page 173.*

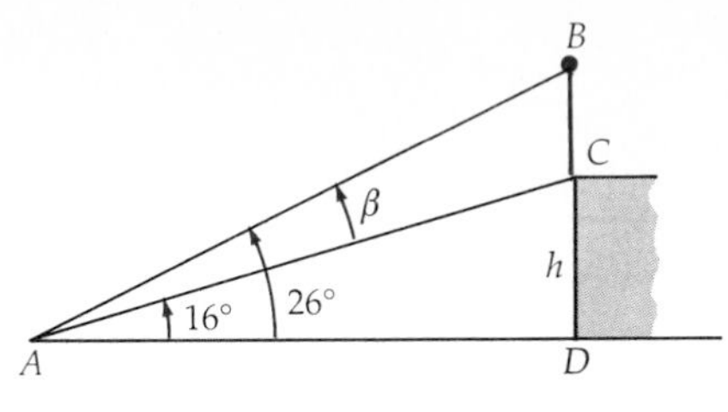

Figure 4.14

EXAMPLE 4 Solve an Application Using the Law of Sines

An 80-foot high radio antenna is located on top of an office building. At a distance d from the base of the building, the angle of elevation to the top of the antenna is 26°, and the angle of elevation to the bottom of the antenna is 16°. Find the height of the building.

Solution Sketch the diagram. Find angles B and β.

$$B = 90^\circ - 26^\circ = 64^\circ$$

$$\beta = 26^\circ - 16^\circ = 10^\circ$$

Since we know the length BC and the measure of β, we can use the Law of Sines to find length AC.

$$\frac{BC}{\sin \beta} = \frac{AC}{\sin B}$$

$$\frac{80}{\sin 10^\circ} = \frac{AC}{\sin 64^\circ}$$

$$AC = \frac{80 \sin 64^\circ}{\sin 10^\circ}$$

Having found AC, we can now find the height of the building.

$$\sin 16^\circ = \frac{h}{AC}$$

$$h = AC \sin 16^\circ$$

$$= \frac{80 \sin 64^\circ}{\sin 10^\circ} \sin 16^\circ \approx 114 \text{ ft} \quad \text{Substitute for } AC.$$

Using the rounding convention, the height of the building to two significant digits is 110 feet.

■ *Try Exercise* **24,** *page 174.*

EXERCISE SET 4.2

In Exercises 1 to 8, solve the triangles.

1. $A = 42^\circ$, $B = 61^\circ$, $a = 12$

2. $B = 25^\circ$, $C = 125^\circ$, $b = 5.0$

3. $A = 110^\circ$, $C = 32^\circ$, $b = 12$

4. $B = 28^\circ$, $C = 78^\circ$, $c = 44$

5. $A = 82.0^\circ$, $B = 65.4^\circ$, $b = 36.5$

6. $B = 54.8^\circ$, $C = 72.6^\circ$, $a = 14.4$

7. $A = 33.8^\circ$, $C = 98.5^\circ$, $c = 102$

8. $B = 36.9^\circ$, $C = 69.2^\circ$, $a = 166$

In Exercises 9 to 18, solve the triangles that exist.

9. $A = 37^\circ$, $c = 40$, $a = 28$

10. $B = 32^\circ$, $c = 14$, $b = 9$

11. $C = 65^\circ$, $b = 10$, $c = 8.0$

12. $A = 42^\circ$, $a = 12$, $c = 18$

13. $A = 30^\circ$, $a = 1.0$, $b = 2.4$

14. $B = 22.6^\circ$, $b = 5.55$, $a = 13.8$

15. $A = 14.8^\circ$, $c = 6.35$, $a = 4.80$

16. $C = 37.9^\circ$, $b = 3.50$, $c = 2.84$

17. $C = 47.2°$, $a = 8.25$, $c = 5.80$

18. $B = 52.7°$, $b = 12.3$, $c = 16.3$

Application Exercises

19. A navigator on a ship sights a lighthouse at a bearing of N36°E. After traveling 8.0 miles at a bearing of N28°W, the ship sights the lighthouse at a bearing of S82°E. How far is the ship from the lighthouse at the second sighting?

20. Two fire lookouts are located on mountains 20 miles apart. Lookout B is at a bearing of S65°E from A. A fire was sighted at a bearing of N50°E from A and N8°E from B. Find the distance of the fire from lookout A.

21. The navigator on a ship traveling at 8 mph due east sights a lighthouse at a heading of 125°. One hour later the lighthouse is sighted at a heading of 205°. Find the closest the ship came to the lighthouse.

22. Two observers are directly in line with a balloon and 220 feet apart. The angle of elevation of the balloon from one observer is 67°, and the angle of elevation of the balloon from the other observer is 31°. How far is the balloon from the observer who sees the balloon at an angle of 67°?

23. To find the distance across a canyon, a surveying team locates points A and B on one side of the canyon and point C on the other side of the canyon. The distance between A and B is 85 yards. The angle CAB is 68°, and the angle CBA is 88°. Find the distance across the canyon.

24. The longer side of a parallelogram is 6.0 meters. Angle A is 56°, and angle α is 35°. Find the length of the longer diagonal.

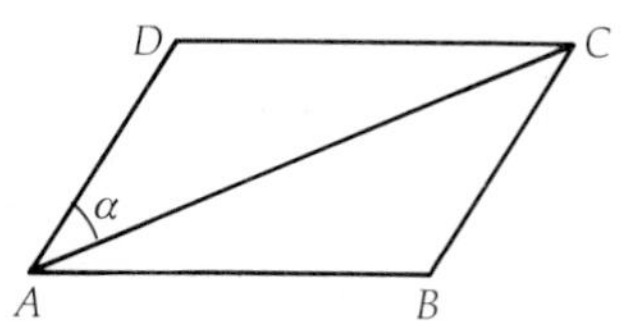

25. Two observers, in a line directly under a kite and a distance of 30 feet apart, observe the kite at an angle of elevation of 62° and 78°, respectively. Find the height of the kite.

26. A 35-foot high telephone pole is situated on an 11 percent slope from A. The angle of elevation from point A to the top of the pole is 32°. Find the length of the guy wire AC.

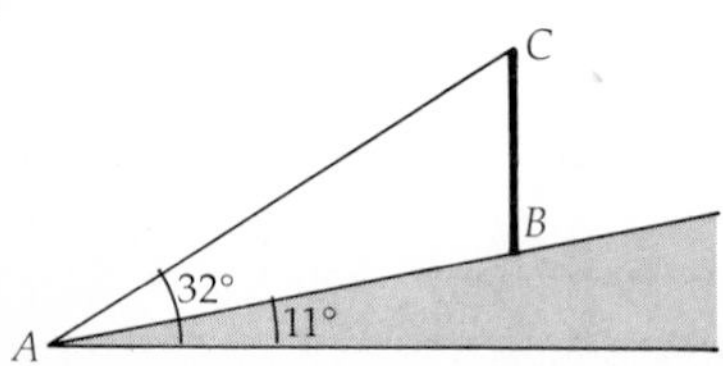

27. A surveying team determines the height of a hill by placing a 12-foot pole at the top of the hill and measuring the angles of elevation to the bottom and the top of the pole. Find the height of the hill.

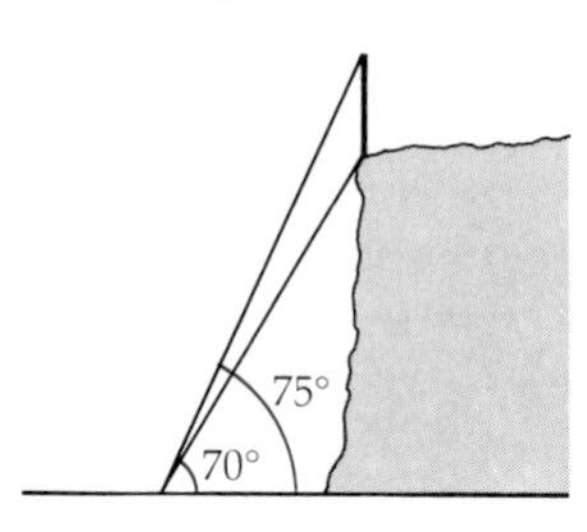

28. Three roads intersect in such a way to form a triangular piece of land. Find the lengths of the other two sides of the land.

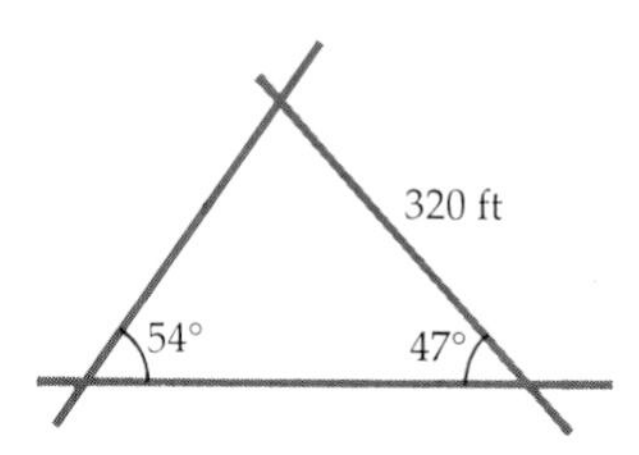

29. An airplane flew 450 miles at a heading of 65° from airport A to airport B. The pilot then flew at a heading of 142° to airport C. Find the distance from A to C if the heading from airport A to airport C is 120°.

Supplemental Exercises

30. A house B is located at a heading of 67° from house A. A house C is 300 meters at a heading of 112° from house A. House B is located at a heading of 349° from house C. Find the distance from house A to house B.

31. For any triangle ABC, show that

$$\frac{a - b}{b} = \frac{\sin A - \sin B}{\sin B}.$$

32. For any triangle ABC, show that

$$\frac{a + b}{b} = \frac{\sin A + \sin B}{\sin B}.$$

33. For any triangle ABC, show that

$$\frac{a - b}{a + b} = \frac{\sin A - \sin B}{\sin A + \sin B}.$$

4.3 The Law of Cosines and Area

The *Law of Cosines* can be used to solve triangles in which two sides and the included angle (angle between the two sides) are known or in which three sides are known. Consider the triangle in Figure 4.15. The altitude BD is drawn from B perpendicular to the x-axis. The triangle BDA is a right triangle, and the coordinates of B are $(a \cos C, a \sin C)$. The coordinates of A are $(b, 0)$. Using the distance formula, we can find the distance c.

$$c = \sqrt{(a \cos C - b)^2 + (a \sin C - 0)^2}$$
$$c^2 = a^2 \cos^2 C - 2ab \cos C + b^2 + a^2 \sin^2 C$$
$$c^2 = a^2(\cos^2 C + \sin^2 C) + b^2 - 2ab \cos C$$
$$c^2 = a^2 + b^2 - 2ab \cos C$$

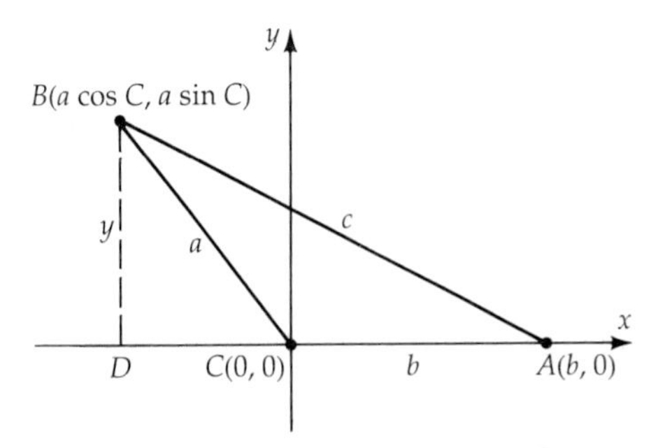

Figure 4.15

The Law of Cosines

> If A, B, and C are the angles of an oblique triangle and a, b, and c are the lengths of sides opposite these angles,
>
> $$c^2 = a^2 + b^2 - 2ab \cos C,$$
> $$a^2 = b^2 + c^2 - 2bc \cos A,$$
> $$b^2 = a^2 + c^2 - 2ac \cos B.$$

EXAMPLE 1 Solve a Triangle Using the Law of Cosines

In triangle ABC, angle $B = 110.0°$, side $a = 10.0$ centimeters, and side $c = 15.0$ centimeters. Find side b.

Solution The Law of Cosines can be used because two sides and the

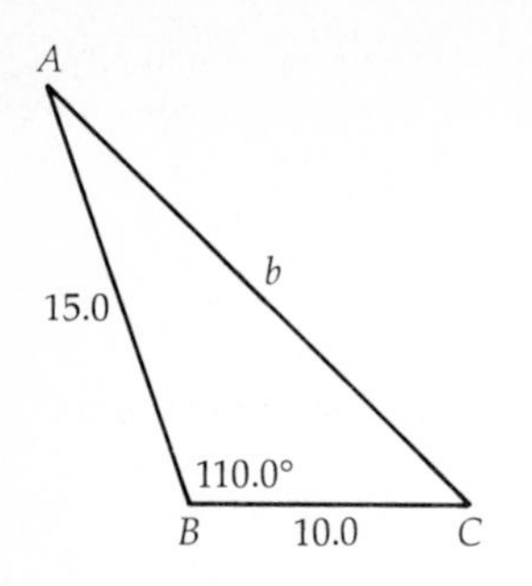

Figure 4.16

included angle are known.

$$b^2 = a^2 + c^2 - 2ac \cos B$$
$$= 10.0^2 + 15.0^2 - 2(10.0)(15.0) \cos 110.0°$$
$$\approx 100 + 225 + 103 = 428$$
$$b \approx 20.7 \text{ cm}$$

■ *Try Exercise* **12,** *page 180.*

EXAMPLE 2 Solve a Triangle Using the Law of Cosines

In triangle ABC, $a = 32$ feet, $b = 20$ feet, and $c = 40$ feet. Find angle B.

Solution $b^2 = a^2 + c^2 - 2ac \cos B$

$$\cos B = \frac{a^2 + c^2 - b^2}{2ac}$$ Solve for cos B.

$$= \frac{32^2 + 40^2 - 20^2}{2(32)(40)} \approx 0.8688$$ Substitute the values in the formula and solve for angle B.

$$B \approx 30°$$

■ *Try Exercise* **18,** *page 180.*

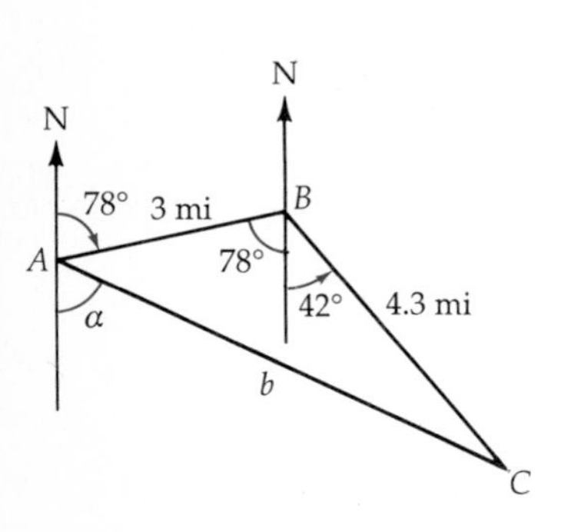

Figure 4.17

EXAMPLE 3 Solve an Application Using the Law of Cosines

A car traveled 3.0 miles at a bearing of N78°E. The road turned and another 4.3 miles was traveled at a bearing of S42°E. Find the distance and the bearing of the car from the starting point.

Solution Sketch a diagram. First find the measure of angle B. Then use the Law of Cosines to find b.

$$B = 78° + 42° = 120°$$
$$b^2 = a^2 + c^2 - 2ac \cos B$$
$$= 3.0^2 + 4.3^2 - 2(3.0)(4.3) \cos 120°$$
$$= 9 + 18.49 + 12.9 = 40.39$$
$$b \approx 6.4 \text{ mi}$$

Find angle A.

$$\cos A = \frac{b^2 + c^2 - a^2}{2bc} = \frac{6.4^2 + 3.0^2 - 4.3^2}{(2)(6.4)(3.0)}$$
$$= \frac{40.96 + 9.0 - 18.49}{38.4} \approx 0.8195$$
$$A \approx 35°$$

The bearing of the present position of the car from the starting point A can be determined by calculating the measure of angle α from Figure 4.17.

$$\alpha = 180° - (78° + 35°) = 67°$$

The distance is approximately 6.4 miles, and the bearing to the nearest degree is S67°E.

■ *Try Exercise* **48,** *page 181.*

Remark The measure of A in Example 3 also can be determined by using the Law of Sines.

Area

We used the formula $A = \frac{1}{2}bh$ for the area of a triangle when the base and height were given. In this section, we will find the area of triangles when the height is not given. We will use K for the area of a triangle since A is often used to represent an angle.

Consider the areas of the acute and obtuse triangles in Figure 4.18.

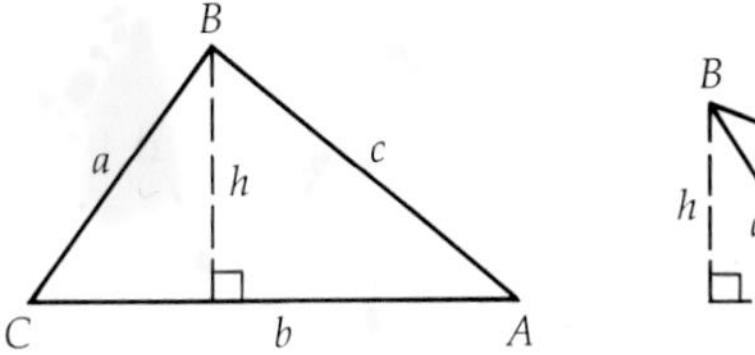

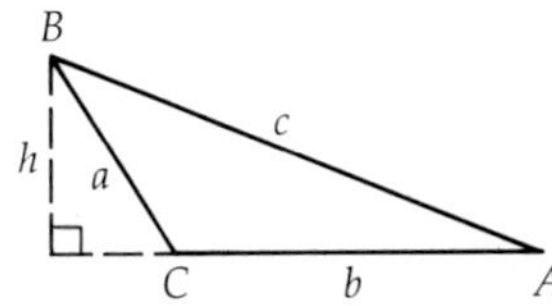

Figure 4.18
a. Acute triangle. b. Obtuse triangle.

Area of each triangle: $K = \dfrac{1}{2}bh$

Height of each triangle: $h = c \sin A$

Substitute for h: $K = \dfrac{1}{2}bc \sin A$

Thus we have established the following theorem.

Area of a Triangle

The area K of triangle ABC is one-half the product of the lengths of any two sides and the included angle. Thus

$$K = \frac{1}{2}bc \sin A,$$

$$K = \frac{1}{2}ab \sin C,$$

$$K = \frac{1}{2}ac \sin B.$$

EXAMPLE 4 **Find the Area of a Triangle**

Given Angle $A = 62°$, $b = 12$ meters, and $c = 5.0$ meters, find the area of triangle ABC.

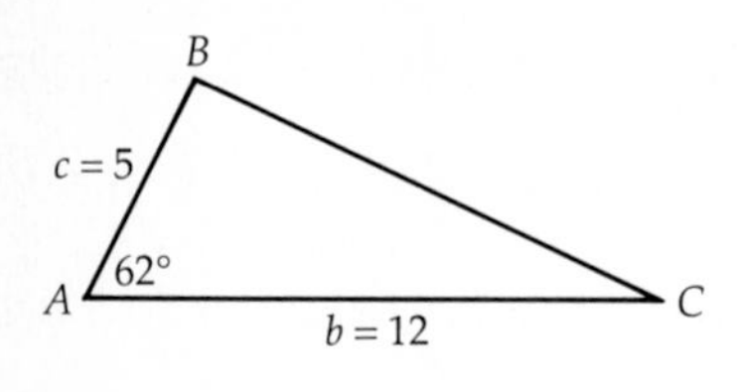

Figure 4.19

Solution Two sides and the included angle of the triangle are given. Using the formula for area, we have

$$K = \frac{1}{2}bc \sin A = \frac{1}{2}(12)(5.0)(\sin 62°) \approx 26 \text{ m}^2.$$

■ *Try Exercise* **28**, *page 180.*

When two angles and a side are given, the Law of Sines is used to derive a formula for the area of a triangle.

First, solve for c in the Law of Sines.

$$\frac{c}{\sin C} = \frac{b}{\sin B}$$

$$c = \frac{b \sin C}{\sin B}$$

Substitute for c in the first formula for area:

$$K = \frac{1}{2}b \cdot c \sin A$$

$$= \frac{1}{2}b \cdot \frac{b \sin C}{\sin B} \sin A$$

$$K = \frac{b^2 \sin C \sin A}{2 \sin B}$$

In like manner, the following two alternate formulas can be derived for the area of a triangle:

$$K = \frac{a^2 \sin B \sin C}{2 \sin A} \quad \text{and} \quad K = \frac{c^2 \sin A \sin B}{2 \sin C}$$

EXAMPLE 5 **Find the Area of a Triangle**

Given angle $A = 32°$, angle $C = 77°$, and side $a = 14$ inches. Find the area of triangle ABC.

Solution First we find the third angle.

$$B = 180° - 32° - 77° = 71°$$

Thus

$$K = \frac{a^2 \sin B \sin C}{2 \sin A} = \frac{14^2 \sin 71° \sin 77°}{2 \sin 32°} \approx 170 \text{ in}^2$$

■ *Try Exercise* **28**, *page 180.*

The Law of Cosines can be used to derive *Heron's formula* for the area of a triangle in which three sides of the triangle are given.

Heron's Formula for Finding the Area of a Triangle

If a, b, and c are the lengths of the sides of a triangle, then the area K of the triangle is

$$K = \sqrt{s(s-a)(s-b)(s-c)}, \quad \text{where } s = \frac{1}{2}(a+b+c)$$

EXAMPLE 6 Find an Area by Heron's Formula

Find the area of the triangle with $a = 7.0$ meters, $b = 15$ meters, and $c = 12$ meters.

Solution Using Heron's formula, we have

$$s = \frac{a+b+c}{2} = \frac{7.0+15+12}{2} = 17$$

$$\begin{aligned} K &= \sqrt{s(s-a)(s-b)(s-c)} \\ &= \sqrt{17(17-7)(17-15)(17-12)} \\ &= \sqrt{1700} \approx 41 \text{ m}^2 \end{aligned}$$

■ *Try Exercise* **36**, *page 180.*

EXAMPLE 7 Solve an Application Using Heron's Formula

A commercial piece of real estate is priced at \$6.50 per square foot. Find the cost of a triangular piece of commercial property measuring 200 feet by 350 feet by 400 feet.

Solution

$$s = \frac{a+b+c}{2} = \frac{200+350+400}{2} = 475$$

$$\begin{aligned} K &= \sqrt{s(s-a)(s-b)(s-c)} \\ &= \sqrt{475(475-200)(475-350)(475-400)} \\ &= \sqrt{475(275)(125)(75)} = \sqrt{1224609375} \\ &\approx 35{,}000 \end{aligned}$$

The area is approximately 35,000 square feet. Find the value of the property by multiplying the cost per foot by the number of square feet.

$$\text{Cost} = 6.50(35{,}000) = 227{,}500$$

The cost of the commercial property is approximately \$227,500.

■ *Try Exercise* **56**, *page 181.*

EXERCISE SET 4.3

In Exercises 1 to 14, find the third side of the triangle.

1. $a = 12$, $b = 18$, $C = 44°$
2. $b = 30$, $c = 24$, $A = 120°$
3. $a = 120$, $c = 180$, $B = 56°$
4. $a = 400$, $b = 620$, $C = 116°$
5. $b = 60$, $c = 84$, $A = 13°$
6. $a = 122$, $c = 144$, $B = 48°$
7. $a = 9.0$, $b = 7.0$, $C = 72°$
8. $b = 12$, $c = 22$, $A = 55°$
9. $a = 4.6$, $b = 7.2$, $C = 124°$
10. $b = 12.3$, $c = 14.5$, $A = 6.5°$
11. $a = 25.9$, $c = 33.4$, $B = 84°$
12. $a = 14.2$, $b = 9.30$, $C = 9.20°$
13. $a = 122$, $c = 55.9$, $B = 44.2°$
14. $b = 444.8$, $c = 389.6$, $A = 78.44°$

In Exercises 15 to 24, given three sides of a triangle, find the specified angle.

15. $a = 25$, $b = 32$, $c = 40$; find A.
16. $a = 60$, $b = 88$, $c = 120$; find B.
17. $a = 8.0$, $b = 9.0$, $c = 12$; find C.
18. $a = 108$, $b = 132$, $c = 160$; find A.
19. $a = 80.0$, $b = 92.0$, $c = 124$; find B.
20. $a = 166$, $b = 124$, $c = 139$; find B.
21. $a = 1025$, $b = 625.0$, $c = 1420$; find C.
22. $a = 4.7$, $b = 3.2$, $c = 5.9$; find A.
23. $a = 32.5$, $b = 40.1$, $c = 29.6$; find B.
24. $a = 112.4$, $b = 96.80$, $c = 129.2$; find C.

In Exercises 25 to 36, find the area of the given triangle.

25. $A = 105°$, $b = 12$, $c = 24$
26. $B = 127°$, $a = 32$, $c = 25$
27. $A = 42°$, $B = 76°$, $c = 12$
28. $B = 102°$, $C = 27°$, $a = 8.5$
29. $a = 10$, $b = 12$, $c = 14$
30. $a = 32$, $b = 24$, $c = 36$
31. $B = 54.3°$, $a = 22.4$, $b = 26.9$
32. $C = 18.2°$, $b = 13.4$, $a = 9.84$
33. $A = 116°$, $B = 34°$, $c = 8.5$
34. $B = 42.8°$, $C = 76.3°$, $c = 17.9$
35. $a = 3.6$, $b = 4.2$, $c = 4.8$
36. $a = 13.3$, $b = 15.4$, $c = 10.2$

Application Exercises

37. A plane leaves airport A and travels 560 miles to airport B at a heading of 32°. The plane leaves airport B and travels to airport C 320 miles away at a heading of 108°. Find the distance from airport A to airport C.

38. A developer has a triangular lot at the intersection of two streets. The streets meet at an angle of 72°, and the lot has 300 feet of frontage along one street and 416 feet of frontage along the other street. Find the length of the third side of the lot.

39. Two ships left a port at the same time. One ship traveled at a speed of 18 mph at a heading of 318°. The other ship traveled at a speed of 22 mph at a heading of 198°. Find the distance between the two ships after 10 hours of travel.

40. Find the distance across a lake using the measurements as shown in the figure.

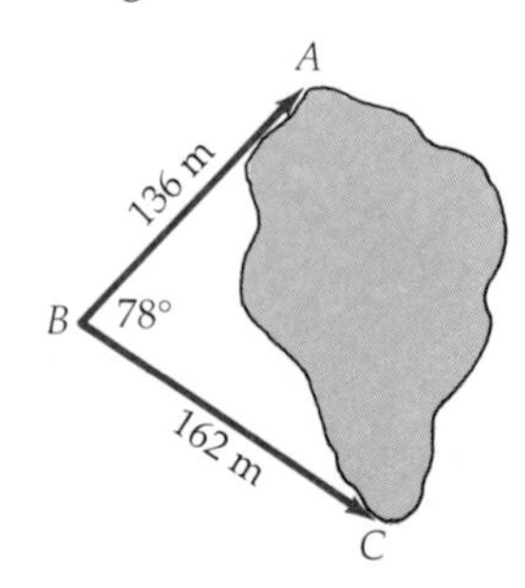

41. A regular hexagon is inscribed in a circle with a radius of 40 centimeters. Find the length of one side of the hexagon.

42. A regular pentagon is inscribed in a circle with a radius of 25 inches. Find the length of one side of the pentagon.

43. The length of the diagonals of a parallelogram are 20 inches and 32 inches. The diagonals intersect at an angle of 35°. Find the lengths of the sides of the parallelogram. (*Hint:* The diagonals of a parallelogram bisect one another.)

44. The sides of a parallelogram are 10 feet and 14 feet. The longer diagonal of the parallelogram is 18 feet. Find the length of the shorter diagonal of the parallelogram. (*Hint:* The diagonals of a parallelogram bisect one another.)

45. The sides of a parallelogram are 30 centimeters and 40 centimeters. The shorter diagonal of the parallelogram is 44 centimeters. Find the length of the longer diagonal of the parallelogram. (*Hint:* The diagonals of a parallelogram bisect one another.)

46. The sides of a triangular city lot have sides of 224 feet, 182 feet, and 165 feet. Find the angle between the longer two sides of the lot.

47. A plane traveling at 180 mph passes 400 feet directly overhead of an observer. The plane is traveling along a path with an angle of elevation of 14°. Find the distance of the plane from the observer 10 seconds after the plane has passed directly overhead.

48. A ship leaves a port at a speed of 16 mph at a bearing of N32°E. One hour later another ship leaves the port at a speed of 22 mph at a bearing of S74°W. Find the distance between the ships 4 hours after the first ship leaves the port.

49. Find the area of a triangular piece of land that is bounded by sides of 236 meters, 620 meters, and 814 meters.

50. Find the area of a parallelogram whose diagonals are 24 inches and 32 inches and the diagonals intersect at an angle of 40°.

51. Find the area of a parallelogram with sides of 12 meters and 18 meters and with one angle of 70°.

52. Find the area of a parallelogram with sides of 8 feet and 12 feet. The shorter diagonal is 10 feet.

53. Find the area of a square inscribed in a circle with a radius of 9 inches.

54. Find the area of a regular hexagon inscribed in a circle with a radius of 24 centimeters.

55. A commercial piece of real estate is priced at $2.20 per square foot. Find the cost of a triangular lot measuring 212 feet by 185 feet by 240 feet.

56. An industrial piece of real estate is priced at $4.15 per square foot. Find the cost of a triangular lot measuring 324 feet by 516 feet by 412 feet.

57. Find the number of acres in a pasture whose shape is a triangle measuring 800 feet by 1020 feet by 680 feet. (An acre is 43,560 square feet.)

58. Find the number of acres in a housing tract whose shape is a triangle measuring 420 yards by 540 yards by 500 yards. (An acre is 4840 square yards.)

Supplemental Exercises

59. Find the angle formed by the sides P_1P_2 and P_1P_3 of a triangle with the vertices at $P_1(-2, 4)$, $P_2(2, 1)$, and $P_3(4, -3)$.

60. The sides of a parallelogram are x and y, and the diagonals are w and z. Show that $w^2 + z^2 = 2x^2 + 2y^2$.

61. A rectangular box has dimensions of length 10 feet, width 4 feet, and height 3 feet. Find the angle between the diagonal of the bottom of the box and the diagonal of the end of the box.

62. A regular pentagon is inscribed in a circle with a radius of 4 inches. Find the perimeter of the pentagon.

63. An equilateral triangle is inscribed in a circle of radius 10 centimeters. Find the perimeter of the triangle.

64. Given a triangle ABC, prove that

$$a^2 = b^2 + c^2 - 2bc \cos A.$$

65. Use the Law of Cosines to show that

$$\cos A = \frac{(b + c - a)(b + c + a)}{2bc} - 1.$$

66. Prove that $K = xy \sin A$ for a parallelogram, where x and y are adjacent sides and A is the angle between x and y.

67. Show that the area of the parallelogram in the figure is $K = 2ab \sin C$.

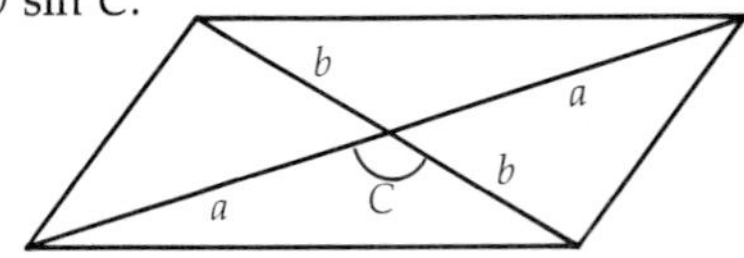

68. Given a regular hexagon inscribed in a circle of 10 inches, find the area of a sector (see the figure).

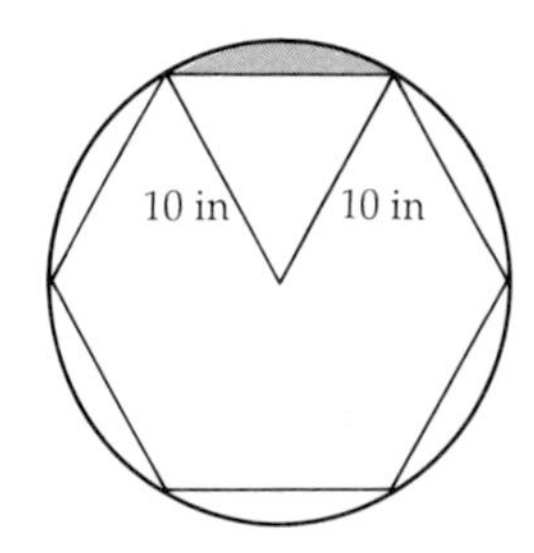

69. Find the volume of the pyramid-shaped piece of aluminum in the figure.

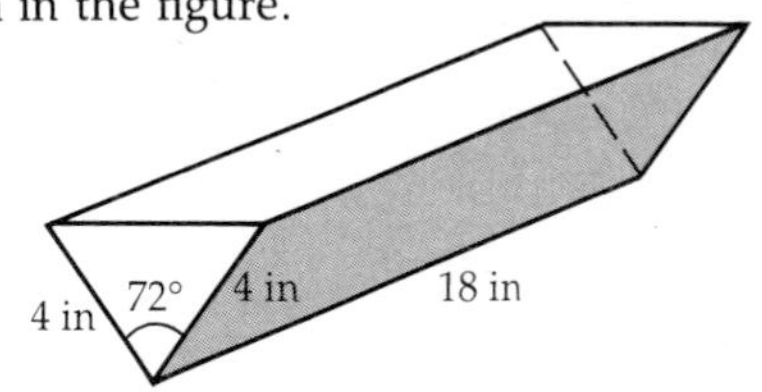

70. Show that the area of the circumscribed triangle in the figure is $K = rs$, where $s = \dfrac{a + b + c}{2}$.

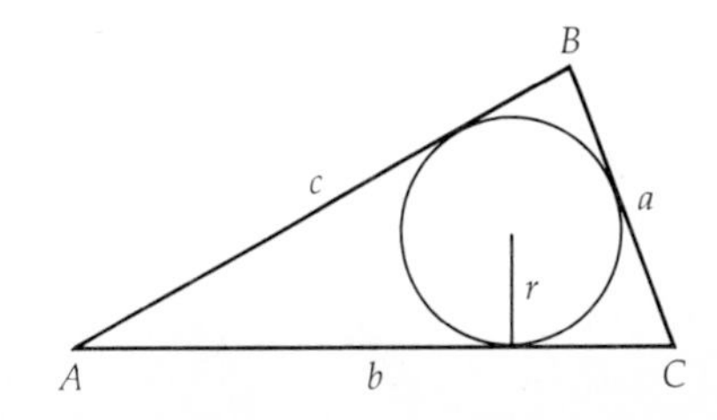

4.4 Vectors

In scientific applications, measurements are made that consist of a number quantity (magnitude) and a unit. Some of these measurements are temperature, force, displacement, and time. The magnitude and unit are sufficient to completely represent some of these measurements, but in other instances it is important to have a direction associated with the measurement. For example, a temperature of 37° does not need a direction to completely represent the quantity. However, a measurement, such as exerting 30 pounds of force on a box, must have the direction specified to completely represent the measurement. For example, is the force being used to lift the box up or push the box along the floor? Also, a displacement (change in location) of 2 miles east is different from a displacement of 2 miles north.

Physical measurements are divided into two classes: scalars and vectors. **Scalars** are quantities that are fully represented by a magnitude only. Temperature is an example of a scalar quantity. **Vector quantities** are represented by a magnitude and a direction. Force and velocity are examples of vector quantities. Force is specified by a magnitude and a direction. Velocity is a vector quantity with magnitude and direction. Speed is a scalar quantity; it is the magnitude part of velocity. For example, 55 meters per second east is a vector. The speed is 55 meters per second.

Definition of a Vector

> A **vector** is a directed line segment. The length of the vector is the magnitude of the vector, and the direction of the vector is measured by an angle.

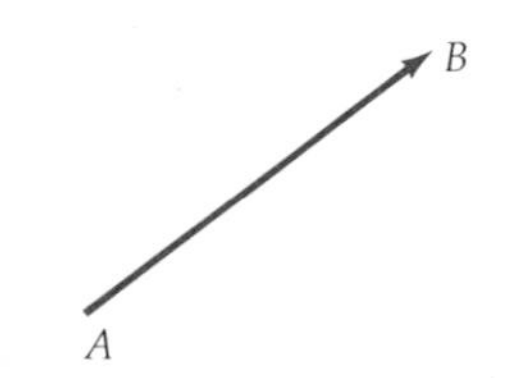

Figure 4.20

The point A of the vector in Figure 4.20 is called the initial point (or tail), and the point B is the terminal point (or head) of the vector. An arrow over the letters $\overrightarrow{AB}$, an arrow over a single letter $\vec{V}$, or boldface type **AB** or **V** is used to represent a vector. The magnitude of the vector is represented by $|\overrightarrow{AB}|$, $|\vec{V}|$, $|\mathbf{V}|$, or $|\mathbf{AB}|$.

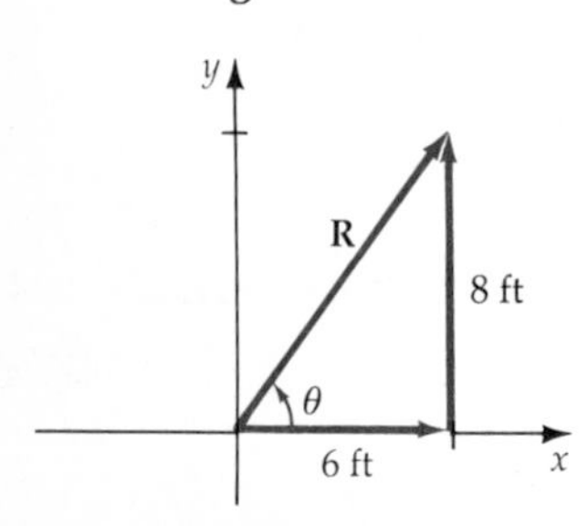

Figure 4.21

The sum of two vectors is the single vector that will have the same effect as the two vectors. For example, the sum of a vector of magnitude 6 feet in the positive x direction and a vector with a magnitude of 8 feet in the positive y direction is equivalent to a vector of magnitude 10 feet at an angle of approximately 53° to the positive x-axis. This vector is called the **resultant** vector. The magnitude of the vector is

$$|\mathbf{V}| = \sqrt{6^2 + 8^2} = 10 \text{ ft}.$$

To find the direction of the vector from the positive x-axis, solve $\tan\theta = \frac{8}{6}$ for θ.

$$\tan \theta = \frac{8}{6}$$

$$\theta = \tan^{-1} \frac{8}{6} \approx 53°$$

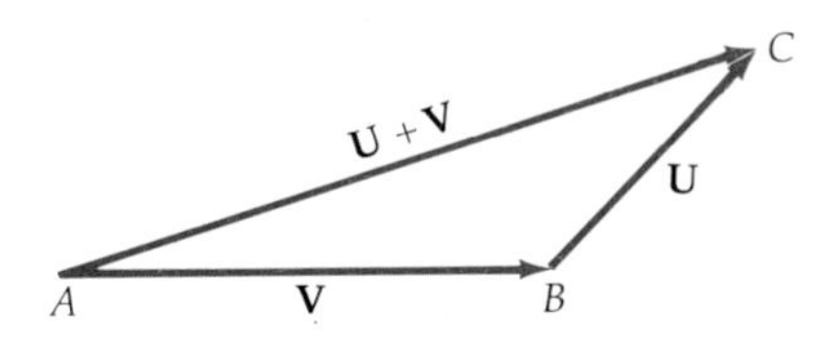

Figure 4.22

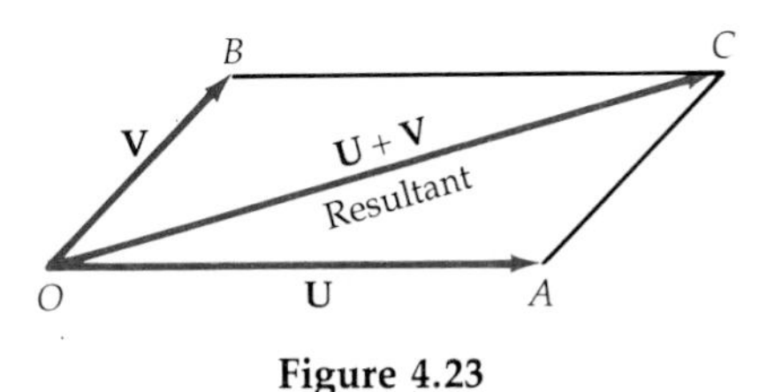

Figure 4.23

The sum of two vectors is another vector that has an equivalent effect of the two vectors. In our example, walking a distance of 10 feet at an angle of 53° with the positive x-axis will put you in the same place as walking 6 feet east and 8 feet north.

One method of adding vectors graphically is to place the tail of one vector **U** at the head of the other vector **V** and complete the triangle, as Figure 4.22 shows. This is called the triangle method of adding vectors.

An alternate method of adding vectors graphically is to place the tails of the two vectors **U** and **V** together, as Figure 4.23 shows. Complete the parallelogram so that **U** and **V** are sides of the parallelogram. The diagonal of the parallelogram is the resultant.

Multiplying a vector by a positive number other than 1 changes the length of the vector but does not affect the direction of the vector. If **V** is any vector, then 2**V** denotes the vector that has the same direction as **V** but twice its magnitude. We say that **V** has been multiplied by the scaler 2. Multiplying a vector by a negative number reverses the direction of the vector.

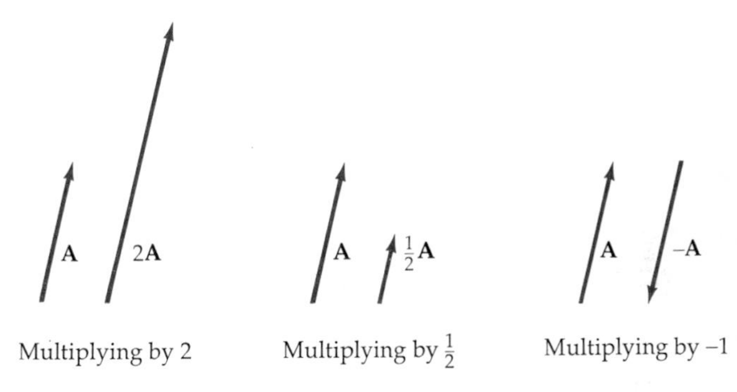

Figure 4.24

To substract **U** from **V** geometrically, add the opposite of **U** to **V**, as Figure 4.25 shows.

$$\mathbf{V} - \mathbf{U} = \mathbf{V} + (-\mathbf{U})$$

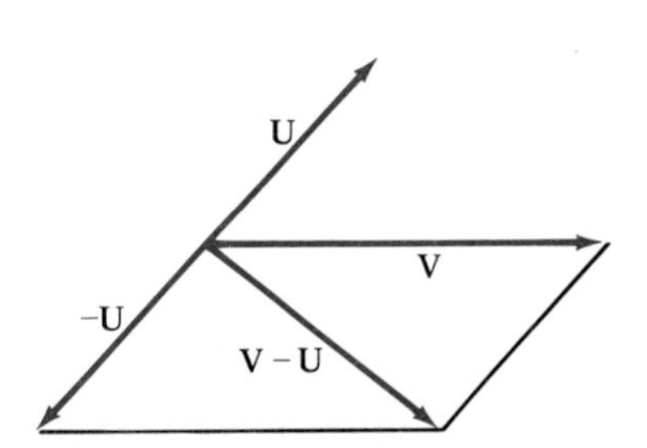

Figure 4.25

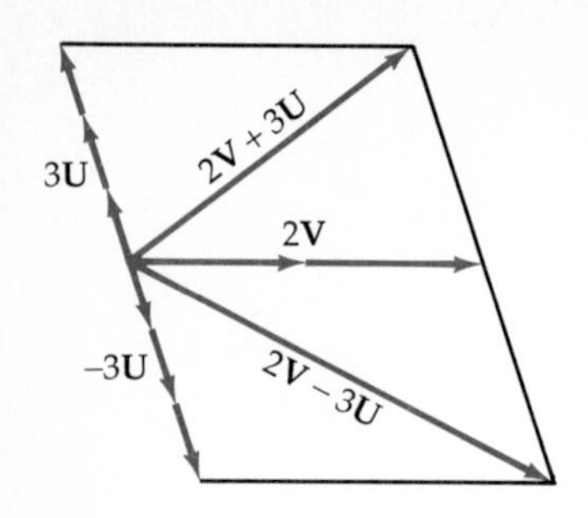

Figure 4.26

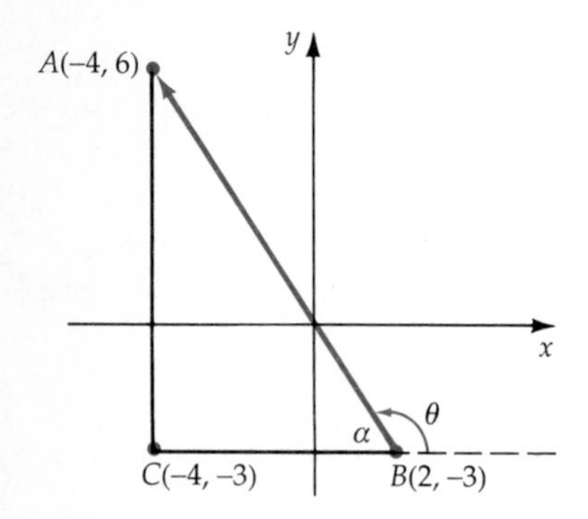

Figure 4.27

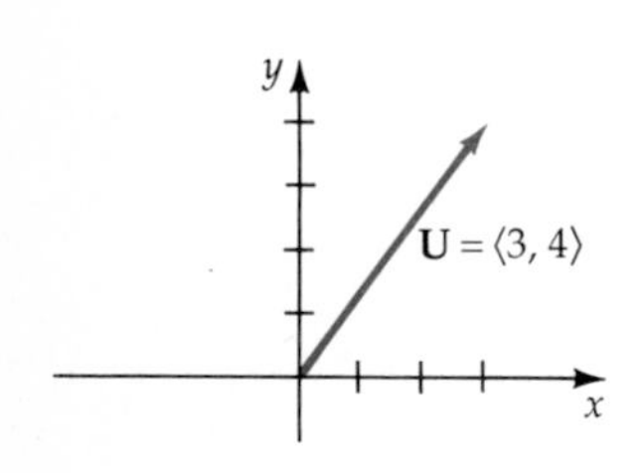

Figure 4.28

Assume that the two vectors **U** and **V** are given as shown in Figure 4.26. The sum $2\mathbf{V} + 3\mathbf{U}$ and the difference $2\mathbf{V} - 3\mathbf{U}$ are shown graphically.

The distance formula and the trigonometric functions can be used to find the magnitude and direction of a vector in the coordinate plane. The direction of a vector is the measure of the angle made by the vector and the positive x-axis.

EXAMPLE 1 Find the Direction and Magnitude of a Vector

Find the magnitude and direction of a vector that has its tail at the point $B(2, -3)$ and its head at the point $A(-4, 6)$.

Solution The vector is sketched as **BA** in Figure 4.27. We use the Pythagorean Theorem to find the magnitude of the vector.

$$|\mathbf{BA}| = \sqrt{[2 - (-4)]^2 + (-3 - 6)^2}$$
$$= \sqrt{36 + 81} = \sqrt{117}$$

Sketch the lines BC and AC to form a right triangle (see Figure 4.27). Find the lengths of BC and AC and then use the tangent function to find the reference angle for the direction angle.

$$a = |-4 - 2| = 6 \quad \text{and} \quad b = |6 - (-3)| = 9$$

$$\tan \alpha = \frac{b}{a} = \frac{9}{6}$$

$\alpha \approx 56°$ The reference angle for the vector is 56°.

$\theta = 180° - 56° = 124°$ The direction angle is the angle made by the vector and the positive x-axis.

The magnitude of the vector is $\sqrt{117}$ and the direction angle is 124°.

■ *Try Exercise* **2**, *page 192.*

The notation $\mathbf{U} = \langle 3, 4\rangle$ is used for a vector with its tail at the origin and its head at the point (3, 4). The distance 3 along the x-axis is called the magnitude of the x-component of the vector, and the distance 4 along the y-axis is called the magnitude of the y-component of the vector. The vector $\langle 0, 0\rangle$ is called the zero (**0**) vector. For any vector $\mathbf{V} + (-\mathbf{V}) = \mathbf{0}$.

Caution The notation $\langle a, b\rangle$ is used for a vector, whereas (a, b) is used for a point. The vector $\langle a, b\rangle$ has its tail at (0, 0) and its head at (a, b).

EXAMPLE 2 Graph Vectors on the Coordinate Plane

Graph each vector. a. $\mathbf{U} = \langle -2, 4\rangle$ b. $\mathbf{V} = \langle 3, -1\rangle$

Solution See Figure 4.29

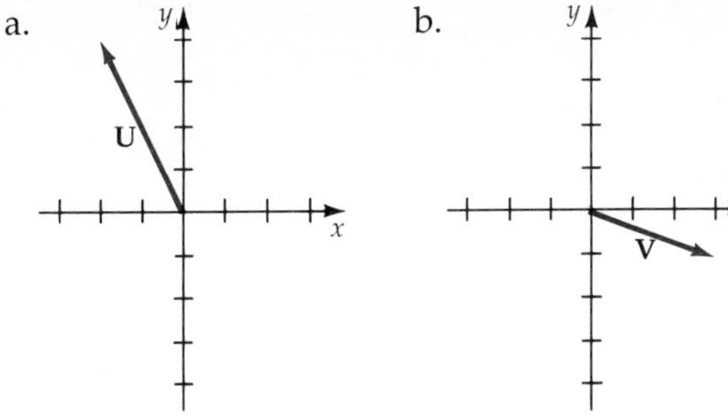

a. b.

Figure 4.29

■ *Try Exercise* **8**, *page 192.*

The vector $\langle 1, 0 \rangle$ has a magnitude of 1, and its direction is in the positive x-direction. It is called the **unit vector i.** The vector $\langle 0, 1 \rangle$ has a magnitude of 1 and a direction in the positive y-direction. It is called the **unit vector j.** See Figure 4.30.

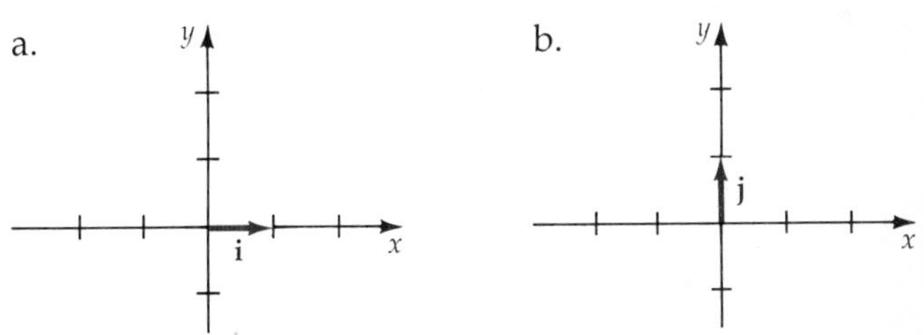

Figure 4.30
a. Unit vector **i**; b. Unit vector **j**.

Any vector in the plane can be written in terms of the unit vectors **i** and **j**. The vector $\mathbf{U} = \langle 3, 4 \rangle$ can be written $\mathbf{U} = 3\mathbf{i} + 4\mathbf{j}$. The vector $3\mathbf{i}$ is called the **x-component** of the vector **U**, and the vector $4\mathbf{j}$ is called the **y-component** of the vector **U**. The magnitude of the vector can be found by the Pythagorean Theorem, and the direction angle can be found by using the tangent function.

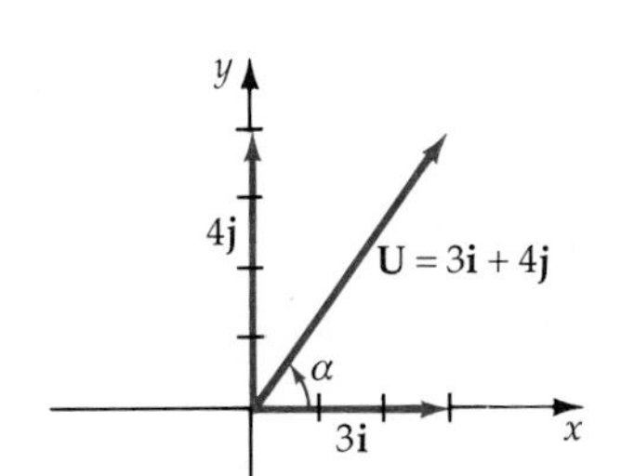

Figure 4.31

$$|\mathbf{U}| = \sqrt{3^2 + 4^2} = 5 \qquad \tan \alpha = \frac{4}{3}$$

$$\alpha \approx 53^\circ$$

The vector $\mathbf{U} = 3\mathbf{i} + 4\mathbf{j}$ has a magnitude of 5 and a direction angle of 53°.

EXAMPLE 3 Find the Magnitude and Direction of Vectors

Find the magnitude and direction angle of each vector.

a. $\mathbf{U} = 5\mathbf{i} - 2\mathbf{j}$ b. $\mathbf{V} = \langle -2, -3 \rangle$

Solution Use the Pythagorean Theorem to find the magnitude of the

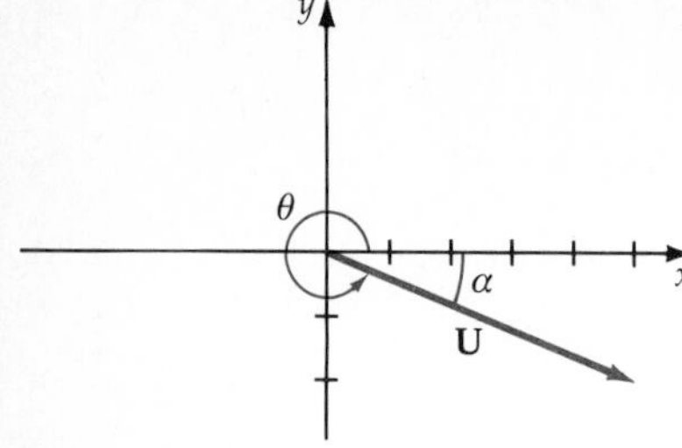

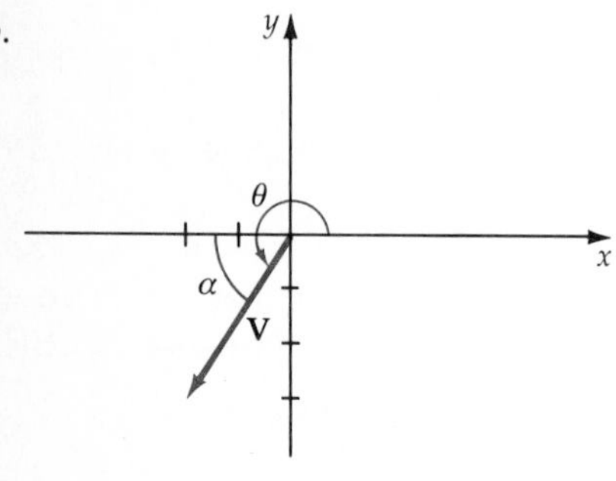

Figure 4.32

vector and the tangent function to find the reference angle for the direction. For each vector, α is the reference angle and θ is the direction angle.

a. $|\mathbf{U}| = \sqrt{5^2 + (-2)^2} \approx 5.4 \qquad \tan \alpha = \left| -\frac{2}{5} \right|$

$$\alpha \approx 22°$$

$$\theta = 360° - 22° = 338°$$

The magnitude is 5.4 and the direction is 338°.

b. $|\mathbf{V}| = \sqrt{(-2)^2 + (-3)^2} \approx 3.6 \qquad \tan = \left| \frac{-3}{-2} \right|$

$$\alpha \approx 56°$$

$$\theta = 180° + 56° = 236°$$

The magnitude is 3.6 and the direction angle is 236°.

■ *Try Exercises* **22**, *page 192.*

Any vector **U** can be written in the form $\mathbf{U} = x\mathbf{i} + y\mathbf{j}$. Figure 8.33 shows that the magnitude of the x- and y-components of vector $\mathbf{U} = x\mathbf{i} + y\mathbf{j}$ are given by $x = |\mathbf{U}| \cos \alpha$ and $y = |\mathbf{U}| \sin \alpha$. Thus

$$\mathbf{U} = |\mathbf{U}| (\cos \alpha)\mathbf{i} + |\mathbf{U}|(\sin \alpha)\mathbf{j}.$$

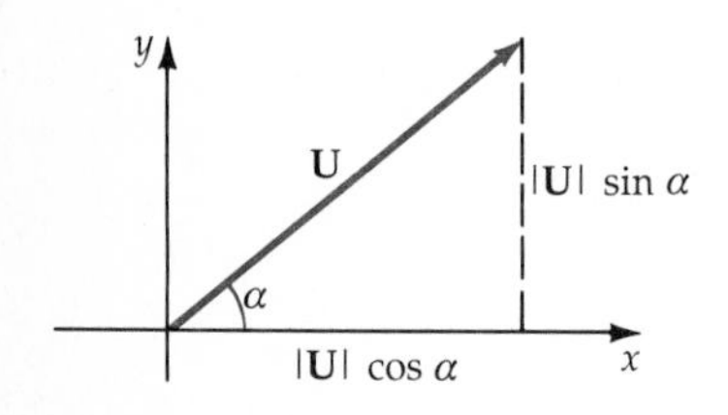

Figure 4.33

EXAMPLE 4 Find the Components of a Vector

Find the x- and y-components of a vector **U** with a magnitude of 15.0 and direction of 125°. Write **U** in terms of unit vectors. (See Figure 4.34.)

Solution Find the magnitude of the x- and y-components.

$$x = 15 \cos 125° \approx -8.60 \qquad y = 15 \sin 125° \approx 12.3$$

Writing in terms of unit vectors, we have $\mathbf{U} \approx -8.60\mathbf{i} + 12.3\mathbf{j}$.

■ *Try Exercise* **28**, *page 192.*

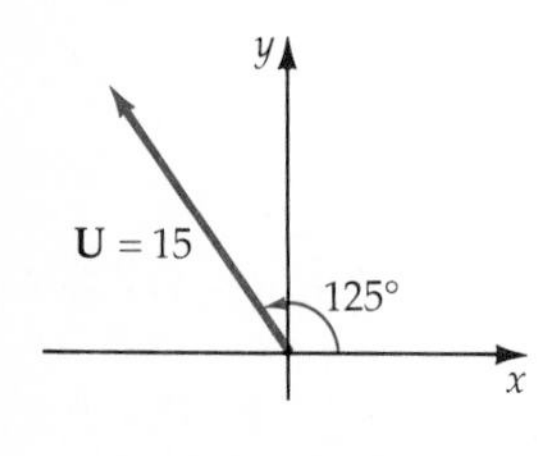

Figure 4.34

Vectors can be added algebraically by adding their x- and y-components.

Definition of the Addition of Vectors

If $\mathbf{U} = a\mathbf{i} + b\mathbf{j}$ and $\mathbf{V} = c\mathbf{i} + d\mathbf{j}$, then

$$\mathbf{U} + \mathbf{V} = (a\mathbf{i} + b\mathbf{j}) + (c\mathbf{i} + d\mathbf{j}) = (a + c)\mathbf{i} + (b + d)\mathbf{j} = \langle a + c, b + d \rangle.$$

EXAMPLE 5 Add Vectors Algebraically

Add the following vectors:

a. $\mathbf{U} = 4\mathbf{i} + 2\mathbf{j}$ and $\mathbf{V} = -3\mathbf{i} + 2\mathbf{j}$ b. $\mathbf{W} = \langle -4, 5 \rangle$ and $\mathbf{Z} = \langle 2, 3 \rangle$

Solution

a. $\mathbf{U} + \mathbf{V} = (4\mathbf{i} + 2\mathbf{j}) + (-3\mathbf{i} + 2\mathbf{j}) = (4 - 3)\mathbf{i} + (2 + 2)\mathbf{j} = \mathbf{i} + 4\mathbf{j}$

b. $\mathbf{W} + \mathbf{Z} = \langle -4, 5\rangle + \langle 2, 3\rangle = \langle -4 + 2, 5 + 3\rangle = \langle -2, 8\rangle$

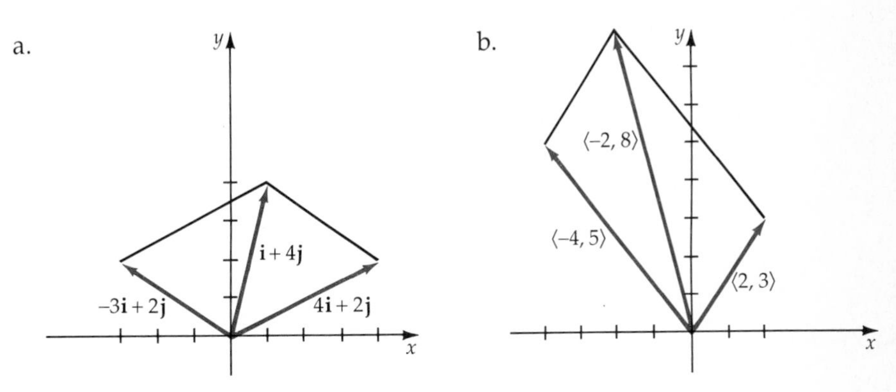

Figure 4.35

■ *Try Exercise* **30,** *page 192.*

To subtract **U** from **V**, add **−U** to **V**.

EXAMPLE 6 Subtract Vectors Algebraically

Subtract $\mathbf{U} = 2\mathbf{i} - 3\mathbf{j}$ from $\mathbf{V} = 4\mathbf{i} - \mathbf{j}$ and find the magnitude and direction of the new vector.

Solution

$$\begin{aligned}\mathbf{V} - \mathbf{U} &= (4\mathbf{i} - \mathbf{j}) - (2\mathbf{i} - 3\mathbf{j})\\ &= (4\mathbf{i} - \mathbf{j}) + (-2\mathbf{i} + 3\mathbf{j})\\ &= (4 - 2)\mathbf{i} + (-1 + 3)\mathbf{j}\\ &= 2\mathbf{i} + 2\mathbf{j}\end{aligned}$$

Use the PythagoreanTheorem and the tangent function to find the magnitude and the direction.

$$|\mathbf{V} - \mathbf{U}| = \sqrt{2^2 + (2)^2} = 2\sqrt{2} \qquad \tan\theta = \frac{2}{2} = 1$$

$$\theta = 45^\circ$$

The magnitude $\mathbf{V} - \mathbf{U}$ is $2\sqrt{2}$ and the direction angle is 45°.

■ *Try Exercise* **32,** *page 192.*

To multiply a vector $x\mathbf{i} + y\mathbf{j}$ by a scalar a, multiply each component of the vector by a.

Definition of the Product of a Scalar and a Vector

> If a is a scalar and $\mathbf{V}$ is the vector $x\mathbf{i} + y\mathbf{j} = \langle x, y\rangle$, then
>
> $$a\mathbf{V} = a(x\mathbf{i} + y\mathbf{j}) = ax\mathbf{i} + ay\mathbf{j} = \langle ax, ay\rangle.$$

EXAMPLE 7 **Perform Vector Operations**

Let $\mathbf{U} = \mathbf{i} - 3\mathbf{j}$ and $\mathbf{V} = -2\mathbf{i} + 2\mathbf{j}$. Find $2\mathbf{U} - 3\mathbf{V}$.

Solution

$$2\mathbf{U} - 3\mathbf{V} = 2(\mathbf{i} - 3\mathbf{j}) - 3(-2\mathbf{i} + 2\mathbf{j}) = 2\mathbf{i} - 6\mathbf{j} + 6\mathbf{i} - 6\mathbf{j} = 8\mathbf{i} - 12\mathbf{j}$$

■ *Try Exercise* **40**, *page 192.*

Dot Product

The *dot product* of two vectors is one way to multiply a vector by a vector. The dot product is a scalar and is useful in certain types of physics and engineering problems involving forces and work.

Definition of the Dot Product

Given $\mathbf{U} = a\mathbf{i} + b\mathbf{j}$ and $\mathbf{V} = c\mathbf{i} + d\mathbf{j}$, the dot product $\mathbf{U} \cdot \mathbf{V}$ is given by

$$\mathbf{U} \cdot \mathbf{V} = (a\mathbf{i} + b\mathbf{j}) \cdot (c\mathbf{i} + d\mathbf{j}) = ac + bd.$$

EXAMPLE 8 **Find the Dot Product**

Find the dot product of $\mathbf{U} = -3\mathbf{i} + 4\mathbf{j}$ and $\mathbf{V} = 2\mathbf{i} + \mathbf{j}$.

Solution $\mathbf{U} \cdot \mathbf{V} = (-3\mathbf{i} + 4\mathbf{j}) \cdot (2\mathbf{i} + \mathbf{j})$ Use the definition of the dot product.

$$= (-3)(2) + (4)(1) = -2$$

■ *Try Exercise* **56**, *page 192.*

Caution Remember that a dot product is a scalar.

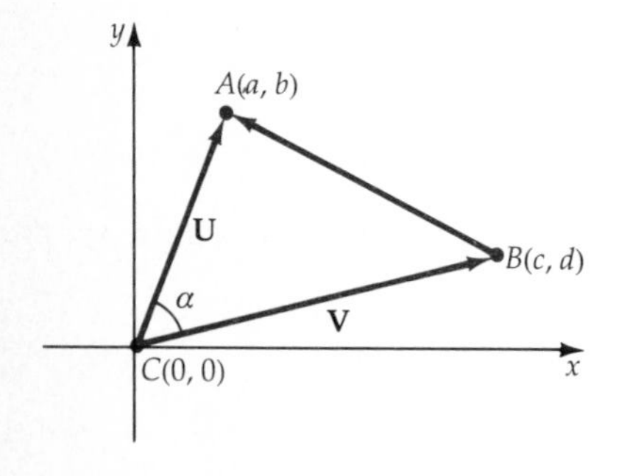

Figure. 4.36

The dot product and the Law of Cosines can be used to derive an alternate expression for the dot product. Given the vectors $\mathbf{U} = a\mathbf{i} + b\mathbf{j}$ and $\mathbf{V} = c\mathbf{i} + d\mathbf{j}$ shown in Figure 4.36, we can derive an expression for the cosine of the angle formed by the two vectors.

Using the Law of Cosines for the triangle OAB, we have

$$|\mathbf{AB}|^2 = |\mathbf{U}|^2 + |\mathbf{V}|^2 - 2|\mathbf{U}||\mathbf{V}| \cos \alpha.$$

By the distance formula, $|\mathbf{AB}|^2 = (a - c)^2 + (b - d)^2$, $|\mathbf{U}|^2 = a^2 + b^2$, and $|\mathbf{V}|^2 = c^2 + d^2$. Thus,

$$(a - c)^2 + (b - d)^2 = (a^2 + b^2) + (c^2 + d^2) - 2|\mathbf{U}||\mathbf{V}| \cos \alpha$$

$$a^2 - 2ac + c^2 + b^2 - 2bd + d^2 = a^2 + b^2 + c^2 + d^2 - 2|\mathbf{U}||\mathbf{V}| \cos \alpha$$

$$-2ac - 2bd = -2|\mathbf{U}||\mathbf{V}| \cos \alpha$$

$$ac + bd = |\mathbf{U}||\mathbf{V}| \cos \alpha$$

$$\mathbf{U} \cdot \mathbf{V} = |\mathbf{U}||\mathbf{V}| \cos \alpha$$ Substitute $\mathbf{U} \cdot \mathbf{V}$ for $ac + bd$

$$\cos \alpha = \frac{\mathbf{U} \cdot \mathbf{V}}{|\mathbf{U}||\mathbf{V}|}$$

Remark The equation for $\cos \alpha$ can be used to find the angle between any two vectors.

The Angle Between Two Vectors

If α is the angle between two nonzero vectors **U** and **V**, $0 \leq \alpha \leq 180°$, then

$$\cos \alpha = \frac{\mathbf{U} \cdot \mathbf{V}}{|\mathbf{U}||\mathbf{V}|}.$$

EXAMPLE 9 Find the Angle Between Two Vectors Using the Dot Product

Find the angle between the vectors $\mathbf{U} = 3\mathbf{i} + 2\mathbf{j}$ and $\mathbf{V} = -5\mathbf{i} + \mathbf{j}$.

Solution Use the equation

$$\begin{aligned} \cos \alpha &= \frac{\mathbf{U} \cdot \mathbf{V}}{|\mathbf{U}||\mathbf{V}|} \\ &= \frac{(3\mathbf{i} + 2\mathbf{j}) \cdot (-5\mathbf{i} + \mathbf{j})}{(\sqrt{3^2 + 2^2})(\sqrt{(-5)^2 + 1^2})} \\ &= \frac{-15 + 2}{\sqrt{13}\sqrt{26}} \\ &= \frac{-13}{\sqrt{338}} \\ &\approx -0.7071 \\ \alpha &\approx 135° \end{aligned}$$

The angle between the vectors is approximately 135°.

■ *Try Exercise* **62,** *page 192.*

Application Problems Using Vectors

We will consider an object on which two vectors are acting simultaneously. This occurs when a boat is moving in a current or an airplane is flying in a wind. We will refer to the airspeed of an airplane as if there were no wind. The actual velocity is the velocity relative to the earth. The magnitude of the actual velocity is called the ground speed.

Remark The bearing of an airplane is the direction in which it is pointed or the direction of the airspeed. The course is the actual direction the airplane is moving relative to the ground or the direction of the ground speed.

Consider an airplane flying due east at an airspeed of 525 mph. The wind is from the south at a speed of 40.0 mph. Suppose we are asked to find the ground speed and course of the airplane.

The airplane is headed east with a speed of 525 mph (airspeed). The airplane is also carried north by the wind. The actual velocity of the airplane (ground speed) is the sum of two vectors.

Graph the two vectors in a coordinate plane and find the resultant.

$$\tan \theta = \frac{40.0}{525}$$

$$\theta \approx 4.36^\circ$$

$$\alpha = 90.00 - 4.36^\circ = 85.64^\circ$$

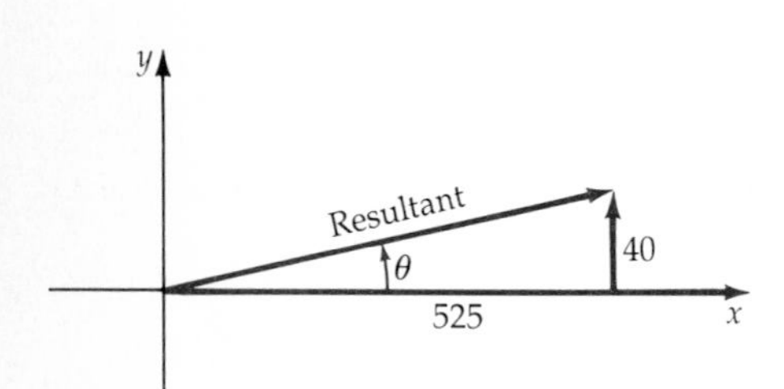

Figure 4.37

The sine function is used to find the ground speed $|\mathbf{V}|$.

$$\sin 4.36^\circ = \frac{40.0}{|\mathbf{V}|}$$

$$|\mathbf{V}| = \frac{40.0}{\sin 4.36^\circ} \approx 526$$

The airplane is on a course N85.6°E traveling at approximately 526 mph.

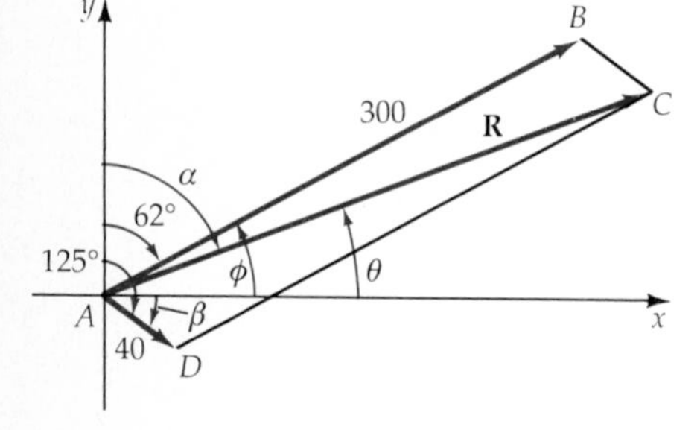

Figure 4.38

EXAMPLE 10 Solve an Application Involving Airspeed

An airplane traveling at an airspeed of 300 mph has a heading of 62.0°. The airplane is traveling through a 40.0-mph wind that has a heading of 125.0°. Find the ground speed and the course of the airplane. (Assume all lengths are accurate to three significant digits.)

Solution Sketch a vector diagram showing the given information. The vector **AB** represents the heading and the airspeed of the airplane. **AD** represents the wind velocity, and **AC** represents the course and the ground speed of the airplane. The angle ϕ (direction of vector **AB**) is equal to 28.0°, and the angle β (direction of vector **AD**) is equal to −35.0°. The resultant vector **AC** can be found by vector addition or by using the Law of Cosines. We will use vector addition.

We find the resultant vector by rewriting the heading and airspeed of the airplane and the wind velocity into component vectors and adding.

$$\mathbf{AB} = 300 \cos 28.^\circ\mathbf{i} + 300 \sin 28.^\circ\mathbf{j}$$

and

$$\mathbf{AD} = 40 \cos(-35^\circ)\mathbf{i} + 40 \sin(-35^\circ)\,.$$

Substitute for **AB** and **AD** in the equation **AC = AB + AD.**

$$\begin{aligned}\mathbf{AC} &= (300 \cos 28.^\circ\mathbf{i} + 300 \sin 28.^\circ\mathbf{j}) + [40 \cos(-35^\circ)\mathbf{i} + 40 \sin(-35^\circ)\mathbf{j}] \\ &\approx (264.9\mathbf{i} + 140.8\mathbf{j}) + (32.76\mathbf{i} - 22.94\mathbf{j}) \\ &= 297.7\mathbf{i} + 117.9\mathbf{j}\end{aligned}$$

The magnitude of the resultant is found by using the Pythagorean Theorem.

$$|\mathbf{R}| = \sqrt{(297.7)^2 + (117.9)^2} \approx 320$$

The angle that the resultant makes with the positive x-axis is found by using the tangent function. The course is found by subtracting this angle from 90°.

$$\tan\theta = \frac{117.9}{297.7}$$

$$\theta = 21.6^\circ$$

$$\alpha = 90.0^\circ - 21.6^\circ$$

$$= 68.4^\circ$$

The airplane is on a course N68.4°E traveling at approximately 320 mph.

■ *Try Exercise* **68,** *page 193.*

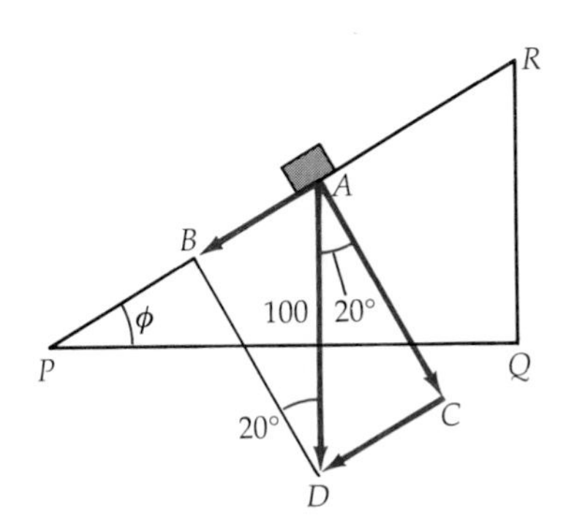

Figure 4.39

EXAMPLE 11 **Solve an Application Involving Force**

A 100-pound box is on a 20° ramp. Find the magnitude of the component of the force vector parallel to the ramp.

Solution We are finding the magnitude of the component of a force in a direction other than that of an axis. The force of gravity vector (100 pounds) can be written as the sum of two components, one parallel to the ramp and the other perpendicular to the ramp, as shown in Figure 4.39. The vector **AB** represents the force that pulls the box down the ramp. The angle ϕ is 20° because two angles are equal if they have mutually perpendicular sides; PQ is perpendicular to AD, and PR is perpendicular to AC. Triangle ADC is a right triangle. The magnitude of the component of the force vector is $|\mathbf{AB}|$.

$$\sin 20^\circ = \frac{|\mathbf{AB}|}{100}$$

$$|\mathbf{AB}| = 34$$

The magnitude of the component of the force vector parallel to the ramp is 34 pounds.

■ *Try Exercise* **70,** *page 193.*

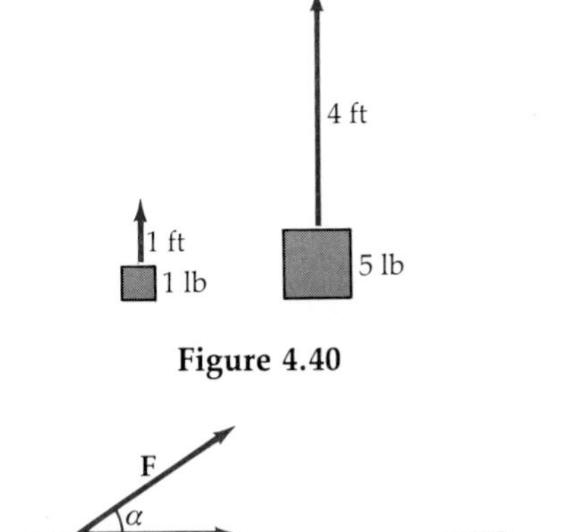

Figure 4.40

Figure 4.41

Work is done only if there is a displacement (or movement through a distance) of an object on which a force acts. **Work** is defined as a force exerted through a distance. Lifting 1 pound a distance of 1 foot is 1 foot pound (ft-lb) of work. Lifting 5 pounds a distance of 4 feet is $5 \cdot 4 = 20$ ft-lb of work. When the force is in the same direction as the displacement, the work is equal to the force times the displacement.

When the force is applied at a nonzero angle to the displacement, as shown in Figure 4.41, the work W is equal to the dot product of the force and the displacement vectors.

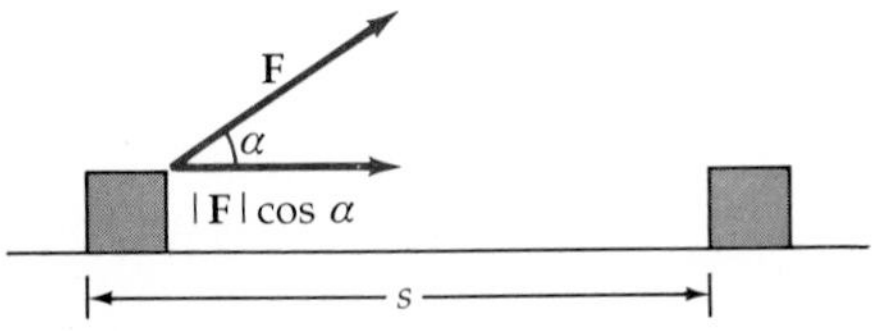

$$W = \mathbf{F} \cdot \mathbf{S}$$

$$= |\mathbf{F}||\mathbf{S}| \cos\alpha \qquad \mathbf{F} \cdot \mathbf{S} = |\mathbf{F}||\mathbf{S}| \cos\alpha.$$

EXAMPLE 12 **Find the Work Done by a Force**

Find the work done when a force of 50 pounds is used to pull a crate 20 feet along a level path if the force is at an angle of 30°. (See Figure 4.42.)

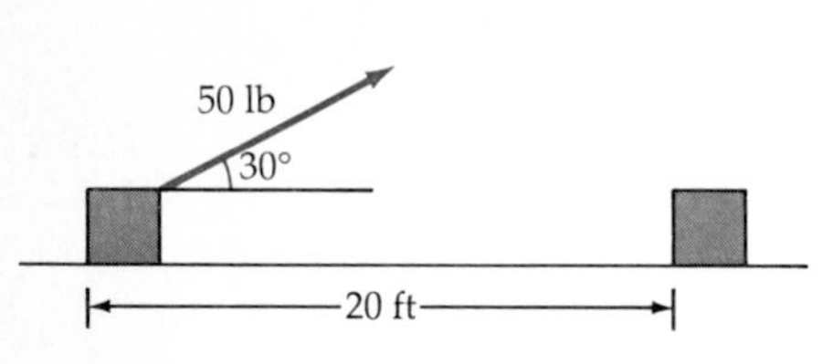

Figure 4.42

Solution Use the definition for work.

$$W = \mathbf{F} \cdot \mathbf{S} = |\mathbf{F}||\mathbf{S}| \cos 30° = 50 \cdot 20 \cos 30° \approx 870$$

The work is 870 foot-pounds (to the nearest 10 ft-lbs).

■ *Try Exercise* **72**, *page 193.*

EXERCISE SET 4.4

In Exercises 1 to 6, find the magnitude and direction of the vectors with the tail at P_1 and the head at P_2.

1. $P_1(2, 3), P_2(-1, 4)$ 2. $P_1(-5, 6), P_2(-4, -7)$
3. $P_1(-4, -6), P_2(5, -6)$ **4.** $P_1(3, -8), P_2(6, 3)$
5. $P_1(20, 30), P_2(-10, 30)$ **6.** $P_1(-45, 10), P_2(25, 25)$

In Exercises 7 to 10, graph each vector.

7. $\mathbf{U} = \langle -1, 4 \rangle$ 8. $\mathbf{V} = \langle 3, -2 \rangle$
9. $\mathbf{V} = \langle -5, -3 \rangle$ **10.** $\mathbf{U} = \langle 3, 0 \rangle$

In Exercises 11 to 18, find the magnitude and direction of the given vector.

11. $\mathbf{U} = \langle -3, 4 \rangle$ **12.** $\mathbf{U} = \langle 6, 10 \rangle$
13. $\mathbf{V} = \langle 20, -40 \rangle$ **14.** $\mathbf{V} = \langle -50, 30 \rangle$
15. $\mathbf{V} = 2\mathbf{i} - 4\mathbf{j}$ **16.** $\mathbf{V} = -5\mathbf{i} + 6\mathbf{j}$
17. $\mathbf{U} = 42\mathbf{i} - 18\mathbf{j}$ **18.** $\mathbf{U} = -22\mathbf{i} - 32\mathbf{j}$

In Exercises 19 to 28, find the magnitude of the x- and y-components of each vector.

19. $\mathbf{U} = \langle -8, 5 \rangle$ **20.** $\mathbf{V} = \langle 4, -3 \rangle$ **21.** $\mathbf{U} = \langle -2, -6 \rangle$
22. $\mathbf{V} = \langle -3, 4 \rangle$ **23.** $\mathbf{V} = -3\mathbf{i} + 4\mathbf{j}$ **24.** $\mathbf{U} = 8\mathbf{i} + 5\mathbf{j}$
25. $\mathbf{V} = 6\mathbf{i} - 4\mathbf{j}$ **26.** $\mathbf{U} = 12\mathbf{i} + 9\mathbf{j}$
27. $|\mathbf{U}| = 6.0, \theta = 50°$
28. $|\mathbf{U}| = 15, \theta = 140°$

In Exercises 29 to 32, perform the indicated operation.

29. $(2\mathbf{i} + 3\mathbf{j}) + (4\mathbf{i} - 2\mathbf{j})$ **30.** $(-2\mathbf{i} + \mathbf{j}) + (-\mathbf{i} + 2\mathbf{j})$
31. $(-3\mathbf{i} + \mathbf{j}) - (2\mathbf{i} + 3\mathbf{j})$ 32. $(\mathbf{i} - 4\mathbf{j}) - (-3\mathbf{i} - 2\mathbf{j})$

In Exercises 33 to 40, perform the indicated operation where $\mathbf{U} = 2\mathbf{i} + 3\mathbf{j}$, $\mathbf{V} = -\mathbf{i} + 2\mathbf{j}$.

33. $3\mathbf{U}$ 34. $-4\mathbf{V}$ **35.** $2\mathbf{U} - \mathbf{V}$
36. $3\mathbf{U} + 2\mathbf{V}$ **37.** $5\mathbf{V} - \frac{1}{2}\mathbf{U}$ **38.** $\frac{3}{4}\mathbf{V} + 2\mathbf{U}$
39. $\frac{2}{3}\mathbf{U} + \frac{1}{2}\mathbf{V}$ 40. $\frac{3}{4}\mathbf{U} - \frac{1}{2}\mathbf{V}$

In Exercises 41 to 48, write the vectors in the form $x\mathbf{i} + y\mathbf{j}$.

41. $\mathbf{U} = \langle -3, 4 \rangle$ **42.** $\mathbf{U} = \langle 7, -10 \rangle$
43. $\mathbf{V} = \langle 1, -3 \rangle$ **44.** $\mathbf{U} = \langle -2, -5 \rangle$
45. $|\mathbf{U}| = 5.0, \theta = 44°$ **46.** $|\mathbf{U}| = 60, \theta = 125°$
47. $|\mathbf{V}| = 8.0, \theta = 320°$ **48.** $|\mathbf{U}| = 12, \theta = 245°$

In Exercises 49 to 56, find the dot product of the vectors.

49. $\mathbf{U} = \langle 2, -3 \rangle, \mathbf{V} = \langle 1, 4 \rangle$
50. $\mathbf{U} = \langle -3, 2 \rangle, \mathbf{V} = \langle 5, 5 \rangle$
51. $\mathbf{U} = \langle 2, 0 \rangle, \mathbf{V} = \langle -5, 4 \rangle$
52. $\mathbf{U} = \langle -5, -2 \rangle, \mathbf{V} = \langle 7, -1 \rangle$
53. $\mathbf{U} = 2\mathbf{i} + 6\mathbf{j}, \mathbf{V} = -3\mathbf{i} + 5\mathbf{j}$
54. $\mathbf{U} = -\mathbf{i} + \mathbf{j}, \mathbf{V} = -4\mathbf{i} + 6\mathbf{j}$
55. $\mathbf{U} = 8\mathbf{i} - 13\mathbf{j}, \mathbf{V} = 7\mathbf{i} - 10\mathbf{j}$
56. $\mathbf{U} = 15\mathbf{i} - 10\mathbf{j}, \mathbf{V} = 18\mathbf{i} + 16\mathbf{j}$

In Exercises 57 to 64, find the angle between the vectors.

57. $\mathbf{U} = \langle -3, 2 \rangle, \mathbf{V} = \langle 1, 4 \rangle$
58. $\mathbf{U} = \langle 5, 5 \rangle, \mathbf{V} = \langle -3, -3 \rangle$
59. $\mathbf{U} = \langle 5, 0 \rangle, \mathbf{V} = \langle 2, -4 \rangle$
60. $\mathbf{U} = \langle 5, -2 \rangle, \mathbf{V} = \langle -4, -1 \rangle$
61. $\mathbf{U} = 3\mathbf{i} - 2\mathbf{j}, \mathbf{V} = 4\mathbf{i} + \mathbf{j}$
62. $\mathbf{U} = -\mathbf{i} + 5\mathbf{j}, \mathbf{V} = 4\mathbf{i} + 7\mathbf{j}$
63. $\mathbf{U} = -2\mathbf{i} - 3\mathbf{j}, \mathbf{V} = 3\mathbf{i} + 5\mathbf{j}$
64. $\mathbf{U} = 5\mathbf{i} - 3\mathbf{j}, \mathbf{V} = 6\mathbf{i} + \mathbf{j}$

65. Two forces of 68.0 pounds and 120 pounds are acting on an object. The angle between the forces is 17.0°. Find the resultant and the angle that the resultant makes with the smaller force.

66. Two forces of 85 pounds and 110 pounds act on an object. The 85-pound force acts at an angle of 9.4° with the positive x-axis, and the 110-pound force acts at an angle of $-12.0°$ with the positive x-axis. Find the resultant vector.

67. A plane is flying at an airspeed of 340 mph at a heading of S56°E. A wind of 45 mph is from the west. Find the ground speed of the plane.

68. A person who can row 2.6 mph in still water wants to row due east across a river. The river is flowing from the north at a rate of 0.8 mph. Determine the heading of the boat to travel due east across the river.

69. Find the magnitude of force necessary to keep a 3000-pound car from sliding down a ramp of 5.6°.

70. A 120-pound force keeps an 800-pound object from sliding down an inclined plane. Find the angle of the inclined plane.

71. A 200-pound object is dragged 18 feet along a level floor. Find the work done if the force used is 75 pounds at an angle of 28° above the horizontal.

72. An 800-pound force is pulling on a sled loaded with 2200 pounds of lumber. The force is at an angle of 35° above the horizontal. Find the work done if the sled is pulled 45 feet.

Supplemental Exercises

73. Show that if $\mathbf{U} \cdot \mathbf{V} = 0$, the vectors $\mathbf{U}$ and $\mathbf{V}$ are orthogonal, that is, the angle between the vectors is a right angle.

74. Show that if $\mathbf{U} \cdot \mathbf{V} = |\mathbf{U}||\mathbf{V}|$, then $\mathbf{U}$ and $\mathbf{V}$ are parallel, that is, the angle between the vectors is 0° or 180°.

75. Show that $\mathbf{i} \cdot \mathbf{i} = 1$.

76. Show that $\mathbf{i} \cdot \mathbf{j} = 0$.

77. Find the sum of the three vectors: $\mathbf{U} = 6\mathbf{i} + 3\mathbf{j}$, $\mathbf{V} = -2\mathbf{i} + 2\mathbf{j}$, $\mathbf{W} = -\mathbf{i} - 4\mathbf{j}$.

In Exercises 78 to 81, find the projection of the vector in the specified direction. The projection of $\mathbf{V}$ in the direction of $\mathbf{U}$ is given by $\dfrac{\mathbf{U} \cdot \mathbf{V}}{|\mathbf{U}|}\mathbf{U}$.

78. Projection of vector $\mathbf{V} = -2\mathbf{i} + 4\mathbf{j}$ in the direction of vector $\mathbf{U} = 3\mathbf{i} - 2\mathbf{j}$.

79. Projection of vector $\mathbf{V} = 4\mathbf{i} - 2\mathbf{j}$ in the direction of vector $\mathbf{U} = -3\mathbf{i} + \mathbf{j}$.

80. Projection of vector $\mathbf{V} = \mathbf{i}$ in the direction of vector $\mathbf{U} = \mathbf{j}$.

81. Projection of vector $\mathbf{V} = -\mathbf{i} + \mathbf{j}$ in the direction of vector $\mathbf{U} = \mathbf{i} - \mathbf{j}$.

Chapter 4 Review

4.1 Right Triangles

To solve a right triangle means to find the lengths of all the sides and the measures of all the angles of the triangle.

4.2 The Law of Sines

The Law of Sines is used to solve general triangles when two angles and the included side are given or in triangles in which two sides and an angle opposite one of them are given.

$$\frac{a}{\sin A} = \frac{b}{\sin B} = \frac{c}{\sin C}$$

4.3 The Law of Cosines and Area

The Law of Cosines $a^2 = b^2 + c^2 - 2bc \cos A$ is used to solve general triangles when two sides and the included angle or three sides of the triangle are given.

Area K of a triangle ABC is

$$K = \frac{1}{2} bc \sin A = \frac{b^2 \sin C \sin A}{2 \sin B}.$$

Area for a triangle in which three sides are given (Heron's formula):

$$K = \sqrt{s(s-a)(s-b)(s-c)}, \quad \text{where } s = \frac{1}{2}(a + b + c).$$

4.4 Vectors

A vector is a quantity with magnitude and direction. Two vectors are equal if they have the same magnitude and direction. The resultant of two or more vectors is the single vector that will have the equivalent effect of the vectors.

Vectors can be added by parallelogram addition, triangle addition, or by adding the x- and y-components.

Multiplication of a vector by a scaler is given by

$$a(x\mathbf{i} + y\mathbf{j}) = ax\mathbf{i} + ay\mathbf{j}.$$

The dot product of two vectors $\mathbf{U} = a\mathbf{i} + b\mathbf{j}$ and $\mathbf{V} = c\mathbf{i} + d\mathbf{j}$ is given by

$$\mathbf{U} \cdot \mathbf{V} = (a\mathbf{i} + b\mathbf{j}) \cdot (c\mathbf{i} + d\mathbf{j}) = ac + bd$$

The cosine of the angle α between two vectors is given by

$$\cos \alpha = \frac{\mathbf{U} \cdot \mathbf{V}}{|\mathbf{U}||\mathbf{V}|}$$

CHALLENGE EXERCISES

For Exercises 1 to 16, answer true or false. If the statement is false, give an example.

1. The Law of Cosines can be used to solve any triangle given two sides and an angle.
2. The law of sines can be used to solve any triangle given two angles and any side.
3. In any triangle, the largest side is opposite the largest angle.
4. If two vectors have the same magnitude, then they are equal.
5. It is possible for the sum of two nonzero vectors to equal zero.
6. The expression $a^2 = b^2 + c^2 + 2bc \cos D$ is true for triangle ABC in which angle D is the supplement of angle A.
7. The measure of angle α formed by two vectors is greater than or equal to 0° and less than or equal to 180°.
8. If A, B, and C are the angles of a triangle, then
$$\sin(A + B + C) = 0.$$
9. Real numbers are complex numbers.
10. Let $\mathbf{V} = a\mathbf{i} + b\mathbf{j}$; then $\mathbf{V} \cdot \mathbf{V} = a^2\mathbf{i} + b^2\mathbf{j}$.
11. If $\mathbf{V}$ and $\mathbf{W}$ are vectors with $\mathbf{V} \cdot \mathbf{W} = 0$, then $\mathbf{V} = 0$ or $\mathbf{W} = 0$.

REVIEW EXERCISES

In Exercises 1 to 10, solve the triangles.

1. $A = 37°, b = 14, C = 90°$

2. $B = 77.4°, c = 11.8, C = 90°$

3. $a = 12, b = 15, c = 20$

4. $a = 24, b = 32, c = 28$

5. $a = 18, b = 22, C = 35°$

6. $b = 102, c = 150, A = 82°$

7. $A = 105°, a = 8, c = 10$

8. $C = 55°, c = 80, b = 110$

9. $A = 55°, B = 80°, c = 25$

10. $B = 25°, C = 40°, c = 40$

In Exercises 11 to 18, find the area of each triangle.

11. $a = 24, b = 30, c = 36$

12. $a = 9.0, b = 7.0, c = 12$

13. $a = 60, b = 44, C = 44°$

14. $b = 8.0, c = 12, A = 75°$

15. $b = 50, c = 75, C = 15°$

16. $b = 18, a = 25, A = 68°$

17. $A = 110°, a = 32, b = 15$

18. $C = 45°, c = 22, b = 18$

In Exercises 19 to 26, find the magnitude and direction of the given vector.

19. $\mathbf{U} = \langle 4, -2 \rangle$

20. $\mathbf{U} = \langle -2, -5 \rangle$

21. $\mathbf{V} = \langle -4, 2 \rangle$

22. $\mathbf{V} = \langle 6, -3 \rangle$

23. $\mathbf{U} = -2\mathbf{i} + 3\mathbf{j}$

24. $\mathbf{U} = -4\mathbf{i} - 7\mathbf{j}$

25. $\mathbf{V} = 5\mathbf{i} + \mathbf{j}$

26. $\mathbf{V}3\mathbf{i} - 5\mathbf{j}$

In Exercises 27 to 30, find the magnitude of the x- and y-components of the following vectors.

27. $\mathbf{U} = \langle -8, 5 \rangle$

28. $\mathbf{U} = \langle 7, -12 \rangle$

29. $\mathbf{V} = 10\mathbf{i} + 6\mathbf{j}$

30. $\mathbf{V} = 8\mathbf{i} - 5\mathbf{j}$

In Exercises 31 to 34, perform the indicated operation. $\mathbf{U} = 3\mathbf{i} + 2\mathbf{j}$, $\mathbf{V} = -4\mathbf{i} - \mathbf{j}$.

31. $\mathbf{V} - \mathbf{U}$

32. $2\mathbf{U} - 3\mathbf{V}$

33. $-\mathbf{U} + \frac{1}{2}\mathbf{V}$

34. $\frac{2}{3}\mathbf{V} - \frac{3}{4}\mathbf{U}$

In Exercises 35 to 38, find the dot product of the vectors.

35. $\mathbf{U} = \langle 3, -2 \rangle, \mathbf{V} = \langle -1, 3 \rangle$

36. $\mathbf{V} = \langle -8, 5 \rangle, \mathbf{U} = \langle 2, -1 \rangle$

37. $\mathbf{V} = -4\mathbf{i} - \mathbf{j}, \mathbf{U} = 2\mathbf{i} + \mathbf{j}$

38. $\mathbf{U} = -3\mathbf{i} + 7\mathbf{j}, \mathbf{V} = -2\mathbf{i} + 2\mathbf{j}$

In Exercises 39 to 42, find the angle between the vectors.

39. $\mathbf{U} = \langle 7, -4 \rangle, \mathbf{V} = \langle 2, 3 \rangle$

40. $\mathbf{V} = \langle -5, 2 \rangle, \mathbf{U} = \langle 2, -4 \rangle$

41. $\mathbf{V} = 6\mathbf{i} - 11\mathbf{j}, \mathbf{U} = 2\mathbf{i} + 4\mathbf{j}$

42. $\mathbf{U} = \mathbf{i} - 5\mathbf{j}, \mathbf{V} = \mathbf{i} + 5\mathbf{j}$

5 Complex Numbers

Clouds are not spheres, mountains are not cones, coastlines are not circles, and bark is not smooth, nor does lightning travel in a straight line.

BENOIT MANDELBROT

THE MANDELBROT SET

The quote on the preceding page, from *The Fractal Geometry of Nature* by Benoit Mandelbrot, appears on the first page of his book and embodies the very essence of fractal geometry.

Mandelbrot says that he coined the word *fractal* from the Latin word *fragere* which means to break into irregular fragments. His studies in geometry have lead to *fractional dimensions* such as 1.26 or 1.67.

Mandelbrot also defined a set of complex numbers, the topic of this chapter, called the Mandelbrot set. The boundary of this set is a fractal curve. To construct the Mandelbrot set, begin with some complex number c. Substitute the value of c for z in the expression $z^2 + c$. The new number is

$$c_1 = c^2 + c\,.$$

Now substitute this number for z in $z^2 + c$. The new number is

$$c_2 = c_1^2 + c = (c^2 + c)^2 + c\,.$$

Repeat this process over and over. For example,

$$c_3 = c_2^2 + c = [(c^2 + c)^2 + c]^2 + c\,.$$

The Mandelbrot set is the set of all complex numbers c, for which $|c_n| < 2$, even after an infinite number of repetitions. The figure below represents a computer investigation into the Mandelbrot set. The white area is the set of complex numbers that belongs to the Mandelbrot set.

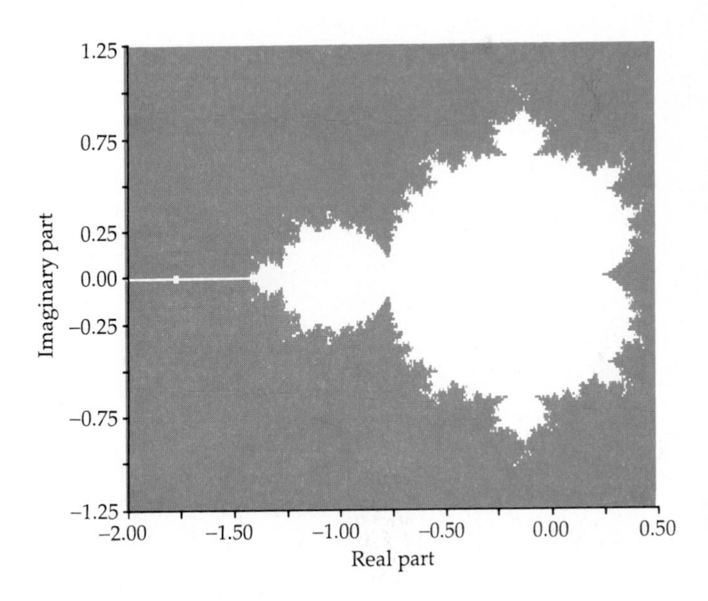

5.1 Complex Numbers

There is no real number whose square is a negative number. For example, there is no real number x such that $x^2 = -1$. In the seventeenth century, a new number, called an **imaginary number,** was defined. The square of an imaginary number is a negative real number. The letter i was chosen to represent an imaginary number whose square is -1.

Definition of i

> The number i, called the **imaginary unit,** is a number such that $i^2 = -1$.

Many of the solutions to equations in the remainder of this text will involve radicals such as $\sqrt{-a}$, where a is a positive real number. The expression $\sqrt{-a}$, with $a > 0$, is defined as follows.

Definition of $\sqrt{-a}$

> For any positive real number a, $\sqrt{-a} = i\sqrt{a}$.

Remark This definition with $a = 1$ implies

$$\sqrt{-1} = i.$$

The above definition is often used to write the square root of a negative real number as the product of the imaginary unit i and a positive real number. For example,

$$\sqrt{-4} = i\sqrt{4} = 2i \qquad \text{and} \qquad \sqrt{-7} = i\sqrt{7}.$$

Definition of a Complex Number

> If a and b are real numbers and i is the imaginary unit, then $a + bi$ is called a **complex number.** The real number a is called the **real part** and the real number b is called the **imaginary part** of the complex number.

Remark Even though b is a real number, it is called the imaginary part of the complex number $a + bi$. For example, the complex number $3 + 8i$ has the real number 8 as its imaginary part.

The real numbers are a subset of the complex numbers. This can be observed by letting $b = 0$. Then $a + bi = a + 0i = a$, which is a real number. Any number that can be written in the form $0 + bi = bi$, where b is a nonzero real number, is an **imaginary number** (or a pure imaginary number). For example, i, $3i$, and $-0.5i$ are all imaginary numbers.

A complex number is in **standard form** when it is written in the form $a + bi$.

EXAMPLE 1 Write Complex Numbers in Standard Form

Write each complex number in standard form.

a. $3 + \sqrt{-4}$ b. $\sqrt{-37} - 3$ c. $-\sqrt{-121}$

Solution Use the definition $\sqrt{-a} = i\sqrt{a}$ and the definition of standard form.

a. $3 + \sqrt{-4} = 3 + i\sqrt{4} = 3 + 2i$

b. $\sqrt{-37} - 3 = i\sqrt{37} - 3 = -3 + i\sqrt{37}$

c. $-\sqrt{-121} = -(i\sqrt{121}) = -11i = 0 - 11i$

■ *Try Exercise* **2**, *page 203.*

Remark The expression $\sqrt{a}\,i$ is often written as $i\sqrt{a}$ so that it is not mistaken for $\sqrt{ai}$.

Arithmetic Operations on Complex Numbers

To compute with complex numbers, we must first define the arithmetic operations of addition, subtraction, multiplication, and division of two complex numbers.

Definition of Addition and Subtraction of Complex Numbers

If $a + bi$ and $c + di$ are complex numbers, then

Addition $(a + bi) + (c + di) = (a + c) + (b + d)i$

Subtraction $(a + bi) - (c + di) = (a - c) + (b - d)i$

Thus to add two numbers, you add their real parts to produce the real part of the sum and you add their imaginary parts to produce the imaginary part of the sum.

EXAMPLE 2 Add Complex Numbers

Write each sum as a complex number in standard form.

a. $(4 + 2i) + (3 + 7i)$ b. $2 + (6 - 5i)$

Solution

a. $(4 + 2i) + (3 + 7i) = (4 + 3) + (2 + 7)i = 7 + 9i$

b. $2 + (6 - 5i) = (2 + 0i) + (6 - 5i) = (2 + 6) + (0 + (-5))i = 8 - 5i$

■ *Try Exercise* **12**, *page 203.*

To subtract two complex numbers, subtract their real parts to produce the real part of the difference and subtract their imaginary parts to produce the imaginary part of the difference.

EXAMPLE 3 Subtract Complex Numbers

Write each difference as a complex number in standard form.

a. $(5 + 7i) - (-3 + 9i)$ b. $i - (3 - 4i)$

Solution

a. $(5 + 7i) - (-3 + 9i) = [5 - (-3)] + (7 - 9)i = 8 - 2i$

b. $i - (3 - 4i) = (0 + 1i) - (3 - 4i)$

$= (0 - 3) + [1 - (-4)]i = -3 + 5i$

■ *Try Exercise* **16,** *page 203.*

Definition of Multiplication of Complex Numbers

If $a + bi$ and $c + di$ are complex numbers, then

$$(a + bi)(c + di) = (ac - bd) + (ad + bc)i$$

Since every complex number can be written as a sum of two terms, it is natural to perform multiplication on complex numbers in a manner consistent with the operation of multiplication defined on binomials and the definition $i^2 = -1$. Thus to multiply complex numbers it is not necessary to memorize the definition of multiplication.

EXAMPLE 4 Multiply Complex Numbers

Write each product as a complex number in standard form.

a. $(3 + 5i)(2 - 4i)$ b. $4i(3 + 7i)$ c. $(2 - 5i)^2$

Solution

a. $(3 + 5i)(2 - 4i) = 6 - 12i + 10i - 20i^2$ The FOIL method

$= 6 - 12i + 10i - 20(-1)$ Substitute (-1) for i^2.

$= 6 + 20 - 12i + 10i$ Simplify.

$= 26 - 2i$

b. $4i(3 + 7i) = 12i + 28i^2$ Distribute the $4i$

$= 12i + 28(-1)$

$= -28 + 12i$

c. $(2 - 5i)^2 = 4 - 20i + 25i^2$ Use $(a - b)^2 = a^2 - 2ab + b^2$.

$= 4 - 20i + 25(-1)$

$= -21 - 20i$

■ *Try Exercise* **24,** *page 203.*

The complex numbers $a + bi$ and $a - bi$ are called **complex conjugates** or **conjugates** of each other. The conjugate of the complex number z is denoted by $\overline{z}$. For example,

$$\overline{3 + 2i} = 3 - 2i \quad \text{and} \quad \overline{7 - 11i} = 7 + 11i.$$

The Product of z and $\overline{z}$

> The product of a complex number and its conjugate is always a real number.

Proof: We can verify this theorem by using the complex number $a + bi$ and its conjugate $a - bi$.

$$(a + bi)(a - bi) = a^2 - abi + abi - b^2i^2$$
$$= a^2 - b^2(-1)$$
$$= a^2 + b^2 \quad \leftarrow \text{a real number}$$

The fact that the product of a complex number and its conjugate is always a real number can be used to find the quotient of two complex numbers. For example, to find the quotient $(a + bi)/(c + di)$, multiply numerator and denominator by the conjugate of the denominator.

EXAMPLE 5 Divide Complex Numbers

Write each quotient as a complex number in standard form.

a. $\dfrac{3 + 2i}{5 - i}$ b. $\dfrac{2 - 7i}{-i}$

Solution

a. $\dfrac{3 + 2i}{5 - i} = \dfrac{(3 + 2i)(5 + i)}{(5 - i)(5 + i)}$ Multiply numerator and denominator by $5 + i$, which is the conjugate of the denominator.

$= \dfrac{15 + 3i + 10i + 2i^2}{25 + 1}$

$= \dfrac{13 + 13i}{26} = \dfrac{1}{2} + \dfrac{1}{2}i$ Write in standard form.

b. $\dfrac{2-7i}{-i} = \dfrac{(2-7i)(i)}{-i(i)}$ Multiply numerator and denominator by i, which is the conjugate of the denominator.

$$= \frac{2i - 7i^2}{-i^2}$$

$$= \frac{7 + 2i}{1} = 7 + 2i$$

■ *Try Exercise* **34,** *page 203.*

There are several theorems about complex numbers and their conjugates. For example, the following theorem concerns the sum of a complex number $z = a + bi$ and its conjugate $\bar{z} = a - bi$.

The Sum of z and $\bar{z}$

> The sum of a complex number z and its conjugate $\bar{z}$ is a real number.

Proof: To verify this, we let $z = a + bi$ and $\bar{z} = a - bi$.

$$z + \bar{z} = (a + bi) + (a - bi) = (a + a) + [b + (-b)]i$$

$$= 2a + 0i = 2a \quad \leftarrow \text{a real number}$$

The following powers of i illustrate a pattern:

$i^1 = i$ $\qquad i^5 = i^4 \cdot i = (1)i = i$

$i^2 = -1$ $\qquad i^6 = i^4 \cdot i^2 = (1)(-1) = -1$

$i^3 = i^2 \cdot i = (-1)i = -i$ $\qquad i^7 = i^4 \cdot i^3 = (1)(-i) = -i$

$i^4 = i^2 \cdot i^2 = (-1)((-1) = 1$ $\qquad i^8 = (i^4)^2 = 1^2 = 1$

Since $(i^4)^n = 1$ for any integer n, it is possible to evaluate powers of i by factoring out powers of i^4, as shown in the following example:

$$i^{25} = (i^4)^6(i) = 1^6(i) = i.$$

The following theorem is often used to evaluate powers of i. Essentially it makes use of division to eliminate powers of i^4.

Powers of i

> If n is a positive integer, then $i^n = i^r$, where r is the remainder of the division of n by 4.

This theorem is particularly useful when evaluating large powers of i.

EXAMPLE 6 **Evaluate Powers of i**

Evaluate each of the following: a. i^{14} b. i^{543}

Solution Use the theorem on powers of i.

a. $i^{14} = i^2 = -1$ Remainder of $14 \div 4$ is 2.

b. $i^{543} = i^3 = -i$ Remainder of $543 \div 4$ is 3.

■ *Try Exercise* **54,** *page 204.*

Caution To compute $\sqrt{a}\sqrt{b}$ when both a and b are negative numbers, write each radical in terms of i before multiplying. For example,

Correct method $\sqrt{-1}\sqrt{-1} = i \cdot i = i^2 = -1$

Incorrect method $\sqrt{-1}\sqrt{-1} = \sqrt{(-1)(-1)} = \sqrt{1} = 1.$

EXAMPLE 7 **Simplify Products Involving Radicals with Negative Radicands**

Simplify each of the following:

a. $\sqrt{-16}\sqrt{-25}$ b. $\sqrt{-9}\sqrt{-7}$ c. $(2 + \sqrt{-5})(2 - \sqrt{-5})$

Solution

a. $\sqrt{-16}\sqrt{-25} = (4i)(5i) = 20i^2 = -20$

b. $\sqrt{-9}\sqrt{-7} = (3i)(i\sqrt{7}) = 3i^2\sqrt{7} = -3\sqrt{7}$

c. $(2 + \sqrt{-5})(2 - \sqrt{-5}) = (2 + i\sqrt{5})(2 - i\sqrt{5}) = 4 - (i^2)5 = 9$

■ *Try Exercise* **68,** *page 204.*

EXERCISE SET 5.1

In Exercises 1 to 10, write the complex number in standard form.

1. $2 + \sqrt{-9}$ **2.** $3 + \sqrt{-25}$

3. $4 - \sqrt{-121}$ **4.** $5 - \sqrt{-144}$

5. $8 + \sqrt{-3}$ **6.** $9 - \sqrt{-75}$

7. $\sqrt{-16} + 7$ **8.** $\sqrt{-49} + 3$

9. $\sqrt{-81}$ **10.** $-\sqrt{-100}$

In Exercises 11 to 28, simplify and write the complex number in standard form.

11. $(2 + 5i) + (3 + 7i)$ **12.** $(1 - 3i) + (6 + 2i)$

13. $(-5 - i) + (9 - 2i)$ **14.** $5 + (3 - 2i)$

15. $(8 - 6i) - (10 - i)$ **16.** $(-3 + i) - (-8 + 2i)$

17. $(7 - 3i) - (-5 - i)$ **18.** $7 - (3 - 2i)$

19. $8i - (2 - 3i)$ **20.** $(4i - 5) - 2$

21. $3(2 + 7i) + 5(2 - i)$ **22.** $8(4 - i) - (4 - 3i)$

23. $(2 + 3i)(4 - 5i)$ **24.** $(5 - 3i)(-2 - 4i)$

25. $(5 + 7i)(5 - 7i)$ **26.** $(-3 - 5i)(-3 + 5i)$

27. $(8i + 11)(-7 + 5i)$ **28.** $(9 - 12i)(15i + 7)$

In Exercises 29 to 48, write each expression as a complex number in standard form.

29. $\dfrac{4 + i}{3 + 5i}$ **30.** $\dfrac{5 - i}{4 + 5i}$ **31.** $\dfrac{1}{7 - 3i}$ **32.** $\dfrac{1}{-8 + i}$

33. $\dfrac{3 + 2i}{3 - 2i}$ **34.** $\dfrac{5 - 7i}{5 + 7i}$ **35.** $\dfrac{2i}{11 + i}$ **36.** $\dfrac{3i}{5 - 2i}$

37. $\dfrac{6 + i}{i}$ **38.** $\dfrac{5 - i}{-i}$

39. $(3 - 5i)^2$ **40.** $(-5 + 7i)^2$

41. $(1 - i) - 2(4 + i)^2$ **42.** $(4 - i) - 5(2 + 3i)^2$

43. $(1 - i)^3$ **44.** $(2 + i)^3$

45. $(2i)(8i)$ **46.** $(-5)(7i)$

47. $(5i)^2(-3i)$ **48.** $(-6i)(-5i)^2$

In Exercises 49 to 64, simplify and write the complex number as either i, $-i$, 1, or -1.

49. i^3 **50.** $-i^3$ **51.** i^5 **52.** $-i^5$

53. i^{10} **54.** i^{28} **55.** $-i^{40}$ **56.** i^{40}

57. i^{223} **58.** i^{553} **59.** i^{2001} **60.** i^{5000}

61. i^{5042} **62.** i^0 **63.** i^{-1} **64.** $i^{10,000}$

In Exercises 65 to 72, simplify each product.

65. $\sqrt{-1}\sqrt{-4}$ **66.** $\sqrt{-16}\sqrt{-49}$

67. $\sqrt{-64}\sqrt{-5}$ **68.** $\sqrt{-3}\sqrt{-121}$

69. $(3 + \sqrt{-2})(3 - \sqrt{-2})$

70. $(4 + \sqrt{-81})(4 - \sqrt{-81})$

71. $(5 + \sqrt{-16})^2$ **72.** $(3 - \sqrt{-144})^2$

In Exercises 73 to 80, evaluate

$$\frac{-b \pm \sqrt{b^2 - 4ac}}{2a}$$

for the given values of a, b, and c. Write your final answer as a complex number in standard form.

73. $a = 3, b = -3, c = 3$ **74.** $a = 1, b = -3, c = 10$

75. $a = 2, b = 4, c = 4$ **76.** $a = 4, b = -4, c = 2$

77. $a = 2, b = 6, c = 6$ **78.** $a = 6, b = -5, c = 5$

79. $a = 2, b = 1, c = 3$ **80.** $a = 3, b = 2, c = 4$

The **absolute value of the complex number** $a + bi$ is denoted by $|a + bi|$ and defined as the real number $\sqrt{a^2 + b^2}$. In Exercises 81 to 88, find the indicated absolute value of each complex number.

81. $|3 + 4i|$ **82.** $|5 + 12i|$ **83.** $|2 - 5i|$ **84.** $|4 - 4i|$

85. $|7 - 4i|$ **86.** $|11 - 2i|$ **87.** $|-3i|$ **88.** $|18i|$

Supplemental Exercises

In Exercises 89 to 92, use the complex number $z = a + bi$ and its conjugate $\overline{z} = a - bi$ to establish each result.

89. Prove that the absolute value of a complex number and the absolute value of its conjugate are equal.

90. Prove that the difference of a complex number and its conjugate is a pure imaginary number.

91. Prove that the conjugate of the sum of two complex numbers equals the sum of the conjugates of the two numbers.

92. Prove that the conjugate of the product of two complex numbers equals the product of the conjugates of the two numbers.

93. Show that if $x = 1 + i\sqrt{3}$, then $x^2 - 2x + 4 = 0$.

94. Show that if $x = 1 - i\sqrt{3}$, then $x^2 - 2x + 4 = 0$.

95. Simplify

$$[(3 + \sqrt{5}) + (7 - \sqrt{3})i][(3 + \sqrt{5}) - (7 - \sqrt{3})i].$$

96. Simplify

$$[2 - (3 - \sqrt{5})i][2 + (3 - \sqrt{5})i].$$

97. Simplify

$$\left(\frac{-1}{2} + \frac{\sqrt{3}}{2}i\right)^3.$$

98. Simplify $(a + bi)^3$, where a and b are real numbers.

99. Simplify $i + i^2 + i^3 + i^4 + \cdots + i^{28}$.

100. Simplify $i + i^2 + i^3 + i^4 + \cdots + i^{100}$.

The product $(a + bi)(a - bi) = a^2 + b^2$ can be used to factor the sum of two squares over the set of complex numbers. *Example:*

$$x^2 + 25 = x^2 + 5^2 = (x + 5i)(x - 5i)$$

In Exercises 101 to 104, factor each polynomial over the set of complex numbers.

101. $x^2 + 9$ **102.** $y^2 + 121$

103. $4x^2 + 81$ **104.** $144y^2 + 625$

In Exercises 105 to 108, evaluate the polynomial for the given value of x.

105. $x^2 + 36$; $x = 6i$

106. $x^2 + 100$; $x = -10i$

107. $x^2 - 6x + 10$; $x = 3 + i$

108. $x^2 + 10x + 29$; $x = -5 + 2i$

5.2 Trigonometric Form of Complex Numbers

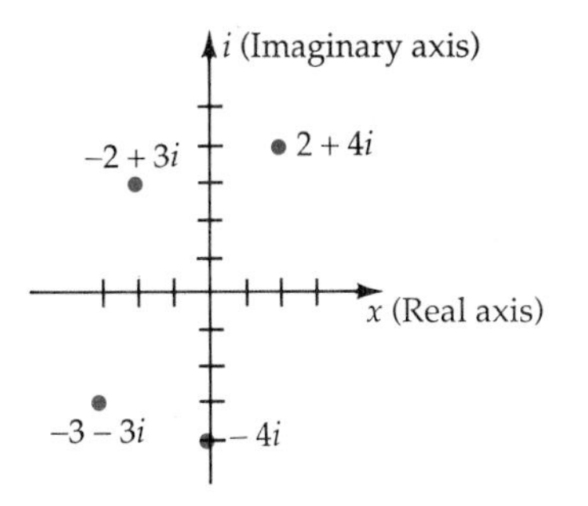

Figure 5.1

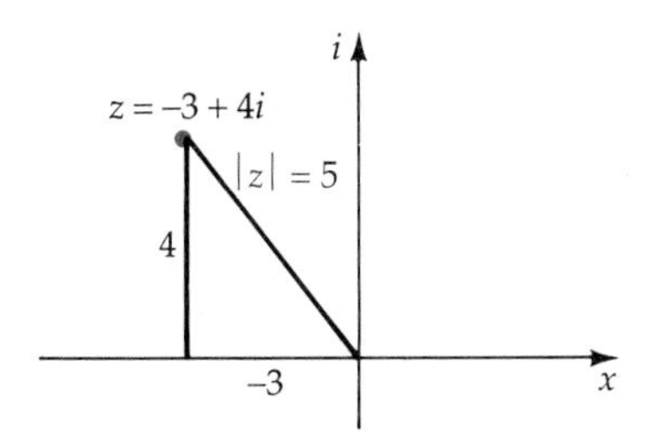

Figure 5.2

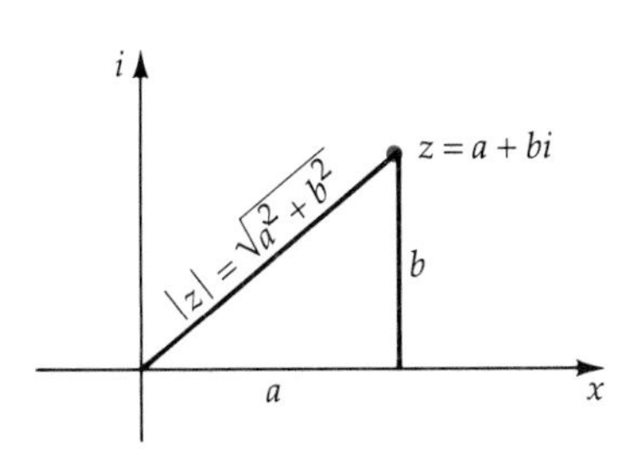

Figure 5.3

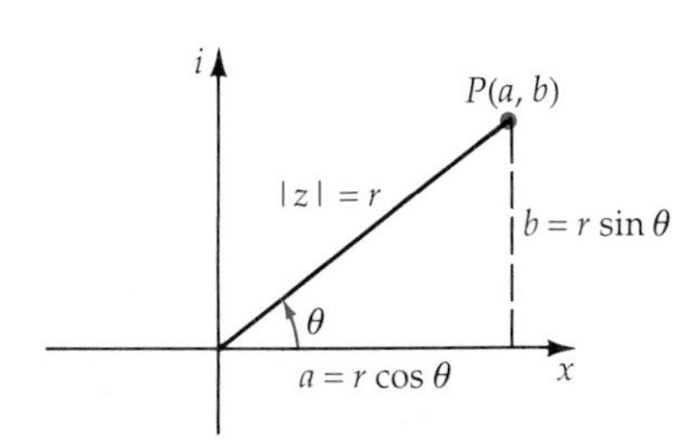

Figure 5.4

Real numbers are graphed as points on a number line. Complex numbers can be graphed on a coordinate grid called an **Argand diagram.** The horizontal axis of the coordinate plane is called the **real axis.** The vertical axis is called the **imaginary axis.** This coordinate system is called the **complex plane.**

A complex number written in the form $z = a + bi$ is said to be written in **standard** or **rectangular form.** The graph of $a + bi$ is associated with the point $P(a, b)$ in the complex plane. Figure 5.1 shows the graphs of several complex numbers.

Figure 5.2 is the graph of the complex number $z = -3 + 4i$. The length of the line segment drawn from the origin to the point $P(-3, 4)$ on the complex plane is the *absolute value* of z. From the Pythagorean Theorem, the absolute value of $z = -3 + 4i$ is

$$\sqrt{(-3)^2 + 4} = \sqrt{25} = 5 .$$

In general, the absolute value of the complex number $z = a + bi$ is

$$|z| = |a + bi| = \sqrt{a^2 + b^2} .$$

Thus $|z|$ is the distance from the origin to z (see Figure 5.3)

A complex number $z = a + bi$ can be written in terms of trigonometric functions. Consider the complex number graphed in Figure 5.4. We can write a and b in terms of the sine and the cosine.

$$\cos\theta = \frac{a}{r} \qquad \sin\theta = \frac{b}{r} \qquad \text{where } r = |z| = \sqrt{a^2 + b^2}$$

$$a = r\cos\theta \qquad b = r\sin\theta$$

Substituting for a and b in $z = a + bi$, we obtain

$$z = r\cos\theta + ir\sin\theta = r(\cos\theta + i\sin\theta) .$$

The expression $z = r(\cos\theta + i\sin\theta)$ is known as the **trigonometric form of a complex number.** The value of r is called the **modulus** of the complex number z, and the angle θ is called the **argument** of z.

The modulus r and the argument θ of a complex number $z = a + bi$ are given by

$$r = \sqrt{a^2 + b^2} \quad \text{and} \quad \cos\theta = \frac{a}{r}, \quad \sin\theta = \frac{b}{r} \quad \text{where } 0 \le \theta < 360°$$

We also can write

$$\alpha = \tan^{-1}\left|\frac{b}{a}\right| ,$$

where α is the reference angle for θ. Because of the periodic nature of the sine function and the cosine function, the trigonometric form of a complex number is not unique.

Since $\cos\theta = \cos(\theta + 2k\pi)$ and $\sin\theta = \sin(\theta + 2k\pi)$, where k is an integer the following complex numbers are equal.

$$r(\cos\theta + i\sin\theta) = r[\cos(\theta + 2k\pi) + i\sin(\theta + 2k\pi)] \quad \text{for } k \text{ an integer}.$$

For example,

$$2\left(\cos\frac{\pi}{6} + i\sin\frac{\pi}{6}\right) = 2\left[\cos\left(\frac{\pi}{6} + 2\pi\right) + i\sin\left(\frac{\pi}{6} + 2\pi\right)\right].$$

EXAMPLE 1 Write a Complex Number in Trigonometric Form

Write $z = -2 - 2i$ in trigonometric form.

Solution Find the modulus and the argument of z. Then substitute these values in the trigonometric form of z.

$$r = \sqrt{(-2)^2 + (-2)^2} = \sqrt{8} = 2\sqrt{2}$$

$\alpha = \tan^{-1}\left|\frac{b}{a}\right|$ — α is the reference angle of angle θ.

$\alpha = \tan^{-1}\left|\frac{-2}{-2}\right|$

$= \tan^{-1}1$

$\alpha = 45°$

$\theta = 180° + 45° = 225°$ — Since z is in the third quadrant, $180° \le \theta \le 270°$.

$z = 2\sqrt{2}(\cos 225° + i\sin 225°)$

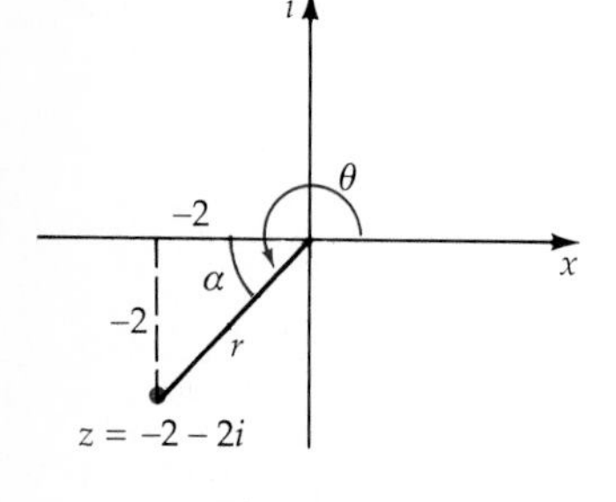

Figure 5.5

■ *Try Exercise* **12,** *page 210.*

EXAMPLE 2 Write a Complex Number in Standard Form

Write $z = 2(\cos 120° + i\sin 120°)$ in standard form.

Solution To change a complex number from its trigonometric form to its standard form, evaluate $\cos\theta$ and $\sin\theta$.

$z = 2(\cos 120° + i\sin 120°)$ — $\cos 120° = -1/2$

$= 2\left(-\frac{1}{2} + i\frac{\sqrt{3}}{2}\right)$ — $\sin 120° = \sqrt{3}/2$

$= -1 + i\sqrt{3}$

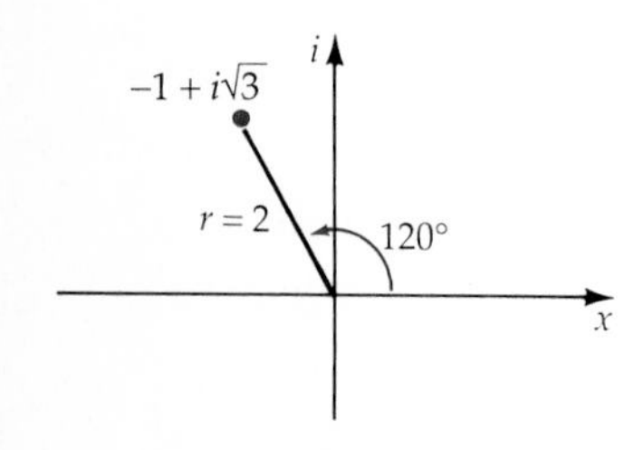

Figure 5.6

■ *Try Exercise* **26,** *page 210.*

Let z_1 and z_2 be two complex numbers written in trigometric form. The product of z_1 and z_2 can be found by using some trigonometric identities.

Let

$$z_1 = r_1(\cos\theta_1 + i\sin\theta_1), \text{ and } z_2 = r_2(\cos\theta_2 + i\sin\theta_2).$$

Then

$$\begin{aligned} z_1z_2 &= r_1(\cos\theta_1 + i\sin\theta_1)\cdot r_2(\cos\theta_2 + i\sin\theta_2) \\ &= r_1r_2(\cos\theta_1\cos\theta_2 + i\cos\theta_1\sin\theta_2 + i\sin\theta_1\cos\theta_2 + i^2\sin\theta_1\sin\theta_2) \\ &= r_1r_2[(\cos\theta_1\cos\theta_2 - \sin\theta_1\sin\theta_2) + i(\cos\theta_1\sin\theta_2 + \sin\theta_1\cos\theta_2)] \end{aligned}$$

Use the sum of two angle identities to simplify the product z_1z_2.

$$\cos\theta_1\cos\theta_2 - \sin\theta_1\sin\theta_2 = \cos(\theta_1 + \theta_2)$$

$$\cos\theta_1\sin\theta_2 + \sin\theta_1\cos\theta_2 = \sin(\theta_1 + \theta_2)$$

$$z_1z_2 = r_1r_2[\cos(\theta_1 + \theta_2) + i\sin(\theta_1 + \theta_2)]$$

The modulus for the product of two complex numbers in trigonometric form is the product of the moduli of the two complex numbers, and the argument of the product is the sum of the arguments of the two numbers.

EXAMPLE 3 Find the Product of Two Complex Numbers in Trigonometric Form

Find the product of

$$z_1 = -1 + i\sqrt{3}$$

and

$$z_2 = -\sqrt{3} + i$$

by using the trigonometric form of the complex numbers. Write the answer in standard form.

Solution

$$z_1 = 2\left(\cos\frac{2\pi}{3} + i\sin\frac{2\pi}{3}\right) \quad \text{See Figure 5.7}$$

$$z_2 = 2\left(\cos\frac{5\pi}{6} + i\sin\frac{5\pi}{6}\right)$$

$$\begin{aligned} z_1z_2 &= 2\left(\cos\frac{2\pi}{3} + i\sin\frac{2\pi}{3}\right)\cdot 2\left(\cos\frac{5\pi}{6} + i\sin\frac{5\pi}{6}\right) \\ &= 2\cdot 2\left[\cos\left(\frac{2\pi}{3} + \frac{5\pi}{6}\right) + i\sin\left(\frac{2\pi}{3} + \frac{5\pi}{6}\right)\right] \\ &= 4\left(\cos\frac{3\pi}{2} + i\sin\frac{3\pi}{2}\right) \\ &= 4(0 - i) \\ z_1z_2 &= -4i \end{aligned}$$

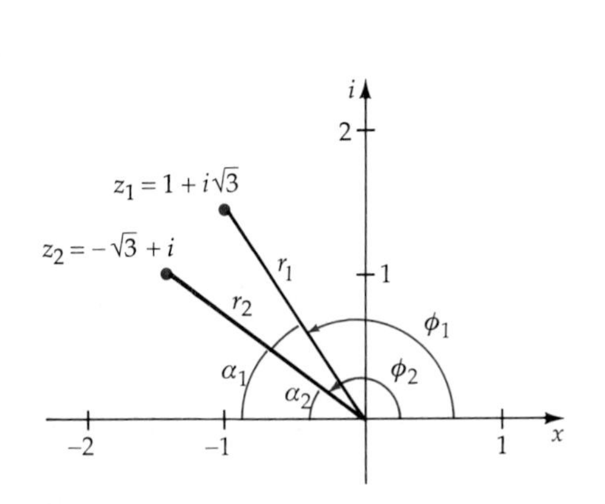

Figure 5.7

■ *Try Exercise* **40**, *page 210.*

EXAMPLE 4 Find the Product of Two Complex Numbers in Trigonometric Form

Find the product of $z_1 = 2 + 3i$ and $z_2 = 1 - 2i$ by using the trigonometric form of the complex numbers. Write the answer in standard form.

Solution Write the complex numbers in trigonometric form.

$$r_1 = \sqrt{2^2 + 3^2} = \sqrt{13} \qquad r_2 = \sqrt{1^2 + (-2)^2} = \sqrt{5}$$

$$\alpha_1 = \tan^{-1}\left|\frac{3}{2}\right| \approx 56.3° \qquad \alpha_2 = \tan^{-1}\left|\frac{-2}{1}\right| \approx 63.4°$$

$$\theta_1 = 56.3° \qquad \theta_2 = 360° - 63.4° = 296.6°$$

$$z_1 = \sqrt{13}(\cos 56.3° + i \sin 56.3°) \qquad z_2 = \sqrt{5}(\cos 296.6° + i \sin 296.6°)$$

$$\begin{aligned} z_1 z_2 &= \sqrt{13}(\cos 56.3° + i \sin 56.3°) \cdot \sqrt{5}(\cos 296.6° + i \sin 296.6°) \\ &= \sqrt{65}[\cos(56.3° + 296.6°) + i \sin(56.3° + 296.6°)] \\ &= \sqrt{65}(\cos 352.9° + i \sin 352.9°) \approx 8.00 - 1.00i \end{aligned}$$

■ *Try Exercise* **42,** *page 210.*

Let z_1 and z_2 be two complex numbers written in trigonometric form. The quotient of z_1 and z_2 can be found by using some trigonometric identities.

$$z_1 = r_1(\cos \theta_1 + i \sin \theta_1) \qquad z_2 = r_2(\cos \theta_2 + i \sin \theta_2)$$

$$\begin{aligned} \frac{z_1}{z_2} &= \frac{r_1(\cos \theta_1 + i \sin \theta_1)}{r_2(\cos \theta_2 + i \sin \theta_2)} \\ &= \frac{r_1(\cos \theta_1 + i \sin \theta_1)(\cos \theta_2 - i \sin \theta_2)}{r_2(\cos \theta_2 + i \sin \theta_2)(\cos \theta_2 - i \sin \theta_2)} \\ &= \frac{r_1(\cos \theta_1 \cos \theta_2 - i \cos \theta_1 \sin \theta_2 + i \sin \theta_1 \cos \theta_2 - i^2 \sin \theta_1 \sin \theta_2)}{r_2(\cos^2\theta_2 - i^2 \sin^2\theta_2)} \\ &= \frac{r_1[(\cos \theta_1 \cos \theta_2 + \sin \theta_1 \sin \theta_2) + i(\sin \theta_1 \cos \theta_2 - \cos \theta_1 \sin \theta_2)]}{r_2(\cos^2 \theta_2 + \sin^2\theta_2)} \end{aligned}$$

$$\frac{z_1}{z_2} = \frac{r_1}{r_2}[(\cos(\theta_1 - \theta_2) + i \sin(\theta_1 - \theta_2)]$$

The modulus for the quotient of two complex numbers in trigonometric form is the quotient of the moduli of the two complex numbers, and the argument of the quotient is the difference of the arguments of the two numbers.

EXAMPLE 5 Divide Two Complex Numbers in Trigonometric Form

Divide $z_1 = 30(\cos 40° - i \sin 40°)$ by $z_2 = 10(\cos 175° + i \sin 175°)$. Write the answer in standard form.

Solution To divide complex numbers, divide the moduli and subtract the arguments.

$$\frac{z_1}{z_2} = \frac{30(\cos 40° + i \sin 40°)}{10(\cos 175° + i \sin 175°)}$$

$$= \frac{30}{10}[\cos(40° - 175°) + i \sin(40° - 175°)]$$

$$= 3[\cos(-135°) + i \sin(-135°)]$$

$$= 3\left(-\frac{\sqrt{2}}{2} - i\frac{\sqrt{2}}{2}\right)$$

$$= -\frac{3\sqrt{2}}{2} - \frac{3i\sqrt{2}}{2}$$

■ *Try Exercise* **50,** *page 210.*

EXAMPLE 6 Divide Two Complex Numbers in Trigonometric Form

Use the trigonometric forms of the complex numbers $z_1 = -1 + i$ and $z_2 = \sqrt{3} - i$ to divide $\frac{z_1}{z_2}$. Write the answer in standard form.

Solution If we write z_1 and z_2 in the form $z_1 = r_1(\cos \theta_1 + i \sin \theta_1)$ and $z_2 = r_2(\cos \theta_2 + i \sin \theta_2)$, then

$$z_1 = \sqrt{2}(\cos 135° + i \sin 135°) \qquad \text{See Figure 5.8}$$

$$z_2 = 2(\cos 330° + i \sin 330°)$$

$$\frac{z_1}{z_2} = \frac{\sqrt{2}(\cos 135° + i \sin 135°)}{2(\cos 330° + i \sin 330°)}$$

$$= \frac{\sqrt{2}}{2}[\cos(135° - 330°) + i \sin(135° - 330°)]$$

$$= \frac{\sqrt{2}}{2}[\cos(-195°) + i \sin(-195°)]$$

$$= \frac{\sqrt{2}}{2}(\cos 195° - i \sin 195°)$$

$$\approx \frac{\sqrt{2}}{2}(-0.9659 + 0.2588i)$$

$$\approx -0.6830 + 0.1830i$$

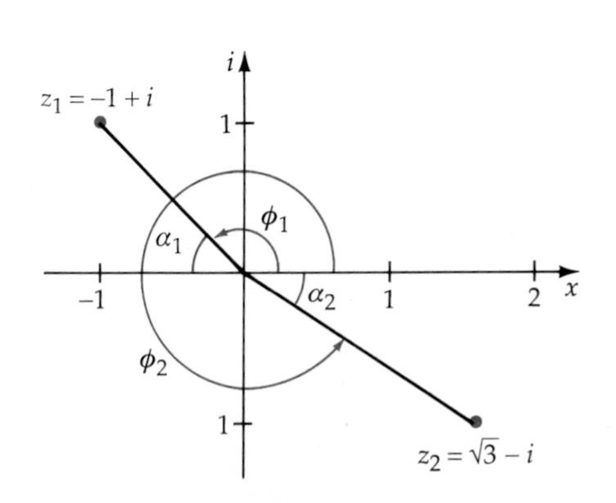

Figure 5.8

■ *Try Exercise* **62,** *page 211.*

EXERCISE SET 5.2

In Exercises 1 to 8, graph the complex numbers. Find the absolute value of each complex number.

1. $z = -2 - 2i$
2. $z = 4 - 4i$
3. $z = \sqrt{3} - i$
4. $z = 1 + i\sqrt{3}$
5. $z = -2i$
6. $z = -5$
7. $z = 3 - 5i$
8. $z = -5 - 4i$

In Exercises 9 to 16, write the complex number in trigonometric form.

9. $z = 1 - i$
10. $z = -4 - 4i$
11. $z = \sqrt{3} - i$
12. $z = 1 + i\sqrt{3}$
13. $z = 3i$
14. $z = -2i$
15. $z = -5$
16. $z = 3$

In Exercises 17 to 38, write the complex number in standard form.

17. $z = 2(\cos 45° + i \sin 45°)$
18. $z = 3(\cos 240° + i \sin 240°)$
19. $z = \cos 315° + i \sin 315°$
20. $z = 5(\cos 120° + i \sin 120°)$
21. $z = 6(\cos 140° + i \sin 140°)$
22. $z = \cos 305° + i \sin 305°$
23. $z = 8(\cos 0° + i \sin 0°)$
24. $z = 5(\cos 90° + i \sin 90°)$
25. $z = 2\left(\cos \frac{5\pi}{6} + i \sin \frac{5\pi}{6}\right)$
26. $z = 4\left(\cos \frac{5\pi}{3} + i \sin \frac{5\pi}{3}\right)$
27. $z = 3\left(\cos \frac{3\pi}{2} + i \sin \frac{3\pi}{2}\right)$
28. $z = 5(\cos \pi + i \sin \pi)$
29. $z = 8\left(\cos \frac{3\pi}{4} + i \sin \frac{3\pi}{4}\right)$
30. $z = 9\left(\cos \frac{4\pi}{3} + i \sin \frac{4\pi}{3}\right)$
31. $z = 9\left(\cos \frac{11\pi}{6} + i \sin \frac{11\pi}{6}\right)$
32. $z = \left(\cos \frac{3\pi}{2} + i \sin \frac{3\pi}{2}\right)$
33. $z = 2(\cos 2 + i \sin 2)$
34. $z = 5(\cos 4 + i \sin 4)$
35. $z = 3(\cos 300° + i \sin 300°)$
36. $z = 2(\cos 225° + i \sin 225°)$
37. $z = 8(\cos 210° + i \sin 210°)$
38. $z = 4(\cos 45° + i \sin 45°)$

In Exercises 39 to 46, multiply the complex numbers. Leave the answer in trigonometric form.

39. $2(\cos 30° + i \sin 30°) \cdot 3(\cos 225° + i \sin 225°)$
40. $4(\cos 120° + i \sin 120°) \cdot 6(\cos 315° + i \sin 315°)$
41. $3(\cos 122° + i \sin 122°) \cdot 4(\cos 213° + i \sin 213°)$
42. $5\left(\cos \frac{2\pi}{3} + i \sin \frac{2\pi}{3}\right) \cdot 2\left(\cos \frac{2\pi}{5} + i \sin \frac{2\pi}{5}\right)$
43. $2\left(\cos \frac{\pi}{4} + i \sin \frac{\pi}{4}\right) \cdot 5\left(\cos \frac{7\pi}{3} + i \sin \frac{7\pi}{3}\right)$
44. $4(\cos 4.5 + i \sin 4.5) \cdot 7(\cos 1.2 + i \sin 1.2)$
45. $3(\cos 45° + i \sin 45°) \cdot 5(\cos 120° + i \sin 120°)$
46. $2(\cos 60° + i \sin 60°) \cdot 6(\cos 150° + i \sin 150°)$

In Exercises 47 to 56, divide the complex numbers. Write the answers in standard form.

47. $\dfrac{32(\cos 30° + i \sin 30°)}{4(\cos 150° + i \sin 150°)}$
48. $\dfrac{15(\cos 240° + i \sin 240°)}{3(\cos 135° + i \sin 135°)}$
49. $\dfrac{12(\cos 2\pi/3 + i \sin 2\pi/3)}{4(\cos 11\pi/6 + i \sin 11\pi/6)}$
50. $\dfrac{10(\cos \pi/3 + i \sin \pi/3)}{5(\cos \pi/4 + i \sin \pi/4)}$
51. $\dfrac{27(\cos 315° + i \sin 315°)}{9(\cos 225° + i \sin 225°)}$
52. $\dfrac{9(\cos 25° + i \sin 25°)}{3(\cos 175° + i \sin 175°)}$
53. $\dfrac{25(\cos 189° + i \sin 189°)}{5(\cos 69° + i \sin 69°)}$
54. $\dfrac{18(\cos 420° + i \sin 420°)}{6(\cos 150° + i \sin 150°)}$
55. $\dfrac{5(\cos 11\pi/12 + i \sin 11\pi/12)}{3(\cos 4\pi/3 + i \sin 4\pi/3)}$
56. $\dfrac{4(\cos 2.4 + i \sin 2.4)}{6(\cos 4.1 + i \sin 4.1)}$

In Exercises 57 to 64, perform the indicated operation in trigonometric form. Write the solution back in standard form.

57. $(1 - i\sqrt{3})(1 + i)$

58. $(\sqrt{3} - i)(1 + i\sqrt{3})$

59. $(3 - 3i)(1 + i)$

60. $(2 + 2i)(\sqrt{3} - i)$

61. $\dfrac{1 + i\sqrt{3}}{1 - i\sqrt{3}}$

62. $\dfrac{1 + i}{1 - i}$

63. $\dfrac{\sqrt{2} - i\sqrt{2}}{1 + i}$

64. $\dfrac{1 + i\sqrt{3}}{4 - 4i}$

Supplemental Exercises

In Exercises 65 to 70, perform the indicated operation in trigonometric form. Write the solution back in standard form.

65. $(\sqrt{3} - i)(2 + 2i)(2 - 2i\sqrt{3})$

66. $(1 - i)(1 + i\sqrt{3})(\sqrt{3} - i)$

67. $\dfrac{\sqrt{3} + i\sqrt{3}}{(1 - i\sqrt{3})(2 - 2i)}$

68. $\dfrac{(2 - 2i\sqrt{3})(1 - i\sqrt{3})}{4\sqrt{3} + 4i}$

69. $(1 - 3i)(2 + 3i)(4 + 5i)$

70. $\dfrac{(2 - 5i)(1 - 6i)}{3 + 4i}$

71. Use the trigonometric forms of the complex numbers z and $\bar{z}$ to find $z \cdot \bar{z}$.

72. Use the trigonometric forms of z and $\bar{z}$ to find $z/\bar{z}$. (*Hint:* The complex form of $z = r(\cos\theta + i\sin\theta)$ can be written as $\bar{z} = r[\cos(-\theta) + i\sin(-\theta)]$.)

5.3 DeMoivre's Theorem

DeMoivre's Theorem is a procedure for finding powers and roots of complex numbers, when the complex numbers are expressed in trigonometric form. This theorem can be illustrated by repeated multiplication of a complex number.

Let $z = r(\cos\theta + i\sin\theta)$. Then z^2 can be written as

$$z \cdot z = r(\cos\theta + i\sin\theta) \cdot r(\cos\theta + i\sin\theta)$$

$$z^2 = r^2(\cos 2\theta + i\sin 2\theta)$$

The product of $z^2 \cdot z$ is

$$z^2 \cdot z = r^2(\cos 2\theta + i\sin 2\theta) \cdot r(\cos\theta + i\sin\theta)$$

$$z^3 = r^3(\cos 3\theta + i\sin 3\theta)$$

In a similar fashion, we can write

$$z^4 = r^4(\cos 4\theta + i\sin 4\theta) \qquad \text{and} \qquad z^5 = r^5(\cos 5\theta + i\sin 5\theta).$$

If we continue this process, the results suggest a formula for the nth power of a complex number that is known as DeMoivre's Theorem.

DeMoivre's Theorem

> If $z = r(\cos\theta + i\sin\theta)$ is a complex number and n is a positive integer, then
>
> $$z^n = r^n(\cos n\theta + i\sin n\theta)$$

EXAMPLE 1 Find the Power of a Complex Number Using DeMoivre's Theorem

Find $(1 + i)^8$ using DeMoivre's Theorem. Write the answer in standard form.

Solution Write $1 + i$ in trigonometric form and then use DeMoivre's Theorem.

$$1 + i = \sqrt{2}(\cos 45° + i \sin 45°)$$

$$(1 + i)^8 = (\sqrt{2})^8[\cos 8(45°) + i \sin 8(45°)$$ DeMoivre's Theorem

$$= 16(\cos 360° + i \sin 360°) = 16(1 + 0i) = 16$$

$$(1 + i)^8 = 16$$

■ *Try Exercise* **10,** *page 216.*

EXAMPLE 2 Find the Power of a Complex Number Using DeMoivre's Theorem

Find $(-\sqrt{3} - i)^{15}$ using DeMoivre's Theorem. Write the answer in standard form.

Solution Write $-\sqrt{3} - i$ in trigonometric form and then use DeMoivre's Theorem.

$$r = \sqrt{(-1)^2 + (-\sqrt{3})^2} = 2$$

$$\alpha = \tan^{-1}\left|\frac{b}{a}\right|$$ α is the reference angle of θ.

$$\alpha = \tan^{-1}\left|\frac{-1}{-\sqrt{3}}\right| = \frac{\pi}{6}$$

$$\theta = \pi + \frac{\pi}{6} = \frac{7\pi}{6}$$ Since z is in the third quadrant, $\pi/2 \le \theta \le \pi$.

$$(-\sqrt{3} - i)^{15} = 2^{15}\left(\cos 15\left(\frac{7\pi}{6}\right) + i \sin 15\left(\frac{7\pi}{6}\right)\right)$$

$$= 32{,}768\left(\cos \frac{105\pi}{6} + i \sin \frac{105\pi}{6}\right)$$

$$= 32{,}768\left(\cos \frac{3\pi}{2} + i \sin \frac{3\pi}{2}\right)$$

$$= 32{,}768(0 + (-i)) = -32{,}768i$$

■ *Try Exercise* **12,** *page 216.*

DeMoivre's Theorem can be extended to include finding the nth roots of any number. Recall that if $w^n = z$, then w is the nth root of z. A complex number to the nth power has n distinct roots. All the nth roots of a complex number can be found by using DeMoivre's Theorem.

Let $w^n = z$, for $z = r(\cos\theta + i\sin\theta)$, and $w = R(\cos\alpha + i\sin\alpha)$. Then by DeMoivre's Theorem,

$$w^n = R^n(\cos n\alpha + i\sin n\alpha). \tag{1}$$

Recall that the sine and cosine functions are periodic functions with a period of 2π or 360°. Therefore, for k an integer,

$$z = r(\cos\theta + i\sin\theta) = r[\cos(\theta + 360°k) + i\sin(\theta + 360°k)]$$

Substituting for w^n and z in the equation $w^n = z$, we have

$$R^n(\cos n\alpha + i\sin n\alpha) = r[\cos(\theta + 360°k) + i\sin(\theta + 360°k)].$$

Two complex numbers written in trigonometric form are equal if and only if the moduli are equal and their arguments are equal. Thus,

$$R^n = r \qquad \text{and} \qquad n\alpha = \theta + 360°k$$

$$R = r^{1/n} \qquad\qquad \alpha = \frac{\theta + 360°k}{n}$$

Since $w = R(\cos\alpha + i\sin\alpha)$, we can write

$$w = r^{1/n}\left[\cos\left(\frac{\theta + 360°k}{n}\right) + i\sin\left(\frac{\theta + 360°k}{n}\right)\right].$$

DeMoivre's Theorem for Finding Roots

If $z = r(\cos\theta + i\sin\theta)$ is a complex number, then there are n distinct nth roots of z given by the formula

$$w = r^{1/n}\left[\cos\left(\frac{\theta + 360°k}{n}\right) + i\sin\left(\frac{\theta + 360°k}{n}\right)\right]$$

for $k = 0, 1, 2, \ldots, n - 1$.

EXAMPLE 3 Find Cube Roots by DeMoivre's Theorem for Finding Roots

Find the three cube roots of 27.

Solution Write 27 in trigonometric form.

$$27 = 27(\cos 0° + i\sin 0°)$$

Then from DeMoivre's Theorem for Finding Roots it follows that the cube roots of 27 are

$$w = 27^{1/3}\left[\cos\left(\frac{0° + 360°k}{3}\right) + i\sin\left(\frac{0° + 360°k}{3}\right)\right], \quad \text{for } k = 0, 1, 2.$$

Find the values of the arguments of the roots for which $k = 0, 1, 2.$

$$k = 0 \qquad \frac{0^\circ}{3} = 0$$

$$k = 1 \qquad \frac{0^\circ + 360^\circ}{3} = 120^\circ$$

$$k = 2 \qquad \frac{0^\circ + 720^\circ}{3} = 240^\circ$$

For $k = 3$, $\dfrac{0^\circ + 1080^\circ}{3} = 360^\circ$. The angles start repeating; thus, there are only three cube roots of 27. These roots are shown below and also graphed in Figure 5.9.

$$w_1 = 3(\cos 0^\circ + i \sin 0^\circ) = 3$$

$$w_2 = 3(\cos 120^\circ + i \sin 120^\circ) = -\frac{3}{2} + \frac{3i\sqrt{3}}{2}$$

$$w_3 = 3(\cos 240^\circ + i \sin 240^\circ) = -\frac{3}{2} - \frac{3i\sqrt{3}}{2}$$

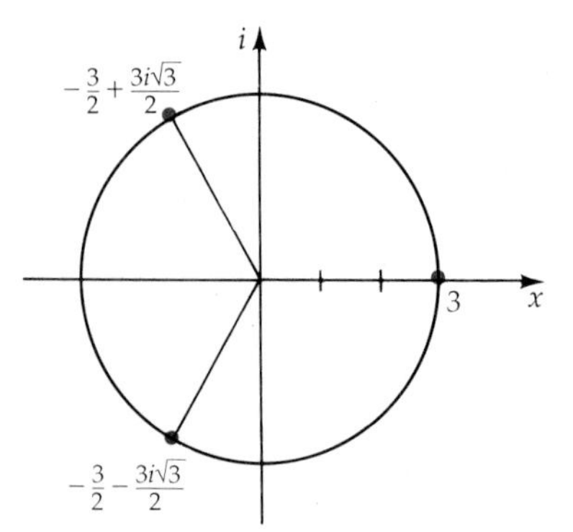

Figure 5.9

Check the roots

$$3^3 = 27$$

$$[3(\cos 120^\circ + i \sin 120^\circ)]^3 = 27(\cos 360^\circ + 1 \sin 360^\circ)$$
$$= 27(1 + 0) = 27$$

$$[3(\cos 240^\circ + i \sin 240^\circ)]^3 = 27(\cos 720^\circ + i \sin 720^\circ)$$
$$= 27(1 + 0) = 27$$

■ *Try Exercise* **24,** *page 216.*

Note from Figure 5.9 that the arguments of the cube roots of 27 are 0°, 120° and 240° and that $|w| = 3$. This means that the complex numbers representing the cube roots of 27 are equally spaced around a circle of radius 3.

EXAMPLE 4 Find the Fifth Roots of a Complex Number

Find the fifth roots of $z = 1 + i\sqrt{3}$.

Solution Write z in trigonometric form.

$$r = \sqrt{1^2 + (\sqrt{3})^2} = 2$$

$$\alpha = \tan^{-1}\left|\frac{\sqrt{3}}{1}\right| = 60^\circ \quad \text{Angle } \theta \text{ is in the first quadrant.}$$

$$\theta = 60^\circ$$

$$z = 2(\cos 60^\circ + i \sin 60^\circ)$$

From DeMoivre's Theorem, the modulus of each root is $\sqrt[5]{2}$, and the arguments are determined by

$$\frac{60^\circ + 360^\circ k}{5}, \quad k = 0, 1, 2, 3, 4.$$

$$w = \sqrt[5]{2}\left(\cos \frac{60 + 360^\circ k}{5} + i \sin \frac{60^\circ + 360^\circ k}{5}\right), \quad k = 0, 1, 2, 3, 4$$

$$w = \sqrt[5]{2}[\cos(12^\circ + 72^\circ k) + i \sin(12^\circ + 72^\circ k)], \quad k = 0, 1, 2, 3, 4$$

Substitute for k to find the five roots of z.

$$k = 0, \quad w_1 = \sqrt[5]{2}(\cos 12^\circ + i \sin 12^\circ)$$

$$k = 1, \quad w_2 = \sqrt[5]{2}(\cos 84^\circ + i \sin 84^\circ)$$

$$k = 2, \quad w_3 = \sqrt[5]{2}(\cos 156^\circ + i \sin 156^\circ)$$

$$k = 3, \quad w_4 = \sqrt[5]{2}(\cos 228^\circ + i \sin 228^\circ)$$

$$k = 4, \quad w_5 = \sqrt[5]{2}(\cos 300^\circ + i \sin 300^\circ)$$

■ *Try Exercise* **26,** *page 216.*

Remark The five roots are graphed in Figure 5.10. The radius of the circle is $\sqrt[5]{2} \approx 1.15$. The complex numbers are spaced equally around the circle.

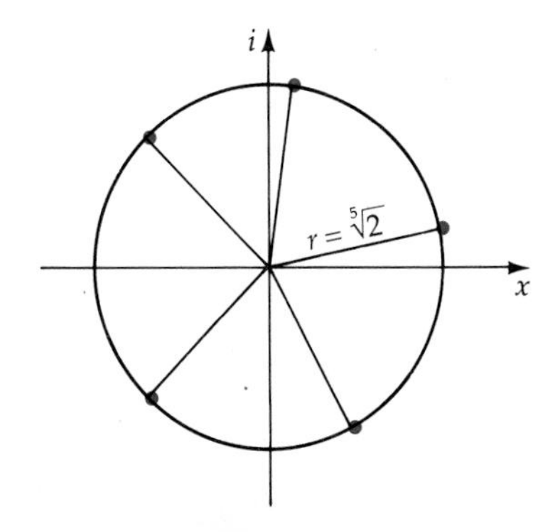

Figure 5.10

EXAMPLE 5 **Find the Complex Roots of an Equation by DeMoivre's Theorem**

Find all the roots of $x^3 + 16 = 0$.

Solution Since $x^3 = -16$, use DeMoivre's Theorem to find the cube roots of -16.

$$-16 = 16(\cos 180° + i \sin 180°)$$

$$w = \sqrt[3]{16}\left(\cos \frac{180° + 360°k}{3} + i \sin \frac{180° + 360°k}{3}\right), \quad k = 0, 1, 2$$

$$k = 0, \quad w_1 = \sqrt[3]{16}(\cos 60° + i \sin 60°) \approx 1.26 + 2.18i$$

$$k = 1, \quad w_2 = \sqrt[3]{16}(\cos 180° + i \sin 180°) \approx -2.52$$

$$k = 2, \quad w_3 = \sqrt[3]{16}(\cos 300° + i \sin 300°) \approx 1.26-2.18i$$

■ *Try Exercise* **34,** *page 216.*

EXERCISE SET 5.3

In Exercises 1 to 14, find the indicated power. Leave the answers in trigonometric form.

1. $[2(\cos 30° + i \sin 30°)]^8$

2. $(\cos 240° + i \sin 240°)^{12}$

3. $[2(\cos 240° + i \sin 240°)]^5$

4. $[2(\cos 45° + i \sin 45°)]^{10}$

5. $[2(\cos 225° + i \sin 225°)]^5$

6. $[3(\cos 330° + i \sin 330°)]^4$

7. $[2(\cos 120° + i \sin 120°)]^6$

8. $[4(\cos 150° + i \sin 150°)]^3$

9. $(1 - i)^{10}$

10. $(1 + i\sqrt{3})^8$

11. $(2 + 2i)^7$

12. $(2\sqrt{3} - 2i)^5$

13. $\left(\frac{\sqrt{2}}{2} + i\frac{\sqrt{2}}{2}\right)^6$

14. $\left(-\frac{\sqrt{2}}{2} + i\frac{\sqrt{2}}{2}\right)^{12}$

In Exercises 15 to 28, find the indicated root. Leave all answers in standard form.

15. The square roots of $9(\cos 135° + i \sin 135°)$

16. The square roots of $16(\cos 45° + i \sin 45°)$

17. The sixth roots of $64(\cos 120° + i \sin 120°)$

18. The fifth roots of $\cos 315° + i \sin 315°$

19. The fifth roots of $\cos 90° + i \sin 90°$

20. The fourth roots of $\cos 0° + i \sin 0°$

21. The cube roots of 1

22. The cube roots of i

23. The fourth roots of $1 + i$

24. The fifth roots of $-1 + i$

25. The cube roots of $2 - 2i\sqrt{3}$

26. The cube roots of $-2 + 2i\sqrt{3}$

27. The square roots of $-16 + 16i\sqrt{3}$

28. The square roots of $-1 + i\sqrt{3}$

In Exercises 29 to 40, find all roots of the equations. Leave the answers in trigonometric form.

29. $x^3 + 8 = 0$

30. $x^5 - 32 = 0$

31. $x^4 + i = 0$

32. $x^3 - 2i = 0$

33. $x^3 - 27 = 0$

34. $x^5 + 32i = 0$

35. $x^4 + 81 = 0$

36. $x^3 - 64i = 0$

37. $x^4 - (1 - i\sqrt{3}) = 0$

38. $x^3 + (2\sqrt{3} - 2i) = 0$

39. $x^3 + (1 + i\sqrt{3}) = 0$

40. $x^6 - (4 - 4i) = 0$

Supplemental Exercises

41. Show that the conjugate of $z = r(\cos\theta + i \sin\theta)$ is equal to $\bar{z} = r(\cos\theta - i \sin\theta)$.

42. If $z = r(\cos\theta - i \sin\theta)$, show that

$$z^{-1} = r^{-1}(\cos\theta - i \sin\theta).$$

43. If $z = r(\cos\theta - i\sin\theta)$, show that

$$z^{-2} = r^{-2}(\cos 2\theta - i\sin 2\theta).$$

(Note that Exercises 42 and 43 indicate that the general expression is

$$z^{-n} = r^{-n}(\cos n\theta - i\sin n\theta).$$

44. Use the results of Exercise 43 to find z^{-4} for $z = 1 - i\sqrt{3}$.

45. Raise $(\cos\theta + i\sin\theta)$ to the second power by using DeMoivre's Theorem. Now square $(\cos\theta + i\sin\theta)$ as a binomial. Equate the real and imaginary parts of the two complex numbers and show that:

a. $\cos 2\theta = \cos^2\theta - \sin^2\theta$,

b. $\sin 2\theta = 2\sin\theta\cos\theta$.

46. Raise $r(\cos\theta + i\sin\theta)$ to the fourth power by using DeMoivre's Theorem. Now find the fourth power of the binomial $(\cos\theta + i\sin\theta)$. Equate the real and the imaginary parts of the complex numbers and show that:

a. $\cos 4\theta = \cos^2 2\theta - 2\sin^2 2\theta$,

b. $\sin 4\theta = 4\cos^3\theta\sin\theta - 4\cos\theta\sin^3\theta$.

Chapter Review

5.1 Complex Numbers

The symbol i, called the imaginary unit, is defined as the number which has the following properties:

$$i^2 = -1, \quad \text{and} \quad \sqrt{-a} = i\sqrt{a} \quad (\text{where } a > 0).$$

If a and b are real numbers, and i is the imaginary unit, then $a + bi$ is called a complex number. The real number a is called the real part and the real number b is called the imaginary part of the complex number.

The complex numbers $a + bi$ and $a - bi$ are called complex conjugates or conjugates of each other,

Powers of i
If n is a positive integer, then:

$$i^n = i^r,$$

where r is the remainder of the division of n by 4.

5.2 Trigonometric Form of Complex Numbers

The standard form of a complex number is $z = a + bi$. The trigonometric form of a complex number is $z = r(\cos\theta + i\sin\theta)$.

The product of two complex numbers in trigonometric form is given by:

$$z_1z_2 = r_1r_2[\cos(\theta_1 + \theta_2) + i\sin(\theta_1 + \theta_2)]$$

for $z_1 = r_1(\cos\theta_1 + i\sin\theta_1)$ and $z_2 = r_2(\cos\theta_2 + i\sin\theta_2)$.

The quotient of two complex numbers in trigonometric form is given by:

$$\frac{z_1}{z_2} = \frac{r_1}{r_2}[\cos(\theta_1 - \theta_2) + i \sin(\theta_1 - \theta_2)]$$

for $z_1 = r_1(\cos \theta_1 + i \sin \theta_1)$ and $z_2 = r_2(\cos \theta_2 + i \sin \theta_2)$.

5.3 DeMoivre's Theorem

DeMoivre's Theorem

If $z = r(\cos \theta + i \sin \theta)$ is a complex number and n is a positive integer, then

$$z^n = r^n(\cos n\theta + i \sin n\theta).$$

If w and z are complex numbers and w is the nth root of z, $w^n = z$, then

$$w = r^{1/n}\left[\cos\left(\frac{\theta + 360°k}{n}\right) + i \sin\left(\frac{\theta + 360°k}{n}\right)\right], \quad \text{where } k = 0, 1, 2, \ldots, n - 1.$$

CHALLENGE EXERCISES

For Exercises 1–12, answer true or false. If the statement is false, given an example.

1. If n is a negative number, then $\sqrt{n^2} = ni$.
2. Every real number a is a complex number.
3. The sum of a complex number z and its conjugate $\bar{z}$ is a real number.
4. The product of a complex number z and its conjugate $\bar{z}$ is a real number.
5. The quotient of a complex number z and its conjugate $\bar{z}$ is a real number.
6. Multiplication of two complex numbers is commutative.
7. The n roots of a complex number can be graphed on a circle and are equally spaced around the circle.
8. Let $z = r(\cos \theta + i \sin \theta)$, then $z^2 = r^2(\cos^2 \theta + i \sin^2 \theta)$.
9. $|a + bi| = \sqrt{a^2 + b^2}$
10. $i = \cos \pi + i \sin \pi$
11. $1 = \cos \pi + i \sin \pi$
12. $z = \cos 45° + i \sin 45°$ is a square root of i.

REVIEW EXERCISES

In Exercises 1 to 4, write the complex number in standard form and give its congugate.

1. $3 - \sqrt{-64}$
2. $\sqrt{-4} + 6$
3. $-2 + \sqrt{-5}$
4. $-5 - \sqrt{-27}$

In Exercises 5 to 20, simplify and write the complex number in standard form.

5. $(\sqrt{-4})(\sqrt{-4})$
6. $(-\sqrt{-27})(\sqrt{-3})$
7. $(3 + 7i) + (2 - 5i)$
8. $(3 - 4i) + (-6 + 8i)$
9. $(6 - 8i) - (9 - 11i)$
10. $(-3 - 5i) - (2 + 10i)$
11. $(5 + 3i)(2 - 5i)$
12. $(-2 - 3i)(-4 + 7i)$
13. $\dfrac{-2i}{3 - 4i}$
14. $\dfrac{4 + i}{7 - 2i}$
15. $i(2i) - (1 + i)^2$
16. $(2 - i)^3$
17. $(3 + \sqrt{-4}) - (-3 - (\sqrt{-16}))$
18. $(-2 + \sqrt{-9}) + (-3 - \sqrt{-81})$
19. $(2 - \sqrt{-3})(2 + \sqrt{-3})$
20. $(3 - \sqrt{-5})(2 + \sqrt{-8})$

In Exercises 21 to 24, simplify and write each complex number as either i, $-i$, 1, or -1.

21. i^{27}

22. i^{105}

23. $\dfrac{i}{i^{17}}$

24. i^{62}

In Exercises 25 to 28, find the indicated absolute value of each complex number.

25. $|-8i|$

26. $|2 - 3i|$

27. $|-4 + 5i|$

28. $|-1 - i|$

In Exercises 29 to 32, write the complex numbers in trigonometric form.

29. $z = 2 - 2i$

30. $z = -\sqrt{3} + i$

31. $z = -3 + 2i$

32. $z = 4 - i$

In Exercises 33 to 36, write the complex number in standard form.

33. $z = 5(\cos 315° + i \sin 315°)$

34. $z = 6\left(\cos \dfrac{4\pi}{3} + i \sin \dfrac{4\pi}{3}\right)$

35. $z = 2(\cos 2 + i \sin 2)$

36. $z = 3(\cos 115° + i \sin 115°)$

In Exercises 37 to 42, multiply the complex numbers. Convert the answers to standard form.

37. $3(\cos 225° + i \sin 225°) \cdot 10(\cos 45° + i \sin 45°)$

38. $5(\cos 162° + i \sin 162°) \cdot 2(\cos 63° + i \sin 63°)$

39. $3(\cos 12° + i \sin 12°) \cdot 4(\cos 126° + i \sin 126°)$

40. $(\cos 23° + i \sin 23°) \cdot 4(\cos 233° + i \sin 233°)$

41. $3(\cos 1.8 + i \sin 1.8) \cdot 5(\cos 2.5 + i \sin 2.5)$

42. $6(\cos 3.1 + i \sin 3.1) \cdot 5(\cos 4.3 + i \sin 4.3)$

In Exercises 43 to 48, divide the complex numbers. Leave the answers in trigonometric form.

43. $\dfrac{6(\cos 50° + i \sin 50°)}{2(\cos 150° + i \sin 150°)}$

44. $\dfrac{30(\cos 165° + i \sin 165°)}{10(\cos 55° + i \sin 55°)}$

45. $\dfrac{40(\cos 66° + i \sin 66°)}{8(\cos 125° + i \sin 125°)}$

46. $\dfrac{2(\cos 150° + i \sin 150°)}{\sqrt{2}(\cos 200° + i \sin 200°)}$

47. $\dfrac{10(\cos 3.7 + i \sin 3.7)}{6(\cos 1.8 + i \sin 1.8)}$

48. $\dfrac{4(\cos 1.2 + i \sin 1.2)}{8(\cos 5.2 + i \sin 5.2)}$

In Exercises 49 to 54, find the indicated power. Leave the answers in standard form.

49. $[3(\cos 45° + i \sin 45°)]^5$

50. $\left[\cos\left(\dfrac{11\pi}{8}\right) + i \sin\left(\dfrac{11\pi}{8}\right)\right]^8$

51. $(1 - i\sqrt{3})^7$

52. $(-2 - 2i)^{10}$

53. $(\sqrt{2} - i\sqrt{2})^5$

54. $(3 - 4i)^5$

In Exercises 55 to 60, find the indicated roots. Leave the answers in trigonometric form.

55. The cube roots of $27i$

56. The fourth roots of $8i$

57. The fourth roots of $256(\cos 120° + i \sin 120°)$

58. The fifth roots of $-1 - i$

59. The fourth roots of 81

60. The cube roots of $\cos 300° + i \sin 300°$

6 Topics in Analytic Geometry

My thesis, then, is that the essence of analytical geometry is the study of loci by means of their equations, and that this was known to the Greeks and was the basis of their study in conic sections.

J. L. COOLIDGE

KIDNEY STONES AND ELLIPSES

The conic sections, some of the curves we will study in this chapter, were studied by the ancient Greeks. The names of the curves are derived by looking at various sections of a right circular cone.

One of the conic sections is an ellipse. Besides applications of the ellipse to astronomy and optics, the ellipse now has an application to medicine. The ellipse is used in a nonsurgical treatment of kidney stones. To understand this application, we will introduce some properties of an ellipse.

An ellipse is an oval shaped curve. Inside the ellipse there are two points called foci of the ellipse. Sound or light waves emitted from one focus are reflected off the surface of the ellipse to the other focus. In an analagous way, think of a billiard table shaped like an ellipse. A ball struck from one focus would pass through the other focus no matter the direction of the initial shot.

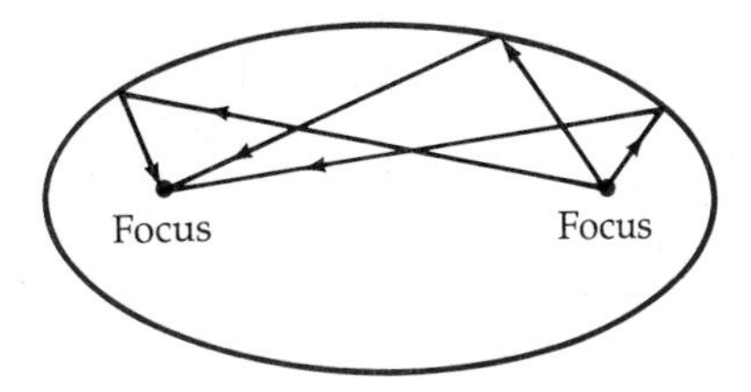

The treatment of kidney stones is based on this reflective property of an ellipse. An electrode is placed at one focus of an ellipse and the patient is placed at the other focus. When the electrode is discharged, ultrasound waves are produced. These waves hit the walls of the ellipse and are reflected to the kidney stone. Because of the reflective property of the ellipse, there is very little energy loss. As a result, it is as if the electrode actually discharges at the kidney stone. The energy of the discharge pulverizes the kidney stone thereby allowing the fragments to be passed through the system.

Research evaluation of this method has been encouraging. Hospitals are reporting success rates of 90% to 95%. Besides successfully treating the condition, the patient spends less time in the hospital and the complete recovery time is shorter.

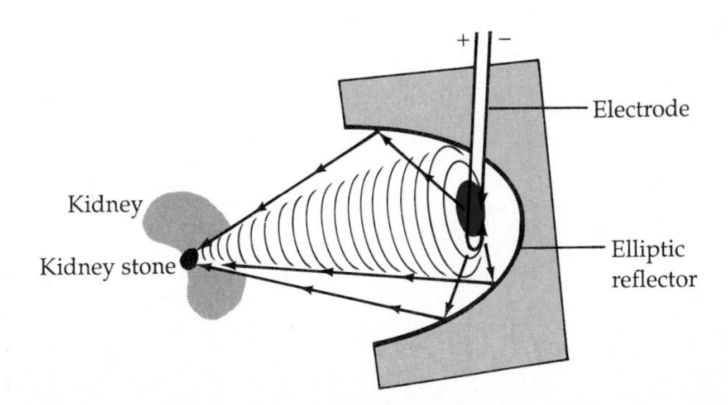

6.1 Conic Sections

The graph of a parabola, circle, ellipse, or a hyperbola can be formed by the intersection of a plane and a double cone. Hence these figures are referred to as **conic sections.**

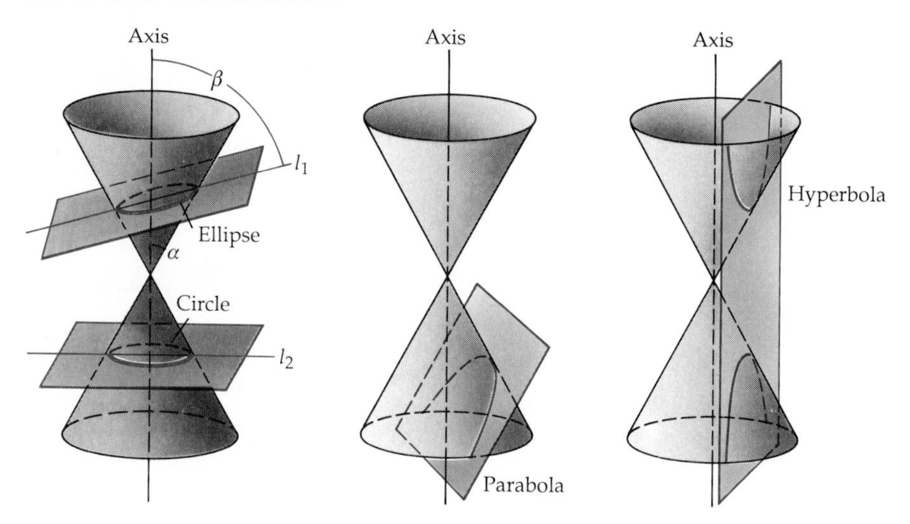

Figure 6.1
Cones intersected by planes.

A plane perpendicular to the axis of the cone intersects the cone in a circle (plane C). The plane E, tilted so that it is not perpendicular to the axis, intersects the cone in an ellipse. When the plane is parallel to a line on the surface of the cone, the plane intersects the cone in a parabola. When the plane intersects both parts of the cone, a hyperbola is formed.

Parabolas with Vertex at (0, 0)

Besides the geometric description of a conic section just given, a conic can be defined as a set of points. This method uses some specified conditions about the curve to determine which points in a coordinate system are points of the graph. For example, a parabola can be defined by the following set of points.

Definition of a Parabola

A **parabola** is the set of points in the plane that are equidistance from a fixed line (the **directrix**) and a fixed point (the **focus**) not on the directrix.

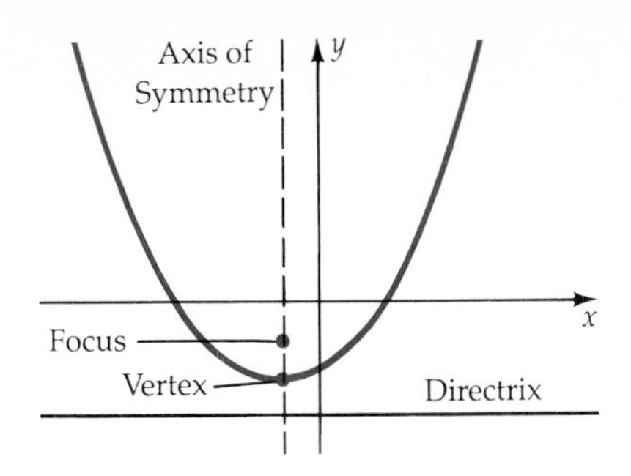

Figure 6.2

The line that passes through the focus and is perpendicular to the directrix is called the **axis of symmetry** of the parabola. The midpoint of the line segment between the focus and directrix on the axis of symmetry is the **vertex** of the parabola as shown in Figure 6.2.

Using this definition of a parabola, we can determine an equation of a parabola. Suppose that the coordinates of the vertex of a parabola are $(0, 0)$ and the axis of symmetry is the y-axis. The equation of the directrix is $y = -p$, $p > 0$. The focus lies on the axis of symmetry and is the same distance from the vertex as the vertex is from the directrix. Thus the coordinates of the focus are $(0, p)$ as shown in Figure 6.3.

Let $P(x, y)$ be any point on the parabola. Then, using the distance formula and the fact that the distance between any point on the parabola and the focus is equal to the distance from the point P to the directrix (why?[1]), we can write the equation

$$d(P, F) = d(P, D).$$

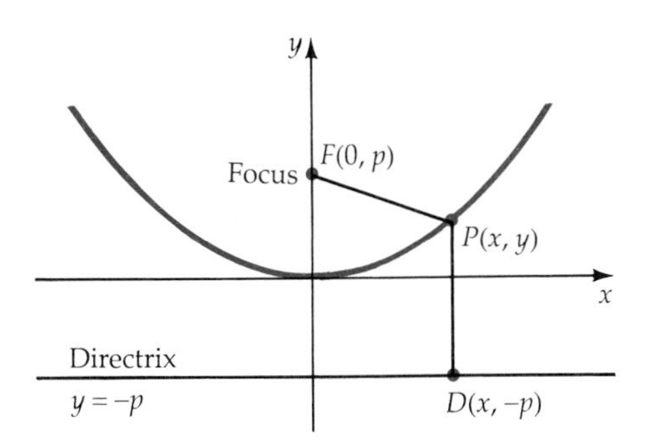

Figure 6.3

By the distance formula,

$$\sqrt{(x-0)^2 + (y-p)^2} = y + p$$

Now squaring each side and simplifying,

$$\left(\sqrt{(x-0)^2 + (y-p)^2}\right)^2 = (y+p)^2$$

$$x^2 + y^2 - 2py + p^2 = y^2 + 2py + p^2$$

$$x^2 = 4py.$$

This is an equation of a parabola with a vertex at the origin and a vertical axis of symmetry. The equation of a parabola with a horizontal axis of symmetry is derived in a similar manner.

Standard Form of the Equation of a Parabola with Vertex at the Origin

Vertical Axis of Symmetry
The standard form of the equation of a parabola with vertex $(0, 0)$ and vertical axis of symmetry is $x^2 = 4py$. The focus is $(0, p)$, and the equation of the directrix is $y = -p$.

Horizontal Axis of Symmetry
The standard form of the equation of a parabola with vertex $(0, 0)$ and horizontal axis of symmetry is $y^2 = 4px$. The focus is $(p, 0)$, and the equation of the directrix is $x = -p$.

Remark In the equation $x^2 = 4py$, $x^2 \geq 0$. Therefore $4py \geq 0$. Thus if $p > 0$, then $y \geq 0$ and the parabola opens up. If $p < 0$, then $y \leq 0$ and the parabola opens down. A similar analysis shows that for $y^2 = 4px$, the parabola opens to the right when $p > 0$ and opens to the left when $p < 0$.

[1] From the definition of a parabola, the distance from a point on the parabola to the focus is the same as the distance from the point to the directrix.

EXAMPLE 1 **Find the Focus and Directrix of a Parabola**

Find the focus and directrix of the parabola given by the equation $y = -\frac{1}{2}x^2$.

Solution Because the x term is squared, the standard form of the equation is $x^2 = 4py$. Write the given equation in standard form:

$$y = -\frac{1}{2}x^2$$
$$x^2 = -2y.$$

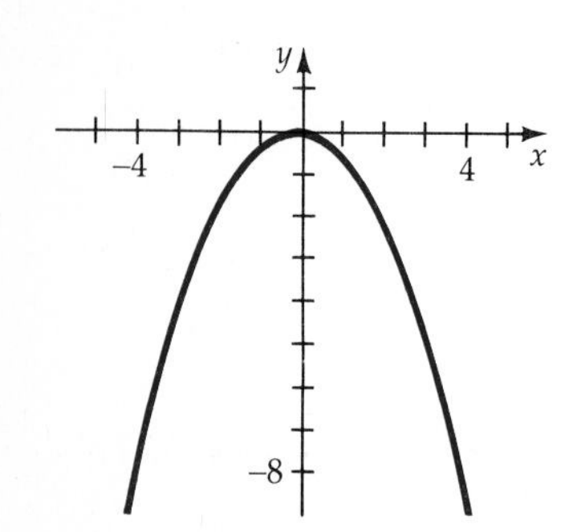

Figure 6.4
$y = -\frac{1}{2}x^2$

Comparing this equation with the equation in standard form gives

$$4p = -2$$

or

$$p = -\frac{1}{2}.$$

Because p is negative, the parabola will open down and the focus will be below the vertex $(0, 0)$.

The coordinates of the focus are $\left(0, -\frac{1}{2}\right)$. The equation of the directrix is $y = \frac{1}{2}$.

■ *Try Exercise* **4**, *page 227.*

EXAMPLE 2 **Find the Equation of a Parabola in Standard Form**

Find the equation of the parabola in standard form with vertex at the origin and focus at $(-2, 0)$.

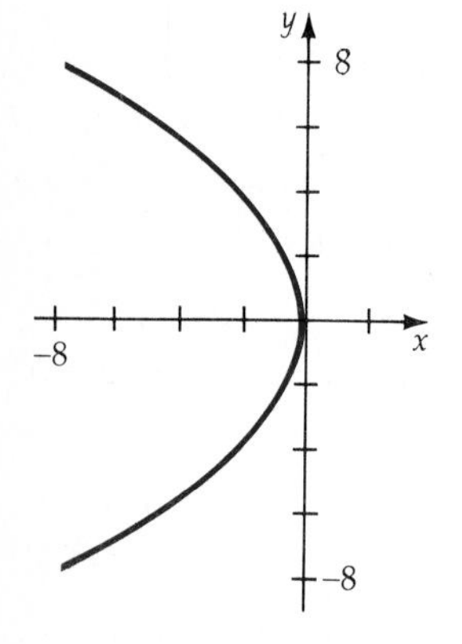

Figure 6.5
$y^2 = -8x$

Solution Because the vertex is at $(0, 0)$ and the focus is at $(-2, 0)$, $p = -2$. The graph of the parabola opens toward the focus and thus, in this case, the parabola is opening to the left. The equation of the parabola in standard form that opens to the left is $y^2 = 4px$. Substitute -2 for p in this equation and simplify:

$$y^2 = 4(-2)x = -8x$$

The equation is $y^2 = -8x$.

■ *Try Exercise* **28**, *page 227.*

Parabolas with the Vertex at (h, k)

The equation of a parabola with vertex at a point (h, k) can be found by using the translations discussed previously. Consider a coordinate system with coordinate axes labeled x' and y' placed so that its origin is at (h, k) of the xy-coordinate system.

The relationship between an ordered pair in the $x'y'$-coordinate sys-

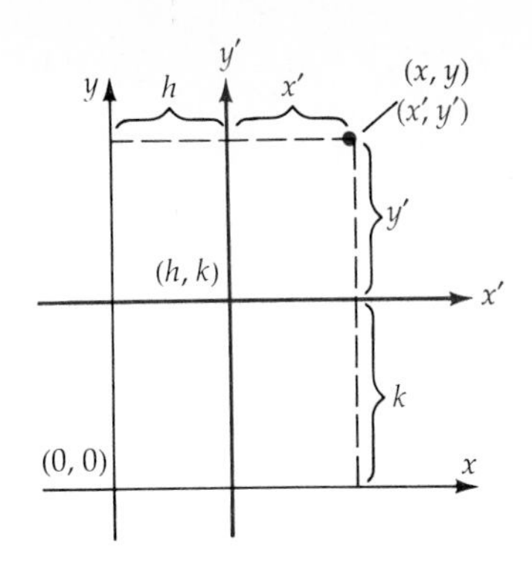

Figure 6.6
$x = x' + h,\ y = y' + k$

tem and the xy-coordinate system is given by the transformation equations

$$\begin{aligned} x' &= x - h \\ y' &= y - k \end{aligned} \tag{1}$$

Now consider a parabola that opens up with vertex at (h, k). Place a new coordinate system labeled x' and y' with its origin at (h, k). The equation of a parabola in the $x'y'$-coordinate system is

$$(x')^2 = 4py', \quad p > 0. \tag{2}$$

Using the transformation equations (1), we can substitute the expressions for x' and y' into Equation (2). The standard form of the equation of the parabola with vertex (h, k) and a vertical axis of symmetry is

$$(x - h)^2 = 4p(y - k)$$

Similarly, we can derive the standard form of the equation of the parabola with vertex (h, k) and a horizontal axis of symmetry.

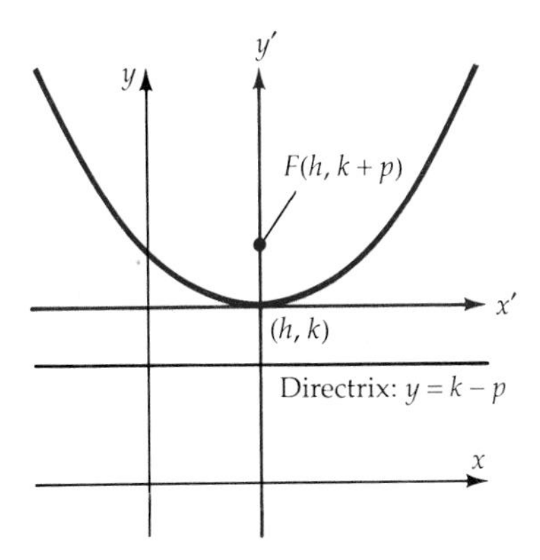

Figure 6.7

Standard Form of the Equation of a Parabola with Vertex at (h, k)

Vertical Axis of Symmetry
The standard form of the equation of the parabola with vertex (h, k) and vertical axis of symmetry is

$$(x - h)^2 = 4p(y - k)$$

The focus is $(h, k + p)$, and the equation of the directrix is $y = k - p$.

Horizontal Axis of Symmetry
The standard form of the equation of the parabola with vertex (h, k) and horizontal axis of symmetry is

$$(y - k)^2 = 4p(x - h)$$

The focus is $(h + p, k)$, and the equation of the directrix is $x = h - p$.

EXAMPLE 3 Find the Focus and Directrix of a Parabola

Find the equation of the directrix and the coordinates of the vertex and focus of the parabola given by the equation $3x + 2y^2 + 8y - 4 = 0$.

Solution Rewrite the equation and then complete the square.

$$3x + 2y^2 + 8y - 4 = 0$$

$$\begin{aligned} 2y^2 + 8y &= -3x + 4 \\ 2(y^2 + 4y) &= -3x + 4 \\ 2(y^2 + 4y + 4) &= -3x + 4 + 8 \\ 2(y + 2)^2 &= -3(x - 4) \end{aligned}$$

Complete the square. Note that $2 \cdot 4 = 8$ is added to each side.

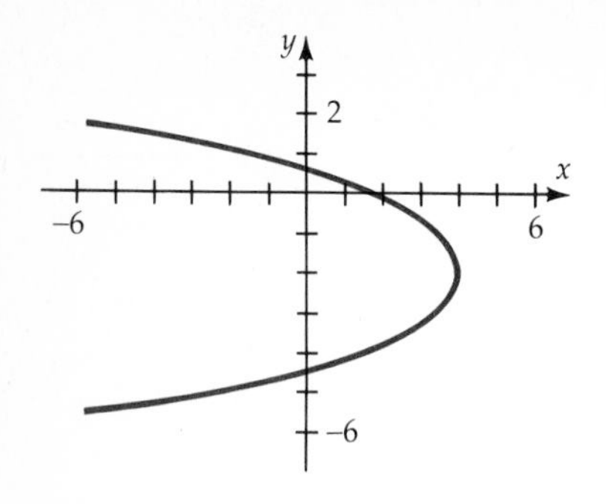

Figure 6.8
$(y+2)^2 = -\frac{3}{2}(x-4)$

Write the equation in standard form.

$$(y+2)^2 = -\frac{3}{2}(x-4)$$

Comparing this equation to $(y-k)^2 = 4p(x-h)$, we have a parabola that opens to the left with vertex $(4, -2)$ and $4p = -\frac{3}{2}$. Thus $p = -\frac{3}{8}$.

The coordinates of the focus are $\left(4 + \left(-\frac{3}{8}\right), -2\right) = \left(\frac{29}{8}, -2\right)$. The equation of the directrix is $x = 4 - \left(-\frac{3}{8}\right) = \frac{35}{8}$.

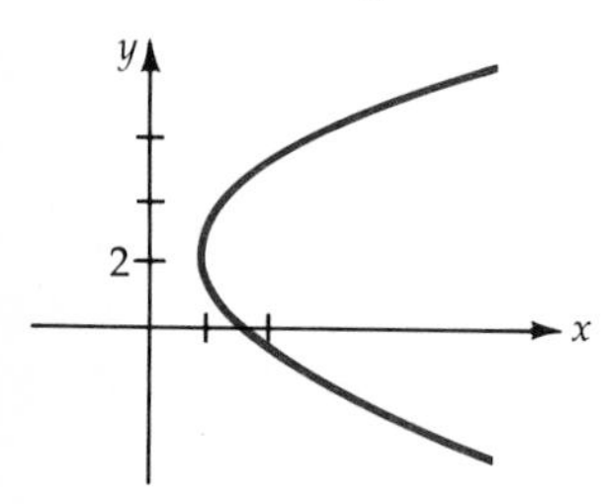

Figure 6.9
$(y-2)^2 = 8(x-1)$

Choosing some values for y and finding the corresponding values for x, we plot a few points. We use the fact that the line $y = -2$ is the axis of symmetry. Thus, for a point on one side of the axis of symmetry, there is a corresponding point on the other side. Two points are $(-2, 1)$ and $(-2, -5)$.

■ *Try Exercise* **20**, *page 227.*

EXAMPLE 4 Find the Equation of a Parabola in Standard Form

Find the equation in standard form of the parabola with directrix $x = -1$ and focus $(3, 2)$.

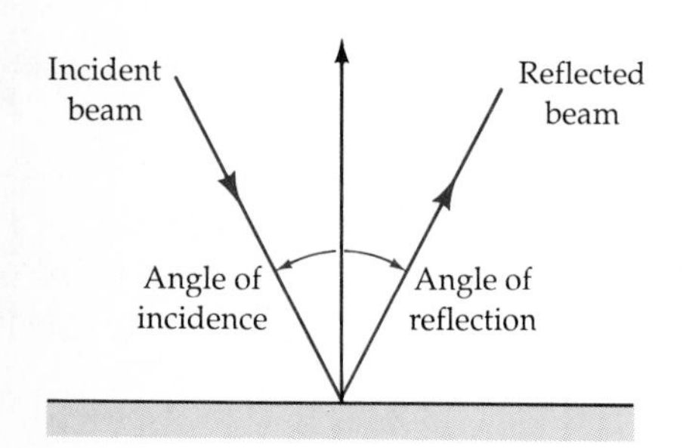

Figure 6.10

Solution The vertex is the midpoint of the line segment joining $(3, 2)$ and the point $(-1, 2)$ on the directrix.

$$(h, k) = \left(\frac{-1+3}{2}, \frac{2+2}{2}\right) = (1, 2).$$

The standard form of the equation will be of the form $(y-k)^2 = 4p(x-h)$. The distance from the vertex to the focus is 2. Thus $4p = 4(2) = 8$ and the equation of the parabola is

$$(y-2)^2 = 8(x-1).$$

■ *Try Exercise* **30**, *page 227.*

A principle of physics states that when light is reflected from a point P on a surface, the angle of incidence (incoming ray) equals the angle of reflection (outgoing ray). This principle applied to parabolas has some useful consequences.

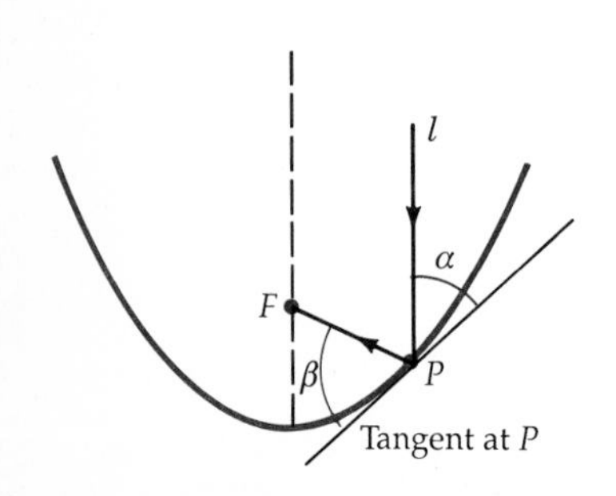

Figure 6.11

Optical Property of a Parabola

> The line tangent to a parabola at a point P makes equal angles with the line through P and parallel to the axis of symmetry and the line through P and the focus of the parabola (see Figure 6.11).

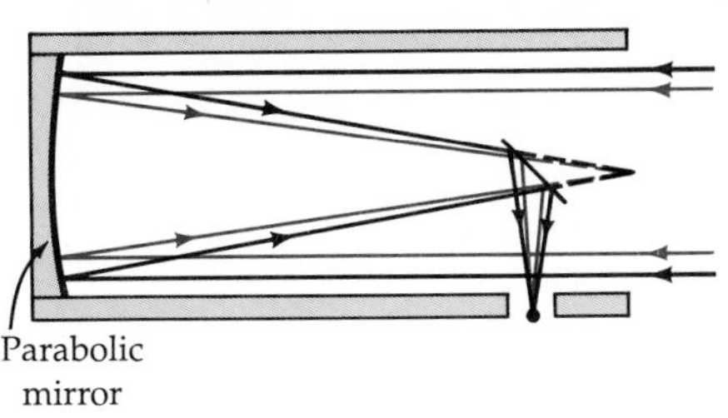

Figure 6.12

Figure 6.13

The reflecting mirror of a telescope is designed in the shape of a parabola. All the incoming light is reflected from the surface of the mirror and to the focus. See Figure 6.12.

Flashlights and car headlights also make use of this property. The light bulb is positioned at the focus of the parabolic reflector, causing the reflected light to be reflected outward in parallel rays. See Figure 6.13.

EXERCISE SET 6.1

In Exercises 1 to 26, find the vertex, focus, and directrix of the parabola given by each equation. Sketch the graph.

1. $x^2 = -4y$

2. $2y^2 = x$

3. $y^2 = \frac{1}{3}x$

4. $x^2 = -\frac{1}{4}y$

5. $(x - 2)^2 = 8(y + 3)$

6. $(y + 1)^2 = 6(x - 1)$

7. $(y + 4)^2 = -4(x - 2)$

8. $(x - 3)^2 = -(y + 2)$

9. $(y - 1)^2 = 2x + 8$

10. $(x + 2)^2 = 3y - 6$

11. $(2x - 4)^2 = 8y - 16$

12. $(3x + 6)^2 = 18y - 36$

13. $x^2 + 8x - y + 6 = 0$

14. $x^2 - 6x + y + 10 = 0$

15. $x + y^2 - 3y + 4 = 0$

16. $x - y^2 - 4y + 9 = 0$

17. $2x - y^2 - 6y + 1 = 0$

18. $3x + y^2 + 8y + 4 = 0$

19. $x^2 + 3x + 3y - 1 = 0$

20. $x^2 + 5x - 4y - 1 = 0$

21. $2x^2 - 8x - 4y + 3 = 0$

22. $6x - 3y^2 - 12y + 4 = 0$

23. $2x + 4y^2 + 8y - 5 = 0$

24. $4x^2 - 12x + 12y + 7 = 0$

25. $3x^2 - 6x - 9y + 4 = 0$

26. $2x - 3y^2 + 9y + 5 = 0$

27. Find the equation in standard form of the parabola with vertex at the origin and focus $(0, -4)$.

28. Find the equation in standard form of the parabola with vertex at the origin and focus $(5, 0)$.

29. Find the equation in standard form of the parabola with vertex at $(-1, 2)$ and focus $(-1, 3)$.

30. Find the equation in standard form of the parabola with vertex at $(2, -3)$ and focus $(0, -3)$.

31. Find the equation in standard form of the parabola with focus $(3, -3)$ and directrix $y = -5$.

32. Find the equation in standard form of the parabola with focus $(-2, 4)$ and directrix $x = 4$.

33. Find the equation in standard form of the parabola with vertex $(-4, 1)$, axis of symmetry parallel to the y-axis, and passing through the point $(-2, 2)$.

34. Find the equation in standard form of the parabola with vertex $(3, -5)$, axis of symmetry parallel to the x-axis, and passing through the point $(4, 3)$.

Supplemental Exercises

In Exercises 35 to 37, use the following definition of latus rectum: the line segment with endpoints on the parabola, through the focus of a parabola and perpendicular to the axis of symmetry is called the **latus rectum** of the parabola.

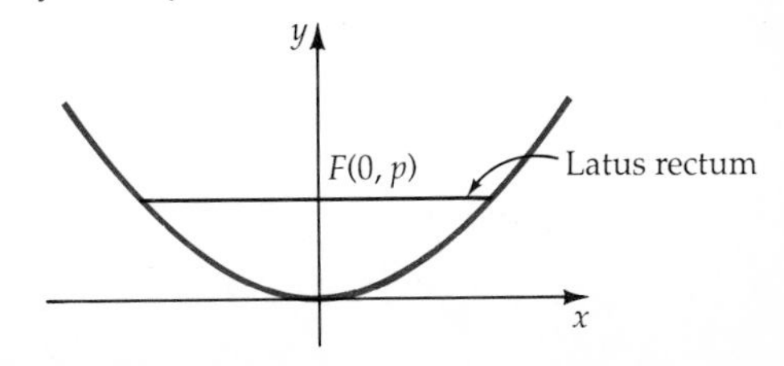

35. Find the length of the latus rectum for the parabola $x^2 = 4y$.

36. Find the length of the latus rectum for the parabola $y^2 = -8x$.

37. Find the length of the latus rectum for any parabola in terms of $|p|$, the distance from the vertex of the parabola to the focus.

The result of Exercise 37 can be stated as the following theorem. **Theorem** Two points on a parabola will be $2|p|$ units on each side of the axis of symmetry on the line through the focus and perpendicular to that axis.

38. Use the theorem to sketch a graph of the parabola given by the equation $(x - 3)^2 = 2(y + 1)$.

39. Use the theorem to sketch a graph of the parabola given by the equation $(y + 4)^2 = -(x - 1)$.

40. Use the theorem to sketch a graph of the parabola given by the equation $4x - y^2 + 8y = 0$.

41. Show that the point on the parabola closest to the focus is the vertex. (*Hint:* Consider the parabola $x^2 = 4py$ and a point on the parabola (a, b). Find the square of the distance between the point (a, b) and the focus. You may want to review the technique of minimizing a quadratic expression.)

42. By using the definition for a parabola, find the equation in standard form of the parabola with $V(0, 0)$, $F(-c, 0)$, and directrix $x = c$.

43. Sketch a graph of $4(y - 2) = (x|x| - 1)$.

44. Find the equation of the directrix of the parabola with vertex at the origin and whose focus is the point $(1, 1)$.

45. Find the equation of the parabola with vertex at the origin and focus at the point $(1, 1)$. (*Hint:* You will need the answer to Exercise 44 and the definition of a parabola.)

6.2 Ellipses

An ellipse is another of the conic sections formed when a plane intersects a right circular cone. If β is the angle at which the plane intersects the axis of the cone and α is the angle shown in Figure 6.14, an ellipse is formed when $\alpha < \beta < 90°$. If $\beta = 90°$, then a circle is formed.

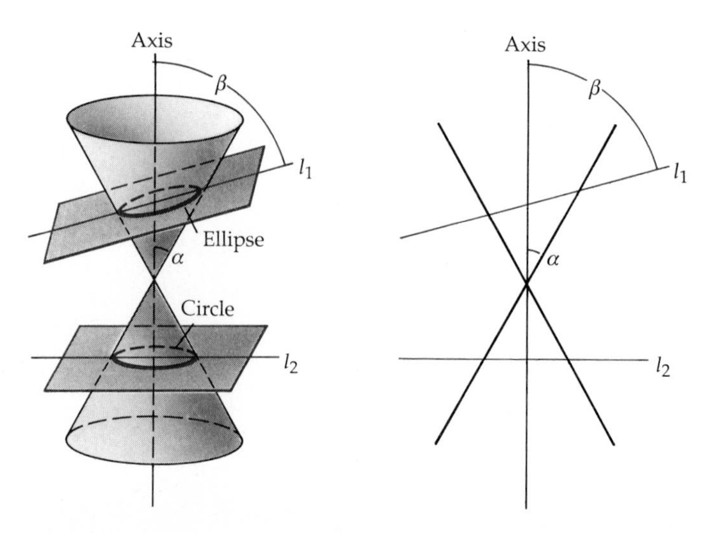

Figure 6.14

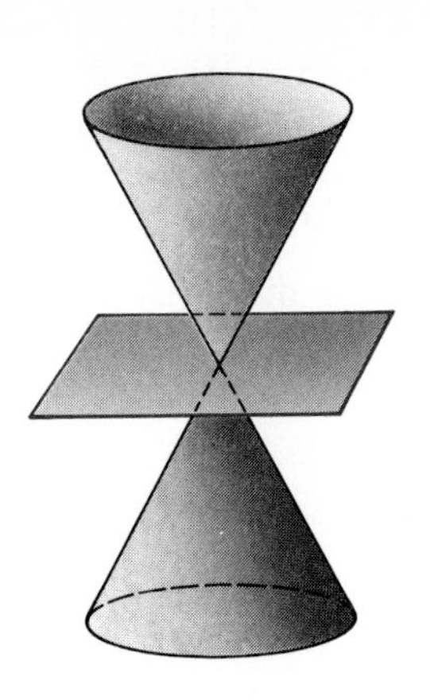

Figure 6.15

Remark If the plane were to intersect the cone at the vertex of the cone so that the resulting figure is a point, the point is a **degenerate ellipse.**

As was the case for a parabola, there is a definition for an ellipse in terms of a certain set of points in the plane.

Definition of an Ellipse

An **ellipse** is the set of all points in the plane, the sum of whose distances from two fixed points (foci) is a positive constant.

This definition can be used to draw an ellipse using a piece of string and two tacks (see Figure 6.16). Tack the ends of the string to the foci, and trace a curve with a pencil held tight against the string. The resulting curve is an ellipse. The positive constant is the length of the string.

Ellipses with Center at (0, 0)

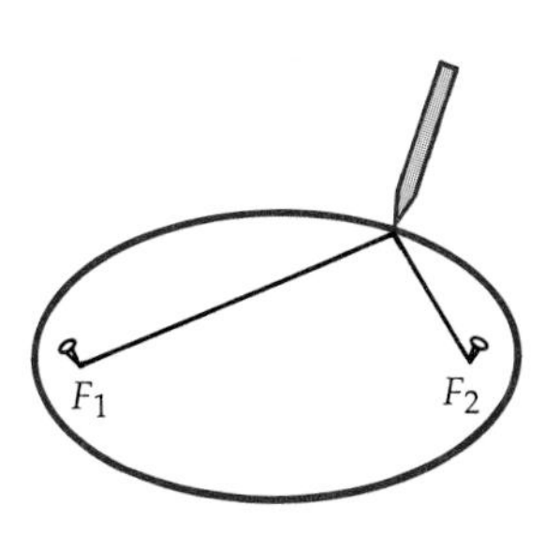

Figure 6.16

The graph of an ellipse is oval-shaped, with two axes of symmetry (see Figure 6.17). The longer axis is called the **major axis.** The foci of the ellipse are on the major axis. The shorter axis is called the **minor axis.** It is customary to denote the length of the major axis as $2a$ and the length of the minor axis as $2b$. The length of the **semiaxes** are one-half the axes. Thus the length of the semimajor axis is denoted by a and the length of the semiminor axis by b. The **center** of the ellipse is the midpoint of the major axis. The endpoints of the major axis are the **vertices** (plural of vertex) of the ellipse.

Consider the point $(a, 0)$, which is one vertex of an ellipse, and the point $(c, 0)$ and $(-c, 0)$, which are the foci of the ellipse shown in Figure 6.18. The distance from $(a, 0)$ to $(c, 0)$ is $a - c$. Similarly, the distance from $(a, 0)$ to $(-c, 0)$ is $a + c$. From the definition of an ellipse, the sum of distances from any point on the ellipse to the foci is a constant. By adding the expressions $a - c$ and $a + c$, we have

$$(a - c) + (a + c) = 2a$$

Thus the constant is precisely the length of the major axis.

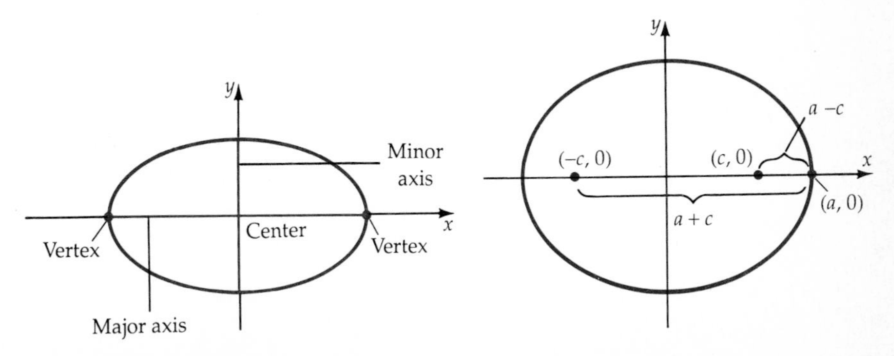

Figure 6.17

Figure 6.18

Now let $P(x, y)$ be any point on the ellipse. By using the definition of an ellipse, we have

$$d(P, F_1) + d(P, F_2) = 2a$$
$$\sqrt{(x + c)^2 + y^2} + \sqrt{(x - c)^2 + y^2} = 2a$$

Subtract the second radical from each side of the equation and then square each side.

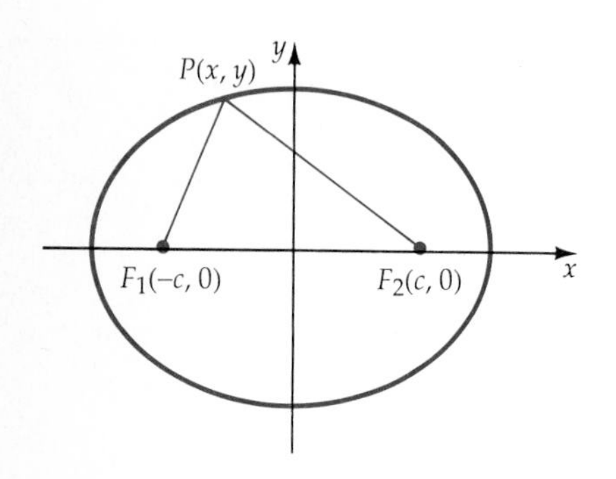

Figure 6.19

$$[\sqrt{(x + c)^2 + y^2}]^2 = [2a - \sqrt{(x - c)^2 + y^2}]^2$$
$$(x + c)^2 + y^2 = 4a^2 - 4a\sqrt{(x - c)^2 + y^2} + (x - c)^2 + y^2$$
$$x^2 + 2cx + c^2 + y^2 = 4a^2 - 4a\sqrt{(x - c)^2 + y^2} + x^2 - 2cx + c^2 + y^2$$
$$4cx - 4a^2 = -4a\sqrt{(x - c)^2 + y^2}$$

Divide each side by -4 and then square each side again.

$$[-cx + a^2]^2 = [a\sqrt{(x - c)^2 + y^2}]^2$$
$$c^2x^2 - 2cxa^2 + a^4 = a^2x^2 - 2cxa^2 + a^2c^2 + a^2y^2$$

Simplify and then rewrite with x and y terms on the left side.

$$-a^2x^2 + c^2x^2 - a^2y^2 = -a^4 + a^2c^2$$
$$-(a^2 - c^2)x^2 - a^2y^2 = -a^2(a^2 - c^2) \quad \text{Factor and let } b^2 = a^2 - c^2.$$
$$-b^2x^2 - a^2y^2 = -a^2b^2 \quad \text{Divide each side by } -a^2b^2.$$
$$\frac{x^2}{a^2} + \frac{y^2}{b^2} = 1 \quad \text{An equation of an ellipse}$$

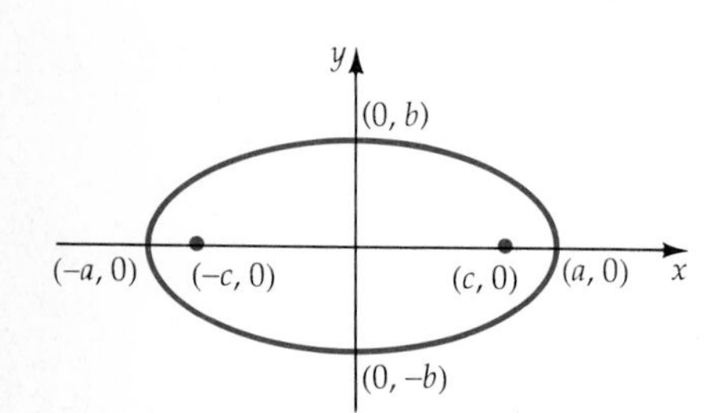

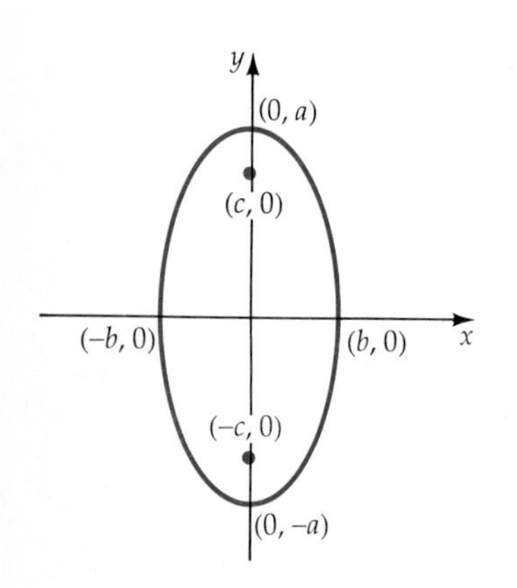

Figure 6.20

Standard Forms of the Equation of an Ellipse with Center at the Origin

Major Axis on the x-axis
The standard form of the equation of an ellipse with the center at the origin and major axis on the x-axis is given by

$$\frac{x^2}{a^2} + \frac{y^2}{b^2} = 1 \quad a > b.$$

The coordinates of the vertices are $(a, 0)$ and $(-a, 0)$, and the coordinates of the foci are $(c, 0)$ and $(-c, 0)$, where $c^2 = a^2 - b^2$.

Major Axis on the y-axis
The standard form of the equation of an ellipse with the center at the origin and major axis on the y-axis is given by

$$\frac{x^2}{b^2} + \frac{y^2}{a^2} = 1 \quad a > b.$$

The coordinates of the vertices are $(0, a)$ and $(0, -a)$, and the coordinates of the foci are $(0, c)$ and $(0, -c)$, where $c^2 = a^2 - b^2$.

Remark By looking at the standard form of the equations of an ellipse and noting that $a > b$, observe that the orientation of the major axis is determined by the larger denominator. When the x^2 term has the larger denominator, the major axis is on the x-axis. When the y^2 term has the larger denominator, the major axis is on the y-axis.

EXAMPLE 1 **Find the Vertices and Foci of an Ellipse**

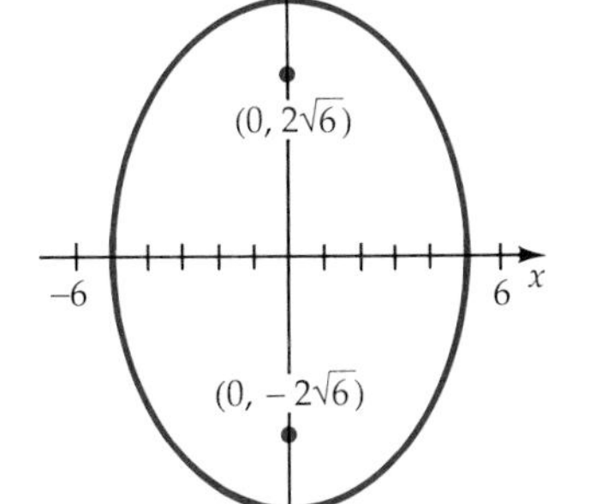

Figure 6.21
$\dfrac{x^2}{25} + \dfrac{y^2}{49} = 1$

Find the vertices and foci of the ellipse given by the equation $\dfrac{x^2}{25} + \dfrac{y^2}{49} = 1$. Sketch the graph.

Solution Because the y^2 term has the larger denominator, the major axis is on the y-axis.

$$a^2 = 49 \qquad b^2 = 25 \qquad c^2 = a^2 - b^2$$
$$a = 7 \qquad b = 5 \qquad = 49 - 25 = 24$$
$$c = \sqrt{24} = 2\sqrt{6}$$

The vertices are $(0, 7)$ and $(0, -7)$. The foci are $(0, 2\sqrt{6})$ and $(0, -2\sqrt{6})$.

■ *Try Exercise* **8,** *page 236.*

EXAMPLE 2 **Find the Equation of an Ellipse**

Find the equation in standard form of the ellipse with foci $(3, 0)$ and $(-3, 0)$ and major axis of length 10. Sketch the graph.

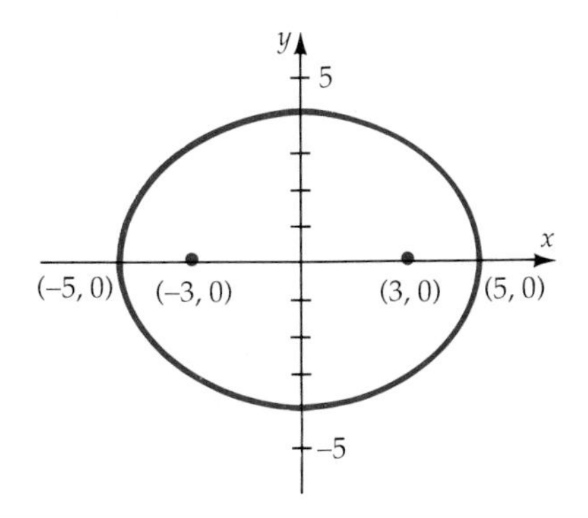

Figure 6.22
$\dfrac{x^2}{25} + \dfrac{y^2}{16} = 1$

Solution Because the foci are on the major axis, the major axis is on the x-axis. The length of the major axis is $2a$.

Thus $2a = 10$. Solving for a, we have $a = 5$.

Because the foci are $(3, 0)$ and $(-3, 0)$ and the center of the ellipse is the midpoint between the two foci, the distance from the center of the ellipse to a focus is 3. Therefore, $c = 3$.

To find b^2, use the equation $c^2 = a^2 - b^2$.

$$9 = 25 - b^2$$
$$b^2 = 16$$

The equation of the ellipse is $\dfrac{x^2}{25} + \dfrac{y^2}{16} = 1$.

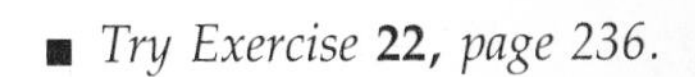
■ *Try Exercise* **22,** *page 236.*

Ellipses with the Center at (h, k)

The equation of an ellipse with center (h, k) and with horizontal or vertical major axes can be found by using a translation of coordinates. Given a

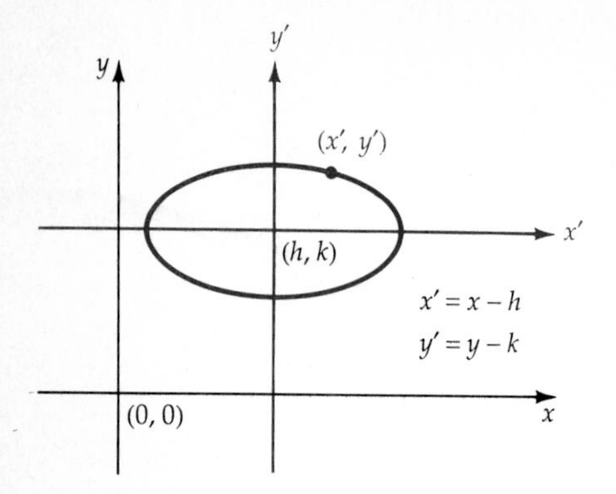

Figure 6.23

coordinate system with axes labeled x' and y', the standard form of the equation of an ellipse with center at the origin is

$$\frac{(x')^2}{a^2} + \frac{(y')^2}{b^2} = 1.$$

Now place the origin of the $x'y'$-coordinate system at (h, k) in an xy-coordinate system.

The relationship between an ordered pair in the $x'y'$-coordinate system and the xy-coordinate system is given by transformation equations

$$x' = x - h$$

$$y' = y - k$$

Substitute the expressions for x' and y' into the equation of an ellipse. The equation of the ellipse with center at (h, k) is

$$\frac{(x-h)^2}{a^2} + \frac{(y-k)^2}{b^2} = 1.$$

Standard Form of Ellipses with Center at (h, k)

Major Axis Parallel to the x-axis

The standard form of the equation of an ellipse with the center at (h, k) and major axis parallel to the x-axis is given by

$$\frac{(x-h)^2}{a^2} + \frac{(y-k)^2}{b^2} = 1 \quad a > b.$$

The coordinates of the vertices are $(h + a, k)$ and $(h - a, k)$, and the coordinates of the foci are $(h + c, k)$ and $(h - c, k)$, where $c^2 = a^2 - b^2$.

Major Axis Parallel to the y-axis

The standard form of the equation of an ellipse with the center at (h, k) and major axis parallel to the y-axis is given by

$$\frac{(x-h)^2}{b^2} + \frac{(y-k)^2}{a^2} = 1 \quad a > b.$$

The coordinates of the vertices are $(h, k + a)$ and $(h, k - a)$, and the coordinates of the foci are $(h, k + c)$ and $(h, k - c)$, where $c^2 = a^2 - b^2$.

EXAMPLE 3 Find the Vertices and Foci of an Ellipse

Find the vertices and foci of the ellipse $4x^2 + 9y^2 - 8x + 36y + 4 = 0$. Sketch the graph.

Solution Write the equation of the ellipse in standard form by completing the square.

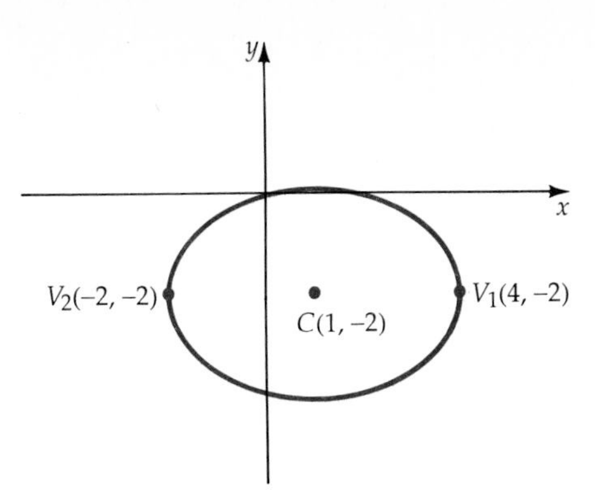

Figure 6.24
$$\frac{(x-1)^2}{9} + \frac{(y+2)^2}{4} = 1$$

$$\begin{aligned}
4x^2 + 9y^2 - 8x + 36y + 4 &= 0 \\
4x^2 - 8x + 9y^2 + 36y &= -4 && \text{Rearrange terms.} \\
4(x^2 - 2x) + 9(y^2 + 4y) &= -4 && \text{Factor.} \\
4(x^2 - 2x + 1) + 9(y^2 + 4y + 4) &= -4 + 4 + 36 && \text{Complete the square} \\
4(x-1)^2 + 9(y+2)^2 &= 36 && \text{Factor.} \\
\frac{(x-1)^2}{9} + \frac{(y+2)^2}{4} &= 1 && \text{Divide by 36.}
\end{aligned}$$

From the equation of the ellipse in standard form, we see that the coordinates of the center of the ellipse are $(1, -2)$. Because the larger denominator is 9, the major axis is parallel to the x-axis and $a^2 = 9$. Thus $a = 3$. The vertices are $(4, -2)$ and $(-2, -2)$.

To find the coordinates of the foci, we find c.

$$c^2 = a^2 - b^2 = 9 - 4 = 5$$

thus
$$c = \sqrt{5}.$$

The foci are $(1 + \sqrt{5}, -2)$ and $(1 - \sqrt{5}, -2)$.

■ *Try Exercise* **14,** *page 236.*

EXAMPLE 4 Find the Equation of an Ellipse

Find the standard form of the equation of the ellipse with center at $(4, -2)$, foci $(4, 1)$ and $(4, -5)$, and minor axis of length 10.

Solution Because the foci are on the major axis, the major axis is parallel to the y-axis. The distance from the center of the ellipse to a focus is c. The distance between $(4, -2)$ and $(4, 1)$ is 3. Therefore $c = 3$.

Recall that the length of the minor axis is $2b$. Thus $2b = 10$. Solving for b, we have $b = 5$.

To find a^2, use the equation $c^2 = a^2 - b^2$.

$$9 = a^2 - 25$$
$$a^2 = 34$$

Thus, the equation is

$$\frac{(x-4)^2}{25} + \frac{(y+2)^2}{34} = 1.$$

■ *Try Exercise* **30,** *page 236.*

Eccentricity of an Ellipse

The graph of an ellipse can be very long and thin, or it can be much like a circle. The **eccentricity** of an ellipse is a measure of its "roundness."

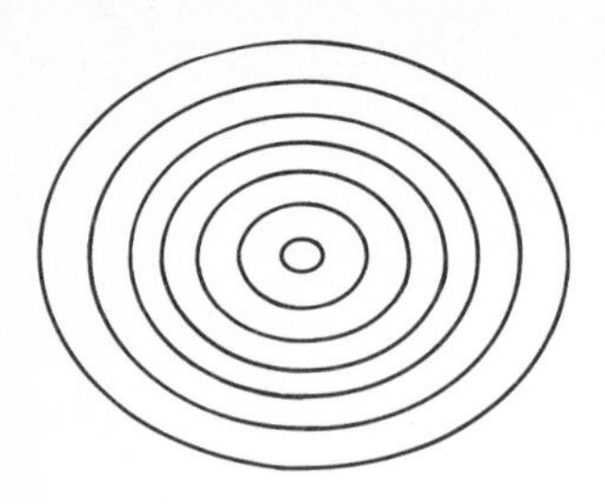

Figure 6.25

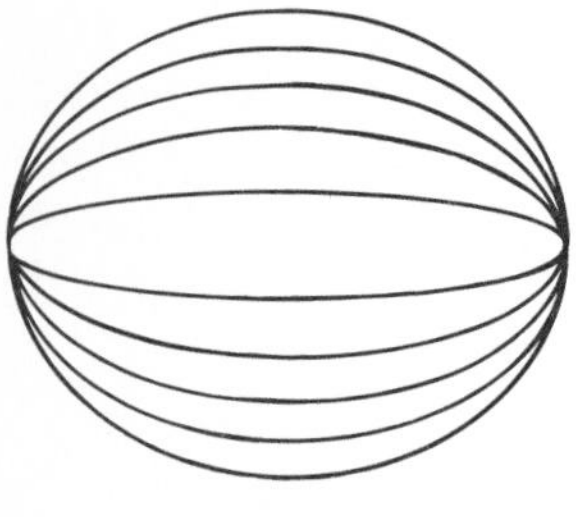

Figure 6.26

Eccentricity (e) of an Ellipse

> The eccentricity e of an ellipse is the ratio of c to a, where c is the distance from the center to the focus and a is the length of the semimajor axis. That is,
>
> $$e = \frac{c}{a}.$$

Because $c < a$, for an ellipse, $0 < e < 1$. If $c \approx 0$, then $e \approx 0$ and the graph will be almost like a circle.

If $c \approx a$, then $e \approx 1$ and the graph will be long and thin.

EXAMPLE 5 Find the Eccentricity of an Ellipse

Find the eccentricity of the ellipse $8x^2 + 9y^2 = 18$.

Solution First, write the equation of the ellipse in standard form. Divide each side of the equation by 18.

$$\frac{8x^2}{18} + \frac{9y^2}{18} = 1$$

$$\frac{4x^2}{9} + \frac{y^2}{2} = 1$$

$$\frac{x^2}{9/4} + \frac{y^2}{2} = 1$$

The last step is necessary because the standard form of the equation has coefficients of 1 in the numerator. Thus we have

$$a^2 = \frac{9}{4}$$

$$a = \frac{3}{2}$$

Use the equation $c^2 = a^2 - b^2$ to find c.

$$c^2 = \frac{9}{4} - 2 = \frac{1}{4}$$

$$c = \sqrt{\frac{1}{4}} = \frac{1}{2}.$$

Now we can find the eccentricity.

$$e = \frac{c}{a} = \frac{1/2}{3/2} = \frac{1}{3}$$

The eccentricity is $\frac{1}{3}$.

■ *Try Exercise* **36,** *page 236.*

The planets revolve around the sun in elliptical orbits. The eccentricities of planets in our solar system are given below.

Which planet has the most circular orbit? Why?[2]

Planet	Eccentricity
Mercury	0.206
Venus	0.007
Earth	0.017
Mars	0.093
Jupiter	0.049
Saturn	0.051
Uranus	0.046
Neptune	0.005
Pluto	0.250

Acoustic Property of an Ellipse

Sound waves, although different from light waves, have a similiar reflective property. When sound is reflected from a point P on a surface, the angle of incidence equals the angle of reflection. Applying this principle to an ellipse results in what are called "whispering galleries." These galleries are based on the following theorem.

The Reflective Property of an Ellipse

> The lines from the foci to a point on the ellipse make equal angles with the tangent line at that point.

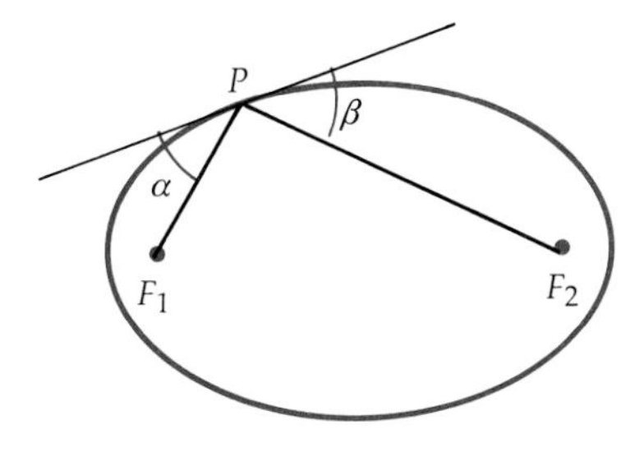

Figure 6.27
$\angle\alpha = \angle\beta$.

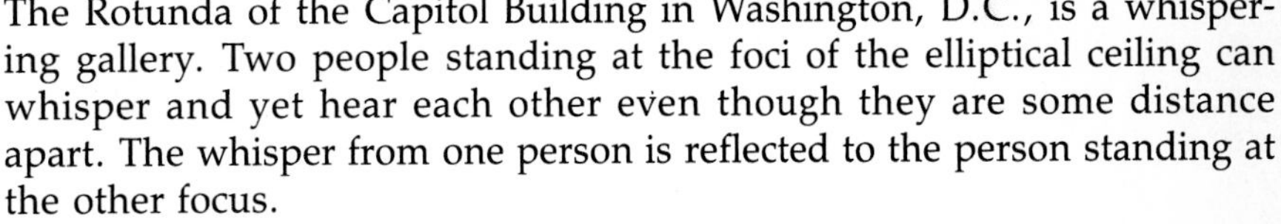
The Rotunda of the Capitol Building in Washington, D.C., is a whispering gallery. Two people standing at the foci of the elliptical ceiling can whisper and yet hear each other even though they are some distance apart. The whisper from one person is reflected to the person standing at the other focus.

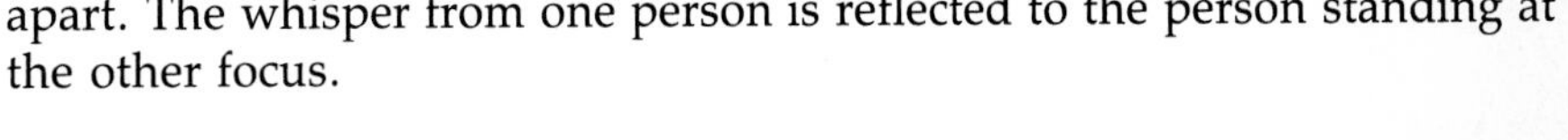

[2] Neptune has the smallest eccentricity, it is therefore the planet with the most circular orbit.

EXERCISE SET 6.2

In Exercises 1 to 20, find the vertices and foci of the ellipse given by each equation. Sketch the graph.

1. $\dfrac{x^2}{16} + \dfrac{y^2}{25} = 1$

2. $\dfrac{x^2}{49} + \dfrac{y^2}{36} = 1$

3. $\dfrac{x^2}{9} + \dfrac{y^2}{4} = 1$

4. $\dfrac{x^2}{64} + \dfrac{y^2}{25} = 1$

5. $3x^2 + 4y^2 = 12$

6. $5x^2 + 4y^2 = 20$

7. $25x^2 + 16y^2 = 400$

8. $25x^2 + 12y^2 = 300$

9. $64x^2 + 25y^2 = 400$

10. $9x^2 + 64y^2 = 144$

11. $4x^2 + y^2 - 24x - 8y + 48 = 0$

12. $x^2 + 9y^2 + 6x - 36y + 36 = 0$

13. $5x^2 + 9y^2 - 20x + 54y + 56 = 0$

14. $9x^2 + 16y^2 + 36x - 16y - 104 = 0$
15. $16x^2 + 9y^2 - 64x - 80 = 0$
16. $16x^2 + 9y^2 + 36y - 108 = 0$
17. $25x^2 + 16y^2 + 50x - 32y - 359 = 0$
18. $16x^2 + 9y^2 - 64x - 54y + 1 = 0$
19. $8x^2 + 25y^2 - 48x + 50y + 47 = 0$
20. $4x^2 + 9y^2 + 24x + 18y + 44 = 0$

In Exercises 21 to 32, find the equation in standard form of each ellipse, given the information provided.

21. Center $(0, 0)$, major axis of length 10, foci at $(4, 0)$ and $(-4, 0)$
22. Center $(0, 0)$, minor axis of length 6, foci at $(0, 4)$ and $(0, -4)$
23. Vertices $(6, 0)$, $(-6, 0)$; ellipse passes through $(0, -4)$, and $(0, 4)$
24. Vertices $(5, 0)$, $(-5, 0)$; ellipse passes through $(0, 7)$, and $(0, -7)$
25. Major axis of length 12 on the x-axis, center at $(0, 0)$, and passing through $(2, -3)$
26. Minor axis of length 8, center at $(0, 0)$, and passing through $(-2, 2)$
27. Center $(-2, 4)$, vertices $(-6, 4)$ and $(2,4)$, foci $(-5, 4)$ and $(1, 4)$
28. Center $(0, 3)$, minor axis of length 4, foci $(0, 0)$ and $(0,6)$
29. Center $(2, 4)$, major axis parallel to the y-axis and of length 10, the ellipse passes through the point $(3, 3)$
30. Center $(-4, 1)$, minor axis parallel to the y-axis of length 8, and the ellipse passes through the point $(0, 4)$
31. Vertices $(5, 6)$ and $(5, -4)$, foci $(5, 4)$ and $(5, -2)$
32. Vertices $(-7, -1)$ and $(5, -1)$, foci $(-5, -1)$ and $(3, -1)$

In Exercises 33 to 40, use the eccentricity of the ellipse to find the equation in standard form of each of the following ellipses.

33. Eccentricity 2/5, major axis on the x-axis of length 10, and center at $(0, 0)$
34. Eccentricity 3/4, foci at $(9, 0)$ and $(-9, 0)$
35. Foci at $(0, -4)$ and $(0, 4)$, eccentricity 2/3
36. Foci at $(0, -3)$ and $(0, 3)$, eccentricity 1/4
37. Eccentricity 2/5, foci $(-1, 3)$ and $(3, 3)$
38. Eccentricity 1/4, foci $(-2, 4)$ and $(-2, -2)$
39. Eccentricity 2/3, major axis of length 24 on the y-axis, centered at $(0, 0)$
40. Eccentricity 3/5, major axis of length 15 on the x-axis, center at $(0, 0)$

Supplemental Exercises

41. Explain why the graph of the equation $4x^2 + 9y^2 - 8x + 36y + 76 = 0$ is or is not an ellipse.
42. Explain why the graph of the equation $4x^2 + 9y - 16x - 2 = 0$ is or is not an ellipse. Sketch the graph of this equation. See section 9.1 for assistance.

In Exercises 43 to 46, find the equation in standard form of an ellipse by using the definition of an ellipse.

43. Find the equation of the ellipse with foci at $(-3, 0)$ and $(3, 0)$ that passes through the point $(3, 9/2)$.
44. Find the equation of the ellipse with foci at $(0, 4)$ and $(0, -4)$ that passes through the point $(9/5, 4)$.
45. Find the equation of the ellipse with foci at $(-1, 2)$ and $(3, 2)$ that passes through the point $(3, 5)$.
46. Find the equation of the ellipse with foci at $(-1, 1)$ and $(-1, 7)$ that passes through the point $(3/4, 1)$.

In Exercises 47 and 48, find the latus rectum of the given ellipse. The line segment with endpoints on the ellipse that is perpendicular to the major axis and passes through the focus is the **latus rectum** of the ellipse.

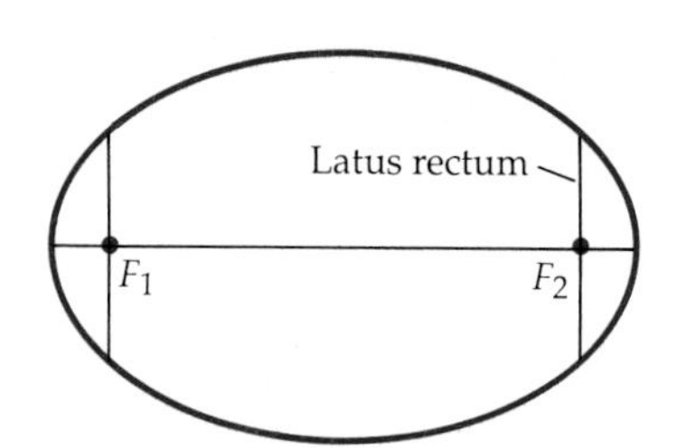

47. Find the length of the latus rectum of the ellipse given by
$$\frac{(x-1)^2}{9} + \frac{(y+1)^2}{16} = 1.$$
48. Find the length of the latus rectum of the ellipse given by
$$9x^2 + 16y^2 - 36x + 96y + 36 = 0.$$
49. Show that for any ellipse, the length of the latus rectum is $\dfrac{2b^2}{a}$.
50. Use the definition of an ellipse to find the equation of an ellipse with center at $(0, 0)$ and foci $(0, c)$ and $(0, -c)$.

Recall that a parabola has a directrix that is a line perpendicular to the axis of symmetry. An ellipse has two directrixes, both of which are perpendicular to the major axis and outside the ellipse. For an ellipse with center at the origin and whose major axis is the x-axis, the equations of the directrixes are $x = a^2/c$ and $x = -a^2/c$.

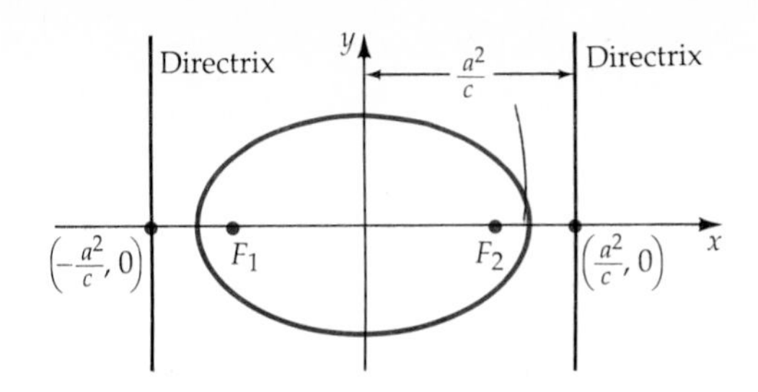

51. Find the directrix of the ellipse in Exercise 3.
52. Find the directrix of the ellipse in Exercise 4.
53. Let $P(x, y)$ be a point on the ellipse $\frac{x^2}{12} + \frac{y^2}{8} = 1$. Show that the distance from the point P to the focus $(2, 0)$ divided by the distance from the point P to the directrix $x = 6$ equals the eccentricity. (*Hint:* Solve the equation of the ellipse for y^2. Substitute this value for y^2 after applying the distance formula.)
54. Let $P(x, y)$ be a point on the ellipse $\frac{x^2}{25} + \frac{y^2}{16} = 1$. Show that the distance from the point P to the focus $(3, 0)$ divided by the distance from the point to the directrix $x = 25/3$ equals the eccentricity. (*Hint:* Solve the equation of the ellipse for y^2. Substitute this value for y^2 after applying the distance formula.)
55. Generalize the results of Exercises 53 and 54. That is, show that if $P(x, y)$ is a point on the ellipse $\frac{x^2}{a^2} + \frac{y^2}{b^2} = 1$, where $F(c, 0)$ is the focus and $x = a^2/c$ is the directrix, then the following equation is true: $e = d(P, F)/d(P, D)$. (*Hint:* Solve the equation of the ellipse for y^2. Substitute this value for y^2 after applying this distance formula.)

6.3 Hyperbolas

The hyperbola is a conic section formed when a plane intersects a right circular cone at a certain angle. If β is the angle at which the plane intersects the axis of the cone and α is the angle shown in Figure 6.28, a hyperbola is formed when $0° < \beta < \alpha$ or when the plane is parallel to the axis of the cone.

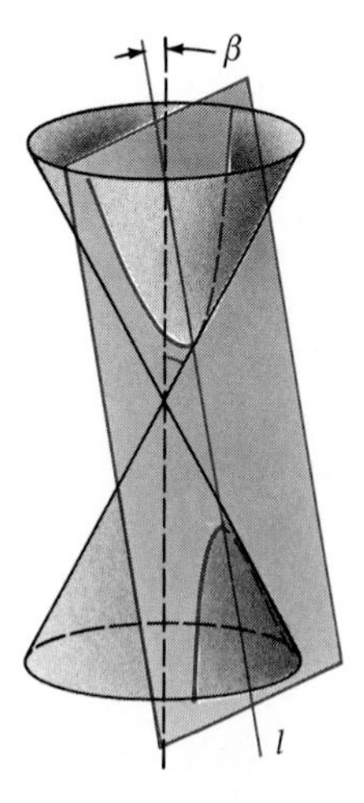

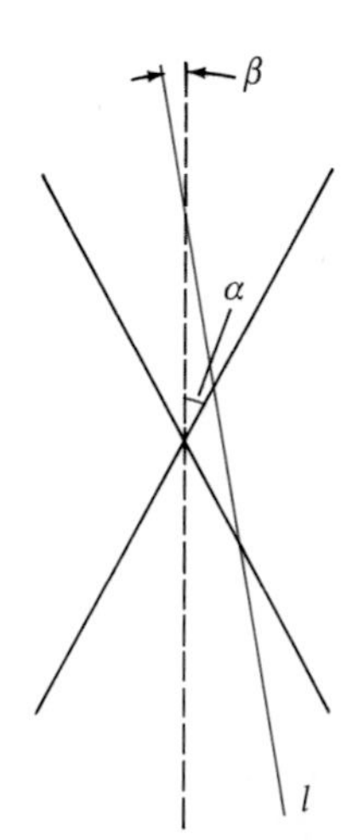

Figure 6.28

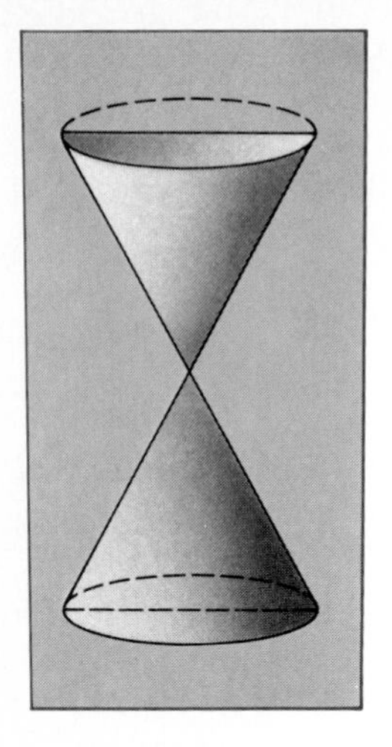

Figure 6.29

Remark If the plane intersects the cone along the axis of the cone, the resulting curve is two intersecting straight lines. This is the **degenerate** form of a hyperbola.

As with the other conic sections, there is a definition of a hyperbola in terms of a certain set of points in the plane.

Definition of a Hyperbola

> A **hyperbola** is the set of all points in the plane, the difference of whose distances from two fixed points (foci) is a positive constant.

Remark This definition differs from that of an ellipse in that the ellipse was defined in terms of the *sum* of the distances, whereas the hyperbola is defined in terms of the *difference* of two distances.

Hyperbolas with Center at (0, 0)

The **transverse axis** is the line segment joining the intercepts through the foci of a hyperbola (see Figure 6.30). The midpoint of the transverse axis is called the **center** of the hyperbola. The **conjugate** axis passes through the center of the hyperbola and is perpendicular to the transverse axis.

The length of the transverse axis is customarily denoted $2a$, and the distance between the two foci is denoted $2c$. The length of the conjugate axis is denoted $2b$.

The **vertices** of a hyperbola are the points where the hyperbola intersects the transverse axis.

To determine the positive constant stated in the definition of a hyperbola, consider the point $V(a, 0)$, which is one vertex of a hyperbola, and the points $F_1(c, 0)$ and $F_2(-c, 0)$, which are the foci of the hyperbola (see Figure 6.31). The difference of the distance from V to F_1, $c - a$, and the distance from $V(a, 0)$ to $F_2(-c, 0)$, $a + c$, must be a constant. By subtracting these distances, we find

$$|(c - a) - (c + a)| = |-2a| = 2a$$

Thus the constant is $2a$ and is the length of the transverse axis. The absolute value was used to ensure that the distance is a positive number.

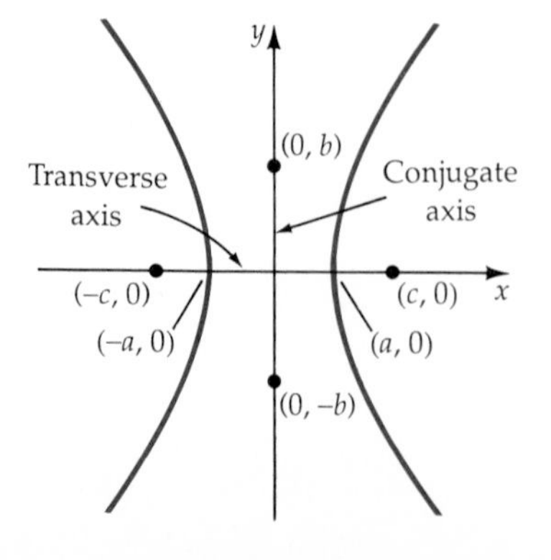

Figure 6.30

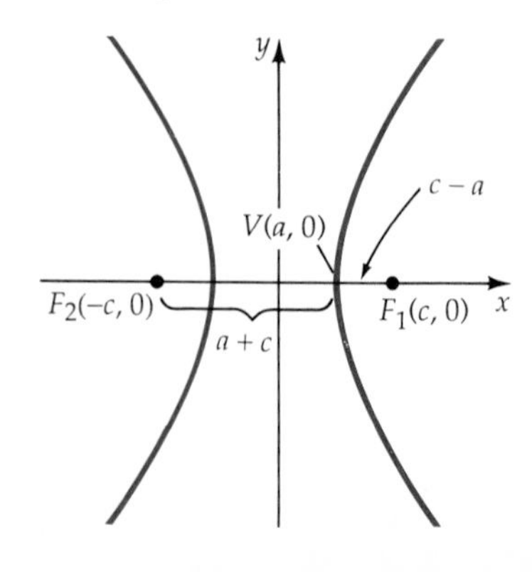

Figure 6.31

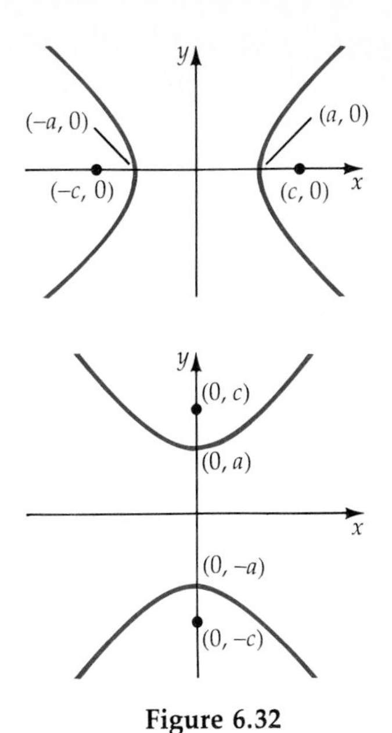

Figure 6.32

Standard Forms of the Equation of a Hyperbola with Center at the Origin

> ***Transverse axis on the x-axis***
> The standard form of the equation of a hyperbola with the center at the origin and transverse axis on the x-axis is given by
>
> $$\frac{x^2}{a^2} - \frac{y^2}{b^2} = 1$$
>
> The coordinates of the vertices are $(a, 0)$ and $(-a, 0)$, and the coordinates of the foci are $(c, 0)$ and $(-c, 0)$, where $c^2 = a^2 + b^2$.
>
> ***Transverse axis on the y-axis***
> The standard form of the equation of a hyperbola with the center at the origin and transverse axis on the y-axis is given by
>
> $$\frac{y^2}{a^2} - \frac{x^2}{b^2} = 1$$
>
> The coordinates of the vertices are $(0, a)$ and $(0, -a)$, and the coordinates of the foci are $(0, c)$ and $(0, -c)$, where $c^2 = a^2 + b^2$.

Remark By looking at the equations, note that it is possible to determine the transverse axis by finding which term in the equation is positive. If the x^2 term is positive, then the transverse axis is on the x-axis. When the y^2 term is positive, the transverse axis is on the y-axis.

Consider the hyperbola given by the equation $\frac{x^2}{16} - \frac{y^2}{9} = 1$. Because the x^2 term is positive, the transverse axis is on the x-axis, $a^2 = 16$, thus $a = 4$. The vertices are $(4, 0)$ and $(-4, 0)$. To find the foci, we determine c.

$$c^2 = a^2 + b^2 = 16 + 9 = 25$$

$$c = \sqrt{25} = 5.$$

The foci are $(5, 0)$ and $(-5, 0)$. The graph is shown in Figure 6.33.

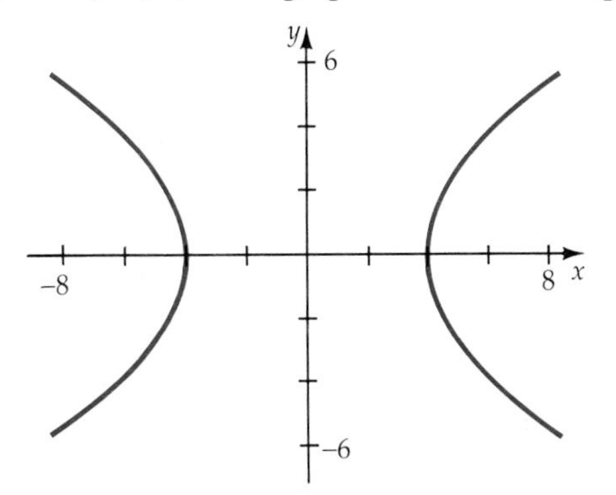

Figure 6.33
$\frac{x^2}{16} - \frac{y^2}{9} = 1$

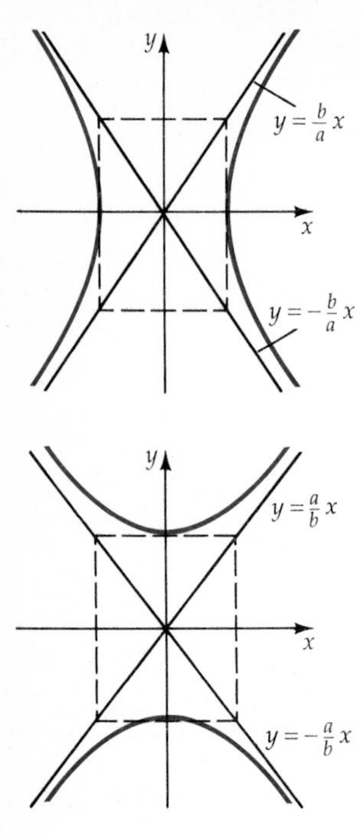

Figure 6.34

The asymptotes of the hyperbola are a useful guide to sketching the graph of the hyperbola. Each hyperbola has two asymptotes that pass through the center of the hyperbola.

Asymptotes of a Hyperbola with Center at the Origin

The **asymptotes** of the hyperbola $\frac{x^2}{a^2} - \frac{y^2}{b^2} = 1$ are given by the equations $y = \frac{b}{a}x$ and $y = -\frac{b}{a}x$. The asymptotes of the hyperbola $\frac{y^2}{a^2} - \frac{x^2}{b^2} = 1$ are given by the equations $y = \frac{a}{b}x$ and $y = -\frac{a}{b}x$.

We can outline a proof for the equations of the asymptotes by using the equation of a hyperbola in standard form.

$$\frac{x^2}{a^2} - \frac{y^2}{b^2} = 1$$

$$y^2 = b^2\left(\frac{x^2}{a^2} - 1\right) \qquad \text{Solve for } y^2.$$

$$= \frac{b^2}{a^2}(x^2 - a^2) \qquad \text{Factor out } 1/a^2.$$

$$= \frac{b^2}{a^2}x^2\left(1 - \frac{a^2}{x^2}\right) \qquad \text{Factor out } x^2.$$

$$y = \pm\frac{b}{a}x\sqrt{1 - \frac{a^2}{x^2}} \qquad \text{Take the square root of each side.}$$

As $|x|$ becomes larger and larger, $1 - \frac{a^2}{x^2}$ approaches 1. (Why?[3]) For large values of $|x|$, $y \approx \pm\frac{b}{a}x$, and thus $y = \pm\frac{b}{a}x$ are asymptotes for the hyperbola. A similar outline of a proof can be given for hyperbolas with the transverse axis on the y-axis.

Remark One method for remembering the equations of the asymptotes is to write the equation of a hyperbola in standard form but replace 1 by 0 and then solve for y.

$$\frac{x^2}{a^2} - \frac{y^2}{b^2} = 0 \quad \text{thus } y^2 = \frac{b^2}{a^2}x^2 \text{ or } y = \pm\frac{b}{a}x$$

$$\frac{y^2}{a^2} - \frac{x^2}{b^2} = 0 \quad \text{thus } y^2 = \frac{a^2}{b^2}x^2 \text{ or } y = \pm\frac{a}{b}x$$

[3] Because a^2 is a constant, a^2/x^2 approaches 0 as $|x|$ becomes very large. Thus $1 - a^2/x^2$ approaches $1 - 0$ or 1.

EXAMPLE 1 Find the Vertices, Foci, and Asymptotes of a Hyperbola

Find the foci, vertices, and asymptotes of the hyperbola given by the equation $\frac{y^2}{9} - \frac{x^2}{4} = 1$. Sketch the graph.

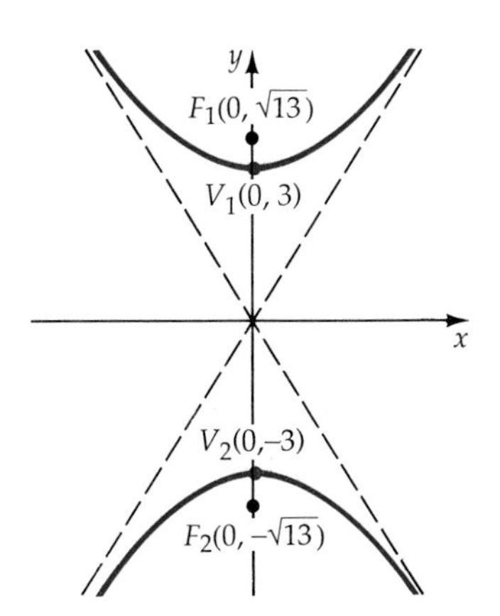

Figure 6.35
$\frac{y^2}{9} - \frac{x^2}{4} = 1$

Solution Because the y^2 term is positive, the transverse axis is the y-axis. We know $a^2 = 9$; thus $a = 3$. The vertices are $V_1(0, 3)$ and $V_2(0, -3)$.

$$c^2 = a^2 + b^2 = 9 + 4$$
$$c = \sqrt{13}$$

The foci are $F_1(0, \sqrt{13})$ and $F_2(0, -\sqrt{13})$.

Because $a = 3$ and $b = 2$ ($b^2 = 4$), the equations of the asymptotes are $y = \frac{3}{2}x$ and $y = -\frac{3}{2}x$.

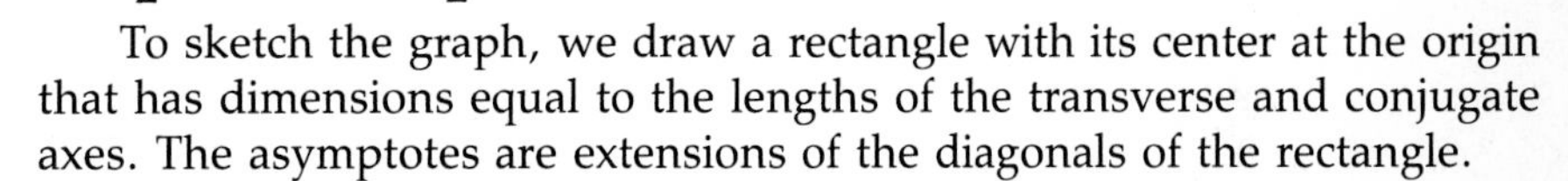

To sketch the graph, we draw a rectangle with its center at the origin that has dimensions equal to the lengths of the transverse and conjugate axes. The asymptotes are extensions of the diagonals of the rectangle.

■ *Try Exercise* **4**, *page 245.*

Hyperbolas with the Center at the Point (h, k)

Using a translation of coordinates similar to that used for ellipses, we can write the equation of a hyperbola with its center at the point (h, k). Given coordinates axes labeled x' and y', an equation of a hyperbola with center at the origin is

$$\frac{(x')^2}{a^2} - \frac{(y')^2}{b^2} = 1. \tag{1}$$

Now place the origin of this coordinate system at the point (h, k) of the xy-coordinate system. The relationship between an ordered pair in

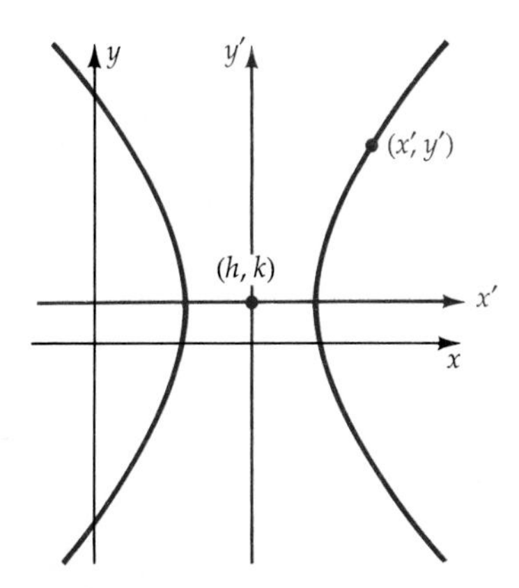

Figure 6.36

the $x'y'$-coordinate system and the xy-coordinate system is given by the transformation equations

$$x' = x - h$$
$$y' = y - k$$

Substitute the expressions for x' and y' into Equation (1). The equation of the hyperbola with center at (h, k) is

$$\frac{(x-h)^2}{a^2} - \frac{(y-k)^2}{b^2} = 1.$$

Standard Form of Hyperbolas with Center at (h, k)

Transverse Axis Parallel to the x-axis
The standard form of the equation of a hyperbola with center (h, k) and transverse axis parallel to the x-axis is given by

$$\frac{(x-h)^2}{a^2} - \frac{(y-k)^2}{b^2} = 1.$$

The coordinates of the vertices are $V_1(h + a, k)$ and $V_2(h - a, k)$. The coordinates of the foci are $F_1(h + c, k)$ and $F_2(h - c, k)$. The equations of the asymptotes are $y - k = \frac{b}{a}(x - h)$ and $y - k = -\frac{b}{a}(x - h)$.

Transverse Axis Parallel to the y-axis
The standard form of the equation of a hyperbola with center (h, k) and transverse axis parallel to the y-axis is given by

$$\frac{(y-k)^2}{a^2} - \frac{(x-h)^2}{b^2} = 1.$$

The coordinates of the vertices are $V_1(h, k + a)$ and $V_2(h, k - a)$. The coordinates of the foci are $F_1(h, k + c)$ and $F_2(h, k - c)$. The equations of the asymptotes are $y - k = \frac{a}{b}(x - h)$ and $y - k = -\frac{a}{b}(x - h)$.

EXAMPLE 2 Find the Vertices, Foci, and Asymptotes of a Hyperbola

Find the vertices, foci, and asymptotes of the hyperbola given by the equation $4x^2 - 9y^2 - 16x + 54y - 29 = 0$. Sketch the graph.

Solution Write the equation of the hyperbola in standard form by completing the square.

$$4x^2 - 9y^2 - 16x + 54y - 29 = 0$$
$$4x^2 - 16x - 9y^2 + 54y = 29 \qquad \text{Rearrange terms.}$$
$$4(x^2 - 4x) - 9(y^2 - 6y) = 29 \qquad \text{Factor.}$$

$$4(x^2 - 4x + 4) - 9(y^2 - 6y + 9) = 29 + 16 - 81 \quad \text{Complete the square.}$$

$$4(x - 2)^2 - 9(y - 3)^2 = -36 \quad \text{Factor.}$$

$$\frac{(y - 3)^2}{4} - \frac{(x - 2)^2}{9} = 1 \quad \text{Divide by } -36.$$

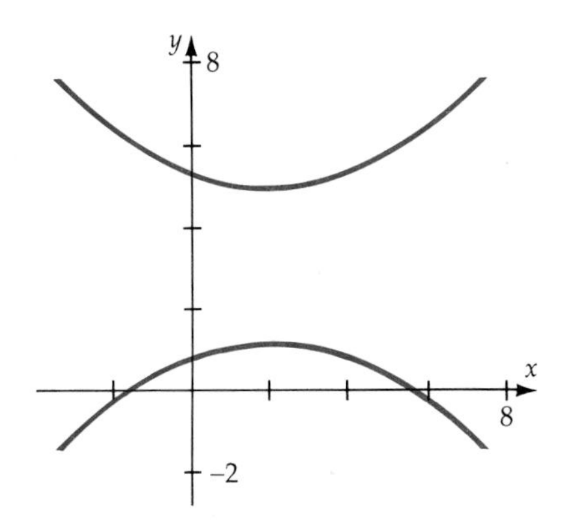

Figure 6.37
$\dfrac{(y - 3)^2}{4} - \dfrac{(x - 2)^2}{9} = 1$

The coordinates of the center are (2, 3). Because the term containing $(y - 3)^2$ is positive, the transverse axis is parallel to the y-axis. We know $a^2 = 4$; thus $a = 2$. The vertices are (2, 5) and (2, 1).

$$c^2 = a^2 + b^2 = 4 + 9$$

$$c = \sqrt{13}.$$

The foci are $(2, 3 + \sqrt{13})$ and $(2, 3 - \sqrt{13})$. We know $b^2 = 9$; thus $b = 3$. The equations of the asymptotes are

$$y = \frac{2}{3}x + \frac{5}{3} \quad \text{and} \quad y = -\frac{2}{3}x + \frac{13}{3}.$$

■ *Try Exercise* **18**, *page 245.*

Eccentricity of a Hyperbola

The graph of a hyperbola can be very wide or very narrow. The **eccentricity** of a hyperbola is a measure of its "wideness."

Eccentricity (e) of a Hyperbola

> The eccentricity e of a hyperbola is the ratio of c to a, where c is the distance from the center to a focus and a is the length of the semi-transverse axis.
>
> $$e = \frac{c}{a}$$

For a hyperbola, $c > a$ and therefore $e > 1$. As the eccentricity of the hyperbola increases, the graph becomes wider and wider.

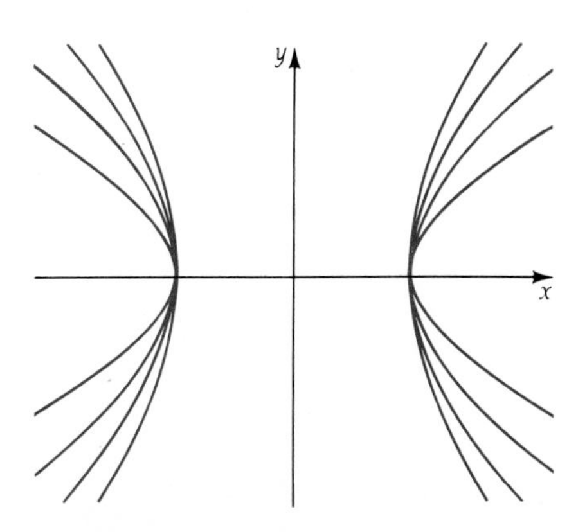

Figure 6.38

EXAMPLE 3 Find the Equation of a Hyperbola Given Its Eccentricity

Find the standard form of the equation of the hyperbola that has eccentricity 3/2, center at the origin, and a focus (6, 0).

Solution Because the focus is located at (6, 0) and the center is at the origin, $c = 6$. An extension of the transverse axis contain the foci, and thus the transverse axis is on the x-axis.

$$e = \frac{3}{2} = \frac{c}{a}$$

$$\frac{3}{2} = \frac{6}{a} \quad \text{Substitute the value for } c.$$

$$a = 4 \quad \text{Solve for } a.$$

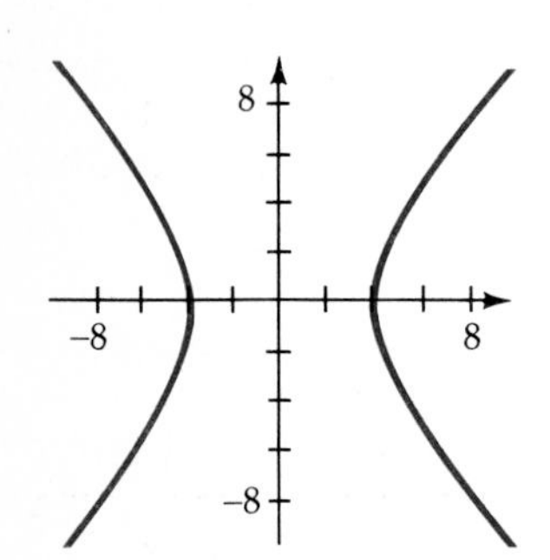

Figure 6.39 $\frac{x^2}{16} - \frac{y^2}{20} = 1$

To find b, use the equation $c^2 = a^2 + b^2$ and the values for c and a.

$$c^2 = a^2 + b^2$$

$$36 = 16 + b^2$$

$$b^2 = 20$$

The equation of the hyperbola is $\frac{x^2}{16} - \frac{y^2}{20} = 1$.

■ *Try Exercise* **40**, *page 246.*

Orbits of Comets

In Section 2 we noted that orbits of the planets are elliptical. Some comets have elliptical orbits also, the most notable being Halley's comet, whose eccentricity is 0.97.

Other comets have hyperbolic orbits with the sun at a focus. These comets pass by the sun only once. The velocity of a comet determines whether its orbit is elliptical or hyperbolic.

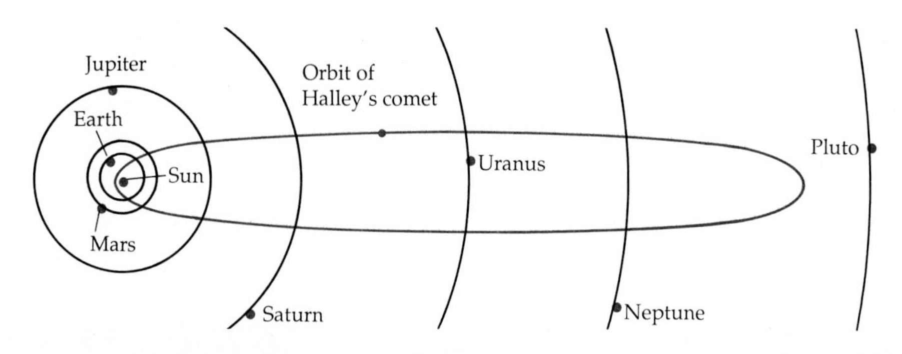

Figure 6.40

Hyperbolas as an Aid to Navigation

Consider two radio transmitters, T_1 and T_2, placed some distance apart. A ship with electronic equipment measures the difference in time it takes signals from the transmitters to reach the ship. Because the difference in time is proportional to the distance of the ship from the transmitter, the ship must be located on the hyperbola with foci at the two transmitters.

Using a third transmitter, T_3, a second hyperbola can be found with foci T_2 and T_3. The ship lies on the intersection of the two hyperbolas.

Figure 6.41

Optical Property of a Hyperbola

> A ray of light directed toward one focus of a hyperbolic mirror is reflected toward the other focus. This property, along with the reflective property of an ellipse, is used in telescopes to focus light.

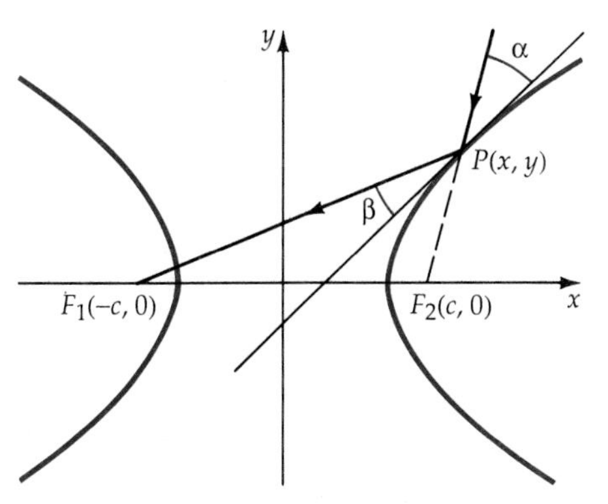

Figure 6.42

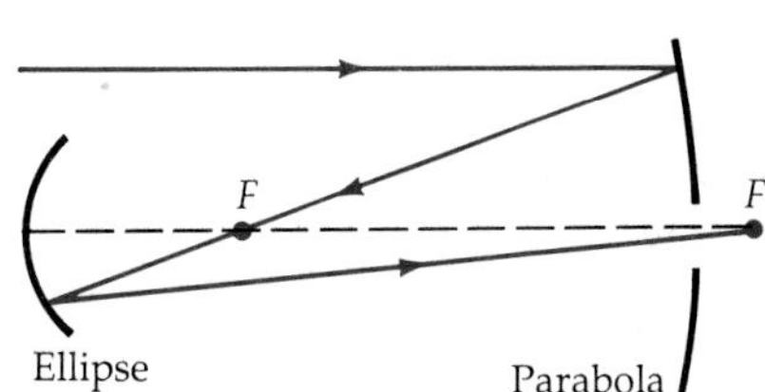

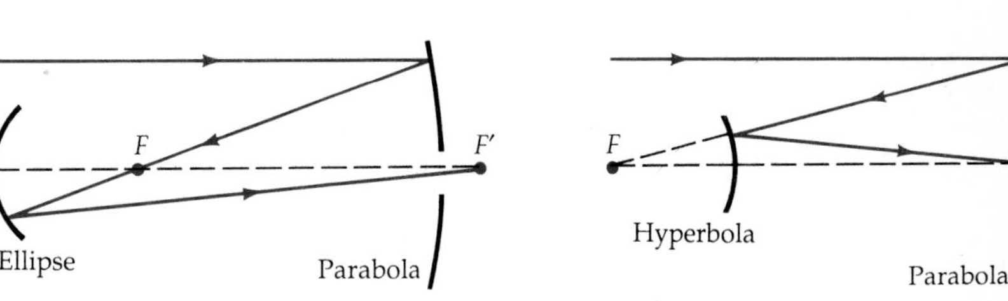

Figure 6.43

EXERCISE SET 6.3

In Exercises 1 to 24, find the center, vertices, foci, and asymptotes for the hyperbola given by each equation. Sketch the graph.

1. $\dfrac{x^2}{16} - \dfrac{y^2}{25} = 1$
2. $\dfrac{x^2}{16} - \dfrac{y^2}{9} = 1$
3. $\dfrac{y^2}{4} - \dfrac{x^2}{25} = 1$
4. $\dfrac{y^2}{25} - \dfrac{x^2}{36} = 1$
5. $\dfrac{(x-3)^2}{16} - \dfrac{(y+4)^2}{9} = 1$
6. $\dfrac{(x+3)^2}{25} - \dfrac{y^2}{4} = 1$
7. $\dfrac{(y+2)^2}{4} - \dfrac{(x-1)^2}{16} = 1$
8. $\dfrac{(y-2)^2}{36} - \dfrac{(x+1)^2}{49} = 1$
9. $x^2 - y^2 = 9$
10. $4x^2 - y^2 = 16$
11. $16y^2 - 9x^2 = 144$
12. $9y^2 - 25x^2 = 225$
13. $9y^2 - 36x^2 = 4$
14. $16x^2 - 25y^2 = 9$
15. $x^2 - y^2 - 6x + 8y - 3 = 0$
16. $4x^2 - 25y^2 + 16x + 50y - 109 = 0$
17. $9x^2 - 4y^2 + 36x - 8y + 68 = 0$
18. $16x^2 - 9y^2 - 32x - 54y + 79 = 0$
19. $4x^2 - y^2 + 32x + 6y + 39 = 0$
20. $x^2 - 16y^2 + 8x - 64y + 16 = 0$
21. $9x^2 - 16y^2 - 36x - 64y + 116 = 0$
22. $2x^2 - 9y^2 + 12x - 18y + 18 = 0$
23. $4x^2 - 9y^2 + 8x - 18y - 6 = 0$
24. $2x^2 - 9y^2 - 8x + 36y - 46 = 0$

In Exercises 25 to 38, find the equation in standard form of the hyperbola satisfying the stated conditions.

25. Vertices $(3, 0)$ and $(-3, 0)$, foci $(4, 0)$ and $(-4, 0)$
26. Vertices $(0, 2)$ and $(0, -2)$, foci $(0, 3)$ and $(0, -3)$

27. Foci $(0, 5)$ and $(0, -5)$, asymptotes $y = 2x$ and $y = -2x$

28. Foci $(4, 0)$ and $(-4, 0)$, asymptotes $y = x$ and $y = -x$

29. Vertices $(0, 3)$ and $(0, -3)$ and passing through $(2, 4)$

30. Vertices $(5, 0)$ and $(-5, 0)$ and passing through $(-1, 3)$

31. Asymptotes $y = \frac{1}{2}x$ and $y = -\frac{1}{2}x$, vertices $(0, 4)$ and $(0, -4)$

32. Asymptotes $y = \frac{2}{3}x$ and $y = -\frac{2}{3}x$, vertices $(6, 0)$ and $(-6, 0)$

33. Vertices $(6, 3)$ and $(2, 3)$, foci $(7, 3)$ and $(1, 3)$

34. Vertices $(-1, 5)$ and $(-1, -1)$, foci $(-1, 7)$ and $(-1, -3)$

35. Foci $(1, -2)$ and $(7, -2)$, slope of an asymptote $5/4$

36. Foci $(-3, -6)$ and $(-3, -2)$, slope of an asymptote 1

37. Passing through $(9, 4)$, slope of an asymptote $1/2$, center $(7, 2)$, transverse axis parallel to the y-axis

38. Passing through $(6, 1)$, slope of an asymptote 2, center $(3, 3)$, transverse axis parallel to the x-axis

In Exercises 39 to 44, use the eccentricity to find the equation in standard form of a hyperbola.

39. Vertices $(1, 6)$ and $(1, 8)$, eccentricity 2

40. Vertices $(2, 3)$ and $(-2, 3)$, eccentricity $5/2$

41. Eccentricity 2, foci $(4, 0)$ and $(-4, 0)$

42. Eccentricity $4/3$, foci $(0, 6)$ and $(0, -6)$

43. Center $(4, 1)$, conjugate axis length 4, eccentricity $4/3$

44. Center $(-3, -3)$, conjugate axis length 6, eccentricity 2

In Exercises 45 to 52, identify the graph of each equation as a parabola, ellipse, or hyperbola. Sketch the graph.

45. $4x^2 + 9y^2 - 16x - 36y + 16 = 0$

46. $2x^2 + 3y - 8x + 2 = 0$

47. $5x - 4y^2 + 24y - 11 = 0$

48. $9x^2 - 25y^2 - 18x + 50y = 0$

49. $x^2 + 2y - 8x = 0$

50. $9x^2 + 16y^2 + 36x - 64y - 44 = 0$

51. $25x^2 + 9y^2 - 50x - 72y - 56 = 0$

52. $(x - 3)^2 + (y - 4)^2 = (x + 1)^2$

53. Find the equation of the path of Halley's comet in astronomical units by letting the sun (one focus) be at the origin and the other focus on the positive x-axis. The length of the major axis of the orbit of Halley's comet is approximately 36 astronomical units (36 AU) and the length of the minor axis 9 AU wide (1 AU = 92,600,000 miles).

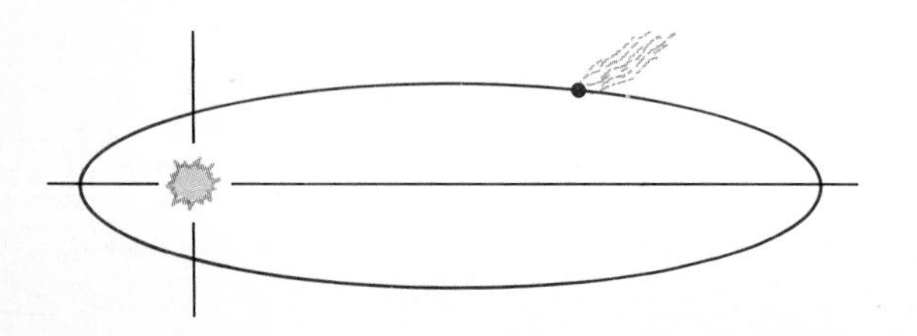

54. A foot suspension bridge is 100 feet long and supported by cables that hang in the shape of a parabola. Find the equation of the parabola if the positive x-axis is along the footpath, as in the figure.

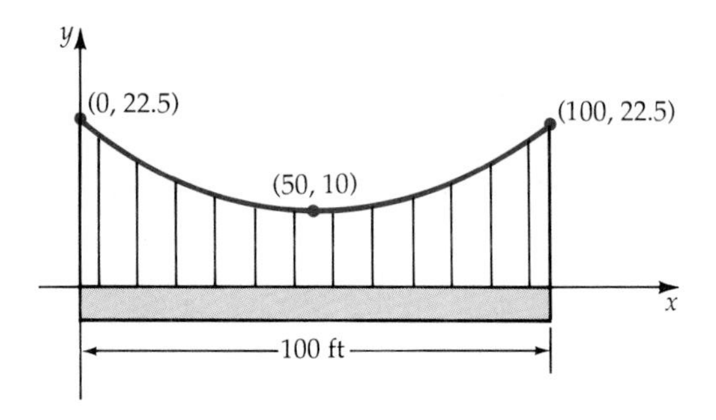

55. Two radio towers are positioned along the coast of California, 200 miles apart. A signal is sent simultaneously from each tower to a ship off the coast. The signal from tower B is received by the ship 500 microseconds after the signal sent by A. If the radio signal travels 0.2 miles per microsecond, find the equation of the hyperbola on which the ship is located.

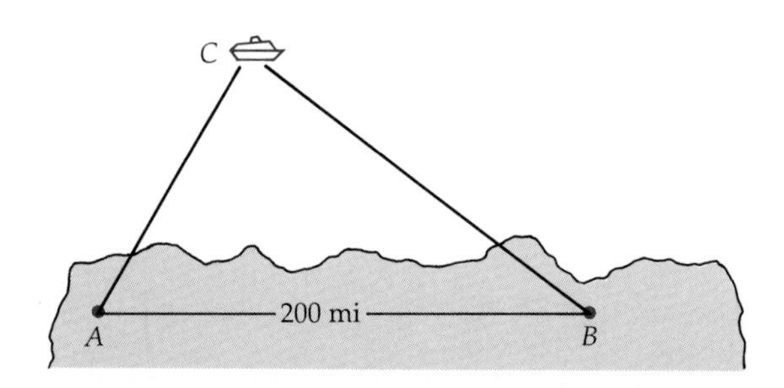

56. A softball player releases a softball at a height of 5 feet above the ground with a speed of 88 feet per second (60 mph). The initial trajectory of the ball is 45°. Neglecting air resistance, the path of the ball is a parabola given by $y = -0.004x^2 + x + 5$. How far will the ball travel before hitting the ground? *Hint:* Let $y = 0$ and solve for x.

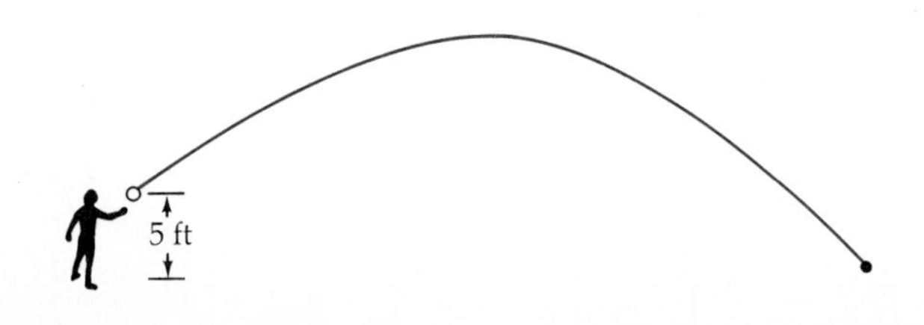

Supplemental Exercises

In Exercises 57 to 60, use the definition for a hyperbola to find the equation of the hyperbola in standard form.

57. Foci $(2, 0)$ and $(-2, 0)$ and passes through the point $(2, 3)$

58. Foci $(0, 3)$ and $(0, -3)$ and passes through the point $(5/2, 3)$

59. Foci $(0, 4)$ and $(0, -4)$ and passes through the point $(7/3, 4)$

60. Foci $(5, 0)$ and $(-5, 0)$ and passes through the point $(5, 9/4)$

Recall that an ellipse has two directrixes that are lines perpendicular to the line containing the foci. A hyperbola also has two directrixes that are perpendicular to the transverse axis and outside the hyperbola. For a hyperbola with center

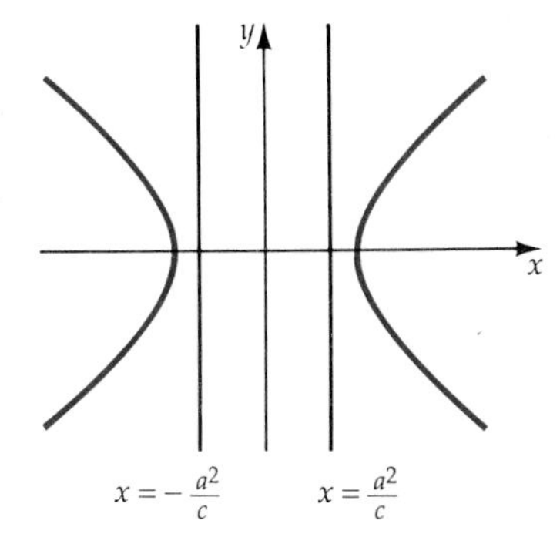

at the origin and transverse axis on the x-axis, the equations of the directrixes are $x = a^2/c$ and $x = -a^2/c$. In Exercises 61 to 65, use this information to solve each exercise.

61. Find the directrixes for the ellipse in Exercise 21.

62. Find the directrixes for the ellipse in Exercise 22.

63. Let $P(x, y)$ be a point on the hyperbola $\frac{x^2}{9} - \frac{y^2}{16} = 1$. Show that the distance from the point P to the focus $(5, 0)$ divided by the distance from the point P to the directrix $x = 9/5$ equals the eccentricity.

64. Let $P(x,y)$ be a point on the hyperbola $\frac{x^2}{7} - \frac{y^2}{9} = 1$. Show that the distance from the point P to the focus $(4, 0)$ divided by the distance from the point to the directrix $x = 7/4$ equals the eccentricity.

65. Generalize the results of Exercises 63 and 64. That is, show that if $P(x, y)$ is a point on the hyperbola $\frac{x^2}{a^2} - \frac{y}{b^2} = 1$, $F(c, 0)$ is the focus, and $x = a^2/c$ is the directrix, then the following equation is true: $e = d(P, F)/d(P, D)$.

66. Derive the equation of a hyperbola with center at the origin, foci at $(0, c)$ and $(0, -c)$, and vertices $(0, a)$ and $(0, -a)$.

67. Sketch a graph of $\frac{x|x|}{16} - \frac{y|y|}{9} = 1$.

68. Sketch a graph of $\frac{x|x|}{16} + \frac{y|y|}{9} = 1$.

6.4 Introduction to Polar Coordinates

Until now, *rectangular coordinate systems* have been used to locate a point in the coordinate plane. An alternate method is to use a *polar coordinate system*. Using this method, a point is located in a manner similar to giving a distance and an angle from some fixed direction.

$P(r, \theta)$ r θ O Pole Polar axis

Figure 6.44

The Polar Coordinate System

A **polar coordinate system** is formed by drawing a horizontal ray. The ray is called the **polar axis**, and the beginning point is called the **pole**. A point $P(r, \theta)$ in the plane is located by specifying a distance r from the pole and an angle θ measured from the polar axis to the line segment OP. The angle can be measured in degrees or radians.

The coordinates of the pole are $(0, \theta)$, where θ is an arbitrary angle. Positive angles are measured counterclockwise from the polar axis. Negative angles are measured clockwise from the axis. Positive values of r are measured along the ray that makes an angle θ from the polar axis.

Negative values of r are measured along the ray that makes an angle of $\theta + 180°$ from the polar axis.

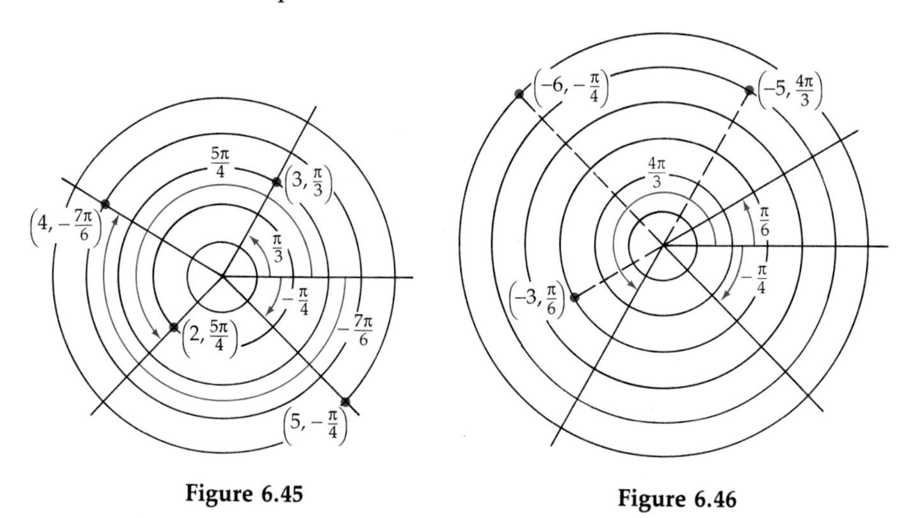

Figure 6.45

Figure 6.46

In a rectangular coordinate system, there is a one-to-one correspondence between the points in the plane and the ordered pairs (x, y). This is not true for a polar coordinate system. For polar coordinates, the relation is one-to-many. For each point $P(r, \theta)$ in a polar coordinate system there corresponds infinitely many ordered pair descriptions of that point.

For example, consider a point whose coordinates are $P(3, 45°)$. Because there are 360° in one complete revolution around a circle, the point P could also be written as $(3, 405°)$, $(3, 765°)$, $(3, 1125°)$, and generally as $(3, 45° + n \cdot 360°)$, where n is an integer. It is also possible to describe the point $P(3, 45°)$ by $(-3, 225°)$, $(-3, -135°)$, and $(3, -315°)$, to name just a few.

Remark The relationship between an ordered pair and a point is not one-to-many. That is, given an ordered pair (r, θ), there is exactly one point in the plane that corresponds to that point.

Graphs of Equations in a Polar Coordinate System

A **polar equation** is an equation in r and θ. A **solution** to a polar equation is an ordered pair (r, θ) that satisfies the equation. The **graph** of a polar equation is the set of all points whose ordered pairs that are solutions of the equation.

Figure 6.47 is the graph of the polar equation $r = 2$. The graph is drawn on a polar coordinate grid that consists of concentric circles and rays beginning at the pole. Because r is independent of θ, r is 2 units from

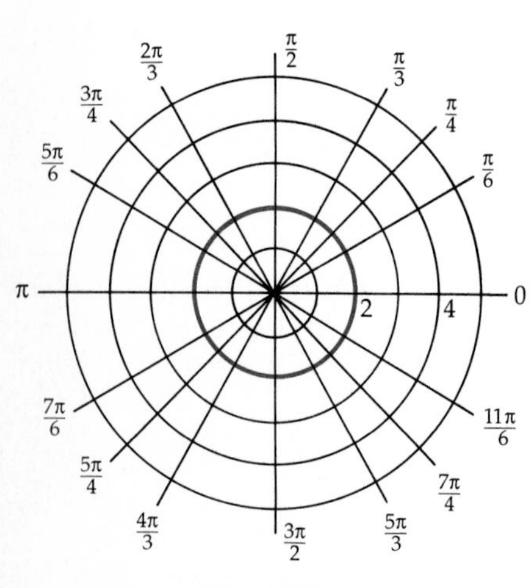

Figure 6.47
$r = 2$

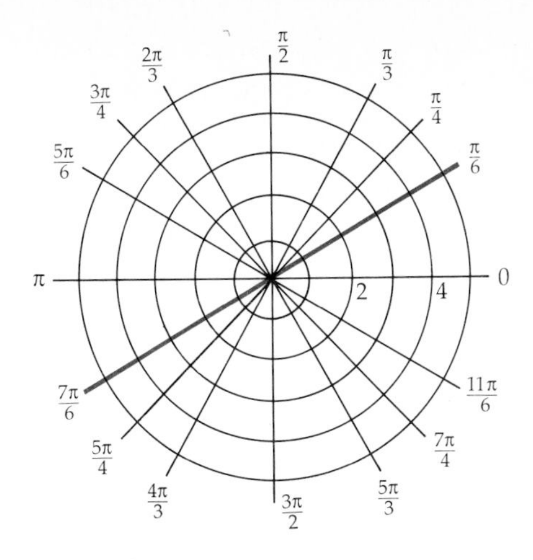

Figure 6.48
$\theta = \dfrac{\pi}{6}$

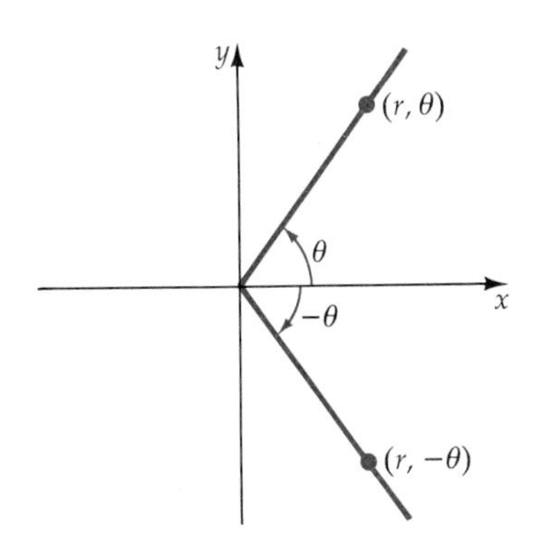

Figure 6.49

the pole for all values of θ. The graph is a circle of radius 2 with center at the pole.

The Graph of $r = a$

> The graph of $r = a$ is a circle with center at the pole and radius a.

The graph of the polar equation $\theta = \pi/6$ is a line. Because θ is independent of r, θ is $\pi/6$ radians from the polar axis for all values of r. The graph is a straight line that makes an angle of $\pi/6$ radians (30°) from the polar axis.

The Graph of $\theta = \alpha$

> The graph of $\theta = \alpha$ is a straight line through the pole at an angle of α from the polar axis.

As an aid to graphing other polar equations, it is useful to examine the equation for symmetries with respect to the pole, the polar axis, or the line $\theta = \pi/2$. By superimposing an xy-coordinate system on the polar coordinate system, this means finding symmetries with respect to the origin, the x-axis, or the y-axis.

Suppose that whenever the ordered pair (r, θ) lies on the graph of a polar equation, $(r, -\theta)$ also lies on the graph. From Figure 9.49, the graph will have symmetry with respect to the polar axis. Thus one test for symmetry is to replace θ by $-\theta$ in the polar equation. If the resulting equation is equivalent to the original equation, the graph is symmetric with respect to the polar axis.

EXAMPLE 1 Graph a Polar Equation That is Symmetric With Respect to the Polar Axis

Show that the graph of $r = 4 \cos \theta$ is symmetric to the polar axis. Graph the equation.

Solution Test for symmetry with respect to the polar axis. Replace θ by $-\theta$.

$$r = 4 \cos(-\theta) = 4 \cos \theta \quad \cos(-\theta) = \cos \theta .$$

Because replacing θ by $-\theta$ results in the same equation, the graph is symmetric with respect to the polar axis.

To graph the equation, begin choosing various values of θ and finding the corresponding values of r. However, before doing so, two further observations will reduce the number of points we must choose.

First, because cosine is a periodic function with period 2π, it is only necessary to choose points between 0 and 2π (0 and 360°). Second, when $\dfrac{\pi}{2} < \theta < \dfrac{3\pi}{2}$, $\cos \theta$ is negative, which means that any θ between these values will produce a negative r. Thus the point will be in the first or

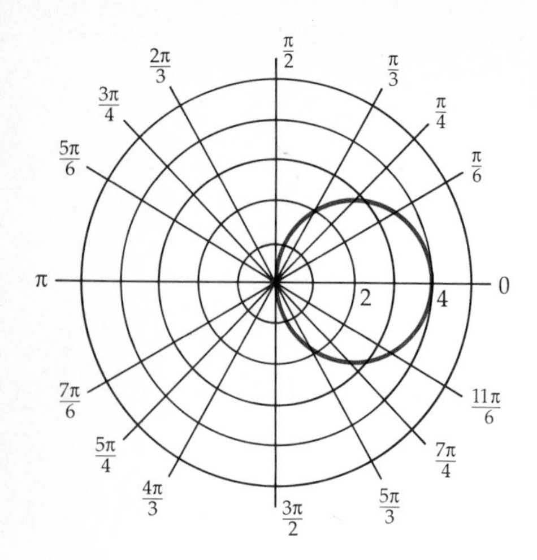

Figure 6.50
$r = 4 \cos \theta$

fourth quadrant. That is, we need consider only angles θ in the first or fourth quadrants. However, since the graph is symmetric with respect to the polar axis, it is only necessary to choose values of θ between 0 and $\pi/2$.

						By symmetry			
θ	0	$\pi/6$	$\pi/4$	$\pi/3$	$\pi/2$	$-\pi/6$	$-\pi/4$	$-\pi/3$	$-\pi/2$
r	4.0	3.5	2.8	2.0	0.0	3.5	2.8	2.0	0.0

As we will show later, the graph of $r = 4 \cos \theta$ is a circle with a center at $(2, 0)$. See Figure 6.50.

■ *Try Exercise* **14**, *page 257.*

The following table shows the types of symmetry and their associated tests. For each type, if the recommended substitution results in an equivalent equation, the graph will have the indicated symmetry.

Tests for Symmetry

Substitution	Symmetry with respect to
$-\theta$ for θ	The line $\theta = 0$
$\pi - \theta$ for θ, $-r$ for r	The line $\theta = 0$
$\pi - \theta$ for θ	The line $\theta = \pi/2$
$-\theta$ for θ, $-r$ for r	The line $\theta = \pi/2$
$-r$ for r	The pole (origin)
$\pi + \theta$ for θ	The pole (origin)

Caution The graph of a polar equation may have a symmetry even though a test for that symmetry fails. For example, as you will see later, the graph of $r = \sin 2\theta$ is symmetric with respect to the line $\theta = 0$. However, using the symmetry test of substituting $-\theta$ for θ, we have

$$\sin 2(-\theta) = -\sin 2\theta \neq r.$$

Thus this test fails to show symmetry with respect to the line $\theta = 0$. The symmetry test of substituting $\pi - \theta$ for θ and $-r$ for r will establish symmetry with respect to the line $\theta = 0$. Why?[4]

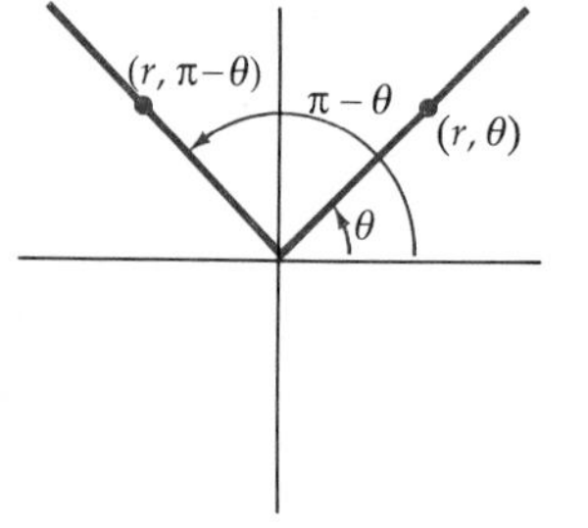

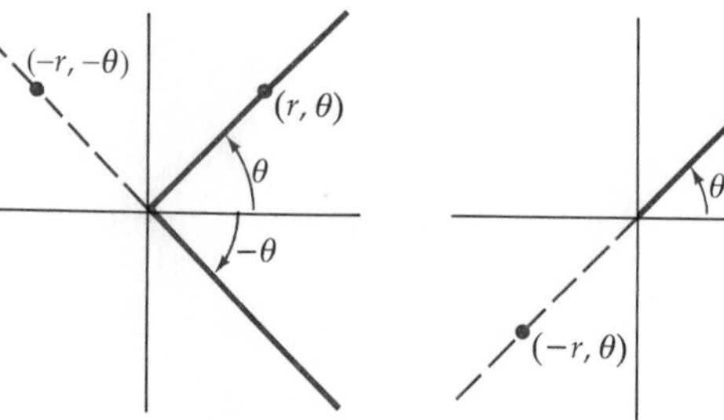

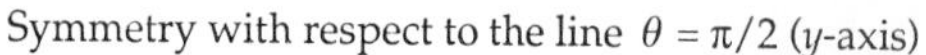
Symmetry with respect to the line $\theta = \pi/2$ (y-axis)

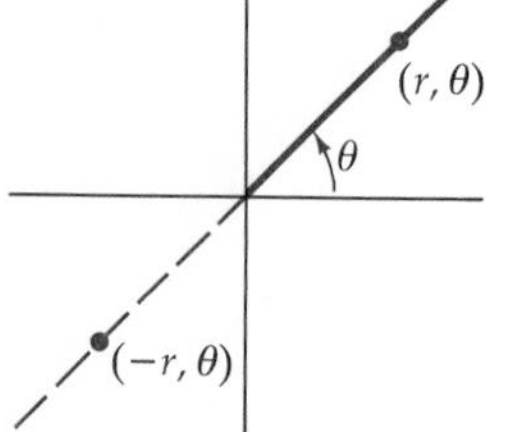

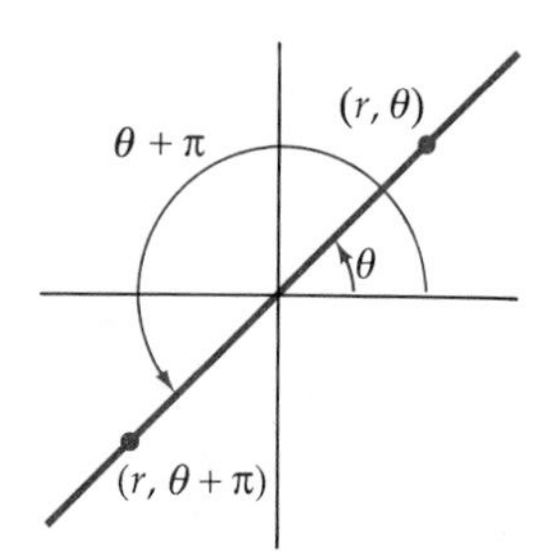

Symmetry with respect to the pole (origin)

Figure 6.51

[4] $\sin(2(\pi - \theta)) = \sin(2\pi - 2\theta) = \sin(-2\theta) = -\sin 2\theta = -r.$

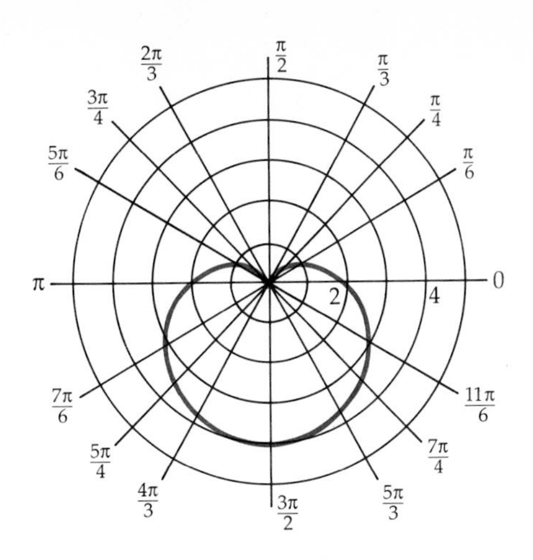

Figure 6.52
$r = 2 - 2\sin\theta$

EXAMPLE 2 Sketch the Graph of a Cardioid

Sketch a graph of $r = 2 - 2\sin\theta$.

Solution Try each of the tests for symmetry (here we will show only the first and third tests). Substitute $-\theta$ for θ.

$$2 - 2\sin(-\theta) = 2 + 2\sin\theta \neq 2 - 2\sin\theta \quad \sin(-\theta) = -\sin\theta$$

This test fails to show symmetry with respect to the line $\theta = 0$.

Substitute $\pi - \theta$ for θ.

$$2 - 2\sin(\pi - \theta) = 2 - 2\sin\theta \quad \sin(\pi - \theta) = \sin\theta$$

The graph is symmetric with respect to the line $\theta = \pi/2$. A check of the other tests for symmetry will show that the graph has no other symmetries.

Choose some values of θ and find the corresponding values for r. Plot these points and use symmetry to sketch the graph.

θ	$-\pi/2$	$-\pi/3$	$-\pi/4$	$-\pi/6$	0	$\pi/6$	$\pi/4$	$\pi/3$	$\pi/2$
r	4.0	3.7	3.4	3.0	2.0	1.0	0.6	0.3	0.0

The graph is called a **cardioid**. See Figure 6.52

■ *Try Exercise* **18**, *page 257.*

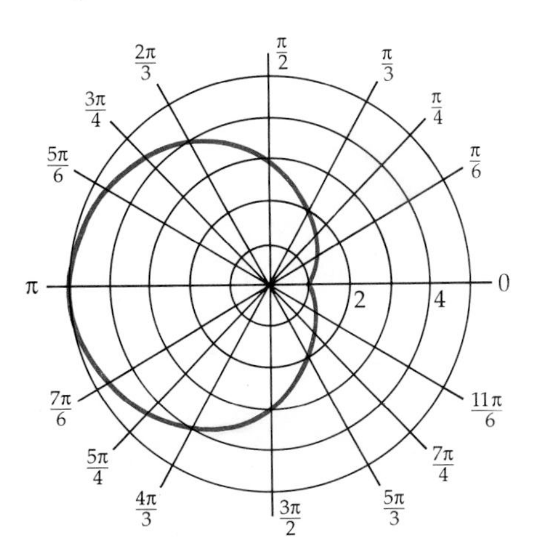

Figure 6.53
$r = 3 - 2\cos\theta$

EXAMPLE 3 Sketch the Graph of a Cardioid

Sketch a graph of $r = 3 - 2\cos\theta$.

Solution Try each of the tests for symmetry (here we will show only the first and sixth tests). Substitute $-\theta$ for θ.

$$3 - 2\cos(-\theta) = 3 - 2\cos\theta \quad \cos(-\theta) = \cos\theta$$

The graph is symmetric with respect to the line $\theta = 0$.

Substitute $\pi + \theta$ for θ.

$$3 - 2\cos(\pi + \theta) = 3 + 2\cos\theta \neq 3 - 2\cos\theta \quad \cos(\pi + \theta) = -\cos\theta$$

This test fails to show symmetry with respect to the line $\theta = \pi/2$. This graph does not have any other symmetries.

Choose some values of θ and find the corresponding values of r. Plot these points and use symmetry to draw the graph.

θ	0	$\pi/6$	$\pi/4$	$\pi/3$	$\pi/2$	$2\pi/3$	$3\pi/4$	$5\pi/6$	π
r	1.0	1.3	1.6	2.0	3.0	4.0	4.4	4.7	5.0

This is also a cardioid. See Figure 6.53.

■ *Try Exercise* **20**, *page 257.*

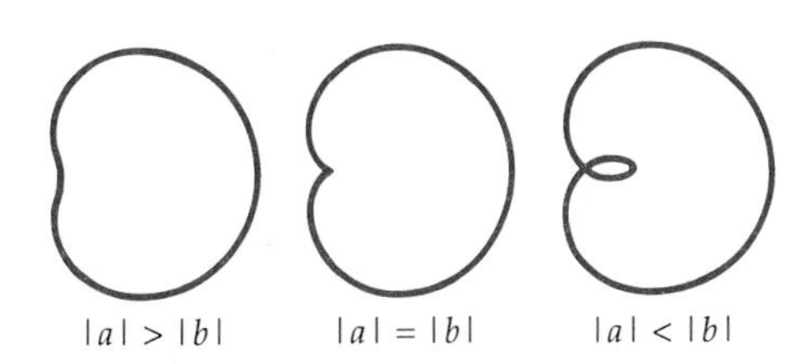

Figure 6.54

Examples 2 and 3 illustrate two possible graphs of a cardioid. There is also a third possibility, shown in Figure 6.54.

General Equations of a Cardioid

> The graph of the equation $r = a + b \cos \theta$ is a cardioid that is symmetric with respect to the line $\theta = 0$. The graph of the equation $r = a + b \sin \theta$ is a cardioid that is symmetric with respect to the line $\theta = \frac{\pi}{2}$.

The graphs illustrate the three forms of a cardioid with symmetry with respect to the line $\theta = 0$. The magnitudes of the absolute values of a and b determine the shape of the graph. All cardioids are reflections or rotations of the three basic graphs.

EXAMPLE 4 Sketch the Graph of a Three-Leaf Rose

Sketch a graph of $r = 3 \sin 3\theta$.

Solution Try each of the tests for symmetry (here we will show only the second and fourth tests).

$$-r = 3 \sin 3(\pi - \theta) \quad \text{Substitute } \pi - \theta \text{ for } \theta \text{ and } -r \text{ for } r.$$
$$r = -3 \sin 3(\pi - \theta) \quad \text{Multiply by } -1.$$
$$= -3 \sin 3\theta \neq 3 \sin 3\theta \quad \sin(\pi - \theta) = \sin \theta.$$

This test fails to show symmetry with respect to the line $\theta = 0$.

$$-r = 3 \sin 3(-\theta) \quad \text{Substitute } -\theta \text{ for } \theta \text{ and } -r \text{ for } r.$$
$$r = -3 \sin 3(-\theta) \quad \text{Multiply by } -1.$$
$$= 3 \sin 3\theta \quad \sin(-\theta) = -\sin \theta.$$

The graph is symmetric with the line $\theta = \dfrac{\pi}{2}$. The graph does not have any other symmetries.

Choose some values for θ and find the corresponding values of r. Use symmetry to sketch the graph.

θ	0	$\pi/18$	$\pi/6$	$5\pi/18$	$\pi/3$	$7\pi/18$	$\pi/2$
r	0.0	1.5	3.0	1.5	0.0	−1.5	−3.0

The graph is a **three-leaf rose.** See Figure 6.55.

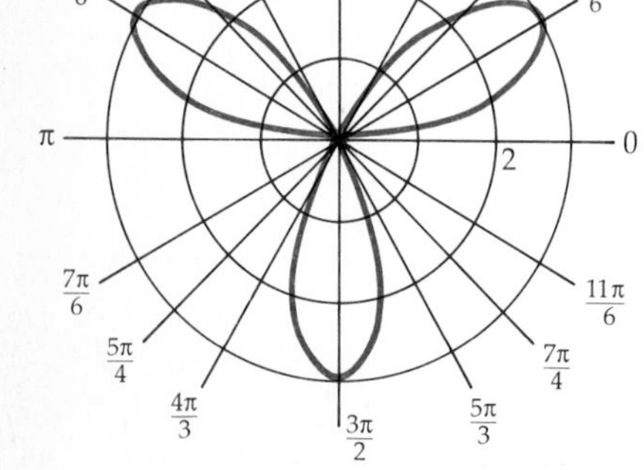

Figure 6.55
$r = 3 \sin 3\theta$

■ *Try Exercise* **28,** *page 257.*

EXAMPLE 5 Sketch the Graph of a Four-Leaf Rose

Sketch a graph of $r = 4 \cos 2\theta$

Solution Try each of the tests for symmetry (here we will show only the first and third tests). Substitute $-\theta$ for θ.

$$4 \cos(-2\theta) = 4 \cos 2\theta \quad \cos(-\theta) = \cos \theta$$

The graph is symmetric with respect to the line $\theta = 0$.

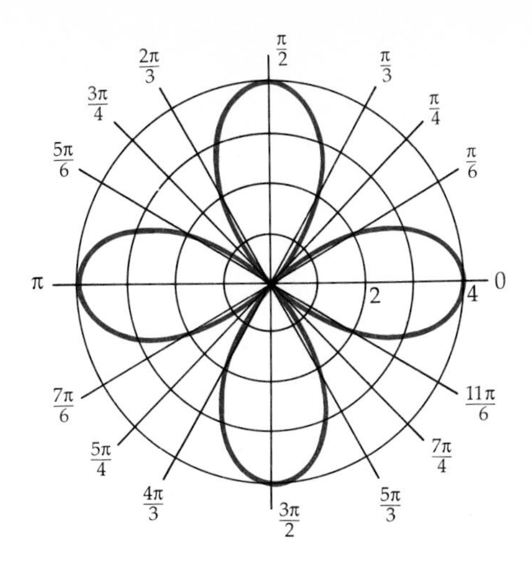

Figure 6.56
$r = 4 \cos 2\theta$

Substitute $\pi - \theta$ for θ.

$$4 \cos(2(\pi - \theta)) = 4 \cos(2\pi - 2\theta) = 4 \cos 2\theta \qquad \cos(2\pi - \alpha) = \cos \alpha$$

The graph is symmetric with respect to the line $\theta = \pi/2$. The graph does not have any other symmetries.

Choose some values of θ and find the corresponding values of r. Use symmetry to sketch the graph. The graph is a **four-leaf rose.**

θ	0	$\pi/12$	$\pi/6$	$\pi/4$	$\pi/3$	$5\pi/12$	$\pi/2$
r	4	3.5	2	0	−2	−3.5	−4

■ *Try Exercise* **32**, *page 257.*

Rose curves have the form $r = a \cos n\theta$ or $r = a \sin n\theta$.

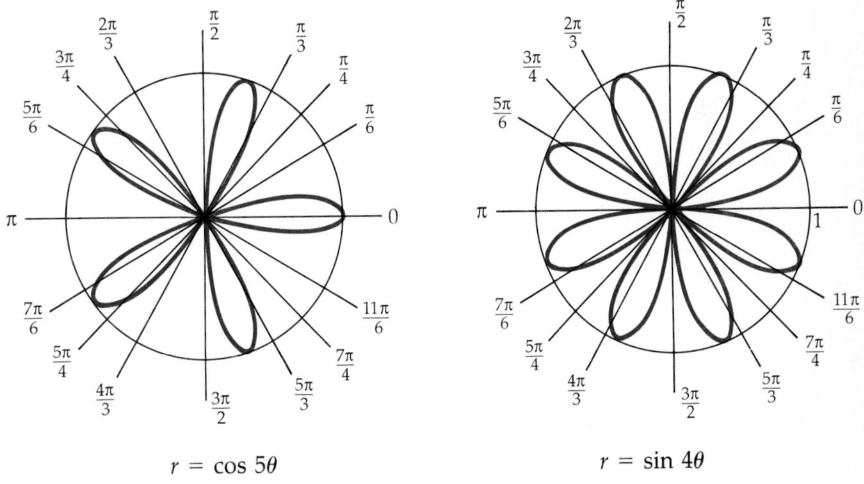

Figure 6.57

General Equations of Rose Curves

> The graphs of the equation $r = a \cos n\theta$ and $r = a \sin n\theta$ are rose curves. When n is an even number, the number of petals is $2n$. When n is an odd number, the number of petals is n.

Transformations Between Rectangular and Polar Coordinates

A transformation between coordinate systems is a set of equations that relate the coordinates of one system with the coordinates in a second system. By superimposing a rectangular coordinate system on a polar system, we can derive the set of transformation equations.

Construct a polar coordinate system and a rectangular system so that the pole coincides with the origin and the polar axis coincides with the positive x-axis. Let a point P have coordinates (x, y) in one system and (r, θ) in the other ($r > 0$).

From the definitions of the sine and cosine of an acute angle in a right triangle, we have

$$\frac{x}{r} = \cos\theta \quad \text{or} \quad x = r\cos\theta,$$

and

$$\frac{y}{r} = \sin\theta \quad \text{or} \quad y = r\sin\theta.$$

It can be shown that these equations are also true when $r < 0$.

Thus given the point (r, θ) in a polar coordinate system, the coordinates of the point in the xy-coordinate system are given by

$$x = r\cos\theta \qquad y = r\sin\theta$$

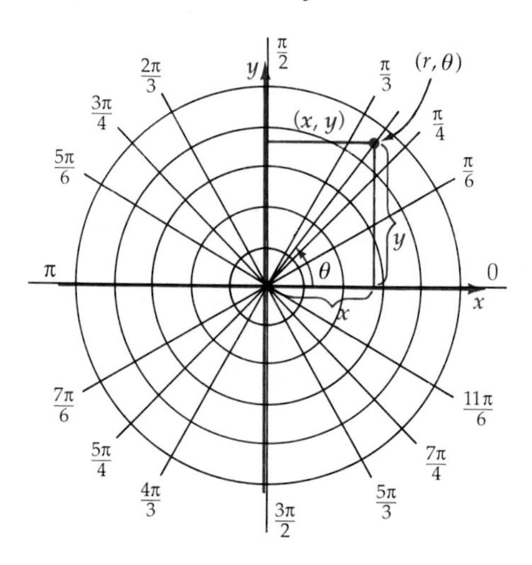

Figure 6.58

For example, to find the point in the xy-coordinate system that corresponds to the point $\left(4, \frac{2\pi}{3}\right)$ in the $r\theta$-coordinate system, substitute into the equations and solve for x and y.

$$x = 4\cos\left(\frac{2\pi}{3}\right) = 4\left(-\frac{1}{2}\right) = -2$$

$$y = 4\sin\left(\frac{2\pi}{3}\right) = 4\left(\frac{\sqrt{3}}{2}\right) = 2\sqrt{3}$$

The point is $(-2, 2\sqrt{3})$.

To find the polar coordinates of a given point in the xy-coordinate system, use the Pythagorean Theorem and the definition of the tangent function. Let $P(x, y)$ be a point in the plane and r the distance from the origin to the point P. Then

$$r^2 = x^2 + y^2 \quad \text{or} \quad r = \sqrt{x^2 + y^2}.$$

From the definition of the tangent function of an angle in a right triangle,

$$\tan\theta = \frac{y}{x}.$$

Thus θ is the angle whose tangent is y/x. The quadrant for θ is chosen according to the following:

- $x > 0$ and $y > 0$, θ is a first-quadrant angle.
- $x < 0$ and $y > 0$, θ is a second-quadrant angle.
- $x < 0$ and $y < 0$, θ is a third-quadrant angle.
- $x > 0$ and $y < 0$, θ is a fourth-quadrant angle.

The equations of transformations between a polar and rectangular coordinate system are now summarized.

Transformations between Polar and Rectangular Coordinates

Given the point (r, θ) in the polar coordinate system, the transformation equations to change from polar to rectangular coordinates are

$$x = r \cos \theta \qquad y = r \sin \theta$$

Given the point (x, y) in the rectangular coordinate system, the transformation equations to change from rectangular to polar coordinates are

$$r = \sqrt{x^2 + y^2} \qquad \tan \theta = \frac{y}{x}, \ x \neq 0$$

where $r \geq 0$, $0 \leq \theta < 2\pi$, and θ is chosen so that the point lies in the appropriate quadrant. If $x = 0$, then $\theta = \pi/2$ or $\theta = 3\pi/2$.

EXAMPLE 6 Transform from Polar to Rectangular Coordinates

Find the rectangular coordinates of the points whose polar coordinates are a. $(6, 3\pi/4)$ b. $(-4, 30°)$.

Solution Use the two transformation equations $x = r \cos \theta$ and $y = r \sin \theta$.

a. $$x = 6 \cos\left(\frac{3\pi}{4}\right) = 6 \cdot -\frac{\sqrt{2}}{2} = -3\sqrt{2}$$

$$y = 6 \sin\left(\frac{3\pi}{4}\right) = 6 \cdot \frac{\sqrt{2}}{2} = 3\sqrt{2}$$

The rectangular coordinates are $(-3\sqrt{2}, 3\sqrt{2})$.

b. $$x = -4 \cos(30°) = -4 \cdot \frac{\sqrt{3}}{2} = -2\sqrt{3}$$

$$y = -4 \sin(30°) = -4 \cdot \frac{1}{2} = -2$$

The rectangular coordinates are $(-2\sqrt{3}, -2)$.

■ *Try Exercise* **44,** *page 257.*

EXAMPLE 7 Transform from Rectangular to Polar Coordinates

Find the polar coordinates of the points whose rectangular coordinates are a. $(-3, 4)$ b. $(-2, -2\sqrt{3})$.

Solution Use the transformation equations $r = \sqrt{x^2 + y^2}$ and $\tan\theta = \dfrac{y}{x}$.

a.
$$r = \sqrt{(-3)^2 + 4^2} = \sqrt{9 + 16} = \sqrt{25} = 5$$
$$\tan\theta = \frac{4}{-3} = -\frac{4}{3}$$

From this and the fact that $(-3, 4)$ lies in the second quadrant, $\theta \approx 127°$. The approximate polar coordinates of the point are $(5, 127°)$.

b.
$$r = \sqrt{(-2)^2 + (-2\sqrt{3})^2} = \sqrt{4 + 12} = \sqrt{16} = 4$$
$$\tan\theta = \frac{-2\sqrt{3}}{-2} = \sqrt{3}$$

From this and the fact that $(-2, -2\sqrt{3})$ lies in the third quadrant, $\theta = 4\pi/3$. The polar coordinates of the point are $(4, 4\pi/3)$.

■ *Try Exercise* **48,** *page 257.*

Using the transformation equations, it is possible to write a polar equation in rectangular form or a rectangular coordinate equation in polar form.

EXAMPLE 8 Write a Rectangular Coordinate Equation in Polar Form

Find a polar form of the equation $x^2 + y^2 - 2x = 3$.

Solution

$$x^2 + y^2 - 2x = 3$$
$$(r\cos\theta)^2 + (r\sin\theta)^2 - 2r\cos\theta = 3 \quad \text{Use the transformation equations.}$$
$$r^2(\cos^2\theta + \sin^2\theta) - 2r\cos\theta = 3 \quad \text{Factor.}$$
$$r^2 - 2r\cos\theta = 3$$

A polar form of the equation is $r^2 - 2r\cos\theta = 3$.

■ *Try Exercise* **56,** *page 257.*

EXAMPLE 9 Write a Polar Coordinate Equation in Rectangular Form.

Find a rectangular form of the equation $r^2 \cos 2\theta = 3$.

Solution

$$r^2 \cos 2\theta = 3$$
$$r^2(1 - 2\sin^2\theta) = 3 \quad \cos 2\theta = 1 - 2\sin^2\theta$$
$$r^2 - 2r^2 \sin^2\theta = 3$$
$$x^2 + y^2 - 2y^2 = 3 \quad \text{Use the transformation equations.}$$
$$x^2 - y^2 = 3$$

A rectangular form of the equation is $x^2 - y^2 = 3$.

■ *Try Exercise* **58**, *page 257.*

EXERCISE SET 6.4

In Exercises 1 to 8, plot the point on a polar coordinate system.

1. $(2, 60°)$ **2.** $(3, -90°)$ **3.** $(1, 315°)$

4. $(2, 400°)$ **5.** $\left(-2, \frac{\pi}{4}\right)$ **6.** $\left(4, \frac{7\pi}{6}\right)$

7. $\left(-3, \frac{5\pi}{3}\right)$ **8.** $(-3, \pi)$

In Exercises 9 to 40, sketch the graphs of the polar equations.

9. $r = 3$ **10.** $r = 5$

11. $\theta = 2$ **12.** $\theta = -\frac{\pi}{3}$

13. $r = 6\cos\theta$ **14.** $r = 4\sin\theta$

15. $r = 3 + 3\cos\theta$ **16.** $r = 4 - 4\sin\theta$

17. $r = 2 - 3\sin\theta$ **18.** $r = 2 - 2\cos\theta$

19. $r = 4 + 3\sin\theta$ **20.** $r = 2 + 4\sin\theta$

21. $r = -2 + 2\cos\theta$ **22.** $r = -1 - \cos\theta$

23. $r = 2 - 4\sin\theta$ **24.** $r = 4(1 - \sin\theta)$

25. $r = 3\sin 2\theta$ **26.** $r = 2\cos 2\theta$

27. $r = 4\cos 3\theta$ **28.** $r = 5\sin 3\theta$

29. $r = 2\sin 5\theta$ **30.** $r = 3\cos 5\theta$

31. $r = 3\cos 4\theta$ **32.** $r = 6\sin 4\theta$

33. $r = 3\sec\theta$ **34.** $r = 4\csc\theta$

35. $r = 5\csc\theta$ **36.** $r = 2\sec\theta$

37. $r = \theta$ **38.** $r = -\theta$

39. $r = 2^\theta, \theta \geq 0$ **40.** $r = \frac{1}{\theta}, \theta > 0$

In Exercises 41 to 48, transform the given coordinates to the indicated ordered pair.

41. $(1, -\sqrt{3})$ to (r, θ) **42.** $(-2\sqrt{3}, 2)$ to (r, θ)

43. $\left(-3, \frac{2\pi}{3}\right)$ to (x, y) **44.** $\left(2, -\frac{\pi}{3}\right)$ to (x, y)

45. $\left(0, -\frac{\pi}{2}\right)$ to (x, y) **46.** $\left(3, \frac{5\pi}{6}\right)$ to (x, y)

47. $(3, 4)$ to (r, θ) **48.** $(12, -5)$ to (r, θ)

In Exercises 49 to 60, find an equation in x and y that has the same graph as the given polar equation.

49. $r = 3\cos\theta$ **50.** $r = 2\sin\theta$

51. $r = 3\sec\theta$ **52.** $r = 4\csc\theta$

53. $r = 4$ **54.** $\theta = \frac{\pi}{4}$

55. $r = \tan\theta$ **56.** $r = \cot\theta$

57. $r = \frac{2}{1 + \cos\theta}$ **58.** $r = \frac{2}{1 - \sin\theta}$

59. $r(\sin\theta - 2\cos\theta) = 6$ **60.** $r(2\cos\theta + \sin\theta) = 3$

In Exercises 61 to 68, find a polar equation that has the same graph as the given equation.

61. $y = 2$ **62.** $x = -4$

63. $x^2 + y^2 = 4$ **64.** $2x - 3y = 6$

65. $x^2 = 8y$ **66.** $y^2 = 4y$

67. $x^2 - y^2 = 25$ **68.** $x^2 + 4y^2 = 16$

Supplemental Exercises

For Exercises 69 to 76, sketch a graph of the polar equation.

69. $r^2 = 4\cos 2\theta$ (lemniscate)

70. $r^2 = -2\sin 2\theta$ (lemniscate)

71. $r = 2(1 + \sec\theta)$ (conchoid)

72. $r = 2\cos 2\theta\sec\theta$ (strophoid)

73. $r\theta = 2$ (spiral)

74. $r = 2 \sin \theta \cos^2 2\theta$ (bifolium)

75. $r = |\theta|$

76. $r = \ln \theta$

77. If $P_1(r_1, \theta_1)$ and $P_2(r_2, \theta_2)$ are two points in the $r\theta$-plane, use the Law of Cosines to show that the distance between the two points $d(P_1, P_2)$ is given by

$$[d(P_1, P_2)]^2 = r_1^2 + r_2^2 - 2r_1r_2 \cos(\theta_1 - \theta_2).$$

78. Prove that the graph of $r = a \sin \theta + b \cos \theta$, $ab \neq 0$, is a circle in polar coordinates.

Chapter 6 Review

6.1 Conic Sections

A parabola is the set of points in the plane that are equidistant from a fixed line (the directrix) and a fixed point (the focus) not on the directrix.

The equations of a parabola with vertex at (h, k) and axis of symmetry parallel to a coordinate axis are given by

$$(x - h)^2 = 4p(y - k) \quad \text{Focus: } (h, k + p), \text{ directrix: } y = k - p$$

$$(y - k)^2 = 4p(x - h) \quad \text{Focus: } (h + p, k), \text{ directrix: } x = h - p$$

6.2 Ellipses

An ellipse is the set of all points in the plane, the sum of whose distances from two fixed points (foci) is a positive constant.

The equations of an ellipse with center at (h, k) and major axis parallel to the coordinate axes are given by

$$\frac{(x - h)^2}{a^2} + \frac{(y - k)^2}{b^2} = 1 \quad \text{Foci: } (h \pm c, k), \text{ vertices: } (h \pm a, k)$$

$$\frac{(x - h)^2}{b^2} + \frac{(y - k)^2}{a^2} = 1 \quad \text{Foci: } (h, k \pm c), \text{ vertices: } (h, k \pm a)$$

For each equation, $a > b$ and $c^2 = a^2 - b^2$.

The eccentricity e of an ellipse is given by $e = c/a$.

6.3 Hyperbolas

A hyperbola is the set of all points in the plane, the difference of whose distances from two fixed points (foci) is a positive constant.

The equations of a hyperbola with center at (h, k) and transverse axis parallel to a coordinate axis are given by

$$\frac{(x-h)^2}{a^2} - \frac{(y-k)^2}{b^2} = 1 \quad \text{Foci: } (h \pm c, k), \text{ vertices: } (h \pm a, k)$$

$$\frac{(y-k)^2}{a^2} + \frac{(x-h)^2}{b^2} = 1 \quad \text{Foci: } (h, k \pm c), \text{ vertices: } (h, k \pm a)$$

For each equation, $c^2 = a^2 + b^2$.

The eccentricity e of a hyperbola is given by $e = c/a$.

6.4 Introduction to Polar Coordinates

A polar coordinate system is formed by drawing a ray (polar axis) and concentric circles with center at the beginning of the ray. The pole is the origin of a polar coordinate system.

A point is specified by coordinates (r, θ), where r is a directed distance from the pole and θ is an angle measured from the polar axis.

The transformation equations between a polar coordinate system and a rectangular coordinate system are

Polar to rectangular: $x = r \cos \theta$ Rectangular to polar: $r = \sqrt{x^2 + y^2}$

$y = r \sin \theta$ $\tan \theta = \dfrac{y}{x}$

CHALLENGE EXERCISES

In Exercises 1 to 10, answer true or false. If the answer is false, give an example.

1. The graph of a parabola is the same shape as one branch of a hyperbola.
2. For the two axes of an ellipse, the major axis and the minor axis, the major axis is always the longer axis.
3. For the two axes of a hyperbola, the transverse axis and the conjugate axis, the transverse axis is always the longer axis.
4. If two ellipses have the same foci, they will have the same graph.
5. A hyperbola is similar to a parabola in that both curves have asymptotes.
6. If a hyperbola with center at the origin and a parabola with vertex at the origin have the same focus, $(0, c)$, the two graphs will always intersect.
7. The graphs of all the conic sections are not the graphs of a function.
8. If F_1 and F_2 are the two foci of an ellipse and P is a point on the ellipse, then $d(P, F_1) + d(P, F_2) = 2a$ where a is the length of the semimajor axis of the ellipse.
9. The eccentricity of a hyperbola is always greater than 1.
10. Each ordered pair (r, θ) in a polar coordinate system specifies exactly one point.

REVIEW EXERCISES

In Exercises 1 to 12, find the foci and vertices of each conic. If the conic is a hyperbola, find the asymptotes. Sketch the graph.

1. $x^2 - y^2 = 4$
2. $y^2 = 16x$
3. $x^2 + 4y^2 - 6x + 8y - 3 = 0$
4. $3x^2 - 4y^2 + 12x - 24y - 36 = 0$
5. $3x - 4y^2 + 8y + 2 = 0$
6. $3x + 2y^2 - 4y - 7 = 0$

7. $9x^2 + 4y^2 + 36x - 8y + 4 = 0$

8. $11x^2 - 25y^2 - 44x - 50y - 256 = 0$

9. $4x^2 - 9y^2 - 8x + 12y - 144 = 0$

10. $9x^2 + 16y^2 + 36x - 16y - 104 = 0$

11. $4x^2 + 28x + 32y + 81 = 0$

12. $x^2 - 6x - 9y + 27 = 0$

In Exercises 13 to 20, find the equation in standard form of the conic that satisfies the given conditions.

13. Ellipse with vertices at $(7, 3)$ and $(-3, 3)$; length of minor axis 8

14. Hyperbola with vertices at $(4, 1)$ and $(-2, 1)$; eccentricity $4/3$

15. Hyperbola with foci $(-5, 2)$ and $(1, 2)$; length of transverse axis 8

16. Parabola with focus $(2, -3)$ and directrix $x = 6$.

17. Parabola with vertex $(0, -2)$ and passing through the point $(3, 4)$

18. Ellipse with eccentricity $2/3$ and foci $(-4, -1)$ and $(0, -1)$

19. Hyperbola with vertices $(\pm 6, 0)$ and asymptotes whose equations are $y = \pm(1/9)x$

20. Parabola passing through the points $(1, 0)$, $(2, 1)$, and $(0, 1)$ with axis of symmetry parallel to the y-axis.

21. Find the equation of the parabola traced by a point $P(x, y)$ that moves so that the distance between $P(x, y)$ and the line $x = 2$ equals the distance between $P(x, y)$ and the point $(-2, 3)$.

22. Find the equation of the parabola traced by a point $P(x, y)$ that moves so that the distance between $P(x, y)$ and the line $y = 1$ equals the distance between $P(x, y)$ and the point $(-1, 2)$.

23. Find the equation of the ellipse traced by a point $P(x, y)$ that moves so that the sum of its distances to $(-3, 1)$ and $(5, 1)$ is 10.

24. Find the equation of the ellipse traced by a point $P(x, y)$ that moves so that the sum of its distances to $(3, 5)$ and $(3, -1)$ is 8.

In Exercises 25 to 34, sketch the graph of the polar equations.

25. $r = 4 \cos 3\theta$

26. $r = 1 + \cos \theta$

27. $r = 2(1 - 2 \sin \theta)$

28. $r = 4 \sin 4\theta$

29. $r = 5 \sin \theta$

30. $r = 3 \sec \theta$

31. $r = 4 \csc \theta$

32. $r = 4 \sin \theta$

33. $r = 3 + 2 \cos \theta$

34. $r = 4 + 2 \sin \theta$

In Exercises 35 to 38, change the equations to polar equations.

35. $y^2 = 16x$

36. $x^2 + y^2 + 4x + 3y = 0$

37. $3x - 2y = 6$

38. $xy = 4$

In Exercises 39 to 42, change the equations to rectangular equations.

39. $r = \dfrac{4}{1 - \cos \theta}$

40. $r = 3 \cos \theta - 4 \sin \theta$

41. $r^2 = \cos 2\theta$

42. $\theta = 1$

7 Exponential and Logarithmic Functions

The science of Pure Mathematics, in its modern developments, may claim to be the most original creation of the human spirit.

A. N. WHITEHEAD (1861–1947)

1, 2, 4, 8, 16, ?

Inductive reasoning is the process of reasoning from particular facts to a general conclusion. Use inductive reasoning to see if you can discover a function between the number of points on a circle and the number of regions in the interior of the circle that are partitioned off by the line segments connecting the points. For example, consider the following five circles. For each circle, count the number of points on the circle and the number of regions that the line segments between the points partition each circle.

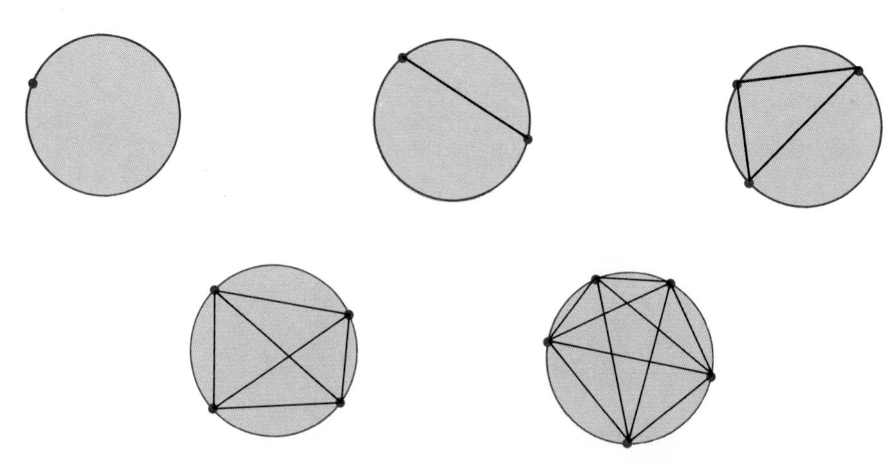

The line segments between points on each circle partition the circle into a number of regions. Can you guess how many regions a circle with six points has?

Your results should agree with the following results:

Number of points on the circle	1	2	3	4	5	6
Maximum number of partitioned regions	1	2	4	8	16	?

Now use inductive reasoning to guess the maximum number of partitioned regions you expect a circle with six points to have. To verify your guess, place six points on a large circle. Connect the six points in all possible ways. Count the regions formed. How does this result compare with your guess?

Completion of this experiment should convince you that thirty-one is the maximum number of partitioned regions that can be formed by the line segments that connect six points on a circle. The purpose of this experiment is to show that inductive reasoning may not lead to valid conclusions.

You will be encouraged to use inductive reasoning to develop mathematical ideas. However, you need to be aware that conclusions based on inductive reasoning may prove to be incorrect.

7.1 Exponential Functions and Their Graphs

In Chapter 1 we defined the real number b^x for every positive base b and every rational number x. For example,

$$2^3 = 8, \qquad 2^{-4} = \frac{1}{2^4} = \frac{1}{16}, \qquad \text{and} \qquad 2^{2/3} = \sqrt[3]{2^2} = \sqrt[3]{4}$$

The exponential key $\boxed{y^x}$ (or on some calculators $\boxed{x^y}$) on a scientific calculator can be used to evaluate a positive number that is raised to a rational power. The following examples illustrate the sequence of key strokes for a calculator that uses algebraic logic.

Power	Key Sequence	Calculator Display
3^2	3 $\boxed{y^x}$ 2 $\boxed{=}$	9.
$2^{1.4}$	2 $\boxed{y^x}$ 1.4 $\boxed{=}$	2.6390158
$1.4^{-3.21}$	1.4 $\boxed{y^x}$ 3.21 $\boxed{+/-}$ $\boxed{=}$	0.3395697
$5^{7/3}$	5 $\boxed{y^x}$ $\boxed{(}$ 7 $\boxed{\div}$ 3 $\boxed{)}$ $\boxed{=}$	42.749398

To define powers of the form b^x, where b is a positive real number and x is a real number, we will require a definition that includes powers with irrational exponents, such as $2^{\sqrt{2}}$, 3^{π}, and $10^{-\sqrt{5}}$.

For our purposes it is convenient to define $2^{\sqrt{2}}$ as the unique real number that can be approximated as closely as desired using an exponent that takes on closer and closer rational approximations of $\sqrt{2}$.

Using a scientific calculator, the key stroke sequence

$$2 \ \boxed{y^x} \ 2 \ \boxed{\sqrt{x}} \ \boxed{=}$$

yields $2^{\sqrt{2}} = 2.6651441$ (to the nearest ten millionth).

Definition of Exponential Functions

> The **exponential function f with base b** is defined by
>
> $$f(x) = b^x$$
>
> where b is a positive constant other than 1 and x is any real number.

The following graphs are representative of the graphs of exponential functions. Each one was graphed by plotting points and then connecting the points with a smooth curve.

EXAMPLE 1 **Graph an Exponential Function**

Sketch the graph of the following exponential functions:

a. $f(x) = 2^x$ and b. $f(x) = \left(\frac{1}{2}\right)^x$

Solution

a.

x	$f(x) = 2^x$
-2	$2^{-2} = \frac{1}{4}$
-1	$2^{-1} = \frac{1}{2}$
0	$2^0 = 1$
1	$2^1 = 2$
2	$2^2 = 4$
3	$2^3 = 8$

Figure 7.1
$f(x) = 2^x$

b.

x	$f(x) = \left(\frac{1}{2}\right)^x = 2^{-x}$
-3	$2^3 = 8$
-2	$2^2 = 4$
-1	$2^1 = 2$
0	$2^0 = 1$
1	$2^{-1} = \frac{1}{2}$
2	$2^{-2} = \frac{1}{4}$

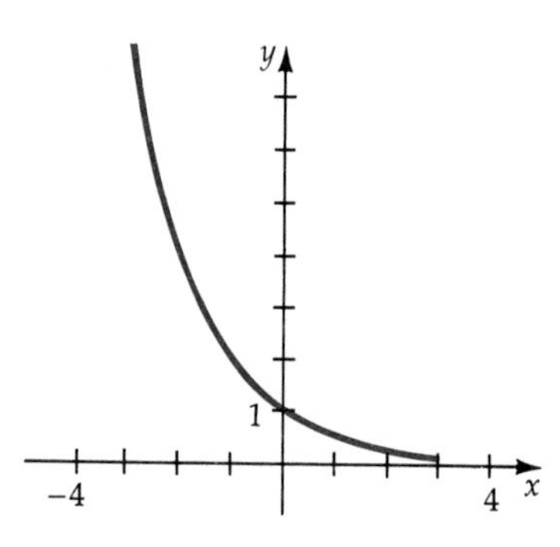

Figure 7.2
$f(x) = \left(\frac{1}{2}\right)^x$

■ *Try Exercise* **26,** *page 268.*

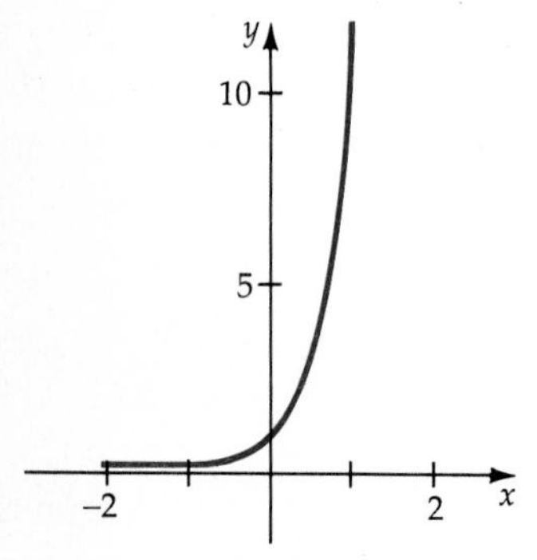

Figure 7.3
$f(x) = 10^x$

To graph an exponential function over a certain portion of its domain, you may need to use different scales on the x- and y-axes. For example, the graph of $f(x) = 10^x$, for $-2 \leq x \leq 1$, is shown in Figure 7.3. Observe that each unit on the y-axis represents a distance of five units and that each unit on the x-axis represents one unit.

For positive real numbers b, $b \neq 1$, the exponential function $f(x) = b^x$ has the following properties:

- f has the set of real numbers as its domain.
- f has the set of positive real numbers as its range.
- f has a graph with a y-intercept of $(0, 1)$.
- f has a graph asymptotic to the x-axis.

- f is a one-to-one function.

The exponential function $f(x) = b^x$ is

- An increasing function if $b > 1$. See Figure 7.4(a).
- A decreasing function if $0 < b < 1$. See Figure 7.4(b).

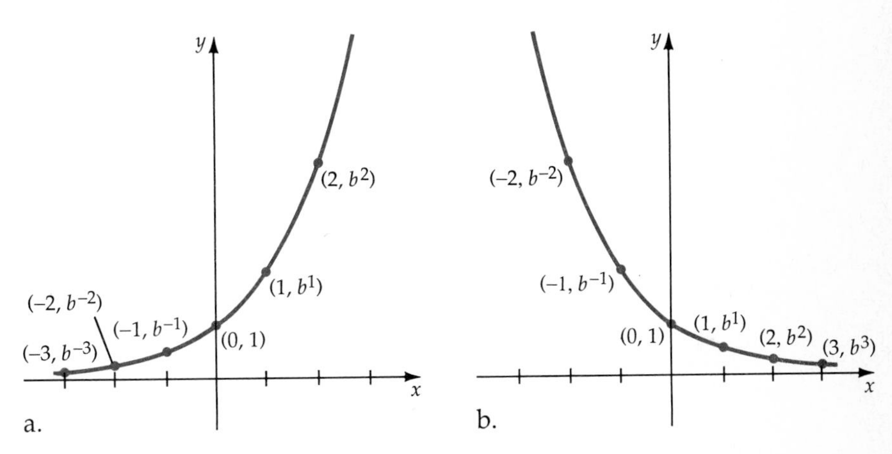

Figure 7.4

Many applications involve functions such as

$$f(x) = ab^p,$$

where a is a constant and p is some expression involving x. Functions of this type can be graphed by plotting points. Be sure to make use of symmetry when possible.

EXAMPLE 2 Graph a Function of the Form $f(x) = ab^p$

Sketch the graph of $f(x) = 3 \cdot 2^{-x^2}$.

Solution Since $f(x) = f(-x)$, this function is an even function. We can sketch its graph by plotting points to the right of the y-axis and then using symmetry to determine points to the left of the y-axis.

x	$f(x) = 3 \cdot 2^{-x^2}$
0	$3 \cdot 2^{-0^2} = 3$
1	$3 \cdot 2^{-1^2} = \frac{3}{2}$
2	$3 \cdot 2^{-2^2} = \frac{3}{16}$

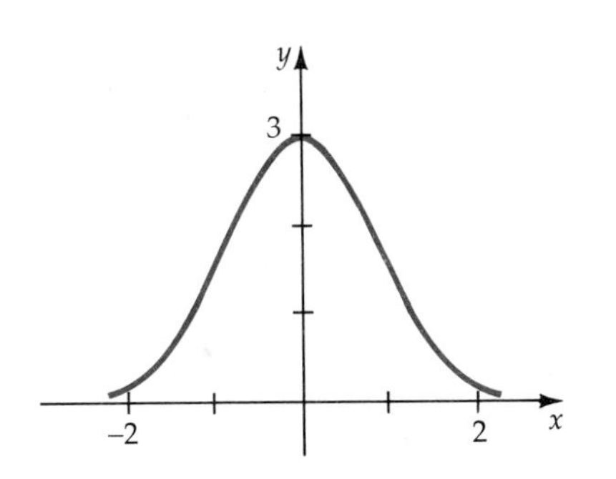

Figure 7.5
$f(x) = 3 \cdot 2^{-x^2}$

■ *Try Exercise* **36,** *page 268.*

Remark A calculator could be used to compute additional points on the graph of f. However, writing f in the fractional form

$$f(x) = 3 \cdot \frac{1}{2^{x^2}}$$

allows us to determine that as $|x|$ increases without bound, the denominator increases without bound while the numerator remains constant. Therefore $f(x)$ approaches 0 as $|x|$ increases without bound. This implies that the graph of f has the x-axis as a horizontal asymptote. Also, the maximum value of f is reached when the denominator 2^{x^2} is its smallest. This occurs when $x = 0$ and $y = 3$. The graph of f is a *bell-shaped* curve similar to the graph of the normal distribution curve, which plays a major role in statistics.

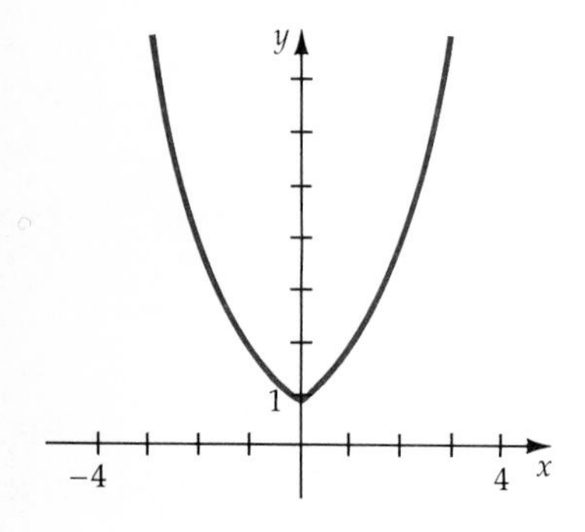

Figure 7.6
$f(x) = 2^{|x|}$

EXAMPLE 3 **Graph a Function of the Form $f(x) = ab^p$**

Sketch the graph of $f(x) = 2^{|x|}$.

Solution

x	$f(x) = 2^{\|x\|}$
0	$2^{\|0\|} = 1$
1	$2^{\|1\|} = 2$
2	$2^{\|2\|} = 4$
3	$2^{\|3\|} = 8$

Because $f(x) = 2^{|x|}$ is an even function, the points to the left of the y-axis can be determined using symmetry.

■ *Try Exercise* **38,** *page 268.*

Remark Note that Figure 7.6 is not the graph of a parabola.

The irrational number π is often used in applications that involve circles. Using techniques developed in a calculus course, we can verify that as n increases without bound,

$$\left(1 + \frac{1}{n}\right)^n$$

approaches an irrational number that is denoted by e. The number e often occurs in applications involving growth or decay. It is denoted by e in honor of the mathematician Leonhard Euler (1707–1783). Although e has a nonterminating and nonrepeating decimal representation, Euler was able to compute e to several decimal places by using large values of n to evaluate $(1 + 1/n)^n$. The entries in Table 7.1 illustrate the process. The value of e accurate to eight decimal places is 2.71828183.

TABLE 7.1

Value of n	Value of $\left(1 + \frac{1}{n}\right)^n$
1	2
10	2.59374246
100	2.704813829
1000	2.716923932
10,000	2.718145927
100,000	2.718268237
1,000,000	2.718280469
10,000,000	2.718281693

The Natural Exponential Function

For all real numbers x, the function defined by

$$f(x) = e^x$$

is called the **natural exponential function.**

To evaluate e^x for specific values of x, you use a calculator with an $\boxed{e^x}$ key, or you can use the $\boxed{y^x}$ key with y as 2.7182818. For example,

Power	Key Sequence	Calculator Display
e^2	2 $\boxed{e^x}$	7.389056099
$e^{4.21}$	4.21 $\boxed{e^x}$	67.35653981
$e^{-1.8}$	1.8 $\boxed{+/-}$ $\boxed{e^x}$	0.165298888

To graph the natural exponential function, use a calculator to approximate e^x for the desired domain values. The resulting points can then be plotted and connected with a smooth curve.

EXAMPLE 4 Graph the Natural Exponential Function

Graph $f(x) = e^x$.

Solution Use a calculator to find the values of e^x. The values in the table have been rounded to the nearest hundredth. Plot the points and then connect the points with a smooth curve.

x	$f(x) = e^x$
−3	0.05
−2	0.14
−1	0.37
0	1.00
1	2.72
2	7.39

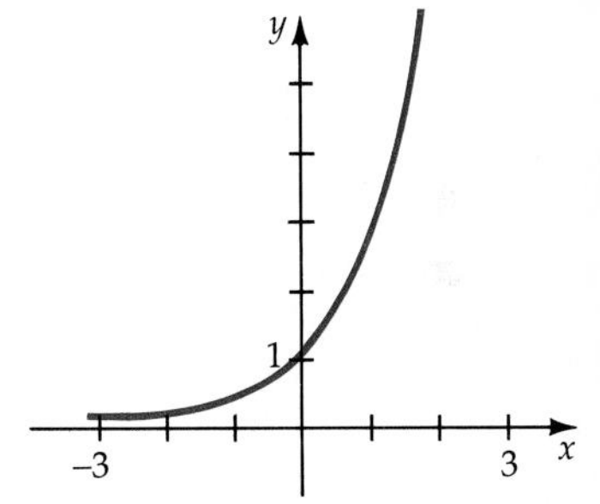

Figure 7.7
$f(x) = e^x$

■ *Try Exercise* **44**, *page 268.*

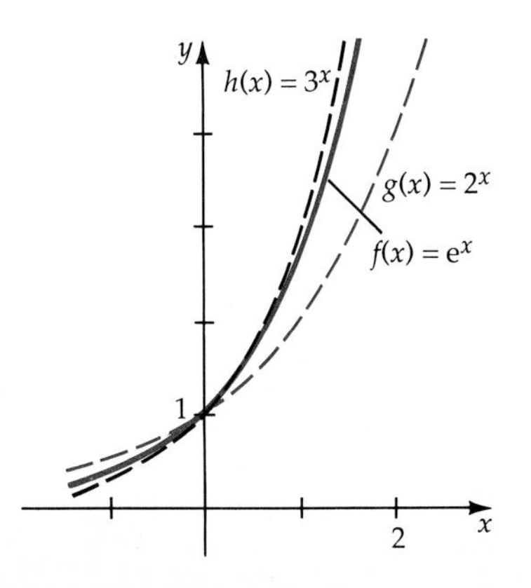

Figure 7.8

Notice in Figure 7.8 how the graph of $f(x) = e^x$ compares with the graphs of $g(x) = 2^x$ and $h(x) = 3^x$. You may have anticipated that the graph of f would be between the graph of g and h because e is between 2 and 3.

Functions of the type

$$f(x) = \frac{b^x + b^{-x}}{2}$$

occur in calculus. These functions are difficult to graph by the method of point-plotting. Example 5 uses the technique of *averaging the y values* of known graphs to sketch the desired graph.

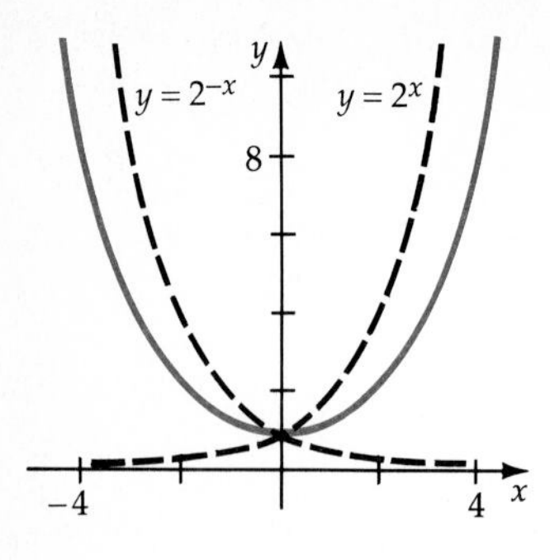

Figure 7.9
$f(x) = \dfrac{2^x + 2^{-x}}{2}$

EXAMPLE 5 **Graph by Averaging *y* Values**

Graph the function $f(x) = \dfrac{2^x + 2^{-x}}{2}$.

Solution Notice that f is the average of 2^x and 2^{-x}. Therefore sketch the graph of f by drawing a curve *halfway between* the graph of $y = 2^x$ and $y = 2^{-x}$. See Figure 7.9.

■ *Try Exercise* **46**, *page 268.*

From the sketch it appears that the graph of f is symmetric with respect to the y-axis, which would imply that f is an even function. This is indeed the case because

$$f(-x) = \frac{2^{-x} + 2^{-(-x)}}{2} = \frac{2^{-x} + 2^x}{2} = f(x).$$

EXERCISE SET 7.1

In Exercises 1 to 12, use a calculator to determine each power accurate to six significant digits.

1. $3^{\sqrt{2}}$ **2.** $5^{\sqrt{3}}$ **3.** $10^{\sqrt{7}}$ **4.** $10^{\sqrt{11}}$
5. $\sqrt{3}^{\sqrt{2}}$ **6.** $\sqrt{5}^{\sqrt{7}}$ **7.** $e^{5.1}$ **8.** $e^{-3.2}$
9. $e^{\sqrt{3}}$ **10.** $e^{\sqrt{5}}$ **11.** $e^{-0.031}$ **12.** $e^{-0.42}$

In Exercises 13 to 24, use a calculator to evaluate each functional value, accurate to six significant digits, given that $f(x) = 3^x$ and $g(x) = e^x$.

13. $f(\sqrt{15})$ **14.** $f(\pi)$ **15.** $f(e)$ **16.** $f(-\sqrt{15})$
17. $g(\sqrt{7})$ **18.** $g(\pi)$ **19.** $g(e)$ **20.** $g(-3.4)$
21. $f[g(2)]$ **22.** $f[g(-1)]$ **23.** $g[f(2)]$ **24.** $g[f(-1)]$

In Exercises 25 to 34, sketch the graph of each exponential function.

25. $f(x) = 3^x$ **26.** $f(x) = 4^x$
27. $f(x) = \left(\frac{3}{2}\right)^x$ **28.** $f(x) = \left(\frac{4}{3}\right)^x$
29. $f(x) = \left(\frac{1}{3}\right)^x$ **30.** $f(x) = \left(\frac{2}{3}\right)^x$
31. $f(x) = \left(\frac{1}{2}\right)^{-x}$ **32.** $f(x) = \left(\frac{1}{3}\right)^{-x}$
33. $f(x) = \frac{5^x}{2}$ **34.** $f(x) = \frac{10^x}{10}$

In Exercises 35 to 48, sketch the graph of each function.

35. $f(x) = \left(\frac{1}{3}\right)^{|x|}$ **36.** $f(x) = 3^{-(x^2)}$
37. $f(x) = 3^{x^2}$ **38.** $f(x) = 2^{-|x|}$
39. $f(x) = 2^{x-3}$ **40.** $f(x) = 2^{x+3}$
41. $f(x) = 3^x - 1$ **42.** $f(x) = 3^x + 1$
43. $f(x) = -e^x$ **44.** $f(x) = \frac{1}{2}e^x$
45. $f(x) = \frac{3^x + 3^{-x}}{2}$ **46.** $f(x) = \frac{e^x + e^{-x}}{2}$
47. $f(x) = -(3^x)$ **48.** $f(x) = -(0.5^x)$

In Exercises 49 to 52, graph each pair of functions on the same set of coordinate axes.

49. $f(x) = 2^x$, $g(x) = 2^{-x}$

50. $f(x) = \left(\frac{2}{3}\right)^x$, $g(x) = \left(\frac{2}{3}\right)^{-x}$

51. $f(x) = 2^{x+1}$, $g(x) = 2^x + 1$

52. $f(x) = 3^x$, $g(x) = \left(\frac{1}{3}\right)^{-x}$

Calculator Exercises

53. Use a calculator to complete the following table:

Value of n	Value of $\left(1 + \frac{1}{n}\right)^n$
5	
50	
500	
5000	
50,000	
500,000	
5,000,000	

54. Use a calculator to complete the following table.

a. How do the entries in this table compare with those in Exercise 53?

b. As n approaches 0, what value does $(1 + n)^{1/n}$ appear to approach?

Value of n	Value of $\left(1 + n\right)^{1/n}$
0.2	
0.02	
0.002	
0.0002	
0.00002	
0.000002	
0.0000002	

Supplemental Exercises

55. Evaluate $h(x) = (-2)^x$, for **a.** $x = 1$, **b.** $x = 2$, and **c.** $x = 1.5$. Explain why h is not an exponential function.

56. Graph $j(x) = 1^x$. Explain why j is not an exponential function.

57. Graph $f(x) = e^x$, and then sketch the graph of f reflected about the graph of the line given by $y = x$.

58. Graph $g(x) = 10^x$, and then sketch the graph of g reflected about the graph of the line given by $y = x$.

59. Prove that

$$F(x) = \frac{e^x - e^{-x}}{2}$$

is an odd function.

60. Prove that $G(x) = e^x$ is neither an odd function nor an even function.

61. The number of bacteria present in a culture is given by $N(t) = 10{,}000(2^t)$, where $N(t)$ is the number of bacteria present after t hours. Find the number of bacteria present when **a.** $t = 1$ hour, **b.** $t = 2$ hours, **c.** $t = 5$ hours.

62. The production function for an oil well is given by the function $B(t) = 100{,}000(e^{-0.2t})$, where $B(t)$ is the number of barrels of oil the well can produce per month after t years. Find the number of barrels of oil the well can produce per month when **a.** $t = 1$ year, **b.** $t = 2$ years, **c.** $t = 5$ years.

63. Which of the following powers, e^π or π^e, is larger?

64. Let $f(x) = x^{(x^x)}$ and $g(x) = (x^x)^x$. Which is larger, $f(3)$ or $g(3)$?

65. Graph $f(x) = \dfrac{2^x - 2^{-x}}{2}$.

66. Graph $f(x) = \dfrac{e^x - e^{-x}}{2}$.

7.2

Logarithms and Logarithmic Properties

Every exponential function is a one-to-one function, and therefore, has an inverse function. Recall from Section 1.5 that we determined the inverse of a function represented by an equation by interchanging the variables and then solving for the dependent variable. If we attempt to use this procedure for the exponential function $f(x) = b^x$, we get

$$f(x) = b^x$$

$$y = b^x$$

$$x = b^y \quad \text{Interchange the variables.}$$

How do we solve $x = b^y$ for the exponent y? None of our previous methods can be used to solve the equation $x = b^y$ for the exponent y.

Thus we must develop a new procedure. One method would be merely to write

$$y = \text{exponent of } b \text{ that produces } x.$$

This procedure would work, but it is not concise. We need compact notation to represent y as the exponent of b that produces x. For historical reasons, we use the notation in the following definition.

Definition of a Logarithm

If $x > 0$ and b is a positive constant ($b \neq 1$), then

$$y = \log_b x \quad \text{if and only if} \quad b^y = x.$$

In the equation $y = \log_b x$, y is referred to as the **logarithm,** b is the **base,** and x is the **argument.**

The notation $\log_b x$ is read as "the logarithm (or log) base b of x." The definition of a logarithm indicates that *a logarithm is an exponent.*

The equations

$$y = \log_b x \qquad \text{and} \qquad b^y = x$$

are different ways of expressing the same thing.

$$y = \log_b x \text{ is the logarithmic form of } b^y = x$$

$$b^y = x \text{ is the exponential form of } y = \log_b x.$$

EXAMPLE 1 Change from Logarithmic to Exponential Form

Write each of the following equations in their exponential form:

a. $2 = \log_7 x$ b. $3 = \log_{10}(x + 8)$ c. $\log_5 125 = x$

Solution Use the definition $y = \log_b x$ if and only if $b^y = x$.

a. Logarithms are exponents

$$2 = \log_7 x \qquad \text{implies} \qquad 7^2 = x$$

Base

b. $3 = \log_{10}(x + 8)$ implies $10^3 = (x + 8)$.

c. $\log_5 125 = x$ implies $5^x = 125$.

■ *Try Exercise* **2,** *page 277.*

EXAMPLE 2 Change from Exponential to Logarithmic Form

Write each of the following equations in their logarithmic form:

a. $x = 25^{1/2}$ b. $\dfrac{1}{16} = x^{-4}$ c. $27^x = 3$

Solution Use $x = b^y$ if and only if $y = \log_b x$.

a.

Exponent

$$x = 25^{1/2} \text{ implies } \tfrac{1}{2} = \log_{25} x$$

Base

b. $\dfrac{1}{16} = x^{-4}$ implies $-4 = \log_x \dfrac{1}{16}$.

c. $27^x = 3$ implies $\log_{27} 3 = x$.

■ *Try Exercise* **12,** *page 277.*

Some logarithms can be evaluated by using the definition of a logarithm and the following theorem.

Equality of Exponents Theorem

If b is a positive real number $(b \neq 1)$ such that $b^x = b^y$, then $x = y$.

EXAMPLE 3 **Evaluate Logarithms**

Evaluate each logarithm.

a. $\log_2 32 = x$ b. $\log_5 125 = x$ c. $\log_{10} \dfrac{1}{100} = x$

Solution

a. $\log_2 32 = x$ implies $2^x = 32$ Change to exponential form.

$2^x = 2^5$ Factor.

$x = 5$ Equality of Exponents Theorem

b. $\log_5 125 = x$ implies $5^x = 125$

$5^x = 5^3$

$x = 3$

c. $\log_{10} \dfrac{1}{100} = x$ implies $10^x = \dfrac{1}{100}$

$10^x = 10^{-2}$

$x = -2$

■ *Try Exercise* **22,** *page 277.*

Since logarithms are exponents, they have many properties that can be established by using the properties of exponents.

Properties of Logarithms

In the following properties, b, M, and N are positive real numbers ($b \neq 1$), and p is any real number.

$\log_b b = 1$	
$\log_b 1 = 0$	
$\log_b(b^p) = p$	
$\log_b(MN) = \log_b M + \log_b N$	Product property
$\log_b\left(\frac{M}{N}\right) = \log_b M - \log_b N$	Quotient property
$\log_b(M^p) = p \log_b M$	Power property
$\log_b M = \log_b N$ implies $M = N$	One-to-one property
$M = N$ implies $\log_b M = \log_b N$	Logarithm of each side property
$b^{\log_b p} = p$ (for $p > 0$)	Inverse property

The first three properties of logarithms can be proven by using the definition of a logarithm. That is,

$$\log_b b = 1 \quad \text{because} \quad b^1 = b.$$

$$\log_b 1 = 0 \quad \text{because} \quad b^0 = 1.$$

$$\log_b(b^p) = p \quad \text{because} \quad b^p = b^p.$$

To prove the product property $\log_b(MN) = \log_b M + \log_b N$, let

$$\log_b M = x \quad \text{and} \quad \log_b N = y.$$

Writing each equation in its equivalent exponential form produces

$$M = b^x \quad \text{and} \quad N = b^y.$$

Forming the product of the respective right and left sides produces

$$MN = b^x b^y \quad \text{or} \quad MN = b^{x+y}.$$

Applying the definition of a logarithm yields the equivalent form

$$\log_b(MN) = x + y.$$

Since $x = \log_b M$ and $y = \log_b N$, this becomes

$$\log_b(MN) = \log_b M + \log_b N.$$

The proofs of the quotient property and the power property are similar to the proof of the product property. They are left as exercises.

The logarithm of each side property will be used several times in this chapter. It states that if two positive real numbers are equal, their logarithms base b are also equal.

Proof: Let $M > 0$, $N > 0$ and $M = N$.

$\log_b M = \log_b M$ The reflexive property

$\log_b M = \log_b N$ Substitute N for M.

The proofs of the one-to-one property and the inverse property are left as exercises.

The properties of logarithms are often used to rewrite logarithms and expressions that involve logarithms.

EXAMPLE 4 Use Logarithmic Properties

Use the properties of logarithms to express the following logarithms in terms of logarithms of x, y, and z:

a. $\log_b xy^2$ b. $\log_b \dfrac{x^2\sqrt{y}}{z^5}$

Solution

a. $\log_b xy^2 = \log_b x + \log_b y^2$ Product property

$= \log_b x + 2\log_b y$ Power property

b. $\log_b \dfrac{x^2\sqrt{y}}{z^5} = \log_b x^2\sqrt{y} - \log_b z^5$ Quotient property

$= \log_b x^2 + \log_b \sqrt{y} - \log_b z^5$ Product property

$= 2\log_b x + \dfrac{1}{2}\log_b y - 5\log_b z$ Power property

■ *Try Exercise* **32,** *page 277.*

Sometimes it is possible to use known logarithmic values and the properties of logarithms to evaluate logarithms.

EXAMPLE 5 Use Logarithmic Properties

Given $\log_8 2 \approx 0.3333$, $\log_8 3 \approx 0.5283$, and $\log_8 5 \approx 0.7740$, evaluate the following:

a. $\log_8 15$ b. $\log_8 \dfrac{5}{2}$ c. $\log_8 \sqrt[3]{9}$

Solution

a. $\log_8 15 = \log_8(3 \cdot 5) = \log_8 3 + \log_8 5$ Product property

$\approx 0.5283 + 0.7740 = 1.3023$

b. $\log_8 \dfrac{5}{2} = \log_8 5 - \log_8 2$ Quotient property

$\approx 0.7740 - 0.3333 = 0.4407$

c. $\log_8 \sqrt[3]{9} = \log_8 3^{2/3} = \dfrac{2}{3} \log_8 3$ Power property

$$\approx \frac{2}{3}(0.5283) = 0.3522$$

■ *Try Exercise* **42**, *page 277.*

The properties of logarithms are also used to rewrite expressions that involve logarithms as a single logarithm.

EXAMPLE 6 Use Logarithmic Properties

Use the properties of logarithms to rewrite the following expressions as a single logarithm:

a. $2 \log_b x + \dfrac{1}{2} \log_b(x + 4)$ b. $4 \log_b(x + 2) - 3 \log_b(x - 5)$

Solution

a. $2 \log_b x + \dfrac{1}{2} \log_b(x + 4) = \log_b x^2 + \log_b(x + 4)^{1/2}$ Power property

$= \log_b x^2(x + 4)^{1/2}$ Product property

b. $4 \log_b(x + 2) - 3 \log_b(x - 5) = \log_b(x + 2)^4 - \log_b(x - 5)^3$ Power property

$= \log_b \dfrac{(x + 2)^4}{(x - 5)^3}$ Quotient property

■ *Try Exercise* **52**, *page 277.*

Definition of Common Logarithm and Natural Logarithm

Logarithms with a base of 10 are called **common logarithms.** It is customary to write $\log_{10} x$ as $\log x$.

Logarithms with a base of e are called **natural logarithms.** They are often used in calculus. It is customary to write $\log_e x$ as $\ln x$.

Most scientific calculators have a key marked [log] for evaluating common logarithms and a key marked [ln] for evaluating natural logarithms. For example,

Logarithm	Key Sequence	Calculator Display
log 24	24 [log]	1.380211242
ln 81	81 [ln]	4.394449155
log 0.58	.58 [log]	-0.236572006

If you use a calculator to try to evaluate the logarithm of a negative number, it will give you an error indication. Recall that the definition of $y = \log_b x$ required x to be a positive real number.

Appendix II contains a table of common logarithms and a table of natural logarithms. Use the tables to evaluate common and natural logarithms if a scientific calculator is not available. Appendix I explains the use of the common logarithmic table. All the properties of logarithms apply to both common and natural logarithms.

Logarithms that are not common logarithms or natural logarithms can be evaluated by using the following theorem.

Change-of-Base Formula

> If x, a, and b are positive real numbers with $a \neq 1$ and $b \neq 1$, then
> $$\log_b x = \frac{\log_a x}{\log_a b}.$$

Proof: Let $\log_b x = y$

Then $b^y = x$ By definition of $\log_b x$

Now take the logarithm with base a of each side.

$\log_a(b^y) = \log_a x$ Logarithm of each side property

$y \log_a b = \log_a x$ Power property

$y = \dfrac{\log_a x}{\log_a b}$ Solve for y by dividing by $\log_a b$.

$\log_b x = \dfrac{\log_a x}{\log_a b}$ Substitute $\log_b x$ for y.

EXAMPLE 7 Use the Change-of-Base Formula

Evaluate each of the following logarithms:

a. $\log_3 18$ b. $\log_{12} 400$ c. $\log_9 4$

Solution In each case we use the change-of-base formula with $a = 10$. That is, we will evaluate these logarithms by using the log key on a scientific calculator.

a. $$\log_3 18 = \frac{\log 18}{\log 3} \approx \frac{1.25527}{0.47712} \approx 2.63093$$

b. $$\log_{12} 400 = \frac{\log 400}{\log 12} \approx \frac{2.60206}{1.07918} \approx 2.41115$$

c. $$\log_9 4 = \frac{\log 4}{\log 9} \approx \frac{0.60206}{0.95424} \approx 0.63093$$

■ *Try Exercise* **62,** *page 277.*

Remark The logarithms could also have been evaluated by using the ln key. For example, for part a,

$$\log_3 18 = \frac{\ln 18}{\ln 3} \approx \frac{2.89037}{1.09861} \approx 2.63093$$

The change-of-base formula can be used to evaluate common logarithms using natural logarithms and to evaluate natural logarithms using common logarithms. For example, if we substitute e for a and 10 for b in the change-of-base formula, we get

$$\log x = \frac{\ln x}{\ln 10} \approx \frac{\ln x}{2.3026} \approx 0.4343 \ln x .$$

Substituting e for b and 10 for a in the change-of-base formula yields

$$\ln x = \frac{\log x}{\log e} \approx \frac{\log x}{0.4343} .$$

Antilogarithms

Given $M = \log N$, it is often necessary to determine the value of N. In this case the number N is called the **antilogarithm of** M.

Definition of Antilogarithms

If M and N are real numbers with $N > 0$, such that

$$\log_b N = M$$

then N is the **antilogarithm of** M for the base b.

Using the definition of a logarithm, we see that N can be evaluated by the formula $N = b^M$. It is convenient to evaluate antilogarithms using a scientific calculator. The following examples demonstrate the method used by some calculators.

Logarithm	Key Sequence	Calculator Display
$\log N = 2.4031$	2.4031 INV log	252.9880456
$\ln N = 3.8067$	3.8067 INV ln	45.00168799

Remark Some calculators have keys marked 10^x and e^x. These keys can be used to find the antilogarithms of a number with base 10 or base e. If $\log N = 2.4031$, you can evaluate N by entering the number 2.4031 and pressing the 10^x key. Also, if $\ln N = 3.8067$, then N can be evaluated by entering the number 3.8067 and pressing the e^x key.

Antilogarithms can also be found using the tables in Appendix II. Appendix I explains the process of using the tables to find logarithms and antilogarithms.

EXERCISE SET 7.2

In Exercises 1 to 10, change each equation to its exponential form.

1. $\log_{10} 100 = 2$

2. $\log_{10} 1000 = 3$

3. $\log_5 125 = 3$

4. $\log_5 \frac{1}{25} = -2$

5. $\log_3 81 = 4$

6. $\log_3 1 = 0$

7. $\log_b r = t$

8. $\log_b(s + t) = r$

9. $-3 = \log_3 \frac{1}{27}$

10. $-1 = \log_7 \frac{1}{7}$

In Exercises 11 to 20, change each equation to its logarithmic form.

11. $2^4 = 16$

12. $3^5 = 243$

13. $7^3 = 343$

14. $7^{-4} = \frac{1}{2401}$

15. $10{,}000 = 10^4$

16. $\frac{1}{1000} = 10^{-3}$

17. $b^k = j$

18. $p = m^n$

19. $b^1 = b$

20. $b^0 = 1$

In Exercises 21 to 30, evaluate each logarithm. Do not use a calculator.

21. $\log_{10} 1{,}000{,}000$

22. $\log_{10} \frac{1}{1000}$

23. $\log_2 32$

24. $\log_3 243$

25. $\log_{3/2} \frac{27}{8}$

26. $\log_{0.5} 16$

27. $\log_5 \frac{1}{25}$

28. $\log_{0.3} \frac{100}{9}$

29. $\log_b 1$

30. $\log_b b$

In Exercises 31 to 40, write the given logarithm in terms of logarithms of x, y, and z.

31. $\log_b xyz$

32. $\log_b x^2y^3$

33. $\log_3 \frac{x}{z^4}$

34. $\log_5 \frac{x^2}{yz^3}$

35. $\log_b \frac{\sqrt{x}}{y^3}$

36. $\log_b \frac{\sqrt{x}}{\sqrt[3]{z}}$

37. $\log_b x\sqrt[3]{\frac{y^2}{z}}$

38. $\log_b \sqrt[3]{x^2z\sqrt{y}}$

39. $\log_7 \frac{\sqrt{x + z^2}}{x^2 - y}$

40. $\log_5\left(\frac{x^2 - y}{z^2}\right)$

In Exercises 41 to 50, evaluate the logarithm using the values $\log_b 2 \approx 0.3562$, $\log_b 3 \approx 0.5646$, and $\log_b 5 \approx 0.8271$, and the properties of logarithms. Do not use a calculator.

41. $\log_b 6$

42. $\log_b 20$

43. $\log_b 9$

44. $\log_b 4$

45. $\log_b \frac{2}{5}$

46. $\log_b \frac{3}{2}$

47. $\log_b 30$

48. $\log_b 45$

49. $\log_b 2b$

50. $\log_b \frac{b^2}{3}$

In Exercises 51 to 60, write each logarithmic expression as a single logarithm.

51. $\log_{10}(x + 5) + 2\log_{10} x$

52. $5\log_3 x - 4\log_3 y + 2\log_3 z$

53. $\frac{1}{2}[3\log_b(x - y) + \log_b(x + y) - \log_b z]$

54. $\log_b(y^3z^2) - 3\log_b(x\sqrt{y}) + 2\log_b\left(\frac{x}{z}\right)$

55. $\log_8(x^2 - y^2) - \log_8(x - y)$

56. $\log_4(x^3 - y^3) - \log_4(x - y)$

57. $4\ln(x - 3) + 2\ln x$

58. $3\ln z - 2\ln(z + 1)$

59. $\ln x - \ln y + \ln z$

60. $\frac{1}{2}\log x + 2\log y$

Calculator Exercises

In Exercises 61 to 70, use the change-of-base formula and a calculator to approximate the following logarithms accurate to five significant digits.

61. $\log_7 20$

62. $\log_5 37$

63. $\log_{11} 8$

64. $\log_{50} 22$

65. $\log_6 0.045$

66. $\log_4 \sqrt{7}$

67. $\log_{0.5} 5$

68. $\log_{0.2} 17$

69. $\log_\pi e$

70. $\log_\pi \sqrt{15}$

In Exercises 71 to 82, use a calculator (or the tables in Appendix II) to approximate the antilogarithm N to three significant digits.

71. $\log N = 0.4857$

72. $\log N = 0.9557$

73. $\log N = 3.5038$

74. $\log N = 7.8476$

75. $\log N = -2.4760$

76. $\log N = -4.3536$

77. $\ln N = 2.001$

78. $\ln N = 2.262$

79. $\ln N = 0.693$

80. $\ln N = 0.531$

81. $\ln N = -1.204$

82. $\ln N = -0.511$

Supplemental Exercises

In Exercises 83 to 88, find all the real numbers that are solutions of the given inequality. Use interval notation to write your answers.

83. $0 \le \log x \le 1000$

84. $-3 \le \log x \le -2$

85. $e \le \ln x \le e^3$

86. $-2 \le \ln x \le 3$

87. $-\log x > 0$

88. $100 - 10\log(x + 1) > 0$

89. Verify the quotient property of logarithms. *Hint:* Use a method similar to the proof of the product property.

90. Verify the power property of logarithms.

91. Give the reason for each step in the proof of the inverse property of logarithms.

$\log_b x = \log_b x$	?
$b^{\log_b x} = x$	?

92. Give the reason for each step in the proof of the one-to-one property of logarithms.

$\log_b M = \log_b N$	Given
$b^{\log_b N} = M$	?
$N = M$	?

7.3 Logarithmic Functions and Their Graphs

Section 2 developed the concept of a logarithm and the properties of logarithms. With this background we can now introduce the concept of a *logarithmic function.*

Definition of a Logarithmic Function

> The **logarithmic function** f **with base** b is defined by
>
> $$f(x) = \log_b x$$
>
> where b is a positive constant $b \neq 1$, and x is any *positive* real number.

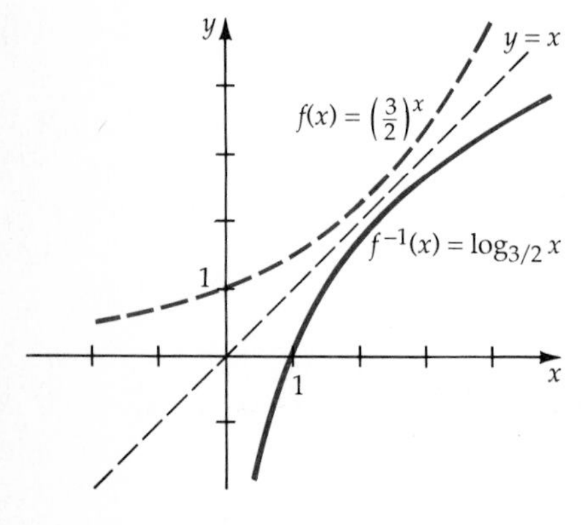

Figure 7.10

Recall that logarithms were defined so that we could write the inverse of $g(x) = b^x$ in a convenient manner. Thus the logarithmic function given by $f(x) = \log_b x$ is the inverse of the exponential function $g(x) = b^x$. The graph of $y = \log_b x$ can be obtained by reflecting the graph of $y = b^x$ across the graph of the line $y = x$. This is illustrated in Figure 7.10 for the exponential function $f(x) = \left(\frac{3}{2}\right)^x$ and its inverse $f^{-1}(x) = \log_{3/2} x$.

If you rewrite a logarithmic function in its equivalent exponential form, then the logarithmic function can be graphed by plotting points.

EXAMPLE 1 Graph a Logarithmic Function

Graph the logarithmic function $f(x) = \log_2 x$.

Solution Changing $y = \log_2 x$ to its exponential form $2^y = x$ allows us to evaluate x for convenient integer values of y.

y	-2	-1	0	1	2	3
$x = 2^y$	$2^{-2} = \frac{1}{4}$	$2^{-1} = \frac{1}{2}$	$2^0 = 1$	$2^1 = 2$	$2^2 = 4$	$2^3 = 8$

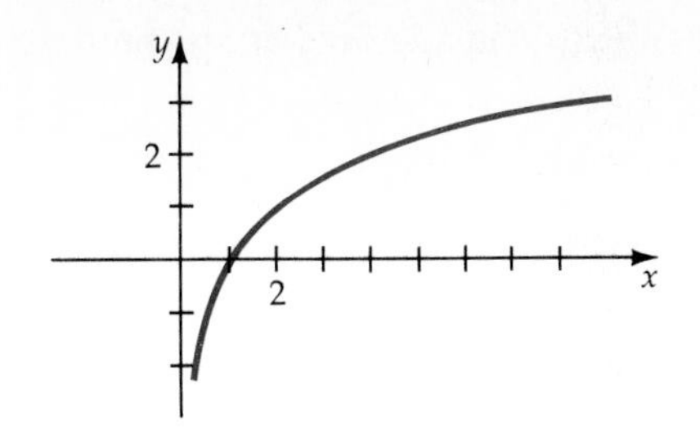

Figure 7.11
$f(x) = \log_2 x$

The above table was constructed by first picking a y value and then computing the corresponding x value. Drawing a smooth curve through the resulting points produces the graph of $f(x) = \log_2 x$ shown in Figure 7.11.

■ *Try Exercise* **2**, *page 282.*

The logarithmic functions $f(x) = \log x$ and $g(x) = \ln x$ are often used in applications and advanced mathematics. Their graphs can be drawn by using the techniques developed in Example 1. However, most scientific calculators have the [log] and the [ln] keys that can be used to determine points on the graphs. For example, the following table was made by entering convenient values of x and then evaluating $\log x$ and $\ln x$ by pressing the [log] and the [ln] keys. Figure 7.12 shows the graph of $f(x) = \log x$ and the graph of $g(x) = \ln x$.

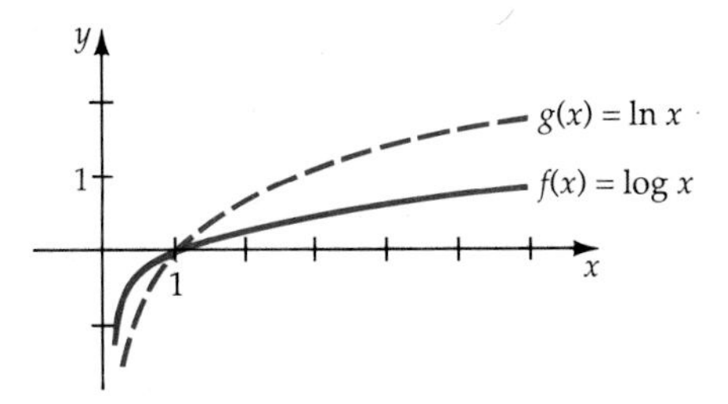

Figure 7.12

x	0.5	1	5	10
$f(x) = \log x$	−0.3010	0	0.6990	1
$g(x) = \ln x$	−0.6931	0	1.6094	2.3026

For all positive real numbers $b \neq 1$, the function $f(x) = \log_b x$ has the following properties:

- f has the set of positive real numbers as its domain.
- f has the set of real numbers as its range.
- f has a graph with an x-intercept of $(1, 0)$.
- f has a graph asymptotic to the y-axis.
- f is a one-to-one function.

The logarithmic function $f(x) = \log_b x$ is:

- an increasing function if $b > 1$. See Figure 7.13a.
- a decreasing function if $0 < b < 1$. See Figure 7.13b.

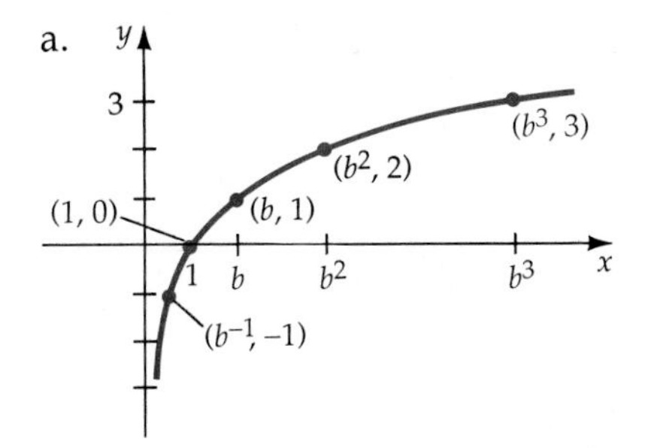

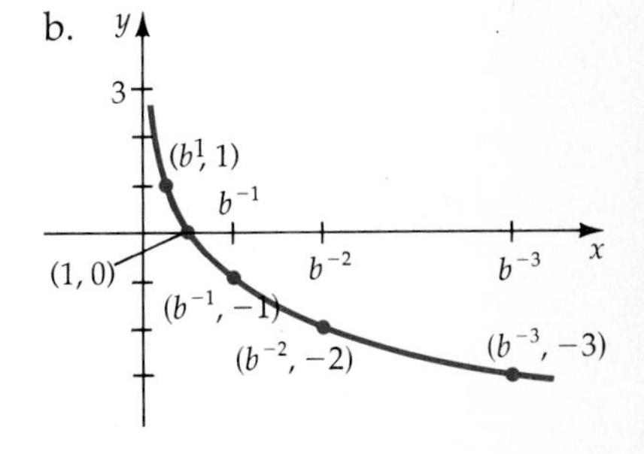

Figure 7.13

Many applied problems involve functions that are defined in terms of

$$f(x) = \log_b c,$$

where c is an algebraic expression that involves x. Example 2 illustrates that the graph of such a function may differ considerably from the graph of the logarithmic function $f(x) = \log_b x$.

EXAMPLE 2 Graph a Function That Is Logarithmic in Form

Graph the function $f(x) = \log_2|x|$.

Solution This graph can be sketched by using the method of plotting points, but the procedure developed in the following discussion allows us to sketch the graph with little or no computation.

The domain of $f(x) = \log_2 x$ is the set of positive real numbers. Since $|x| > 0$ for all $x \neq 0$, the domain of $f(x) = \log_2|x|$ is the set of all nonzero real numbers. If $x > 0$, then $|x| = x$, and the graph of $f(x) = \log_2|x|$ will be the same as the graph of $f(x) = \log_2 x$ shown in Figure 7.11.

For $x < 0$, we use the fact that $f(x) = \log_2|x|$ is an even function. Its graph is symmetrical to the y-axis; thus we reflect the right-hand part of the graph across the y-axis to produce the graph in Figure 7.14.

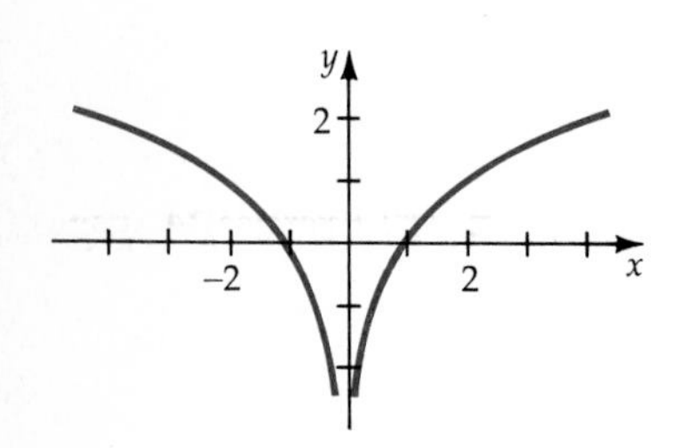

Figure 7.14
$f(x) = \log_2 |x|$

■ *Try Exercise* **10**, *page 282.*

EXAMPLE 3 Graph a Function That Is Logarithmic in Form

Graph the function $f(x) = \log_4(-2x)$.

Solution Writing $y = \log_4(-2x)$ in its exponential form produces $4^y = -2x$ or $(-\frac{1}{2})4^y = x$. Choosing convenient values for y yields the following table:

y	-2	-1	0	1	2
$x = \left(-\frac{1}{2}\right)4^y$	$\left(-\frac{1}{2}\right)4^{-2} = -\frac{1}{32}$	$\left(-\frac{1}{2}\right)4^{-1} = -\frac{1}{8}$	$\left(-\frac{1}{2}\right)4^{0} = -\frac{1}{2}$	$\left(-\frac{1}{2}\right)4^{1} = -2$	$\left(-\frac{1}{2}\right)4^{2} = -8$

The domain of $f(x) = \log_4(-2x)$ is the set of all negative real numbers because $-2x > 0$ if and only if $x < 0$.

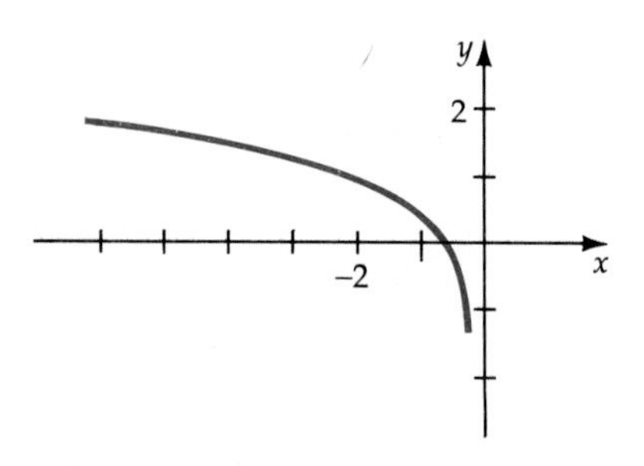

Figure 7.15
$f(x) = \log_4(-2x)$

■ *Try Exercise* **12**, *page 282.*

EXAMPLE 4 Graph a Function That Is Logarithmic in Form

Graph the function $f(x) = \log\sqrt{x}$.

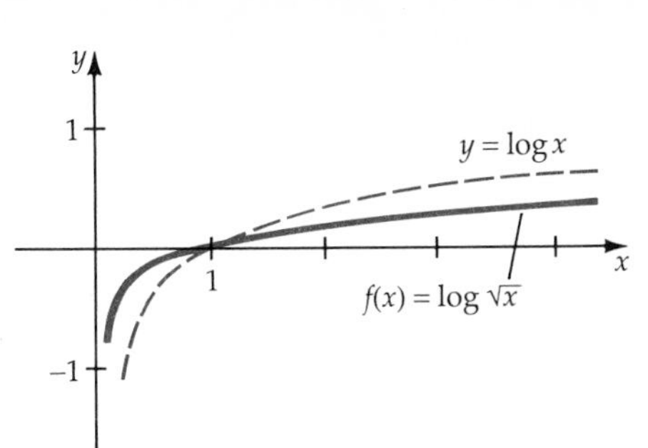

Figure 7.16

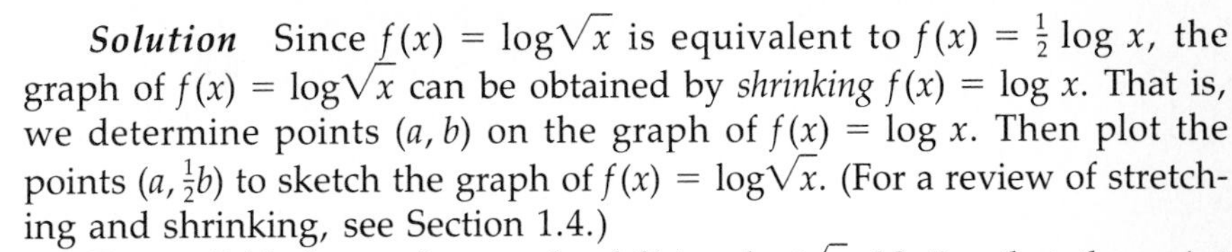

Solution Since $f(x) = \log\sqrt{x}$ is equivalent to $f(x) = \frac{1}{2}\log x$, the graph of $f(x) = \log\sqrt{x}$ can be obtained by *shrinking* $f(x) = \log x$. That is, we determine points (a, b) on the graph of $f(x) = \log x$. Then plot the points $(a, \frac{1}{2}b)$ to sketch the graph of $f(x) = \log\sqrt{x}$. (For a review of stretching and shrinking, see Section 1.4.)

Figure 7.16 shows the graph of $f(x) = \log\sqrt{x}$. Notice that the point $(a, \frac{1}{2}b)$ is on the graph of $f(x) = \log\sqrt{x}$ if and only if the point (a, b) is on the graph of $f(x) = \log x$.

■ *Try Exercise* **14,** *page 282.*

Horizontal and/or vertical translations of the graph of the logarithmic function $f(x) = \log_b x$ sometimes can be used to obtain the graph of functions that involve logarithms.

EXAMPLE 5 Use Translations to Graph

Graph a. $f(x) = \log_4(x + 3)$ b. $f(x) = \log_4 x + 3$.

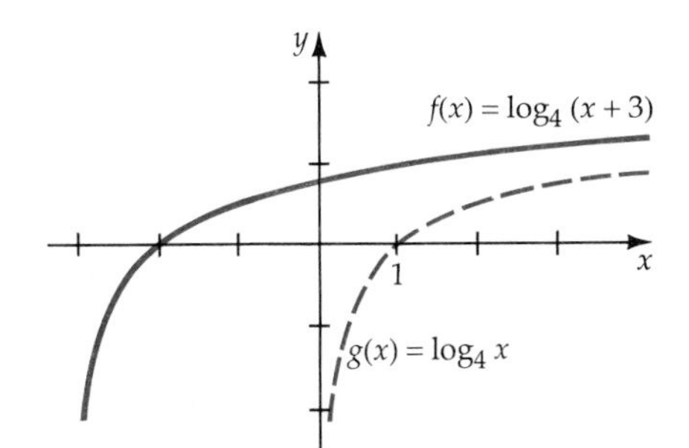

Figure 7.17

Solution

a. The graph of $f(x) = \log_4(x + 3)$ can be obtained by shifting the graph of $g(x) = \log_4 x$ three units to the left. (For a review of *horizontal translations*, see Section 1.4.) Figure 7.17 shows the graph of $g(x) = \log_4 x$ and the graph of $f(x) = \log_4(x + 3)$.

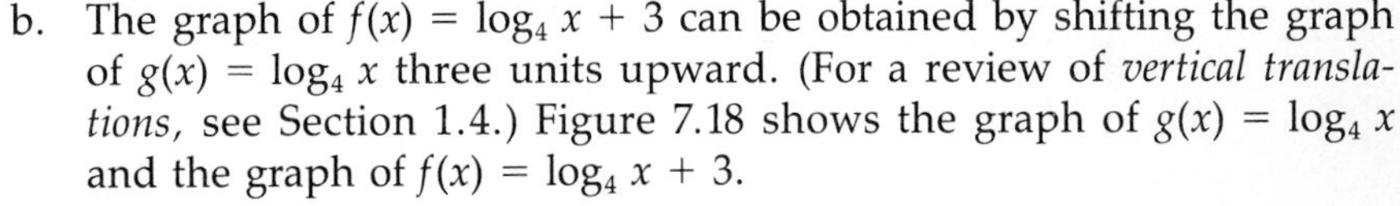

b. The graph of $f(x) = \log_4 x + 3$ can be obtained by shifting the graph of $g(x) = \log_4 x$ three units upward. (For a review of *vertical translations*, see Section 1.4.) Figure 7.18 shows the graph of $g(x) = \log_4 x$ and the graph of $f(x) = \log_4 x + 3$.

■ *Try Exercise* **20,** *page 282.*

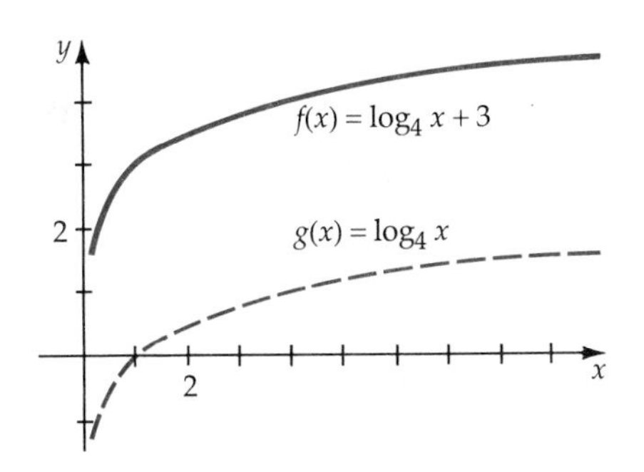

Figure 7.18

Seismologists measure the magnitude of earthquakes using the Richter scale. On this scale, the magnitude R of an earthquake is determined by the function.

$$R = \log\frac{I}{I_0}$$

where I_0 is the measure of a *zero-level earthquake*, and I is the intensity of the earthquake being measured.

EXAMPLE 6 Find the Richter Scale Measure of an Earthquake

Find the Richter scale measure of

a. an earthquake that had an intensity of $I = 10{,}000\ I_0$;

b. the San Francisco earthquake of 1906, which had an intensity of $I = 199{,}526{,}000I_0$.

Solution

a. $R = \log \dfrac{10{,}000I_0}{I_0} = \log 10{,}000 = \log 10^4 = 4$ (on the Richter scale).

b. $R = \log \dfrac{199{,}526{,}000I_0}{I_0} = \log 199{,}526{,}000 \approx 8.3$ (on the Richter scale).

■ *Try Exercise* **42**, *page 283.*

EXAMPLE 7 Compare the Richter Scale Measures of Earthquakes

If an earthquake has an intensity 100 times the intensity of a second earthquake, then how much larger is the Richter scale measure of the larger earthquake than that of the smaller?

Solution Let I represent the intensity of the smaller earthquake and $100I$ represent the intensity of the larger earthquake. The Richter scale measures of the small earthquake R_1 and the larger earthquake R_2 are given by

$$R_1 = \log \frac{I}{I_0} \quad \text{and} \quad R_2 = \log \frac{100I}{I_0}.$$

Using the properties of logarithms, we can write R_2 as

$$R_2 = \log 100\frac{I}{I_0} = \log 100 + \log \frac{I}{I_0} = \log 10^2 + R_1 = 2 + R_1$$

Thus an earthquake that is 100 times as intense as a smaller earthquake will have a Richter scale measure that is 2 more than that of the smaller earthquake.

■ *Try Exercise* **44**, *page 283.*

EXERCISE SET 7.3

In Exercises 1 to 28, graph the given logarithmic functions. If possible, make use of translations.

1. $f(x) = \log_3 x$
2. $f(x) = \log_5 x$
3. $f(x) = \log_{1/2} x$
4. $f(x) = \log_{1/4} x$
5. $f(x) = -2 \ln x$
6. $f(x) = -\log x$
7. $f(x) = \log_4 x^2$
8. $f(x) = 2 \log_5 x$
9. $f(x) = |\ln x|$
10. $f(x) = \ln|x|$
11. $f(x) = -|\ln x|$
12. $f(x) = -|\log_2 x^3|$
13. $f(x) = \log \sqrt[3]{x}$
14. $f(x) = \ln \sqrt{x}$
15. $f(x) = 3 + \log_2 x$
16. $f(x) = -2 + \log_4 x$
17. $f(x) = \log(x + 10)$
18. $f(x) = \ln(x + 3)$
19. $f(x) = \ln(x - 5)$
20. $f(x) = \log(x - 2)$
21. $f(x) = \log_5(x + 2)^2$
22. $f(x) = \log_5(x - 1)^2$
23. $f(x) = -\ln(x - 4)$
24. $f(x) = \ln(x + 3)^{-1}$
25. $f(x) = 4 - 2 \ln x$
26. $f(x) = x + \ln x$
27. $f(x) = \log(4 - x)$
28. $f(x) = -\ln(e - x)$

In Exercises 29 to 34, determine the domain of the given logarithmic function.

29. $J(x) = \dfrac{1}{\ln x}$
30. $K(x) = \log(x^2 + 4)$

31. $L(x) = \log(x^2 - 9)$

32. $f(x) = \dfrac{1}{\ln(4x + 10)}$

33. $h(x) = \log(x^2) + 4$

34. $Q(x) = \log(|x| + 1)$

In Exercises 35 to 40, find the range of the given logarithmic function.

35. $J(x) = \dfrac{1}{\ln x}$

36. $K(x) = \log(x^2 + 4)$

37. $L(x) = \log(x^2 - 9)$

38. $V(x) = |\ln x|$

39. $R(x) = \ln|x|$

40. $B(x) = 4 + \log(x^2)$

41. A furniture outlet finds that the number of sofas it can sell in a month is given by

$$S(d) = 15 + 10 \ln(0.01d + 1),$$

where $S(d)$ is the number of sofas sold and d is the dollar amount spent for advertising.

a. How many sofas will be sold with no advertising expenditure?

b. How many sofas (to the nearest unit) will be sold with an advertising expenditure of $5000?

c. How much advertising expenditure is necessary to sell 195 sofas?

42. What will an earthquake measure on the Richter scale if it has an intensity of $I = 100{,}000I_0$?

43. The Colombia earthquake of 1906 had an intensity of $I = 398{,}107{,}000I_0$. What did it measure on the Richter scale?

44. If an earthquake has an intensity 1000 times the intensity of a second earthquake, then how much larger is the Richter scale measure of the larger earthquake than that of the smaller?

Supplemental Exercises

45. Given $f(x) = \ln x$, evaluate

a. $f(e^3)$, b. $f(e^{\ln 4})$, c. $f(e^{3\ln 3})$.

46. Given $f(x) = \log_5 x$, evaluate

a. $f(5^2)$, b. $f(5^{\log_5 4})$, c. $f(5^{3\log_5 3})$.

47. Explain why the graph of $F(x) = \log_b x^2$ and the graph of $G(x) = 2\log_b x$ are not identical.

48. Explain why the graph of $F(x) = |\log_b x|$ and the graph of $G(x) = \log_b|x|$ are not identical.

49. Graph $f(x) = e^{-x}(\ln x)$ for $1 \le x \le e^2$.

50. Graph $g(x) = \log[\![x]\!]$ for $1 \le x \le 10$. Recall that $[\![x]\!]$ represents the greatest integer function.

In Exercises 51 to 56, determine the domain and the range of the given function.

51. $f(x) = \sqrt{\log x}$

52. $f(x) = \sqrt{\ln x^3}$

53. $f(x) = 100 - \ln\sqrt{1 - x^2}$

54. $f(x) = 10 + |\ln(x - e)|$

55. $f(x) = \log(\log x)$

56. $f(x) = |\ln(-\ln x)|$

57. The Coalinga, California, earthquake of May 2, 1983, had a Richter scale measure of 6.5. Find the Richter scale measure of an earthquake that has an intensity 200 times the intensity of the Coalinga quake.

58. The earthquake just south of Concepción, Chile, on May 22, 1960, had a Richter scale measure of 9.5. Find the Richter scale measure of an earthquake that has an intensity one-half the intensity of this quake.

7.4 Exponential and Logarithmic Equations

If a variable appears as an exponent in a term of an equation, then the equation is called an **exponential equation.** Example 1 uses the Equality of Exponents Theorem to solve exponential equations.

EXAMPLE 1 Solve Exponential Equations

Solve the following exponential equations: a. $5^{x-1} = 125$ b. $49^{2x} = \dfrac{1}{7}$

Solution Write each side of the equation as a power of the same base and then equate the exponents.

a. $5^{x-1} = 125$

$5^{x-1} = 5^3$ Write each side as a power of 5.

$x - 1 = 3$ Equate the exponents.

$x = 4$

b. $49^{2x} = \dfrac{1}{7}$

$(7^2)^{2x} = 7^{-1}$ Write each side as a power of 7.

$7^{4x} = 7^{-1}$

$4x = -1$ Equate the exponents.

$x = -\dfrac{1}{4}$

■ *Try Exercise* **2**, *page 288.*

Some exponential equations can be solved by taking the logarithm of both sides of the equation.

EXAMPLE 2 Use the Logarithm of Each Side Theorem to Solve an Equation

Solve the following exponential equations: a. $5^x = 40$ b. $3^{2x-1} = 5^{x+2}$

Solution Start by taking the logarithm of each side of the equation.

a. $5^x = 40$

$\log(5^x) = \log 40$

$x \log 5 = \log 40$

$x = \dfrac{\log 40}{\log 5}$ Exact solution

$x \approx \dfrac{1.60206}{0.69897} \approx 2.29203$ Decimal approximation

b. $3^{2x-1} = 5^{x+2}$

$\log 3^{2x-1} = \log 5^{x+2}$

$(2x - 1) \log 3 = (x + 2) \log 5$ Power property

$2x \log 3 - \log 3 = x \log 5 + 2 \log 5$ Distributive property

Collecting terms involving the variable x on the left side yields

$$2x \log 3 - x \log 5 = 2 \log 5 + \log 3$$

$$x(2 \log 3 - \log 5) = 2 \log 5 + \log 3$$

$$x = \frac{2 \log 5 + \log 3}{2 \log 3 - \log 5}$$

Using a calculator, a decimal approximation to the solution can be obtained:

$$x \approx \frac{1.87506}{0.25527} \approx 7.34540$$

■ *Try Exercise* **12**, *page 288.*

Logarithmic Equations

Equations that involve logarithms are called **logarithmic equations.** The properties of logarithms, along with the definition of a logarithm, are valuable aids to solving a logarithmic equation.

EXAMPLE 3 Solve a Logarithmic Equation

Solve the logarithmic equation $\log 2x - \log(x - 3) = 1$.

Solution

$$\log 2x - \log(x - 3) = 1$$

$$\log \frac{2x}{x-3} = 1 \qquad \text{Quotient property}$$

$$\frac{2x}{x-3} = 10^1 \qquad \text{Definition of logarithm}$$

$$2x = 10x - 30$$

$$-8x = -30$$

$$x = \frac{15}{4}$$

Check the solution by substituting 15/4 into the original equation.

■ *Try Exercise* **24**, *page 288.*

Example 4 uses the one-to-one property of logarithms to solve a logarithmic equation.

EXAMPLE 4 Solve a Logarithmic Equation

Solve the logarithmic equation $\ln(3x + 8) = \ln(2x + 2) + \ln(x - 2)$.

Solution

$$\ln(3x + 8) = \ln(2x + 2) + \ln(x - 2)$$

$$\ln(3x + 8) = \ln(2x + 2)(x - 2) \qquad \text{Product property}$$

$$\ln(3x + 8) = \ln(2x^2 - 2x - 4)$$

$$3x + 8 = 2x^2 - 2x - 4 \qquad \text{One-to-one property of logarithms}$$

$$0 = 2x^2 - 5x - 12$$

$$0 = (2x + 3)(x - 4)$$

Thus $-3/2$ and 4 are possible solutions. The number $-3/2$ does not check in the original equation. Why?[1] It can be shown that 4 checks and thus the only solution is $x = 4$.

■ *Try Exercise* **32**, *page 288.*

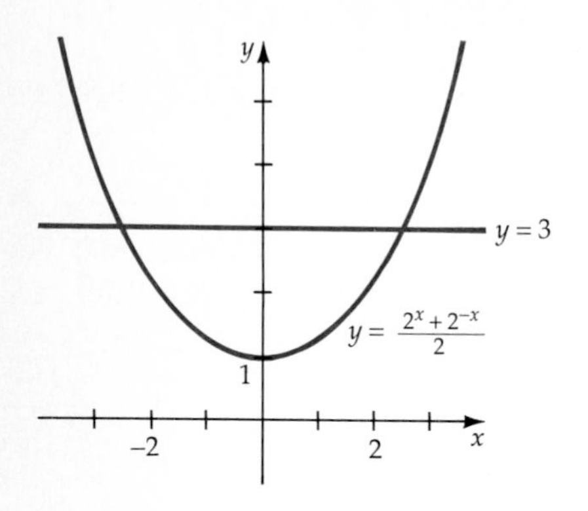

Figure 7.19
$y = \dfrac{2^x + 2^{-x}}{2}$

Equations That Involve $b^x \pm b^{-x}$

Recall from Section 1 that the graph of $f(x) = (2^x + 2^{-x})/2$ can be sketched by drawing a curve *halfway* between the graph of $y = 2^x$ and $y = 2^{-x}$ since $(2^x + 2^{-x})/2$ is the average of 2^x and 2^{-x}. The solutions of the equation $(2^x + 2^{-x})/2 = 3$ are represented by the x-coordinates of the points of intersection of the graph of $y = 3$ and the graph of $y = (2^x + 2^{-x})/2$ as shown in Figure 7.19.

Example 5 uses an algebraic method to solve $(2^x + 2^{-x})/2 = 3$.

EXAMPLE 5 Solve an Equation Involving $b^x + b^{-x}$

Solve the exponential equation $\dfrac{2^x + 2^{-x}}{2} = 3$.

Solution Multiplying each side by 2 produces

$$2^x + 2^{-x} = 6$$

$$2^{2x} + 2^0 = 6 \cdot (2^x) \qquad \text{Multiply by } 2^x \text{ to clear negative exponents.}$$

$$(2^x)^2 - 6(2^x) + 1 = 0 \qquad \text{Write in quadratic form.}$$

Substituting u for 2^x produces the quadratic equation

$$(u)^2 - 6(u) + 1 = 0.$$

By the quadratic formula,

$$u = \frac{6 \pm \sqrt{36 - 4}}{2} = \frac{6 \pm 4\sqrt{2}}{2} = 3 \pm 2\sqrt{2}.$$

Replacing u with 2^x produces

$$2^x = 3 \pm 2\sqrt{2}.$$

Now take the common logarithm of each side.

$$\log 2^x = \log(3 \pm 2\sqrt{2})$$

$$x \log 2 = \log(3 \pm 2\sqrt{2}) \qquad \text{Power property of logarithms}$$

$$x = \frac{\log(3 \pm 2\sqrt{2})}{\log 2} \approx \pm 2.54$$

■ *Try Exercise* **42**, *page 288.*

[1] If $x = -3/2$, the original equation becomes $\ln(7/2) = \ln(-1) + \ln(-7/2)$. This cannot be true because the function $f(x) = \ln x$ is not defined for negative values of x.

Remark If natural logarithms had been used in Example 5, then the exact solutions would be

$$x = \frac{\ln(3 \pm 2\sqrt{2})}{\ln 2}.$$

The pH of a Solution

Whether an aqueous solution is acidic or basic depends on its hydronium-ion concentration. Thus acidity is a function of hydronium-ion concentration. Since these hydronium-ion concentrations may be very small, it is convenient to measure acidity in terms of **pH**, which is defined as the negative of the common logarithm of the molar hydronium-ion concentration. As a mathematical formula, this is stated as

$$\text{pH} = -\log[H_3O^+].$$

EXAMPLE 6 Find the pH of a Solution

Find the pH of the following: a. Orange juice with $[H_3O^+] = 2.80 \times 10^{-4}$ M b. milk with $[H_3O^+] = 3.97 \times 10^{-7}$ M

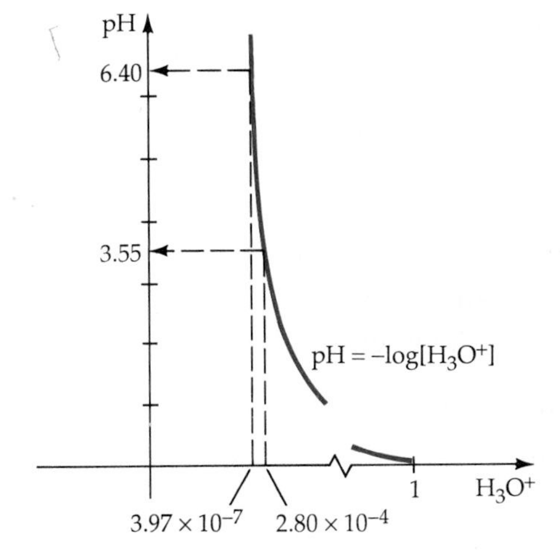

Figure 7.20

Solution

a. $\text{pH} = -\log[H_3O^+] = -\log(2.80 \times 10^{-4}) \approx -(-3.55) = 3.55$

The orange juice has a pH of 3.55 (to the nearest hundredth).

b. $\text{pH} = -\log[H_3O^+] = -\log(3.97 \times 10^{-7}) \approx -(-6.40) = 6.40$

The milk has a pH of 6.40 (to the nearest hundredth).

■ *Try Exercise* **50,** *page 288.*

Figure 7.20 shows how the pH function *maps* small positive numbers that are relatively close together on the hydronium-ion concentration axis into numbers that are farther apart on the pH axis. The pH of pure water is 7.0.

EXAMPLE 7 Solve an Application

Determine the hydronium-ion concentration of a sample of blood with pH = 7.41.

Solution Substitute 7.41 for the pH and solve for H_3O^+.

$$\text{pH} = -\log[H_3O^+]$$

$$7.41 = -\log[H_3O^+] \quad \text{Substitute 7.41 for pH.}$$

$$-7.41 = \log[H_3O^+] \quad \text{Multiply both sides by } -1.$$

$$10^{-7.41} = [H_3O^+] \quad \text{Definition of } y = \log_b x.$$

$$3.9 \times 10^{-8} \approx [H_3O^+]$$

The hydronium-ion concentration of the blood sample is 3.9×10^{-8} M.

■ *Try Exercise* **52,** *page 288.*

EXERCISE SET 7.4

In Exercises 1 to 40, solve for x.

1. $2^x = 64$
2. $3^x = 243$
3. $8^x = 512$
4. $25^x = 3125$
5. $49^x = \frac{1}{343}$
6. $9^x = \frac{1}{243}$
7. $2^{5x+3} = \frac{1}{8}$
8. $3^{4x-7} = \frac{1}{9}$
9. $\left(\frac{2}{5}\right)^x = \frac{8}{125}$
10. $\left(\frac{2}{5}\right)^x = \frac{25}{4}$
11. $5^x = 70$
12. $6^x = 50$
13. $3^{-x} = 120$
14. $7^{-x} = 63$
15. $\left(\frac{3}{5}\right)^x = 0.92$
16. $\left(\frac{7}{3}\right)^x = 22$
17. $10^{2x+3} = 315$
18. $10^{6-x} = 550$
19. $e^x = 10$
20. $e^{x+1} = 20$
21. $\left(1 + \frac{0.08}{12}\right)^{12x} = 1.5$
22. $\left(1 + \frac{0.05}{365}\right)^{365x} = 2$
23. $\log_2 x + \log_2(x - 4) = 2$
24. $\log_3 x + \log_3(x + 6) = 3$
25. $\log(5x - 1) = 2 + \log(x - 2)$
26. $1 + \log(3x - 1) = \log(2x + 1)$
27. $\log \sqrt{x^3 - 17} = \frac{1}{2}$
28. $\log(x^3) = (\log x)^2$
29. $\log(\log x) = 1$
30. $2 \ln \frac{e}{\sqrt{3}} = 3 - \ln x$
31. $\frac{1}{3} \ln 125 + \frac{1}{2} \ln x = \ln x$
32. $\ln x = \frac{1}{2} \ln\left(2x + \frac{5}{2}\right) + \frac{1}{2} \ln 2$
33. $\ln(e^{3x}) = 6$
34. $\log_b(b^{5x+2}) = 4$
35. $\ln x^2 = \ln 9$
36. $\log_x 9 = 3$
37. $\log_x 8 = 2$
38. $4 \log_x 2 - \frac{1}{2} \log_x 4 = 2 - \frac{1}{3} \log_x 8$
39. $e^{\ln(x-1)} = 4$
40. $10^{\log(2x+7)} = 8$

In Exercises 41 to 44, use common logarithms to solve for x.

41. $\frac{10^x - 10^{-x}}{2} = 20$
42. $\frac{10^x + 10^{-x}}{2} = 8$
43. $\frac{10^x + 10^{-x}}{10^x - 10^{-x}} = 5$
44. $\frac{10^x - 10^{-x}}{10^x + 10^{-x}} = \frac{1}{2}$

In Exercises 45 to 48, use natural logarithms to solve for x.

45. $\frac{e^x + e^{-x}}{2} = 15$
46. $\frac{e^x - e^{-x}}{2} = 15$
47. $\frac{1}{e^x - e^{-x}} = 4$
48. $\frac{e^x + e^{-x}}{e^x - e^{-x}} = 3$

49. Find the pH of a sample of lemon juice that has a hydronium-ion concentration of 6.3×10^{-3}.

50. An *acidic solution* has a pH of less than 7, whereas a *basic solution* has a pH of greater than 7. Household ammonia has an hydronium-ion concentration of 1.26×10^{-12}. Determine the pH of the ammonia, and state whether it is an acid or a base.

51. Find the hydronium-ion concentration of beer, which has a pH of 4.5.

52. Normal rain has a pH of 5.6. A recent acid rain had a pH of 3.1. Find the hydronium concentration of this rain.

53. The population P of a city grows exponentially according to the function

$$P(t) = 8500(1.1)^t, \quad 0 \le t \le 8$$

where t is measured in years.

   a. Find the population at time $t = 0$ and also at time $t = 2$.

   b. When, to the nearest year, will the population reach 15,000?

54. After a race, a runner's pulse rate R in beats per minute decreases according to the function $R(t) = 145e^{-0.092t}$, $(0 \le t \le 15)$ where t is measured in minutes.

   a. Find the runner's pulse rate at the end of the race and also 1 minute after the end of the race.

   b. How long, to the nearest minute, after the end of the race will the runner's pulse rate be 80 beats per minute?

55. A can of soda at 70°F is placed into a refrigerator that maintains a constant temperature of 36°F. The temperature T of the soda t minutes after it is placed into the refrigerator is

$$T(t) = 36 + 43e^{-0.058t}$$

   a. Find the temperature of the soda 10 minutes after it is placed into the refrigerator.

   b. When, to the nearest minute, will the temperature of the soda be 45°F?

56. During surgery, a patient's circulatory system requires at least 50 milligrams of an anesthetic. The amount of anesthetic present t hours after 80 milligrams of anesthetic are administered is

$$A(t) = 80(0.727)^t.$$

a. How much of the anesthetic is present in the patient's circulatory system 30 minutes after the anesthetic is administered?

b. How long, to the nearest minute can the operation last if the patient does not receive an additional dosage of anesthetic?

Supplemental Exercises

57. The following argument shows that $0.125 > 0.25$. Find the incorrect step.

$$3 > 2$$
$$3(\log 0.5) > 2(\log 0.5)$$
$$\log 0.5^3 > \log 0.5^2$$
$$0.5^3 > 0.5^2$$
$$0.125 > 0.25$$

58. The following argument shows that $4 = 6$. Find the incorrect step.

$$\begin{aligned} 4 &= \log_2 16 \\ &= \log_2(8 + 8) \\ &= \log_2 8 + \log_2 8 \\ &= 3 + 3 \\ &= 6 \end{aligned}$$

In Exercises 59 to 66, determine the *number* of solutions of the given equation. (*Hint:* Sketch the graph of the function on the left side of the equation and the graph of the function on the right side of the equation on the same coordinate axis. Then determine the *number* of intersections of their graphs to determine the *number* of solutions to the equation.)

59. $2^x = \log x$

60. $10^{-x} = \log x$

61. $e^x - 4 = \ln x$

62. $x = -\log x$

63. $\ln x = x^2 - 5$

64. $\log x = x^3$

65. $\dfrac{2^x + 2^{-x}}{2} - 2 = \log|x|$

66. $\dfrac{2^x + 2^{-x}}{2} = \dfrac{3^x + 3^{-x}}{2}$

67. A common mistake that students make is to write $\log(x + y)$ as $\log x + \log y$. For what values of x and y does $\log(x + y) = \log x + \log y$? (*Hint:* Solve for x in terms of y.)

68. Which is larger, 500^{501} or 506^{500}? (*Hint:* Let $x = 500^{501}$ and $y = 506^{500}$ and then compare $\ln x$ with $\ln y$.)

69. Explain why the functions $F(x) = (1.4)^x$ and $G(x) = e^{0.336x}$ essentially represent the same functions.

70. Find the constant k that will make $f(t) = (2.2)^t$ and $g(t) = e^{-kt}$ essentially represent the same function.

71. Solve $e^{1/x} > 2$. Write your answer using interval notation.

72. Solve $\log(x^2) > (\log x)^2$. Write your answer using interval notation.

7.5 Applications of Exponential and Logarithmic Functions

In many applications, a quantity N grows or decays according to the function $N(t) = N_0e^{kt}$. In this function, N is a function of time t and N_0 is the value of N at time $t = 0$. If k is a *positive* constant, then $N(t) = N_0e^{kt}$ is called an exponential **growth function.** If k is a *negative* constant, then $N(t) = N_0e^{kt}$ is called an exponential **decay function.** The following examples will give you an understanding of how growth and decay functions arise naturally in the investigation of certain phenomena.

Interest is money paid for the use of money. The interest I is called **simple interest** if it is a fixed percent r per time period t of the amount of money invested. The amount of money invested is called the **principal** P. Simple interest is computed using the formula $I = Prt$. For example, if \$1000 is invested at 12% for 3 years, the simple interest is

$$I = Prt = \$1000(0.12)(3) = \$360.$$

The balance after t years is $B = P + I = P + Prt$. In the previous example, the \$1000 invested for 3 years produced \$360 interest. Thus the balance after 3 years is \$1360.

Compound Interest

In many financial transactions, interest is added to the principal at regular intervals so that interest is paid on interest as well as the principal. Interest earned in this manner is called **compound interest.** For example, if \$1000 is invested at 12% annual interest compounded annually for 3 years, then the total interest after 3 years is

First-year interest	$\$1000(0.12) =$	\$120.00
Second-year interest	$\$1120(0.12) =$	\$134.40
Third-year interest	$\$1254.40(0.12) \approx$	\$150.53
		\$404.93 ←Total interest

This method of computing the balance can be tedious and time-consuming. A *compound interest formula* that can be used to determine the balance due after n periods of compounding can be developed as follows.

Note that if P dollars is invested at an interest rate of r per period, then the balance after one period is $B_1 = P + Pr = P(1 + r)$, where Pr represents the interest earned for the period. Observe that B_1 is the product of the original principal P and $(1 + r)$. If the amount B_1 is reinvested for another period, then the balance after the second period is

$$B_2 = (B_1)(1 + r) = P(1 + r)(1 + r) = P(1 + r)^2.$$

Successive reinvestments lead to the following results:

Number of periods	Balance
3	$B_3 = P(1 + r)^3$
4	$B_4 = P(1 + r)^4$
$\vdots$	$\vdots$
n	$B_n = P(1 + r)^n$

The equation $B_n = P(1 + r)^n$ is valid if r is the interest rate paid during each of the n compounding periods. If r is an annual interest rate and n is the number of compounding periods per year, then the interest rate each period is r/n and the number of compounding periods after t years is nt. Thus the compound interest formula is expressed as follows.

The Compound Interest Formula

A principal P invested at an annual interest rate r compounded n times per year for t years produces the balance

$$B = P\left(1 + \frac{r}{n}\right)^{nt}.$$

EXAMPLE 1 Solve an Application

Find the balance if \$1000 is invested at an annual interest rate of 10%, for 2 years compounded a. annually b. daily c. hourly

Solution

a. Use the compound interest formula, with $P = 1000$, $r = 0.1$, $t = 2$, and $n = 1$.

$$B = \$1000\left(1 + \frac{0.1}{1}\right)^{1\cdot 2} = \$1000(1.1)^2 = \$1210.00$$

b. Since there are 365 days in a year, use $n = 365$.

$$B = \$1000\left(1 + \frac{0.1}{365}\right)^{365\cdot 2} \approx \$1000(1.000273973)^{730} \approx \$1221.37$$

c. Since there are 8760 hours in a year, use $n = 8760$.

$$B = \$1000\left(1 + \frac{0.1}{8760}\right)^{8760\cdot 2} \approx \$1000(1.000011416)^{17520} \approx \$1221.41$$

■ *Try Exercise* **4**, *page 297.*

Remark As the number of compounding periods increases, the balance seems to approach some upper limit. Even if the interest is compounded each *second,* the balance to the nearest cent remains \$1221.40.

To **compound continuously** means to increase the number of compounding periods without bound. We can better understand the concept of compounding continuously if we evaluate the balance B in the compound interest formula $B = P(1 + r/n)^{nt}$ with a principal of \$1, a yearly interst rate of $100\% = 1$, for 1 year, as n increases without bound. That is, evaluate

$$\left(1 + \frac{1}{n}\right)^n \text{ as } n \to \infty .$$

Recall that in Section 1 e was defined as the number that $(1 + 1/n)^n$ approaches as n increases without bound. Thus we see that the number e plays an important role in continuous compounding.

To derive a continuous interest formula, substitute $1/m$ for r/n in the compound interest formula

$$B = P\left(1 + \frac{r}{n}\right)^{nt} \tag{1}$$

to produce

$$B = P\left(1 + \frac{1}{m}\right)^{nt} \tag{2}$$

This substitution is motivated by the desire to express $(1 + r/n)^n$ as $[(1 + 1/m)^m]^r$, which will approach e^r as m gets large without bound.

Solving the equation $1/m = r/n$ for n yields $n = mr$, and thus the exponent nt can be written as mrt. Therefore Equation (2) can be expressed as

$$B = P\left(1 + \frac{1}{m}\right)^{mrt} = P\left[\left(1 + \frac{1}{m}\right)^{m}\right]^{rt} \tag{3}$$

By the definition of e, we know that as m gets larger without bound,

$$\left(1 + \frac{1}{m}\right)^{m} \text{ approaches } e.$$

Thus, using continuous compounding, Equation (3) simplifies to $B = Pe^{rt}$.

Continuous Compounding Interest Formula

> If an account with principal P and annual interest rate r is compounded continuously for t years, then the balance is $B = Pe^{rt}$.

EXAMPLE 2 Compound Continuously

Find the balance after 4 years if $800 is invested at an annual rate of 6% compounded continuously.

Solution Use the continuous compounding formula.

$$B = Pe^{rt} = 800e^{0.06(4)} = 800e^{0.24}$$

$$\approx \$800(1.27124915) = \$1017.00 \quad \text{To nearest cent}$$

■ *Try Exercise* **6**, *page 297.*

Exponential Growth

Given any two points on the graph of a growth function $N(t) = N_0e^{kt}$, you can use the given data to solve for the constants N_0 and k.

EXAMPLE 3 Find the Exponential Growth Function That Models Given Data

Find the exponential growth function for a town whose population was 16,400 in 1970 and 20,200 in 1980.

Solution We need to determine N_0 and k in $N(t) = N_0e^{kt}$. If we represent the year 1970 by $t = 0$, then our given data are $N(0) = 16{,}400$ and $N(10) = 20{,}200$. Because N_0 is defined to be $N(0)$, we know $N_0 = 16{,}400$. To determine k, substitute $t = 10$, and $N_0 = 16{,}400$ in $N(t) = N_0e^{kt}$ to produce

$$N(10) = 16{,}400e^{k \cdot 10}$$

$$20{,}200 = 16{,}400e^{10k} \quad \text{Substitute 20,200 for } N(10).$$

$$\frac{20{,}200}{16{,}400} = e^{10k}$$

To solve this equation for k, take the natural logarithm of each side.

$$\ln\left(\frac{20{,}200}{16{,}400}\right) = \ln e^{10k}$$

$$\ln\left(\frac{20{,}200}{16{,}400}\right) = 10k \quad \text{Use } \log_b(b^p) = p.$$

$$\tfrac{1}{10}\ln\left(\frac{20{,}200}{16{,}400}\right) = k$$

$$0.0208 \approx k$$

The exponential growth function is $N(t) = 16{,}400e^{0.0208t}$

■ *Try Exercise* **10,** *page 297.*

EXAMPLE 4 Solve an Application

Use the exponential growth function from Example 3 to

a. estimate the population of the town in the year 1995 and
b. estimate when the population will be double its 1970 population.

Solution

a. Let $t = 25$.

$$N(t) = 16{,}400e^{0.0208t}$$

$$N(25) = 16{,}400e^{0.0208 \cdot 25}$$

$$\approx 27{,}600 \quad \text{To nearest 100}$$

b. We need to solve for t when $N(t) = 2 \cdot 16{,}400 = 32{,}800$.

$$32{,}800 = 16{,}400e^{0.0208t}$$

$$2 = e^{0.0208t} \quad \text{Divide each side by 16,400.}$$

$$\ln 2 = 0.0208t \quad \text{Take the natural logarithm of each side.}$$

$$\frac{\ln 2}{0.0208} = t$$

$$33 \approx t \quad \text{To nearest year.}$$

■ *Try Exercise* **12,** *page 297.*

Exponential Decay

Many radioactive materials *decrease* exponentially. This decrease, called radioactive decay, is measured in terms of **half-life,** which is defined as the time required for the disintegration of half the atoms in a sample of a radioactive substance. Following are the half-lives of selected radioactive isotopes:

Isotope	Half-Life
Carbon (C^{14})	5730 years
Radium (Ra^{226})	1660 years
Polonium (Po^{210})	138 days
Phosphorus (P^{32})	14 days
Polonium (Po^{214})	1/10,000th of a second

EXAMPLE 5 Find the Exponential Decay Function That Models Given Data

Find the exponential decay function for the amount of phosphorus (P^{32}) that remains in a sample after t days.

Solution When $t = 0$, $N(0) = N_0e^{k(0)} = N_0$. Thus, we know that $N(0) = N_0$. Also, because the phosphorus has a half-life of 14 days, $N(14) = 0.5N_0$. To find k, substitute $t = 14$ in $N(t) = N_0e^{kt}$ and solve for k.

$$N(14) = N_0 \cdot e^{k \cdot 14}$$

$$0.5N_0 = N_0e^{14k} \qquad \text{Substitute } 0.5N_0 \text{ for } N(14).$$

$$0.5 = e^{14k} \qquad \text{Divide each side by } N_0.$$

Taking the natural logarithm of each side produces

$$\ln 0.5 = \ln e^{14k}$$

$$\ln 0.5 = 14k$$

$$\frac{1}{14}\ln 0.5 = k$$

$$-0.0495 \approx k$$

The exponential decay function is $N(t) = N_0e^{-0.0495t}$.

■ *Try Exercise* **14**, *page 297.*

Remark Since $e^{-0.0495} \approx (0.5)^{1/14}$, the decay function $N(t) = N_0e^{-0.0495t}$ can also be written as $N(t) = N_0(0.5)^{t/14}$. In this form it is easy to see that if t is increased by 14, N will decrease by a factor of 0.5.

EXAMPLE 6 Solve an Application

Use $N(t) = N_0(0.5)^{t/14}$ to estimate the amount of phosphorus (P^{32}) that remains in a sample after 50 days.

Solution

$$N(t) = N_0(0.5)^{t/14}$$

$$N(50) = N_0(0.5)^{50/14} \approx 0.0841N_0$$

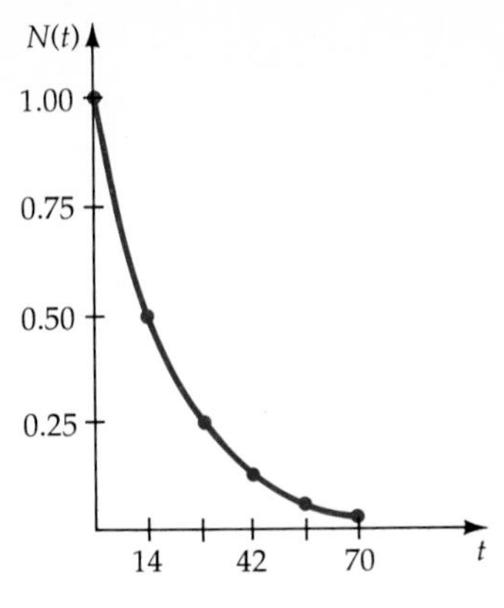

Figure 7.21
$N(t) = N_0(0.5)^{t/14}$

After 50 days, approximately 8% of the original phosphorus (P^{32}) remains.

■ *Try Exercise* **16**, *page 297.*

The following tables and Figure 7.21 show the percent of phosphorus(P^{32}) that remains after 0, 14, 28, 42, 56, and 70 days.

Days t	0	14	28	42	56	70
Percent (P^{32}) remaining $N(t)$	100	50	25	12.5	6.25	3.125

Carbon Dating

The bone tissue in all living animals contains both carbon-12, which is nonradioactive, and carbon-14, which is radioactive with a half-life of approximately 5730 years. As long as the animal is alive, the ratio of carbon-14 to carbon-12 remains constant. When the animal dies ($t = 0$), the carbon-14 begins to decay. Thus a bone that has a smaller ratio of carbon-14 to carbon-12 is older than a bone that has a larger ratio. The amount of carbon-14 present at time t is

$$N(t) = N_0(0.5)^{t/5730}$$

where N_0 is the amount of carbon-14 present in the bone at time $t = 0$.

EXAMPLE 7 Application

Determine the age of a bone if it now contains 85 percent of the carbon-14 it had when $t = 0$.

Solution Let t be the time at which $N(t) = 0.85N_0(0.5)^{t/5730}$.

$$0.85N_0 = N_0(0.5)^{t/5730}$$

$$0.85 = (0.5)^{t/5730} \qquad \text{Divide each side by } N_0.$$

Taking the natural logarithm of each side produces

$$\ln 0.85 = \ln (0.5)^{t/5730}$$

$$\ln 0.85 = \frac{t}{5730} \ln 0.5 \qquad \text{Power property}$$

$$\frac{\ln 0.85}{\ln 0.5} = \frac{t}{5730}$$

$$5730\frac{\ln 0.85}{\ln 0.5} = t$$

$$1340 \approx t \qquad \text{To nearest 10 years}$$

The bone is about 1340 years old.

■ *Try Exercise* **18**, *page 297.*

The Decibel Scale

The range of sound intensities that the human ear can detect is so large that a special *decibel scale* (named after the inventor of the telephone, Alexander Graham Bell) is used to measure and compare different sound intensities. Specifically, the *intensity level* N of sound measured in decibels is directly proportional to the *power* I of the sound measured in watts per square centimeter. That is,

$$N(I) = 10 \log\left(\frac{I}{I_0}\right)$$

where I_0 is the power of sound that is barely audible to the human ear. By international agreement, I_0 is the constant 10^{-16} watts per square centimeter.

EXAMPLE 8 Solve an Application

The power of normal conversation is 10^{-10} watts per square centimeter. What is the intensity level N in decibels of normal conversation?

Solution Evaluate $N(10^{-10})$.

$$N(I) = 10 \log\left(\frac{I}{10^{-16}}\right)$$

$$N(10^{-10}) = 10 \log\left(\frac{10^{-10}}{10^{-16}}\right) \qquad \text{Substitute } 10^{-10} \text{ for } I.$$

$$= 10 \log(10^6) \qquad \text{Since } \frac{10^{-10}}{10^{-16}} = 10^{-10-(-16)} = 10^6$$

$$= 10(6) = 60$$

The intensity level of normal conversation is 60 decibels.

■ *Try Exercise* **22**, *page 297.*

You might be tempted to conclude that three people each conversing at the same time, each with the power of normal conversation, would produce a combined intensity level of $3 \cdot 60$ decibels $= 180$ decibels; however, since the decibel scale is a logarithmic scale, this is not the case. In fact, three people conversing at the same time will produce a combined conversation with power of $3 \cdot (10^{-10})$ watts per square centimeter. To find the intensity level in decibels of their combined conversation, substitute $3 \cdot (10^{-10})$ for I in the intensity function.

$$N(I) = 10 \log\left(\frac{I}{10^{-16}}\right)$$

$$N(3 \cdot 10^{-10}) = 10 \log\left(\frac{3 \cdot 10^{-10}}{10^{-16}}\right) = 10 \log(3 \cdot 10^6) \approx 65$$

Thus the intensity of sound is not *multiplicative.* That is, although the power of the conversation of the three people is three times larger than the power of the conversation of a single person, the intensity of sound in decibels of the three people conversing (65 decibels) is not three times larger than the decibel rating of a single normal conversation (60 decibels).

EXERCISE SET 7.5

1. If \$8000 is invested at an annual interest rate of 5 percent and compounded annually, find the balance after **a.** 4 years, **b.** 7 years.
2. If \$22,000 is invested at an annual interest rate of 4.5 percent and compounded annually, find the balance after **a.** 2 years, **b.** 10 years.
3. If \$38,000 is invested at an annual interest rate of 6.5 percent for 4 years, find the balance if the interest is compounded **a.** annually, **b.** daily, **c.** hourly.
4. If \$12,500 is invested at an annual interest rate of 8 percent for 10 years, find the balance if the interest is compounded **a.** annually, **b.** daily, **c.** hourly.
5. Find the balance if \$15,000 is invested at an annual rate of 10 percent for 5 years, compounded continuously.
6. Find the balance if \$32,000 is invested at an annual rate of 8 percent for 3 years, compounded continuously.
7. The number of bacteria $N(t)$ present in a culture at time t hours is given by $N(t) = 2200(2)^t$. Find the number of bacteria present when **a.** $t = 0$ hours, **b.** $t =$ 3 hours.
8. The population of a town grows exponentially according to the function

$$f(t) = 12{,}400(1.14)^t \quad \text{for } 0 \leq t \leq 5$$

Find the population of the town when t is **a.** 3 years, **b.** 4.25 years.

9. Find the growth function for a town whose population was 22,600 in 1980 and 24,200 in 1985. Use $t = 0$ to represent the year 1980.
10. Find the growth function for a town whose population was 53,700 in 1982 and 58,100 in 1988. Use $t = 0$ to represent the year 1982.
11. The function $P(t) = 9700(e^{0.08t})$ estimates the population of a city at time t years after 1985.
    - **a.** Estimate the population of the city in 1993.
    - **b.** Estimate the year the population will be double its 1985 population.
12. The function $P(t) = 15{,}600(e^{0.09t})$ estimates the population of a city at time t years after 1984.
    - **a.** Estimate the population of the city in 1994.
    - **b.** Estimate the year the population will be double its 1984 population.
13. Radium (Ra^{226}) has a half-life of 1660 years. Find the decay function for the percent of radium (Ra^{226}) that remains in a sample after t years.
14. Polonium (Po^{210}) has a half-life of 138 days. Find the decay function for the percent of polonium (Po^{210}) that remains in a sample after t days.
15. Use $N(t) = N_0(0.5)^{t/1660}$ to estimate the percentage of radium (Ra^{226}) that remains in a sample after 2250 years.
16. Use $N(t) = N_0(0.5)^{t/138}$ to estimate the percentage of polonium (Po^{210}) that remains in a sample after 2 years.
17. Determine the age of a bone if it now contains 77 percent of its original amount of carbon-14.
18. Determine the age of a bone if it now contains 65 percent of its original amount of carbon-14.
19. Newton's Law of Cooling states that if an object at temperature T_0 is placed into an environment at constant temperature A, then the temperature of the object will be $T(t)$ after t minutes according to the function given by $T(t) = A + (T_0 - A)e^{-kt}$, where k is a constant that depends on the object.
    - **a.** Determine the constant k (to the nearest thousandth) for a canned soda drink that takes 5 minutes to cool from 75°F to 65°F after being placed in a refrigerator that maintains a constant temperature of 34°F.
    - **b.** What will be the temperature (to the nearest degree) of the soda drink after 30 minutes?
    - **c.** When (to the nearest minute) will the temperature of the soda drink be 36°F?
    - **d.** When will the temperature of the soda drink be exactly 34°F?
20. Solve the sound intensity equation $N = 10 \log\left(\frac{I}{I_0}\right)$ for I.
21. How much more powerful is a sound that measures 120 decibels than a sound (at the same frequency) that measures 110 decibels?
22. The power of a band is 3.4×10^{-5} watts per square centimeter. What is the band's intensity level N in decibels?
23. If the power of a sound is doubled, what is the increase in the intensity level. (*Hint:* Find $N(2I) - N(I)$.)
24. According to a software company, the users of their typing tutorial can expect to type $N(t)$ words per minute

after t hours of practice with their product according to the function $N(t) = 100(1.04 - 0.99^t)$.

a. How many words per minute can a student expect to type after 2 hours of practice?

b. How many words per minute can a student expect to type after 40 hours of practice?

c. According to the function N, how many hours (to the nearest 1 hour) of practice will be required before a student can expect to type 60 words per minute?

25. In the city of Whispering Palms, the number of people $P(t)$ exposed to a rumor in t hours is given by the function $P(t) = 80{,}000(1 - e^{-0.0005t})$.

a. Find the number of hours until 10 percent of the population have heard the rumor.

b. Find the number of hours until 50 percent of the population have heard the rumor.

26. A lawyer has determined that the number of people $P(t)$ who have been exposed to a news item after t days is given by the function $P(t) = 1{,}200{,}000(1 - e^{-0.03t})$.

a. Find the number of days after a major crime has been reported that 40 percent of the population have heard of the crime.

b. A defense lawyer knows that it will be very difficult to pick an unbiased jury after 80 percent of the population have heard of the crime. How many days are required until 80 percent of the population will have heard of the crime?

27. An automobile depreciates according to the function $V(t) = V_0(1 - r)^t$, where $V(t)$ is the value in dollars after t years, V_0 is the original value, and r is the yearly depreciation rate. If a car has a yearly depreciation rate of 20 percent, determine in how many years the car will depreciate to half its original value.

28. The current $I(t)$ (measured in amperes) of a circuit is given by the function $I(t) = 6(1 - e^{-2.5t})$, where t is the number of seconds after the switch is closed.

a. Find the current when $t = 0$.

b. Find the current when $t = 0.5$.

c. Solve the equation for t.

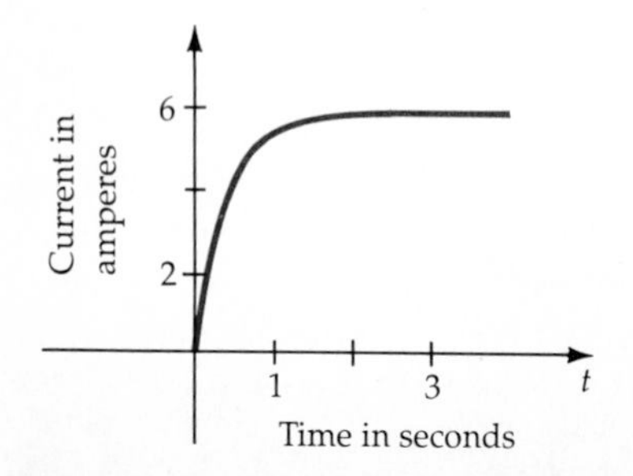

Supplemental Exercises

29. The Prime Number Theorem states that the number of prime numbers $P(n)$ less than a number n can be approximated by the function

$$P(n) = \frac{n}{\ln n}.$$

a. The actual number of prime numbers less than 100 is 25. Compute $P(100)$ and $P(100)/25$.

b. The actual number of prime numbers less than 10,000 is 1229. Compute $P(10{,}000)$ and $P(10{,}000)/1229$.

c. The actual number of prime numbers less than 1,000,000 is 78,498. Compute $P(1{,}000{,}000)$, and then compute the ratio $P(1{,}000{,}000)/78{,}498$.

30. The number $n!$ (read "n factorial") is defined as

$$n! = n(n - 1)(n - 2) \cdots 1$$

for all positive integers n. Thus, $4! = 4 \cdot 3 \cdot 2 \cdot 1 = 12$. *Stirling's Formula* (after James Stirling, 1692–1770)

$$n! \approx \left(\frac{n}{e}\right)^n \sqrt{2\pi n}$$

is often used to approximate very large factorials. Use Stirling's Formula to approximate $10!$, and then compute the ratio of Stirling's approximation of $10!$ divided by the actual value of $10!$, which is 3,628,800.

31. A population that grows or decays according to the function $P(t) = P_0e^{kt}$ is called a **Malthusian model.** This formula models the growth or decay of many populations with unlimited resources. However, if such factors as limited food supply and other limited resources affect the growth of the population, then it may be necessary to model the population growth using the following function, which is known as the **logistic law:**

$$P(t) = \frac{mP_0}{P_0 + (m - P_0)e^{-kt}}$$

where m is the maximum possible population and k is a positive constant. Notice that the graph of P as shown in the figure starts out with an exponential type of growth

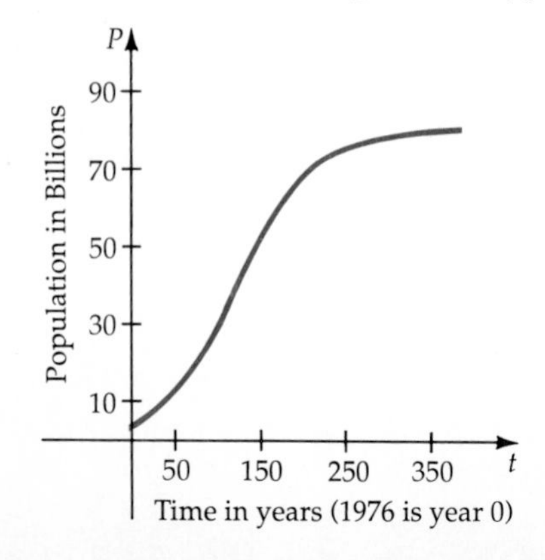

but eventually levels off and approaches the graph of $y = m$ asymptotically. Assume that the world's population growth satisfies the logistics law with $m = 80$ billion. If P was 4 billion in 1976 (think of this as the year $t = 0$) and 5 billion in 1986 ($t = 10$), find the

a. constant k;

b. world's predicted population (according to the logistic law) for the year 2000, and

c. world's predicted population for the year 3000.

32. The world's population reached 3 billion in 1961. How does this compare with the result obtained by using the logistic law and the value of k obtained in Exercise 31?

33. The population of squirrels in a nature reserve satisfies the logistics law, with $P_0 = 1500$, $k = 0.29$, and $P(2) = 2500$.

a. What is the maximum number of squirrels (to the nearest thousand) that the reserve can support? (*Hint:* compute m.)

b. Find the number of squirrels when $t = 10$.

34. The population of walruses in a colony satisfies the logistics law, with $P_0 = 800$, $P(1) = 900$, and $k = 0.14$.

a. What is the maximum number of walruses (to the nearest hundred) that the colony can support? (*Hint:* compute m.)

b. Find the number of walruses when $t = 5$ years.

35. Solve the logistic law in Exercise 31 for t.

36. A farmer knows that planting the same crop in the same field year after year reduces the yield. If the yield on each succeeding years crop is 90 percent of the preceding years yield, then the yield $Y(t)$ at any time t is given by the function $Y(t) = Y_0(0.90)^t$, where Y_0 is the yield when $t = 0$. In how many years (to the nearest year) will the yield be 60 percent of Y_0?

37. Crude oil leaks from a tank at a rate that depends on the amount of oil that remains in the tank. Since 1/8 of the oil in the tank leaks out every 2 hours, the volume of oil $V(t)$ in the tank at time t is given by the function $V(t) = V_0(0.875)^{t/2}$, where $V_0 = 350{,}000$ gallons is the number of gallons in the tank at the time the tank started to leak ($t = 0$).

a. How many gallons does the tank hold after 3 hours?

b. How many gallons does the tank hold after 5 hours?

c. How long will it take until 90 percent of the oil has leaked from the tank?

38. How many times stronger is an earthquake that measures 6 on the Richter scale than one that measures 3 on the Richter scale?

39. How many times stronger was the Chile earthquake of 1960, which measured 9.5 on the Richter scale, than the San Francisco earthquake of 1906, which measured 8.3 on the Richter scale?

40. How many years will it take the price of goods to double if the annual rate of inflation is 5 percent per year? Use continuous compounding.

41. The current rate of inflation will cause the price of goods to double in the next 10 years. Determine the current rate of inflation. Use continuous compounding.

42. The height h in feet of any point P on the cable shown is a function of the horizontal distance in feet from point P to the origin given by the function

$$h(x) = \frac{20}{2}(e^{x/20} + e^{-x/20}) \quad -40 \le x \le 40.$$

a. What is the height of the cable at point P if P is directly above the origin?

b. What is the height of the cable at point P if P is 25 feet to the right of the origin?

c. How far to the right or left of the origin is the cable 30 feet in height? (*Hint:* Use the method developed in Example 5, Section 4.)

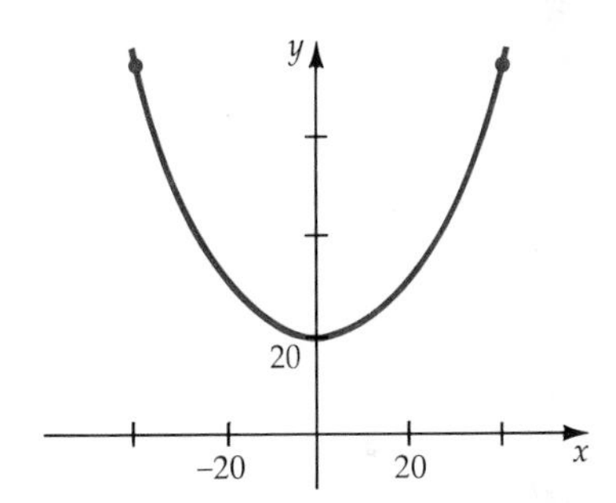

43. Logarithms and a function called the integer function (denoted by *INT*) can be used to determine the number of digits in a number written in exponential notation. The *INT* function is illustrated in the following examples:

$$INT(8.75) = 8 \qquad INT(102.003) = 102$$
$$INT(55) = 55 \qquad INT(e) = 2$$

Note that the *INT* function removes the decimal portion of a real number and returns the integer part of the real number as its output. The number of digits $N(x)$ in the number b^x, with $0 < b < 10$ and x a positive integer, is given by the function $N(x) = INT(x \log b) + 1$.

a. Find the number of digits in 3^{200}.

b. Find the number of digits in 7^{4005}.

c. The largest known prime number in 1980 was the number $2^{44{,}497} - 1$. Find the number of digits in this prime number.

d. The largest known prime number as of 1983 was $2^{132{,}049} - 1$. Find the number of digits in this prime number.

Chapter 7 Review

7.1 Exponential Functions and Their Graphs

For all positive real numbers b ($b \neq 1$), the exponential function $f(x) = b^x$ has the following properties:

- f has the set of real numbers as its domain.
- f has the set of positive real numbers as its range.
- f has a graph with a y-intercept of $(0, 1)$.
- f has a graph asymptotic to the x-axis.
- f is a one-to-one function.

The exponential function $f(x) = b^x$ is

- an increasing function if $b > 1$.
- a decreasing function if $0 < b < 1$.

As n increases without bound, $(1 + 1/n)^n$ approaches an irrational number denoted by e. The value of e accurate to eight decimal places is 2.71828183. The function $f(x) = e^x$ is called the natural exponential function.

7.2 Logarithms and Logarithmic Properties

Definition of a Logarithm
If $x > 0$ and b is a positive constant ($b \neq 1$), then

$$y = \log_b x \quad \text{if and only if} \quad b^y = x.$$

In the equation $y = \log_b x$, y is referred to as the logarithm, b is the base, and x is the argument.

The properties of logarithms are used to simplify logarithmic expressions and to solve both exponential and logarithmic equations. Logarithms with a base of 10 are called common logarithms. It is customary to write $\log_{10} x$ as $\log x$. Logarithms with a base of e are called natural logarithms. It is customary to write $\log_e x$ as $\ln x$.

Change-of-Base Formula If $x > 0$, $a > 0$, $b > 0$, and neither a nor b equals 1, then

$$\log_b x = \frac{\log_a x}{\log_a b}.$$

7.3 Logarithmic Functions and Their Graphs

For all positive real numbers b, $b \neq 1$, the function $f(x) = \log_b x$ has the following properties:

- f has the set of positive real numbers as its domain.
- f has the set of real numbers as its range.
- f has a graph with an x-intercept of $(1, 0)$.

- f has a graph asymptotic to the y-axis.
- f is a one-to-one function.

The logarithmic function $f(x) = \log_b x$ is:

- an increasing function if $b > 1$.
- a decreasing function if $0 < b < 1$.

7.4 Exponential and Logarithmic Equations

To solve exponential equations and logarithmic equations, we use the properties of exponents, the properties of logarithms, the Equality of Exponents Theorem, and the one-to-one property of logarithms.

Equality of Exponents Theorem If b is a positive real number ($b \neq 1$) such that $b^x = b^y$, then $x = y$.

If M, N, and b are positive real numbers with $b \neq 1$, then $\log_b M = \log_b N$ if and only if $M = N$.

7.5 Applications of Exponential and Logarithmic Functions

The function $N(t) = N_0 e^{kt}$ is called an exponential growth function if k is positive, and it is called an exponential decay function if k is negative.

The Compound Interest Formula
A principal P invested at an annual interest rate r compounded n times per year, for t years produces the balance

$$B = P\left(1 + \frac{r}{n}\right)^{nt}.$$

Continuous Compounding Interest Formula
If an account with principal P and annual interest rate r is compounded continuously for t years, then the balance is $B = Pe^{rt}$.

CHALLENGE EXERCISES

In Exercises 1 to 14, answer true or false. If the statement is false, give an example.

1. If $7^x = 40$, then $\log_7 40 = x$.
2. If $\log_4 x = 3.1$, then $4^{3.1} = x$.
3. If $f(x) = \log x$ and $g(x) = 10^x$, then $f[g(x)] = x$ for all real numbers x.
4. If $f(x) = \log x$ and $g(x) = 10^x$, then $g[f(x)] = x$ for all real numbers x.
5. The exponential function $h(x) = b^x$ is an increasing function.
6. The logarithmic function $j(x) = \log_b x$ is an increasing function.
7. The exponential function $h(x) = b^x$ is a one-to-one function.
8. The logarithmic function $j(x) = \log_b x$ is a one-to-one function.

9. The graph of

$$f(x) = \frac{2^x + 2^{-x}}{2}$$

is symmetric with respect to the y-axis.

10. The graph of

$$f(x) = \frac{2^x - 2^{-x}}{2}$$

is symmetric with respect to the origin.

11. If $x > 0$ and $y > 0$, $\log(x + y) = \log x + \log y$.

12. If $x > 0$, $\log x^2 = 2 \log x$.

13. If M and N are positive real numbers, then

$$\ln\left(\frac{M}{N}\right) = \ln M - \ln N.$$

14. For all $p > 0$, $e^{\ln p} = p$.

REVIEW EXERCISES

In Exercises 1 to 12, solve each equation. Do not use a calculator.

1. $\log_5 25 = x$ **2.** $\log_3 81 = x$

3. $\ln e^3 = x$ **4.** $\ln e^{\pi} = x$

5. $3^{2x+7} = 27$ **6.** $5^{x-4} = 625$

7. $2^x = \dfrac{1}{8}$ **8.** $27(3^x) = 3^{-1}$

9. $\log x^2 = 6$ **10.** $\dfrac{1}{2}\log|x| = 5$

11. $10^{\log 2x} = 14$ **12.** $e^{\ln x^2} = 64$

In Exercises 13 to 18, use a calculator to evaluate each power. Give your answers accurate to six significant digits.

13. $7^{\sqrt{2}}$ **14.** $3^{\sqrt{5}}$

15. $e^{1.7}$ **16.** $e^{-2.2}$

17. $10^{1.135}$ **18.** $10^{-\sqrt{10}}$

In Exercises 19 to 32, sketch the graph of each function.

19. $f(x) = (2.5)^x$ **20.** $f(x) = \left(\dfrac{1}{4}\right)^x$

21. $f(x) = 3^{|x|}$ **22.** $f(x) = 4^{-|x|}$

23. $f(x) = 2^x - 3$ **24.** $f(x) = 2^{(x-3)}$

25. $f(x) = \dfrac{4^x + 4^{-x}}{2}$ **26.** $f(x) = \dfrac{3^x - 3^{-x}}{2}$

27. $f(x) = \dfrac{1}{3}\log x$ **28.** $f(x) = 3 \log x^{1/3}$

29. $f(x) = -x + \log x$ **30.** $f(x) = 2^{-x}\log x$

31. $f(x) = -\dfrac{1}{2}\ln x$ **32.** $f(x) = -\ln|x|$

In Exercises 33 to 36, change each logarithmic equation to its exponential form.

33. $\log_4 64 = 3$ **34.** $\log_{1/2} 8 = -3$

35. $\log_{\sqrt{2}} 4 = 4$ **36.** $\ln 1 = 0$

In Exercises 37 to 40, change each exponential equation to its logarithmic form.

37. $5^3 = 125$ **38.** $2^{10} = 1024$

39. $10^0 = 1$ **40.** $8^{1/2} = 2\sqrt{2}$

In Exercises 41 to 44, write the given logarithm in terms of logarithms of x, y, and z.

41. $\log_b \dfrac{x^2y^3}{z}$ **42.** $\log_b \dfrac{\sqrt{x}}{y^2z}$

43. $\ln xy^3$ **44.** $\ln \dfrac{\sqrt{xy}}{z^4}$

In Exercises 45 to 48, write each logarithmic expression as a single logarithm.

45. $2\log x + \dfrac{1}{3}\log(x + 1)$ **46.** $5 \log x - 2\log(x + 5)$

47. $\dfrac{1}{2}\ln 2xy - 3 \ln z$ **48.** $\ln x - (\ln y - \ln z)$

In Exercises 49 to 52, use the change-of-base formula and a calculator to approximate each logarithm accurate to six significant digits.

49. $\log_5 101$ **50.** $\log_3 40$

51. $\log_4 0.85$ **52.** $\log_8 0.3$

In Exercises 53 to 56, use a calculator to approximate N to three significant digits.

53. $\log N = 247$ **54.** $\log N = -0.48$

55. $\ln N = 51$ **56.** $\ln N = -0.09$

In Exercises 57 to 72, solve each equation for x. Give exact answers. Do not use a calculator.

57. $4^x = 30$ **58.** $5^{x+1} = 41$

59. $\ln 3x - \ln(x - 1) = \ln 4$ **60.** $\ln 3x + \ln 2 = 1$

61. $e^{\ln(x+2)} = 6$

62. $10^{\log(2x+1)} = 31$

63. $\dfrac{4^x + 4^{-x}}{4^x - 4^{-x}} = 2$

64. $\dfrac{5^x + 5^{-x}}{2} = 8$

65. $\log(\log x) = 3$

66. $\ln(\ln x) = 2$

67. $\log\sqrt{x - 5} = 3$

68. $\log x + \log(x - 15) = 1$

69. $\log_4(\log_3 x) = 1$

70. $\log_7(\log_5 x^2) = 0$

71. $\log_5 x^3 = \log_5 16x$

72. $25 = 16^{\log_4 x}$

73. Find the pH of tomatoes that have a hydronium-ion concentration of 6.28×10^{-5}.

74. Find the hydronium-ion concentration of rainwater that has a pH of 5.4.

75. Find the balance if $16,000 is invested at an annual rate of 8 percent for 3 years if the interest is compounded **a.** monthly, **b.** continuously.

76. Find the balance if $19,000 is invested at an annual rate of 6 percent for 5 years if the interest is compounded **a.** daily, **b.** continuously.

77. The scrap value S of a product with an expected life span of n years is given by $S(n) = P(1 - r)^n$, where P is the original purchase price of the product and r is the annual rate of depreciation. A taxicab is purchased for $12,400 and is expected to last 3 years. What is its scrap value if it depreciates at a rate of 29 percent per year?

78. A skin wound heals according to the function given by $N(t) = N_0e^{-0.12t}$, where N is the number of square centimeters of unhealed skin t days after the injury, and N_0 is the number of square centimeters covered by the original wound.

a. What percentage of the wound will be healed after 10 days?

b. How many days will it take for 50 percent of the wound to heal?

c. How long will it take for 90 percent of the wound to heal?

In Exercises 79 to 82, find the exponential growth/decay function $N(t) = N_0e^{kt}$ that satisfies the given conditions.

79. $N(0) = 1, \quad N(2) = 5$

80. $N(0) = 2, \quad N(3) = 11$

81. $N(1) = 4, \quad N(5) = 5$

82. $N(-1) = 2, \quad N(0) = 1$

Appendix I

Using Tables

Logarithmic Tables

If a calculator is not available, then the values of logarithmic functions can be found by using the tables in this book.

Let x be a number written in scientific notation. That is, let $x = a \cdot 10^k$ where $1 \le a < 10$ and k is an integer. Applying some properties of logarithms, we have

$$\begin{aligned} \log x &= \log(a \cdot 10^k) \\ &= \log a + \log 10^k \quad \text{The Product Property} \\ &= \log a + k \qquad \log_b b^p = p. \end{aligned}$$

TABLE A1 Portion of a Common Logarithm Table

x	0	1	2	3
2.0	.3010	.3032	.3054	.3075
2.1	.3222	.3243	.3263	.3284
2.2	.3424	.3444	.3464	.3483
2.3	.3617	.3636	.3655	.3674
2.4	.3802	.3820	.3838	.3856
2.5	.3979	.3997	.4014	.4031
2.6	.4150	.4166	.4183	.4200

This last equation states that given any real number x, the real number $\log x$ is the sum of the logarithm of a number between one and ten, and an integer k. The real number $\log a$ is called the **mantissa** of $\log x$ and k is called the **characteristic** of $\log x$.

The Table of Common Logarithms in Appendix II gives the common logarithms accurate to 4 decimal places of numbers between 1.00 and 9.99 in increments of 0.01. For example, from the section of the Table of Common Logarithms at the left, $\log 2.43 \approx 0.3856$.

EXAMPLE 1 Evaluate Logarithms

Find a. $\log 54{,}600$ b. $\log 54.6$ c. $\log 0.00546$

Solution

a. $$\begin{aligned} \log 54{,}600 &= \log(5.46 \times 10^4) && \text{Write 54,600 in scientific notation.} \\ &= \log 5.46 + \log 10^4 && \text{The Product Property} \\ &\approx 0.7372 + 4 = 4.7372 && \log 10^4 = 4 \end{aligned}$$

b. $$\begin{aligned} \log 54.6 &= \log(5.46 \times 10^1) \\ &= \log 5.46 + \log 10^1 \\ &\approx 0.7372 + 1 = 1.7372 \end{aligned}$$

c. $$\begin{aligned} \log 0.00546 &= \log(5.46 \times 10^{-3}) \\ &= \log 5.46 + \log 10^{-3} \\ &\approx 0.7372 + (-3) \end{aligned}$$

In Example 1c, adding the characteristic to the mantissa would result in a negative logarithm. This form of a logarithm is inconvenient to use with logarithmic tables because these tables contain only positive mantissas. The value of log 0.00546 can be written in many forms.

For example, $\log 0.00546 \approx 0.7372 + (-3)$

$$\log 0.00546 \approx 0.7372 + (7 - 10) = 7.7372 - 10$$

$$\log 0.00546 \approx 0.7372 + (1 - 4) = 1.7372 - 4$$

Of these forms, $\log 0.00546 = 7.7372 - 10$ is the most common.

Caution When a logarithm has a negative characteristic like $0.7372 + (-3)$, sometimes $0.7372 + (-3)$ is incorrectly written as -3.7372. The correct result is -2.2628.

As a final note, when using a calculator, log 0.00546 will be displayed as -2.26280736. By adding and subtracting 10 to this number, we can write the number with a positive mantissa.

$$\begin{aligned}\log 0.00546 &\approx -2.26280736\\ &= (-2.26280736 + 10) - 10\\ &= 7.73719264 - 10\end{aligned}$$

Adding and subtracting 10 is a somewhat arbitrary choice. Any integer that will produce a positive mantissa could be used. For example,

$$\begin{aligned}\log 0.00546 &\approx -2.26280736\\ &= (-2.26280736 + 3) - 3\\ &= 0.73719264 - 3\end{aligned}$$

We now examine the process of finding x given $\log x$. For example, given $\log x = 0.3522$, locate 0.3522 in the *body* of the Table of Common Logarithms. Find the number that corresponds to this mantissa. The number is 2.25. Thus $\log 2.25 \approx 0.3522$ and 2.25 is called the antilogarithm of 0.3522. We write antilog $2.25 \approx 0.3522$.

TABLE A2 Portion of a Common Logarithm Table

x	3	4	5	6
2.0	.3075	.3096	.3118	.3139
2.1	.3284	.3304	.3324	.3345
2.2	.3483	.3502	.3522	.3541
2.3	.3674	.3692	.3711	.3729
2.4	.3856	.3874	.3892	.3909
2.5	.4031	.4048	.4065	.4082
2.6	.4200	.4216	.4232	.4249

EXAMPLE 2 Evaluate Antilogarithms

Find a. antilog 3.4639 b. antilog -1.4881

Solution

a. antilog $3.4639 = \text{antilog}(0.4639 + 3)$ — Write the number as the sum of the mantissa and the characteristic.

$\approx 2.91 \times 10^3$ — Use the table to find the number that corresponds to the mantissa. The characteristic is the exponent on 10.

$= 2910$

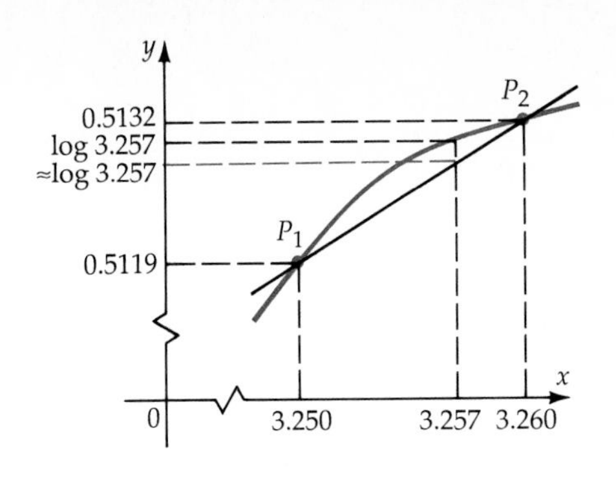

Figure A.1

b. The first step is to write the antilogarithm with a positive mantissa.

$$\begin{aligned}\text{antilog } -1.4881 &= \text{antilog}(-1.4881 + 2 - 2)\\ &= \text{antilog}(0.5119 - 2)\\ &\approx 3.25 \times 10^{-2} = 0.0325\end{aligned}$$

Linear Interpolation

Many functions can be approximated, over small intervals, by a straight line. Using this technique, we can approximate logarithms and antilogarithms of numbers that are not given in the Table of Common Logarithms.

A portion of the graph of $y = \log x$ is shown in Figure A.1. The points $P_1(3.250, 0.5119)$ and $P_2(3.260, 0.5132)$ are two points on the graph of $y = \log x$. The value of $\log 3.257$ is on the curve. An approximation for this value is on the straight line connecting P_1 and P_2. Because the slopes are the same between any two points on a straight line, we can write

$$\frac{y_2 - y_1}{x_2 - x_1} = \frac{y - y_1}{x - x_1}.$$

Substituting the coordinates of P_1, P_2, and using $x = 3.257$, we have

$$\frac{0.5132 - 0.5119}{3.260 - 3.250} = \frac{y - 0.5119}{3.257 - 3.250}.$$

Solving for y, we have

$$\begin{aligned}\frac{0.0013}{0.01} &= \frac{y - 0.5199}{0.007}\\ 0.00091 &= y - 0.5119\\ 0.51281 &= y\end{aligned}$$

$\log 3.257 \approx 0.5128$ rounded to the nearest ten thousandth.

A more convenient method of using linear interpolation is illustrated in the following example.

EXAMPLE 3 Evaluate Logarithms Using Linear Interpolation

Find log 0.02903.

Solution Arrange the numbers and their corresponding mantissas in a table. Indicate the differences between the numbers and the mantissas as shown using d to represent the unknown difference between the mantissa of 2.900 and 2.910.

$$\log 0.02903 = \log(2.903 \times 10^{-2}) = \log 2.903 + \log 10^{-2}$$

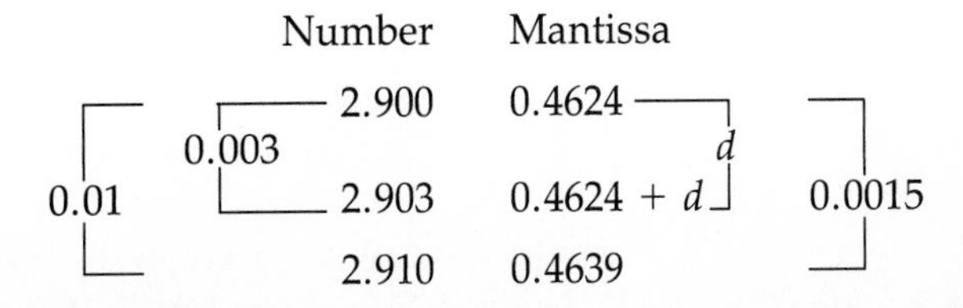

		Number	Mantissa		
0.01	0.003	2.900	0.4624	d	0.0015
		2.903	$0.4624 + d$		
		2.910	0.4639		

Write a proportion and solve for d.

$$\frac{0.003}{0.01} = \frac{d}{0.0015}$$

$$0.0005 = d$$

Add $d = 0.0005$ to the mantissa 0.4624.

$$\begin{aligned}\log 0.02903 &= \log 2.903 + \log 10^{-2}\\ &\approx (0.4624 + 0.0005) + (-2)\\ &= 0.4629 + (-2) = 8.4629 - 10\end{aligned}$$

Therefore $\log 0.02903 \approx 8.4629 - 10$.

Linear interpolation can also be used to find an antilogarithm.

EXAMPLE 4 Evaluate Antilogarithms Using Linear Interpolation

Find antilog(9.8465 − 10).

Solution

		Mantissa	Antilog		
0.0007	0.0002	0.8463	7.02	d	0.01
		0.8465	$7.02 + d$		
		0.8470	7.03		

Write a proportion and solve for d.

$$\frac{0.0002}{0.0007} = \frac{d}{0.01}$$

$$0.003 \approx d$$

Thus,

$$\begin{aligned}\text{antilog}(9.8465 - 10) &\approx (7.02 + d) \times 10^{-1} = (7.02 + 0.003) \times 10^{-1}\\ &= 7.023 \times 10^{-1} = 0.7023\end{aligned}$$

Using Trigonometric Tables

The values of trigonometric functions can be found by using the Table of Values of Trigonometric Functions found in Appendix III. These tables give the values of the trigonometric functions, accurate to four significant digits, in increments of 10′.

To use the tables of trigonometric functions, first note that the table has degrees listed in both the right- and left-hand columns and the trigonometric functions are listed in both the top and bottom rows of the tables. For angles listed at the left, use the top row of functions. For angles listed at the right, use the bottom row of functions. For example, to determine the value of sin 27°40′, locate 27°40′ in the left-hand column of the table and read across until you locate the column at the top headed by "sin." Thus, $\sin 27°40' \approx 0.4643$.

To determine the value of cos 72°20′, locate 72°20′ in the right-hand column of the table and read across until you locate the column at the bottom headed by "cos." Thus, $\cos 72°20' \approx 0.3035$.

The Table of Trigonometric Functions only gives the values of the trigonometric functions from 0° to 90°. To find the value of a trigonometric function for other angles, first find the reference angle. Then attach the appropriate sign depending on the quadrant where the terminal side of the angle lies.

EXAMPLE 1 Evaluate Trigonometric Functions Using the Reference Angle

Find a. $\tan 153°40'$ b. $\sec(-87°10')$ c. $\sin 3.9910$

Solution

a. $\tan 153°40' = -\tan 26°20'$ $\quad 180° - 153°40' = 26°20'$

≈ -0.4950

b. $\sec(-87°10') = \sec 87°10' \approx 20.23$

c. In this case, we are finding the sine of a real number. Find the reference number in the column headed by θ.

$$\sin 3.9910 \approx -\sin 0.8494 \approx -0.7509 \quad 3.9910 - \pi \approx 0.8494$$

We now examine the process of finding θ given a trigonometric function of θ. For example, given $\cos \theta = 0.5736$ with θ an angle in the first quadrant, locate 0.5736 in the body of the table in the cosine column. Thus $\theta \approx 55°$ and $\cos 55° \approx 0.5736$. Because the trigonometric functions are periodic, it is necessary to specify some bounds on θ.

EXAMPLE 2 Solve a Trigonometric Equation

Solve the equation for $0° \le \theta < 360°$.

a. $\sin \theta = 0.5225$ b. $\tan \theta = -1.611$

Solution

a. Let θ' be the reference angle so that $\sin \theta' = 0.5225$. From the table, $\theta' \approx 31°30'$. Because $\sin \theta$ is positive in quadrants I and II,

$$\theta \approx 31°30' \quad \text{or} \quad \theta \approx 180° - 31°30' = 148°30'.$$

b. Let θ' be the reference angle so that $\tan \theta' = 1.611$. From the table, $\theta' \approx 58°10'$. Because $\tan \theta$ is negative in quadrants II and IV,

$$\theta \approx 180° - 58°10' = 121°50' \quad \text{or} \quad \theta \approx 360° - 58°10' = 301°50'.$$

Using the method of linear interpolation, we can approximate trigonometric function values that are not given in the tables.

EXAMPLE 3 Evaluate Trigonometric Functions

Find a. $\cos 58°28'$ b. $\tan(-0.5828)$ c. $\csc 231°12'$

Solution

a.

$$10'\left[\;8'\left[\begin{array}{l}\cos 58°20' \approx 0.5250\\ \cos 58°28' \approx 0.5250 + d\end{array}\right]d\;\right.\quad\begin{array}{l}\cos 58°30' \approx 0.5225\end{array}\;\Bigg]-0.0025$$

Write a proportion and solve for d.

$$\frac{8}{10} = \frac{d}{-0.0025}$$

$$d = -0.0020$$

$\cos 58°28' \approx 0.5250 + (-0.0020) = 0.5230.$

b. Let $\theta' = 0.5828$ be the reference number. Look for 0.5828 in the θ column.

$$0.0029\left[\;0.0010\left[\begin{array}{l}\tan 0.5818 \approx 0.6577\\ \tan 0.5828 \approx 0.6577 + d\end{array}\right]d\;\right.\quad\begin{array}{l}\tan 0.5847 \approx 0.6619\end{array}\;\Bigg]0.0042$$

Write a proportion and solve for d.

$$\frac{0.0010}{0.0029} = \frac{d}{0.0042}$$

$$d \approx 0.0014$$

$\tan(-0.5828) = -\tan 0.5828 \approx -(0.6577 + 0.0014) = -0.6591.$

c. Let $\theta' = 51°12'$ be the reference angle for $231°12'$.

$$10'\left[\;2'\left[\begin{array}{l}\csc 51°10' \approx 1.284\\ \csc 51°12' \approx 1.284 + d\end{array}\right]d\;\right.\quad\begin{array}{l}\csc 51°20' \approx 1.281\end{array}\;\Bigg]-0.0030$$

Write and solve a proportion for d.

$$\frac{2}{10} = \frac{d}{-0.0030}$$

$$d \approx -0.0006$$

$\csc 231°12' = -\csc 51°12' = -(1.284 - 0.0006) = -1.283.$

We can also use interpolation to find θ given the value of the trigonometric function of θ.

EXAMPLE 4 Solve a Trigonometric Equation

Solve the equation $\cot\theta = 1.044$ where $0° \le \theta < 360°$.

Solution Search the body of the trigonometric table to find the two values of θ', where θ' is the reference angle, for which $\cot\theta$ is closest to 1.044.

$$\left.\begin{array}{l} \left.\begin{array}{r} \cot 43°40' \approx 1.048 \\ \cot(43°40' + d) \approx 1.044 \end{array}\right\} d \\ \cot 43°50' \approx 1.042 \end{array}\right\} 10' \qquad \left.\begin{array}{l} \left.\begin{array}{l} 1.048 \\ 1.044 \end{array}\right\} -0.004 \\ 1.042 \end{array}\right\} -0.006$$

Write a proportion and solve for d.

$$\frac{d}{10} = \frac{-0.004}{-0.0060}$$

$$d \approx 7'$$

Thus $\theta' \approx 43°40' + 7' = 43°47'$. Because $\cot\theta$ is positive in the first and third quadrants,

$$\theta \approx 43°47' \quad \text{or} \quad \theta \approx 180° + 43°47' = 223°47'.$$

EXAMPLE 5 Solve a Trigonometric Equation

Solve the equation $\sin\theta = -0.5672$ where $0 \le \theta < 2\pi$.

Solution Find two values of θ', where θ' is the reference number for θ, for which $\sin\theta$ is closest to 0.5672.

$$\left.\begin{array}{l} \left.\begin{array}{r} \sin 0.6021 \approx 0.5664 \\ \sin(0.6021 + d) \approx 0.5672 \end{array}\right\} d \\ \sin 0.6050 \approx 0.5688 \end{array}\right\} 0.0029 \qquad \left.\begin{array}{l} \left.\begin{array}{l} 0.5664 \\ 0.5672 \end{array}\right\} 0.0008 \\ 0.5688 \end{array}\right\} 0.0024$$

Write a proportion and solve for d.

$$\frac{d}{0.0029} = \frac{0.0008}{0.0024}$$

$$d \approx 0.0010$$

Thus $\theta' \approx 0.6021 + 0.0010 = 0.6031$. Because $\sin\theta$ is negative in the third and fourth quadrants,

$$\theta \approx \pi + 0.6031 \approx 3.7447 \quad \text{or} \quad \theta \approx 2\pi - 0.6031 \approx 5.6801.$$

Appendix II Logarithmic and Exponential Tables

Common Logarithms

x	0	1	2	3	4	5	6	7	8	9
1.0	.0000	.0043	.0086	.0128	.0170	.0212	.0253	.0294	.0334	.0374
1.1	.0141	.0453	.0492	.0531	.0569	.0607	.0645	.0682	.0719	.0755
1.2	.0792	.0828	.0864	.0899	.0934	.0969	.1004	.1038	.1072	.1106
1.3	.1139	.1173	.1206	.1239	.1271	.1303	.1335	.1367	.1399	.1430
1.4	.1461	.1492	.1523	.1553	.1584	.1614	.1644	.1673	.1703	.1732
1.5	.1761	.1790	.1818	.1847	.1875	.1903	.1931	.1959	.1987	.2014
1.6	.2041	.2068	.2095	.2122	.2148	.2175	.2201	.2227	.2253	.2279
1.7	.2304	.2330	.2355	.2380	.2405	.2430	.2455	.2480	.2504	.2529
1.8	.2553	.2577	.2601	.2625	.2648	.2672	.2695	.2718	.2742	.2765
1.9	.2788	.2810	.2833	.2856	.2878	.2900	.2923	.2945	.2967	.2989
2.0	.3010	.3032	.3054	.3075	.3096	.3118	.3139	.3160	.3181	.3201
2.1	.3222	.3243	.3263	.3284	.3304	.3324	.3345	.3365	.3385	.3404
2.2	.3424	.3444	.3464	.3483	.3502	.3522	.3541	.3560	.3579	.3598
2.3	.3617	.3636	.3655	.3674	.3692	.3711	.3729	.3747	.3766	.3784
2.4	.3802	.3820	.3838	.3856	.3874	.3892	.3909	.3927	.3945	.3962
2.5	.3979	.3997	.4014	.4031	.4048	.4065	.4082	.4099	.4116	.4133
2.6	.4150	.4166	.4183	.4200	.4216	.4232	.4249	.4265	.4281	.4298
2.7	.4314	.4330	.4346	.4362	.4378	.4393	.4409	.4425	.4440	.4456
2.8	.4472	.4487	.4502	.4518	.4533	.4548	.4564	.4579	.4594	.4609
2.9	.4624	.4639	.4654	.4669	.4683	.4698	.4713	.4728	.4742	.4757
3.0	.4771	.4786	.4800	.4814	.4829	.4843	.4857	.4871	.4886	.4900
3.1	.4914	.4928	.4942	.4955	.4969	.4983	.4997	.5011	.5024	.5038
3.2	.5051	.5065	.5079	.5092	.5105	.5119	.5132	.5145	.5159	.5172
3.3	.5185	.5198	.5211	.5224	.5237	.5250	.5263	.5276	.5289	.5302
3.4	.5315	.5328	.5340	.5353	.5366	.5378	.5391	.5403	.5416	.5428
3.5	.5441	.5453	.5465	.5478	.5490	.5502	.5514	.5527	.5539	.5551
3.6	.5563	.5575	.5587	.5599	.5611	.5623	.5635	.5647	.5658	.5670
3.7	.5682	.5694	.5705	.5717	.5729	.5740	.5752	.5763	.5775	.5786
3.8	.5798	.5809	.5821	.5832	.5843	.5855	.5866	.5877	.5888	.5899
3.9	.5911	.5922	.5933	.5944	.5955	.5966	.5977	.5988	.5999	.6010
4.0	.6021	.6031	.6042	.6053	.6064	.6075	.6085	.6096	.6107	.6117
4.1	.6128	.6138	.6149	.6160	.6170	.6180	.6191	.6201	.6212	.6222
4.2	.6232	.6243	.6253	.6263	.6274	.6284	.6294	.6304	.6314	.6325
4.3	.6335	.6345	.6355	.6365	.6375	.6385	.6395	.6405	.6415	.6425
4.4	.6435	.6444	.6454	.6464	.6474	.6484	.6493	.6503	.6513	.6522
4.5	.6532	.6542	.6551	.6561	.6571	.6580	.6590	.6599	.6609	.6618
4.6	.6628	.6637	.6646	.6656	.6665	.6675	.6684	.6693	.6702	.6712
4.7	.6721	.6730	.6739	.6749	.6758	.6767	.6776	.6785	.6794	.6803
4.8	.6812	.6821	.6830	.6839	.6848	.6857	.6866	.6875	.6884	.6893
4.9	.6902	.6911	.6920	.6928	.6937	.6946	.6955	.6964	.6972	.6981

Common Logarithms (continued)

x	0	1	2	3	4	5	6	7	8	9
5.0	.6990	.6998	.7007	.7016	.7024	.7033	.7042	.7050	.7059	.7067
5.1	.7076	.7084	.7093	.7101	.7110	.7118	.7126	.7135	.7143	.7152
5.2	.7160	.7168	.7177	.7185	.7193	.7202	.7210	.7218	.7226	.7235
5.3	.7243	.7251	.7259	.7267	.7275	.7284	.7292	.7300	.7308	.7316
5.4	.7324	.7332	.7340	.7348	.7356	.7364	.7372	.7380	.7388	.7396
5.5	.7404	.7412	.7419	.7427	.7435	.7443	.7451	.7459	.7466	.7474
5.6	.7482	.7490	.7497	.7505	.7513	.7520	.7528	.7536	.7543	.7551
5.7	.7559	.7566	.7574	.7582	.7589	.7597	.7604	.7612	.7619	.7627
5.8	.7634	.7642	.7649	.7657	.7664	.7672	.7679	.7686	.7694	.7701
5.9	.7709	.7716	.7723	.7731	.7738	.7745	.7752	.7760	.7767	.7774
6.0	.7782	.7789	.7796	.7803	.7810	.7818	.7825	.7832	.7839	.7846
6.1	.7853	.7860	.7868	.7875	.7882	.7889	.7896	.7903	.7910	.7917
6.2	.7924	.7931	.7938	.7945	.7952	.7959	.7966	.7973	.7980	.7987
6.3	.7993	.8000	.8007	.8014	.8021	.8028	.8035	.8041	.8048	.8055
6.4	.8062	.8069	.8075	.8082	.8089	.8096	.8102	.8109	.8116	.8122
6.5	.8129	.8136	.8142	.8149	.8156	.8162	.8169	.8176	.8182	.8189
6.6	.8195	.8202	.8209	.8215	.8222	.8228	.8235	.8241	.8248	.8254
6.7	.8261	.8267	.8274	.8280	.8287	.8293	.8299	.8306	.8312	.8319
6.8	.8325	.8331	.8338	.8344	.8351	.8357	.8363	.8370	.8376	.8382
6.9	.8388	.8395	.8401	.8407	.8414	.8420	.8426	.8432	.8439	.8445
7.0	.8451	.8457	.8463	.8470	.8476	.8482	.8488	.8494	.8500	.8506
7.1	.8513	.8519	.8525	.8531	.8537	.8543	.8549	.8555	.8561	.8567
7.2	.8573	.8579	.8585	.8591	.8597	.8603	.8609	.8615	.8621	.8627
7.3	.8633	.8639	.8645	.8651	.8657	.8663	.8669	.8675	.8681	.8686
7.4	.8692	.8698	.8704	.8710	.8716	.8722	.8727	.8733	.8739	.8745
7.5	.8751	.8756	.8762	.8768	.8774	.8779	.8785	.8791	.8797	.8802
7.6	.8808	.8814	.8820	.8825	.8831	.8837	.8842	.8848	.8854	.8859
7.7	.8865	.8871	.8876	.8882	.8887	.8893	.8899	.8904	.8910	.8915
7.8	.8921	.8927	.8932	.8938	.8943	.8949	.8954	.8960	.8965	.8971
7.9	.8976	.8982	.8987	.8993	.8998	.9004	.9009	.9015	.9020	.9025
8.0	.9031	.9036	.9042	.9047	.9053	.9058	.9063	.9069	.9074	.9079
8.1	.9085	.9090	.9096	.9101	.9106	.9112	.9117	.9122	.9128	.9133
8.2	.9138	.9143	.9149	.9154	.9159	.9165	.9170	.9175	.9180	.9186
8.3	.9191	.9196	.9201	.9206	.9212	.9217	.9222	.9227	.9232	.9238
8.4	.9243	.9248	.9253	.9258	.9263	.9269	.9274	.9279	.9284	.9289
8.5	.9294	.9299	.9304	.9309	.9315	.9320	.9325	.9330	.9335	.9340
8.6	.9345	.9350	.9355	.9360	.9365	.9370	.9375	.9380	.9385	.9390
8.7	.9395	.9400	.9405	.9410	.9415	.9420	.9425	.9430	.9435	.9440
8.8	.9445	.9450	.9455	.9460	.9465	.9469	.9474	.9479	.9484	.9489
8.9	.9494	.9499	.9504	.9509	.9513	.9518	.9523	.9528	.9533	.9538
9.0	.9542	.9547	.9552	.9557	.9562	.9566	.9571	.9576	.9581	.9586
9.1	.9590	.9595	.9600	.9605	.9609	.9614	.9619	.9624	.9628	.9633
9.2	.9638	.9643	.9647	.9652	.9657	.9661	.9666	.9671	.9675	.9680
9.3	.9685	.9689	.9694	.9699	.9703	.9708	.9713	.9717	.9722	.9727
9.4	.9731	.9736	.9741	.9745	.9750	.9754	.9759	.9763	.9768	.9773
9.5	.9777	.9782	.9786	.9791	.9795	.9800	.9805	.9809	.9814	.9818
9.6	.9823	.9827	.9832	.9836	.9841	.9845	.9850	.9854	.9859	.9863
9.7	.9868	.9872	.9877	.9881	.9886	.9890	.9894	.9899	.9903	.9908
9.8	.9912	.9917	.9921	.9926	.9930	.9934	.9939	.9943	.9948	.9952
9.9	.9956	.9961	.9965	.9969	.9974	.9978	.9983	.9987	.9991	.9996

Natural Exponential Function

x	e^x	e^{-x}	x	e^x	e^{-x}	x	e^x	e^{-x}
.00	1.00000	1.00000	.40	1.49182	.67032	.80	2.22554	.44032
.01	1.01005	.99005	.41	1.50682	.66365	.85	2.33965	.42741
.02	1.02020	.98020	.42	1.52196	.65705	.90	2.45960	.40657
.03	1.03045	.97045	.43	1.53726	.65051	.95	2.58571	.38674
.04	1.04081	.96079	.44	1.55271	.64404	1.00	2.71828	.36788
.05	1.05127	.95123	.45	1.56831	.63763	1.10	3.00416	.33287
.06	1.06184	.94176	.46	1.58407	.63128	1.20	3.32011	.30119
.07	1.07251	.93239	.47	1.59999	.62500	1.30	3.66929	.27253
.08	1.08329	.92312	.48	1.61607	.61878	1.40	4.05519	.24659
.09	1.09417	.91393	.49	1.63232	.61263	1.50	4.48168	.22313
.10	1.10517	.90484	.50	1.64872	.60653	1.60	4.95302	.20189
.11	1.11628	.89583	.51	1.66529	.60050	1.70	5.47394	.18268
.12	1.12750	.88692	.52	1.68203	.59452	1.80	6.04964	.16529
.13	1.13883	.87810	.53	1.69893	.58860	1.90	6.68589	.14956
.14	1.15027	.86936	.54	1.71601	.58275	2.00	7.38905	.13533
.15	1.16183	.86071	.55	1.73325	.57695	2.10	8.16616	.12245
.16	1.17351	.85214	.56	1.75067	.57121	2.20	9.02500	.11080
.17	1.18530	.84366	.57	1.76827	.56553	2.30	9.97417	.10025
.18	1.19722	.83527	.58	1.78604	.55990	2.40	11.02316	.09071
.19	1.20925	.82696	.59	1.80399	.55433	2.50	12.18248	.08208
.20	1.22140	.81873	.60	1.82212	.54881	3.00	20.08551	.04978
.21	1.23368	.81058	.61	1.84043	.54335	3.50	33.11545	.03020
.22	1.24608	.80252	.62	1.85893	.53794	4.00	54.59815	.01832
.23	1.25860	.79453	.63	1.87761	.53259	4.50	90.01713	.01111
.24	1.27125	.78663	.64	1.89648	.52729	5.00	148.41316	.00674
.25	1.28403	.77880	.65	1.91554	.52205	5.50	224.69193	.00409
.26	1.29693	.77105	.66	1.93479	.51685	6.00	403.42879	.00248
.27	1.30996	.76338	.67	1.95424	.51171	6.50	665.14163	.00150
.28	1.32313	.75578	.68	1.97388	.50662	7.00	1096.63316	.00091
.29	1.33643	.74826	.69	1.99372	.50158	7.50	1808.04241	.00055
.30	1.34986	.74082	.70	2.01375	.49659	8.00	2980.95799	.00034
.31	1.36343	.73345	.71	2.03399	.49164	8.50	4914.76884	.00020
.32	1.37713	.72615	.72	2.05443	.48675	9.00	8130.08392	.00012
.33	1.39097	.71892	.73	2.07508	.48191	9.50	13359.72683	.00007
.34	1.40495	.71177	.74	2.09594	.47711	10.00	22026.46579	.00005
.35	1.41907	.70469	.75	2.11700	.47237			
.36	1.43333	.69768	.76	2.13828	.46767			
.37	1.44773	.69073	.77	2.15977	.46301			
.38	1.46228	.68386	.78	2.18147	.45841			
.39	1.47698	.67706	.79	2.20340	.45384			

Natural Logarithms

x	$\ln x$	x	$\ln x$	x	$\ln x$
		4.5	1.5041	9.0	2.1972
0.1	−2.3026	4.6	1.5261	9.1	2.2083
0.2	−1.6094	4.7	1.5476	9.2	2.2192
0.3	−1.2040	4.8	1.5686	9.3	2.2300
0.4	−0.9163	4.9	1.5892	9.4	2.2407
0.5	−0.6931	5.0	1.6094	9.5	2.2513
0.6	−0.5108	5.1	1.6292	9.6	2.2618
0.7	−0.3567	5.2	1.6487	9.7	2.2721
0.8	−0.2231	5.3	1.6677	9.8	2.2824
0.9	−0.1054	5.4	1.6864	9.9	2.2925
1.0	0.0000	5.5	1.7047	10	2.3026
1.1	0.0953	5.6	1.7228	11	2.3979
1.2	0.1823	5.7	1.7405	12	2.4849
1.3	0.2624	5.8	1.7579	13	2.5649
1.4	0.3365	5.9	1.7750	14	2.6391
1.5	0.4055	6.0	1.7918	15	2.7081
1.6	0.4700	6.1	1.8083	16	2.7726
1.7	0.5306	6.2	1.8245	17	2.8332
1.8	0.5878	6.3	1.8405	18	2.8904
1.9	0.6419	6.4	1.8563	19	2.9444
2.0	0.6931	6.5	1.8718	20	2.9957
2.1	0.7419	6.6	1.8871	25	3.2189
2.2	0.7885	6.7	1.9021	30	3.4012
2.3	0.8329	6.8	1.9169	35	3.5553
2.4	0.8755	6.9	1.9315	40	3.6889
2.5	0.9163	7.0	1.9459	45	3.8067
2.6	0.9555	7.1	1.9601	50	3.9120
2.7	0.9933	7.2	1.9741	55	4.0073
2.8	1.0296	7.3	1.9879	60	4.0943
2.9	1.0647	7.4	2.0015	65	4.1744
3.0	1.0986	7.5	2.0149	70	4.2485
3.1	1.1314	7.6	2.0281	75	4.3175
3.2	1.1632	7.7	2.0412	80	4.3820
3.3	1.1939	7.8	2.0541	85	4.4427
3.4	1.2238	7.9	2.0669	90	4.4998
3.5	1.2528	8.0	2.0794	100	4.6052
3.6	1.2809	8.1	2.0919	110	4.7005
3.7	1.3083	8.2	2.1041	120	4.7875
3.8	1.3350	8.3	2.1163	130	4.8676
3.9	1.3610	8.4	2.1282	140	4.9416
4.0	1.3863	8.5	2.1401	150	5.0106
4.1	1.4110	8.6	2.1518	160	5.0752
4.2	1.4351	8.7	2.1633	170	5.1358
4.3	1.4586	8.8	2.1748	180	5.1930
4.4	1.4816	8.9	2.1861	190	5.2470

Appendix III Trigonometric Tables

Trigonometric Functions in Degrees and Radians

θ (degrees)	θ (radians)	sin θ	cos θ	tan θ	cot θ	sec θ	csc θ		
0°00′	.0000	.0000	1.0000	.0000	—	1.000	—	1.5708	**90°00′**
10	.0029	.0029	1.0000	.0029	343.8	1.000	343.8	1.5679	50
20	.0058	.0058	1.0000	.0058	171.9	1.000	171.9	1.5650	40
30	.0087	.0087	1.0000	.0087	114.6	1.000	114.6	1.5621	30
40	.0116	.0116	.9999	.0116	85.94	1.000	85.95	1.5592	20
50	.0145	.0145	.9999	.0145	68.75	1.000	68.76	1.5563	10
1°00′	.0175	.0175	.9998	.0175	57.29	1.000	57.30	1.5533	**89°00′**
10	.0204	.0204	.9998	.0204	49.10	1.000	49.11	1.5504	50
20	.0233	.0233	.9997	.0233	42.96	1.000	42.98	1.5475	40
30	.0262	.0262	.9997	.0262	38.19	1.000	38.20	1.5446	30
40	.0291	.0291	.9996	.0291	34.37	1.000	34.38	1.5417	20
50	.0320	.0320	.9995	.0320	31.24	1.001	31.26	1.5388	10
2°00′	.0349	.0349	.9994	.0349	28.64	1.001	28.65	1.5359	**88°00′**
10	.0378	.0378	.9993	.0378	26.43	1.001	26.45	1.5330	50
20	.0407	.0407	.9992	.0407	24.54	1.001	24.56	1.5301	40
30	.0436	.0436	.9990	.0437	22.90	1.001	22.93	1.5272	30
40	.0465	.0465	.9989	.0466	21.47	1.001	21.49	1.5243	20
50	.0495	.0494	.9988	.0495	20.21	1.001	20.23	1.5213	10
3°00′	.0524	.0523	.9986	.0524	19.08	1.001	19.11	1.5184	**87°00′**
10	.0553	.0552	.9985	.0553	18.07	1.002	18.10	1.5155	50
20	.0582	.0581	.9983	.0582	17.17	1.002	17.20	1.5126	40
30	.0611	.0610	.9981	.0612	16.35	1.002	16.38	1.5097	30
40	.0640	.0640	.9980	.0641	15.60	1.002	15.64	1.5068	20
50	.0669	.0669	.9978	.0670	14.92	1.002	14.96	1.5039	10
4°00′	.0698	.0698	.9976	.0699	14.30	1.002	14.34	1.5010	**86°00′**
10	.0727	.0727	.9974	.0729	13.73	1.003	13.76	1.4981	50
20	.0756	.0756	.9971	.0758	13.20	1.003	13.23	1.4952	40
30	.0785	.0785	.9969	.0787	12.71	1.003	12.75	1.4923	30
40	.0814	.0814	.9967	.0816	12.25	1.003	12.29	1.4893	20
50	.0844	.0843	.9964	.0846	11.83	1.004	11.87	1.4864	10
5°00′	.0873	.0872	.9962	.0875	11.43	1.004	11.47	1.4835	**85°00′**
10	.0902	.0901	.9959	.0904	11.06	1.004	11.10	1.4806	50
20	.0931	.0929	.9957	.0934	10.71	1.004	10.76	1.4777	40
30	.0960	.0958	.9954	.0963	10.39	1.005	10.43	1.4748	30
40	.0989	.0987	.9951	.0992	10.08	1.005	10.13	1.4719	20
50	.1018	.1016	.9948	.1022	9.788	1.005	9.839	1.4690	10
6°00′	.1047	.1045	.9945	.1051	9.514	1.006	9.567	1.4661	**84°00′**
10	.1076	.1074	.9942	.1080	9.255	1.006	9.309	1.4632	50
20	.1105	.1103	.9939	.1110	9.010	1.006	9.065	1.4603	40
30	.1134	.1132	.9936	.1139	8.777	1.006	8.834	1.4573	30
40	.1164	.1161	.9932	.1169	8.556	1.007	8.614	1.4544	20
50	.1193	.1190	.9929	.1198	8.345	1.007	8.405	1.4515	10
		cos θ	sin θ	cot θ	tan θ	csc θ	sec θ	θ (radians)	θ (degrees)

θ (degrees)	θ (radians)	sin θ	cos θ	tan θ	cot θ	sec θ	csc θ		
7°00′	.1222	.1219	.9925	.1228	8.144	1.008	8.206	1.4486	**83°00′**
10	.1251	.1248	.9922	.1257	7.953	1.008	8.016	1.4457	50
20	.1280	.1276	.9918	.1287	7.770	1.008	7.834	1.4428	40
30	.1309	.1305	.9914	.1317	7.596	1.009	7.661	1.4399	30
40	.1338	.1334	.9911	.1346	7.429	1.009	7.496	1.4370	20
50	.1376	.1363	.9907	.1376	7.269	1.009	7.337	1.4341	10
8°00′	.1396	.1392	.9903	.1405	7.115	1.010	7.185	1.4312	**82°00′**
10	.1425	.1421	.9899	.1435	6.968	1.010	7.040	1.4283	50
20	.1454	.1449	.9894	.1465	6.827	1.011	6.900	1.4254	40
30	.1484	.1478	.9890	.1495	6.691	1.011	6.765	1.4224	30
40	.1513	.1507	.9886	.1524	6.561	1.012	6.636	1.4195	20
50	.1542	.1536	.9881	.1554	6.435	1.012	6.512	1.4166	10
9°00′	.1571	.1564	.9877	.1584	6.314	1.012	6.392	1.4137	**81°00′**
10	.1600	.1593	.9872	.1614	6.197	1.013	6.277	1.4108	50
20	.1629	.1622	.9868	.1644	6.084	1.013	6.166	1.4079	40
30	.1658	.1650	.9863	.1673	5.976	1.014	6.059	1.4050	30
40	.1687	.1679	.9858	.1703	5.871	1.014	5.955	1.4021	20
50	.1716	.1708	.9853	.1733	5.769	1.015	5.855	1.3992	10
10°00′	.1745	.1736	.9848	.1763	5.671	1.015	5.759	1.3963	**80°00′**
10	.1774	.1765	.9843	.1793	5.576	1.016	5.665	1.3934	50
20	.1804	.1794	.9838	.1823	5.485	1.016	5.575	1.3904	40
30	.1833	.1822	.9833	.1853	5.396	1.017	5.487	1.3875	30
40	.1862	.1851	.9827	.1883	5.309	1.018	5.403	1.3846	20
50	.1891	.1880	.9822	.1914	5.226	1.018	5.320	1.3817	10
11°00′	.1920	.1908	.9816	.1944	5.145	1.019	5.241	1.3788	**79°00′**
10	.1949	.1937	.9811	.1974	5.066	1.019	5.164	1.3759	50
20	.1978	.1965	.9805	.2004	4.989	1.020	5.089	1.3730	40
30	.2007	.1994	.9799	.2035	4.915	1.020	5.016	1.3701	30
40	.2036	.2022	.9793	.2065	4.843	1.021	4.945	1.3672	20
50	.2065	.2051	.9787	.2095	4.773	1.022	4.876	1.3643	10
12°00′	.2094	.2079	.9781	.2126	4.705	1.022	4.810	1.3614	**78°00′**
10	.2123	.2108	.9775	.2156	4.638	1.023	4.745	1.3584	50
20	.2153	.2136	.9769	.2186	4.574	1.024	4.682	1.3555	40
30	.2182	.2164	.9763	.2217	4.511	1.024	4.620	1.3526	30
40	.2211	.2193	.9757	.2247	4.449	1.025	4.560	1.3497	20
50	.2240	.2221	.9750	.2278	4.390	1.026	4.502	1.3468	10
13°00′	.2269	.2250	.9744	.2309	4.331	1.026	4.445	1.3439	**77°00′**
10	.2298	.2278	.9737	.2339	4.275	1.027	4.390	1.3410	50
20	.2327	.2306	.9730	.2370	4.219	1.028	4.336	1.3381	40
30	.2356	.2334	.9724	.2401	4.165	1.028	4.284	1.3352	30
40	.2385	.2363	.9717	.2432	4.113	1.029	4.232	1.3323	20
50	.2414	.2391	.9710	.2462	4.061	1.030	4.182	1.3294	10
		cos θ	sin θ	cot θ	tan θ	csc θ	sec θ	θ (radians)	θ (degrees)

Trigonometric Functions in Degrees and Radians (continued)

θ (degrees)	θ (radians)	sin θ	cos θ	tan θ	cot θ	sec θ	csc θ		
14°00′	.2443	.2419	.9703	.2493	4.011	1.031	4.134	1.3265	76°00′
10	.2473	.2447	.9696	.2524	3.962	1.031	4.086	1.3235	50
20	.2502	.2476	.9689	.2555	3.914	1.032	4.039	1.3206	40
30	.2531	.2504	.9681	.2586	3.867	1.033	3.994	1.3177	30
40	.2560	.2532	.9674	.2617	3.821	1.034	3.950	1.3148	20
50	.2589	.2560	.9667	.2648	3.776	1.034	3.906	1.3119	10
15°00′	.2618	.2588	.9659	.2679	3.732	1.035	3.864	1.3090	75°00′
10	.2647	.2616	.9652	.2711	3.689	1.036	3.822	1.3061	50
20	.2676	.2644	.9644	.2742	3.647	1.037	3.782	1.3032	40
30	.2705	.2672	.9636	.2773	3.606	1.038	3.742	1.3003	30
40	.2734	.2700	.9628	.2805	3.566	1.039	3.703	1.2974	20
50	.2763	.2728	.9621	.2836	3.526	1.039	3.665	1.2945	10
16°00′	.2793	.2756	.9613	.2867	3.487	1.040	3.628	1.2915	74°00′
10	.2822	.2784	.9605	.2899	3.450	1.041	3.592	1.2886	50
20	.2851	.2812	.9596	.2931	3.412	1.042	3.556	1.2857	40
30	.2880	.2840	.9588	.2962	3.376	1.043	3.521	1.2828	30
40	.2909	.2868	.9580	.2994	3.340	1.044	3.487	1.2799	20
50	.2938	.2896	.9572	.3026	3.305	1.045	3.453	1.2770	10
17°00′	.2967	.2924	.9563	.3057	3.271	1.046	3.420	1.2741	73°00′
10	.2996	.2952	.9555	.3089	3.237	1.047	3.388	1.2712	50
20	.3025	.2979	.9546	.3121	3.204	1.048	3.356	1.2683	40
30	.3054	.3007	.9537	.3153	3.172	1.049	3.326	1.2654	30
40	.3083	.3035	.9528	.3185	3.140	1.049	3.295	1.2625	20
50	.3113	.3062	.9520	.3217	3.108	1.050	3.265	1.2595	10
18°00′	.3142	.3090	.9511	.3249	3.078	1.051	3.236	1.2566	72°00′
10	.3171	.3118	.9502	.3281	3.047	1.052	3.207	1.2537	50
20	.3200	.3145	.9492	.3314	3.018	1.053	3.179	1.2508	40
30	.3229	.3173	.9483	.3346	2.989	1.054	3.152	1.2479	30
40	.3258	.3201	.9474	.3378	2.960	1.056	3.124	1.2450	20
50	.3287	.3228	.9465	.3411	2.932	1.057	3.098	1.2421	10
19°00′	.3316	.3256	.9455	.3443	2.904	1.058	3.072	1.2392	71°00′
10	.3345	.3283	.9446	.3476	2.877	1.059	3.046	1.2363	50
20	.3374	.3311	.9436	.3508	2.850	1.060	3.021	1.2334	40
30	.3403	.3338	.9426	.3541	2.824	1.061	2.996	1.2305	30
40	.3432	.3365	.9417	.3574	2.798	1.062	2.971	1.2275	20
50	.3462	.3393	.9407	.3607	2.773	1.063	2.947	1.2246	10
20°00′	.3491	.3420	.9397	.3640	2.747	1.064	2.924	1.2217	70°00′
10	.3520	.3448	.9387	.3673	2.723	1.065	2.901	1.2188	50
20	.3549	.3475	.9377	.3706	2.699	1.066	2.878	1.2159	40
30	.3578	.3502	.9367	.3739	2.675	1.068	2.855	1.2130	30
40	.3607	.3529	.9356	.3772	2.651	1.069	2.833	1.2101	20
50	.3636	.3557	.9346	.3805	2.628	1.070	2.812	1.2072	10
		cos θ	sin θ	cot θ	tan θ	csc θ	sec θ	θ (radians)	θ (degrees)

θ (degrees)	θ (radians)	sin θ	cos θ	tan θ	cot θ	sec θ	csc θ		
21°00′	.3665	.3584	.9336	.3839	2.605	1.071	2.790	1.2043	69°00′
10	.3694	.3611	.9325	.3872	2.583	1.072	2.769	1.2014	50
20	.3723	.3638	.9315	.3906	2.560	1.074	2.749	1.1985	40
30	.3752	.3665	.9304	.3939	2.539	1.075	2.729	1.1956	30
40	.3782	.3692	.9293	.3973	2.517	1.076	2.709	1.1926	20
50	.3811	.3719	.9283	.4006	2.496	1.077	2.689	1.1897	10
22°00′	.3840	.3746	.9272	.4040	2.475	1.079	2.669	1.1868	68°00′
10	.3869	.3773	.9261	.4074	2.455	1.080	2.650	1.1839	50
20	.3898	.3800	.9250	.4108	2.434	1.081	2.632	1.1810	40
30	.3927	.3827	.9239	.4142	2.414	1.082	2.613	1.1781	30
40	.3956	.3854	.9228	.4176	2.394	1.084	2.595	1.1752	20
50	.3985	.3881	.9216	.4210	2.375	1.085	2.577	1.1723	10
23°00′	.4014	.3907	.9205	.4245	2.356	1.086	2.559	1.1694	67°00′
10	.4043	.3934	.9194	.4279	2.337	1.088	2.542	1.1665	50
20	.4072	.3961	.9182	.4314	2.318	1.089	2.525	1.1636	40
30	.4102	.3987	.9171	.4348	2.300	1.090	2.508	1.1606	30
40	.4131	.4014	.9159	.4383	2.282	1.092	2.491	1.1577	20
50	.4160	.4041	.9147	.4417	2.264	1.093	2.475	1.1548	10
24°00′	.4189	.4067	.9135	.4452	2.246	1.095	2.459	1.1519	66°00′
10	.4218	.4094	.9124	.4487	2.229	1.096	2.443	1.1490	50
20	.4247	.4120	.9112	.4522	2.211	1.097	2.427	1.1461	40
30	.4276	.4147	.9100	.4557	2.194	1.099	2.411	1.1432	30
40	.4305	.4173	.9088	.4592	2.177	1.100	2.396	1.1403	20
50	.4334	.4200	.9075	.4628	2.161	1.102	2.381	1.1374	10
25°00′	.4363	.4226	.9063	.4663	2.145	1.103	2.366	1.1345	65°00′
10	.4392	.4253	.9051	.4699	2.128	1.105	2.352	1.1316	50
20	.4422	.4279	.9038	.4734	2.112	1.106	2.337	1.1286	40
30	.4451	.4305	.9026	.4770	2.097	1.108	2.323	1.1257	30
40	.4480	.4331	.9013	.4806	2.081	1.109	2.309	1.1228	20
50	.4509	.4358	.9001	.4841	2.066	1.111	2.295	1.1199	10
26°00′	.4538	.4384	.8988	.4877	2.050	1.113	2.281	1.1170	64°00′
10	.4567	.4410	.8975	.4913	2.035	1.114	2.268	1.1141	50
20	.4596	.4436	.8962	.4950	2.020	1.116	2.254	1.1112	40
30	.4625	.4462	.8949	.4986	2.006	1.117	2.241	1.1083	30
40	.4654	.4488	.8936	.5022	1.991	1.119	2.228	1.1054	20
50	.4683	.4514	.8923	.5059	1.977	1.121	2.215	1.1025	10
27°00′	.4712	.4540	.8910	.5095	1.963	1.122	2.203	1.0996	63°00′
10	.4741	.4566	.8897	.5132	1.949	1.124	2.190	1.0966	50
20	.4771	.4592	.8884	.5169	1.935	1.126	2.178	1.0937	40
30	.4800	.4617	.8870	.5206	1.921	1.127	2.166	1.0908	30
40	.4829	.4643	.8857	.5243	1.907	1.129	2.154	1.0879	20
50	.4858	.4669	.8843	.5280	1.894	1.131	2.142	1.0850	10
		cos θ	sin θ	cot θ	tan θ	csc θ	sec θ	θ (radians)	θ (degrees)

Trigonometric Functions in Degrees and Radians (continued)

θ (degrees)	θ (radians)	sin θ	cos θ	tan θ	cot θ	sec θ	csc θ		
28°00′	.4887	.4695	.8829	.5317	1.881	1.133	2.130	1.0821	**62°00′**
10	.4916	.4720	.8816	.5354	1.868	1.134	2.118	1.0792	50
20	.4945	.4746	.8802	.5392	1.855	1.136	2.107	1.0763	40
30	.4974	.4772	.8788	.5430	1.842	1.138	2.096	1.0734	30
40	.5003	.4797	.8774	.5467	1.829	1.140	2.085	1.0705	20
50	.5032	.4823	.8760	.5505	1.816	1.142	2.074	1.0676	10
29°00′	.5061	.4848	.8746	.5543	1.804	1.143	2.063	1.0647	**61°00′**
10	.5091	.4874	.8732	.5581	1.792	1.145	2.052	1.0617	50
20	.5120	.4899	.8718	.5619	1.780	1.147	2.041	1.0588	40
30	.5149	.4924	.8704	.5658	1.767	1.149	2.031	1.0559	30
40	.5178	.4950	.8689	.5696	1.756	1.151	2.020	1.0530	20
50	.5207	.4975	.8675	.5735	1.744	1.153	2.010	1.0501	10
30°00′	.5236	.5000	.8660	.5774	1.732	1.155	2.000	1.0472	**60°00′**
10	.5265	.5025	.8646	.5812	1.720	1.157	1.990	1.0443	50
20	.5294	.5050	.8631	.5851	1.709	1.159	1.980	1.0414	40
30	.5323	.5075	.8616	.5890	1.698	1.161	1.970	1.0385	30
40	.5352	.5100	.8601	.5930	1.686	1.163	1.961	1.0356	20
50	.5381	.5125	.8587	.5969	1.675	1.165	1.951	1.0327	10
31°00′	.5411	.5150	.8572	.6009	1.664	1.167	1.942	1.0297	**59°00′**
10	.5440	.5175	.8557	.6048	1.653	1.169	1.932	1.0268	50
20	.5469	.5200	.8542	.6088	1.643	1.171	1.923	1.0239	40
30	.5498	.5225	.8526	.6128	1.632	1.173	1.914	1.0210	30
40	.5527	.5250	.8511	.6168	1.621	1.175	1.905	1.0181	20
50	.5556	.5275	.8496	.6208	1.611	1.177	1.896	1.0152	10
32°00′	.5585	.5299	.8480	.6249	1.600	1.179	1.887	1.0123	**58°00′**
10	.5614	.5324	.8465	.6289	1.590	1.181	1.878	1.0094	50
20	.5643	.5348	.8450	.6330	1.580	1.184	1.870	1.0065	40
30	.5672	.5373	.8434	.6371	1.570	1.186	1.861	1.0036	30
40	.5701	.5398	.8418	.6412	1.560	1.188	1.853	1.0007	20
50	.5730	.5422	.8403	.6453	1.550	1.190	1.844	.9977	10
33°00′	.5760	.5446	.8387	.6494	1.540	1.192	1.836	.9948	**57°00′**
10	.5789	.5471	.8371	.6536	1.530	1.195	1.828	.9919	50
20	.5818	.5495	.8355	.6577	1.520	1.197	1.820	.9890	40
30	.5847	.5519	.8339	.6619	1.511	1.199	1.812	.9861	30
40	.5876	.5544	.8323	.6661	1.501	1.202	1.804	.9832	20
50	.5905	.5568	.8307	.6703	1.492	1.204	1.796	.9803	10
34°00′	.5934	.5592	.8290	.6745	1.483	1.206	1.788	.9774	**56°00′**
10	.5963	.5616	.8274	.6787	1.473	1.209	1.781	.9745	50
20	.5992	.5640	.8258	.6830	1.464	1.211	1.773	.9716	40
30	.6021	.5664	.8241	.6873	1.455	1.213	1.766	.9687	30
40	.6050	.5688	.8225	.6916	1.446	1.216	1.758	.9657	20
50	.6080	.5712	.8208	.6959	1.437	1.218	1.751	.9628	10
		cos θ	sin θ	cot θ	tan θ	csc θ	sec θ	θ (radians)	θ (degrees)

θ (degrees)	θ (radians)	sin θ	cos θ	tan θ	cot θ	sec θ	csc θ		
35°00′	.6109	.5736	.8192	.7002	1.428	1.221	1.743	.9599	**55°00′**
10	.6138	.5760	.8175	.7046	1.419	1.223	1.736	.9570	50
20	.6167	.5783	.8158	.7089	1.411	1.226	1.729	.9541	40
30	.6196	.5807	.8141	.7133	1.402	1.228	1.722	.9512	30
40	.6225	.5831	.8124	.7177	1.393	1.231	1.715	.9483	20
50	.6254	.5854	.8107	.7221	1.385	1.233	1.708	.9454	10
36°00′	.6283	.5878	.8090	.7265	1.376	1.236	1.701	.9425	**54°00′**
10	.6312	.5901	.8073	.7310	1.368	1.239	1.695	.9396	50
20	.6341	.5925	.8056	.7355	1.360	1.241	1.688	.9367	40
30	.6370	.5948	.8039	.7400	1.351	1.244	1.681	.9338	30
40	.6400	.5972	.8021	.7445	1.343	1.247	1.675	.9308	20
50	.6429	.5995	.8004	.7490	1.335	1.249	1.668	.9279	10
37°00′	.6458	.6018	.7986	.7536	1.327	1.252	1.662	.9250	**53°00′**
10	.6487	.6041	.7969	.7581	1.319	1.255	1.655	.9221	50
20	.6516	.6065	.7951	.7627	1.311	1.258	1.649	.9192	40
30	.6545	.6088	.7934	.7673	1.303	1.260	1.643	.9163	30
40	.6574	.6111	.7916	.7720	1.295	1.263	1.636	.9134	20
50	.6603	.6134	.7898	.7766	1.288	1.266	1.630	.9105	10
38°00′	.6632	.6157	.7880	.7813	1.280	1.269	1.624	.9076	**52°00′**
10	.6661	.6180	.7862	.7860	1.272	1.272	1.618	.9047	50
20	.6690	.6202	.7844	.7907	1.265	1.275	1.612	.9018	40
30	.6720	.6225	.7826	.7954	1.257	1.278	1.606	.8988	30
40	.6749	.6248	.7808	.8002	1.250	1.281	1.601	.8959	20
50	.6778	.6271	.7790	.8050	1.242	1.284	1.595	.8930	10
39°00′	.6807	.6293	.7771	.8098	1.235	1.287	1.589	.8901	**51°00′**
10	.6836	.6316	.7753	.8146	1.228	1.290	1.583	.8872	50
20	.6865	.6338	.7735	.8195	1.220	1.293	1.578	.8843	40
30	.6894	.6361	.7716	.8243	1.213	1.296	1.572	.8814	30
40	.6923	.6383	.7698	.8292	1.206	1.299	1.567	.8785	20
50	.6952	.6406	.7679	.8342	1.199	1.302	1.561	.8756	10
40°00′	.6981	.6428	.7660	.8391	1.192	1.305	1.556	.8727	**50°00′**
10	.7010	.6450	.7642	.8441	1.185	1.309	1.550	.8698	50
20	.7039	.6472	.7623	.8491	1.178	1.312	1.545	.8668	40
30	.7069	.6494	.7604	.8541	1.171	1.315	1.540	.8639	30
40	.7098	.6517	.7585	.8591	1.164	1.318	1.535	.8610	20
50	.7127	.6539	.7566	.8642	1.157	1.322	1.529	.8581	10
41°00′	.7156	.6561	.7547	.8693	1.150	1.325	1.524	.8552	**49°00′**
10	.7185	.6583	.7528	.8744	1.144	1.328	1.519	.8523	50
20	.7214	.6604	.7509	.8796	1.137	1.332	1.514	.8494	40
30	.7243	.6626	.7490	.8847	1.130	1.335	1.509	.8465	30
40	.7272	.6648	.7470	.8899	1.124	1.339	1.504	.8436	20
50	.7301	.6670	.7451	.8952	1.117	1.342	1.499	.8407	10
		cos θ	sin θ	cot θ	tan θ	csc θ	sec θ	θ (radians)	θ (degrees)

Trigonometric Functions in Degrees and Radians (continued)

θ (degrees)	θ (radians)	sin θ	cos θ	tan θ	cot θ	sec θ	csc θ		
42°00′	.7330	.6691	.7431	.9004	1.111	1.346	1.494	.8378	**48°00′**
10	.7359	.6713	.7412	.9057	1.104	1.349	1.490	.8348	50
20	.7389	.6734	.7392	.9110	1.098	1.353	1.485	.8319	40
30	.7418	.6756	.7373	.9163	1.091	1.356	1.480	.8290	30
40	.7447	.6777	.7353	.9217	1.085	1.360	1.476	.8261	20
50	.7476	.6799	.7333	.9271	1.079	1.364	1.471	.8232	10
43°00′	.7505	.6820	.7314	.9325	1.072	1.367	1.466	.8203	**47°00′**
10	.7534	.6841	.7294	.9380	1.066	1.371	1.462	.8174	50
20	.7563	.6862	.7274	.9435	1.060	1.375	1.457	.8145	40
30	.7592	.6884	.7254	.9490	1.054	1.379	1.453	.8116	30
40	.7621	.6905	.7234	.9545	1.048	1.382	1.448	.8087	20
50	.7650	.6926	.7214	.9601	1.042	1.386	1.444	.8058	10
44°00′	.7679	.6947	.7193	.9657	1.036	1.390	1.440	.8029	**46°00′**
10	.7709	.6967	.7173	.9713	1.030	1.394	1.435	.7999	50
20	.7738	.6988	.7153	.9770	1.024	1.398	1.431	.7970	40
30	.7767	.7009	.7133	.9827	1.018	1.402	1.427	.7941	30
40	.7796	.7030	.7112	.9884	1.012	1.406	1.423	.7912	20
50	.7825	.7050	.7092	.9942	1.006	1.410	1.418	.7883	10
45°00′	.7854	.7071	.7071	1.000	1.000	1.414	1.414	.7854	**45°00′**
		cos θ	sin θ	cot θ	tan θ	csc θ	sec θ	θ (radians)	θ (degrees)

Solutions to Selected Exercises

Exercise Set 1.1, page 9

6.
$$\begin{aligned} 5x - 2 &= 3x + 4 \\ 5x - 3x &= 4 + 2 \\ 2x &= 6 \\ x &= 3 \end{aligned}$$

The solution is 3.

16. $x^2 - 6x - 4 = 0$

$$\begin{aligned} x &= \frac{-(-6) \pm \sqrt{(-6)^2 - 4(1)(-4)}}{2(1)} \\ &= \frac{6 \pm \sqrt{36 + 16}}{2} = \frac{6 \pm \sqrt{52}}{2} \\ &= \frac{6 \pm 2\sqrt{13}}{2} = 3 \pm \sqrt{13} \end{aligned}$$

The solutions are $3 - \sqrt{13}$ and $3 + \sqrt{13}$.

28.
$$\begin{aligned} d &= \sqrt{(5 - 2)^2 + [-2 - (-4)]^2} \\ &= \sqrt{9 + 4} \\ &= \sqrt{13} \end{aligned}$$

38. $\left(\frac{1 + 3}{2}, \frac{-4 + (-5)}{2}\right) = \left(2, -\frac{9}{2}\right)$

Exercise Set 1.2, page 15

16. $y = -|x| + 3$

x	y
−3	0
−2	1
−1	2
0	3
1	2
2	1
3	0

26. $y = -2x^2 + 4x - 5$

x	y
−1	−11
0	−5
1	−3
2	−5
3	−11

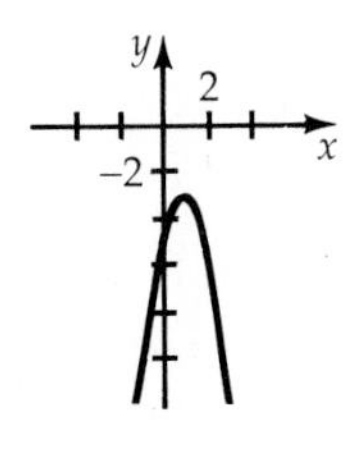

32.
$$2x + 3y = -9 \qquad 2(0) + 3y = -9$$
$$2x + 3(0) = -9 \qquad 3y = -9$$
$$2x = -9 \qquad y = -3$$
$$x = -\frac{9}{2}$$

The intercepts are $\left(-\frac{9}{2}, 0\right)$ and $(0, -3)$.

40. $y = x^2 - 4x + 1$

$$x = \frac{4 \pm \sqrt{16 - 4}}{2} = \frac{4 \pm \sqrt{12}}{2} \qquad y = 0^2 - 4(0) + 1 = 1$$
$$= \frac{4 \pm 2\sqrt{3}}{2} = 2 \pm \sqrt{3}$$

The intercepts are $(2 + \sqrt{3}, 0)$, $(2 - \sqrt{3}, 0)$ and $(0, 1)$.

52.
$$x^2 + y^2 - 6x - 4y - 12 = 0$$
$$(x^2 = 6x + 9) + (y^2 - 4y + 4) = 12 + 9 + 4$$
$$(x - 3)^2 + (y - 2)^2 = 25$$

center $(3, 2)$, radius 5

60.
$$r = \sqrt{(-1 - 2)^2 + (-2 - (-1))^2}$$
$$= \sqrt{9 + 1} = \sqrt{10}$$

The center is $(2, -1)$.

$$(x - 2)^2 + [y - (-1)]^2 = (\sqrt{10})^2$$
$$(x - 2)^2 + (y + 1)^2 = 10$$

Exercise Set 1.3, page 22

4. $2x^2 + y = 3$

$y = -2x^2 + 3$

Because each value of x produces one value of y, $2x^2 + y = 3$ defines y as a function of x.

24.
$$k(x) = 3x + 1$$
$$k\left(-\frac{1}{2}\right) = 3\left(-\frac{1}{2}\right) + 1$$
$$= -\frac{3}{2} + 1 = -\frac{1}{2}$$

30.
$$(h - k)(x) = h(x) - k(x)$$
$$= (4 - 7x) - (3x + 1)$$
$$= -10x + 3$$
$$(h - k)(4) = -10(4) + 3 = -40 + 3 = -37$$

34.
$$f(x) = x^3 - 6$$
$$f(c) = c^3 - 6 = 2$$
$$c^3 = 8$$
$$c = 2$$

38.
$$d = \sqrt{50^2 + (3t)^2}$$
$$= \sqrt{2500 + 9t^2}$$

Exercise Set 1.4, page 31

6.

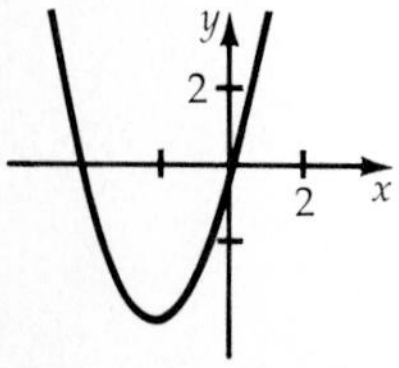

16. This is the graph of a function because every vertical line intersects the graph in exactly one point.

28. This is not the graph of a one-to-one function because some horizontal lines intersect the graph in more than one point.

36.
$$f(x) = |x - 2|$$
$$f(-x) = |-x - 2|$$

Thus $f(-x) \neq f(x)$ and $f(-x) \neq -f(x)$. The function is neither even nor odd.

46.

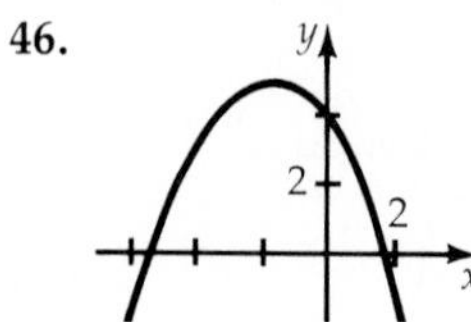

Exercise Set 1.5, page 39

4. $g(2) = \dfrac{1}{2}$

$k(g(2)) = 2\left(\dfrac{1}{2}\right) - 3$

$= 1 - 3 = -2$

14. $H(r) = 5r + 10$

$y = 5r + 10$

$r = 5y + 10$ Interchange y and r.

$r - 10 = 5y$

$\dfrac{r - 10}{5} = y$

$H^{-1}(r) = \dfrac{r - 10}{5} = \dfrac{1}{5}r - 2$

18. $G(s) = \dfrac{3}{s - 2}, \quad s \neq 2$

$y = \dfrac{3}{s - 2}$

$s = \dfrac{3}{y - 2}$

$sy - 2s = 3$

$sy = 2s + 3$

$y = \dfrac{2s + 3}{s}$

$G^{-1}(s) = \dfrac{2s + 3}{s}, \quad s \neq 0$

30. $G(s) = s^2 - 4, \quad s \geqslant 0$

$G^{-1}(s) = \sqrt{s + 4}, \quad s \geq -4$

$G(G^{-1}(s)) = (\sqrt{s + 4})^2 - 4$

$= s + 4 - 4 = s$

$G^{-1}(G(s)) = \sqrt{(s^2 - 4) + 4}$

$= \sqrt{s^2} = s$

Exercise Set 2.1, page 54

22. $35°42' = 35° + 42'' = 35° + 42''\left(\dfrac{1°}{3600''}\right)$

$= 35° + 0.012° = 35.012°$

30. $55.44° = 55° + 0.44°\left(\dfrac{60'}{1°}\right) = 55° + 26.4'$

$= 55° + 26' + 0.4'\left(\dfrac{60''}{1'}\right)$

$= 55° + 26' + 24'' = 55°26'24''$

42. $585° = 585°\left(\dfrac{\pi}{180°}\right) = \dfrac{13}{4}\pi$

52. $\dfrac{11\pi}{18} = \dfrac{11\pi}{18}\left(\dfrac{180°}{\pi}\right) = 110°$

74. $\theta = \dfrac{s}{r} = \dfrac{4}{7} = 0.57$

$= 0.57\left(\dfrac{180°}{\pi}\right) \approx 32.7°$

80. $s = r\theta = 5\,(144°)\left(\dfrac{\pi}{180°}\right) \approx 12.57$ meters

82. $\theta = \dfrac{3}{8}(2\pi) = \dfrac{3}{4}\pi$

84. $\theta_2 = \dfrac{1.2}{0.8}(240°)\left(\dfrac{\pi}{180}\right) = 2\pi$

86. $\omega = \dfrac{\theta}{t} = \dfrac{2\pi}{86400} = 7.27 \times 10^{-5}$ radian/s

90. $v = \omega r$

$= \dfrac{450 \cdot 2\pi \cdot 60 \cdot 15}{12 \cdot 520} \approx 40.16$ mph

Exercise Set 2.2, page 61

8. $a = 10,\ b = 6,\ r = \sqrt{10^2 + 6^2} = 2\sqrt{34}.$

$$\sin\theta = \frac{b}{r} = \frac{6}{2\sqrt{34}} = \frac{3\sqrt{34}}{34} \qquad \csc\theta = \frac{\sqrt{34}}{3}$$

$$\cos\theta = \frac{a}{r} = \frac{10}{2\sqrt{34}} = \frac{5\sqrt{34}}{34} \qquad \sec\theta = \frac{\sqrt{34}}{5}$$

$$\tan\theta = \frac{b}{a} = \frac{6}{10} = \frac{3}{5} \qquad \cot\theta = \frac{5}{3}$$

18. Since $\tan\theta = \frac{b}{a} = \frac{4}{3}$, let $b = 4$ and $a = 3$.

$$r = \sqrt{3^2 + 4^2} = 5$$

$$\sec\theta = \frac{r}{a} = \frac{5}{3}$$

48. $\sec\frac{3\pi}{8} = 2.6131$

70. $\sin\frac{\pi}{3}\cos\frac{\pi}{4} - \tan\frac{\pi}{4} = \frac{\sqrt{3}}{2}\cdot\frac{\sqrt{2}}{2} - 1$

$$= \frac{\sqrt{6}}{4} - 1 = \frac{\sqrt{6} - 4}{4}$$

Exercise Set 2.3, page 68

6. $a = -6,\ b = -9,\ r = \sqrt{(-6)^2 + (-9)^2} = \sqrt{117} = 3\sqrt{13}.$

$$\sin\theta = \frac{b}{r} = \frac{-9}{3\sqrt{13}} = -\frac{3}{\sqrt{13}} = -\frac{3\sqrt{13}}{13} \qquad \csc\theta = -\frac{\sqrt{13}}{3}$$

$$\cos\theta = \frac{a}{r} = \frac{-6}{3\sqrt{13}} = -\frac{2}{\sqrt{13}} = -\frac{2\sqrt{13}}{13} \qquad \sec\theta = -\frac{\sqrt{13}}{2}$$

$$\tan\theta = \frac{b}{a} = \frac{-9}{-6} = \frac{3}{2} \qquad \cot\theta = \frac{2}{3}$$

10. The reference angle for 225° is 45°.

$$\sin 225° = -\frac{\sqrt{2}}{2} \qquad \csc 225° = -\sqrt{2}$$

$$\cos 225° = -\frac{\sqrt{2}}{2} \qquad \sec 225° = -\sqrt{2}$$

$$\tan 225° = 1 \qquad \cot 225° = 1$$

26. $\cos\frac{7\pi}{4}\tan\frac{4\pi}{3} + \cos\frac{7\pi}{6} = \frac{\sqrt{2}}{2}\cdot(\sqrt{3}) + \left(-\frac{\sqrt{3}}{2}\right)$

$$= \frac{\sqrt{6}}{2} - \frac{\sqrt{3}}{2} = \frac{\sqrt{6} - \sqrt{3}}{2}$$

40. $\sec\phi = \frac{2\sqrt{3}}{3} = \frac{r}{a},\ r = 2\sqrt{3},\ a = 3,$

$b = \pm\sqrt{(2\sqrt{3})^2 - 3^2} = \pm\sqrt{3},$

$b = -\sqrt{3}$ in quadrant IV.

$$\sin\phi = \frac{-\sqrt{3}}{2\sqrt{3}} = -\frac{1}{2}$$

54. $\sin 398° \approx 0.6157$

56. $\sin 740° \approx 0.3420$

Exercise Set 2.4, page 76

48. $\sin\frac{37\pi}{4} = \sin\left(\frac{5\pi}{4} + \frac{32\pi}{4}\right) = \sin\left(\frac{5\pi}{4} + 8\pi\right)$

$$= \sin\left(\frac{5\pi}{4} + 4(2\pi)\right) = \sin\frac{5\pi}{4} = -\frac{\sqrt{2}}{2}$$

58.

$$\tan\phi = \frac{y}{x}$$

$$\tan(\phi - 180°) = \frac{-y}{-x} = \frac{y}{x}$$

$$\tan\phi = \tan(\phi - 180°)$$

80. $\cos\phi \sec^2\phi - \dfrac{\cos\phi}{\cot^2\phi} = \cos\phi \dfrac{1}{\cos^2\phi} - \dfrac{\cos\phi}{\dfrac{\cos^2\phi}{\sin^2\phi}}$

$$= \frac{1}{\cos\phi} - \cos\phi \frac{\sin^2\phi}{\cos^2\phi}$$

$$= \frac{1}{\cos\phi} - \frac{\sin^2\phi}{\cos\phi}$$

$$= \frac{1-\sin^2\phi}{\cos\phi}$$

$$= \frac{\cos^2\phi}{\cos\phi} = \cos\phi$$

96. $\sin\phi = \dfrac{1}{\csc\phi} = \dfrac{1}{\sqrt{2}} = \dfrac{\sqrt{2}}{2}$

98. $\cos\phi = -\sqrt{1-\sin^2\phi}$ since ϕ is in quadrant II.

$$\cos\phi = -\sqrt{1-\left(\frac{1}{2}\right)^2} = -\sqrt{1-\frac{1}{4}} = -\frac{\sqrt{3}}{2}$$

$$\tan\phi = \frac{\sin\phi}{\cos\phi} = \frac{\frac{1}{2}}{-\frac{\sqrt{3}}{2}} = -\frac{1}{\sqrt{3}} = -\frac{\sqrt{3}}{3}$$

106. $\tan\phi = \dfrac{\sin\phi}{\cos\phi} = \dfrac{\sin\phi}{\pm\sqrt{1-\sin^2\phi}} = \pm\dfrac{\sin\phi}{\sqrt{1-\sin^2\phi}}$

Exercise Set 2.5, page 87

20. $f(x) = \dfrac{3}{2}\sin x$, $a = \left|\dfrac{3}{2}\right| = \dfrac{3}{2}$, $p = 2\pi$.

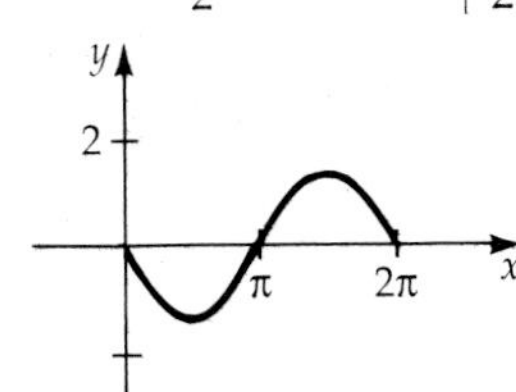

28. $f(x) = 3\cos\dfrac{\pi x}{3}$, $a = 3$, $p = \dfrac{2\pi}{\pi/3} = 6$.

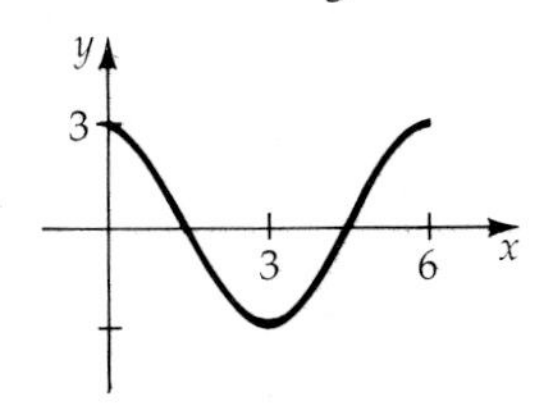

32. $f(x) = \dfrac{1}{2}\sin 2.5x$, $a = \dfrac{1}{2}$, $p = \dfrac{2\pi}{2.5} = \dfrac{4\pi}{5}$.

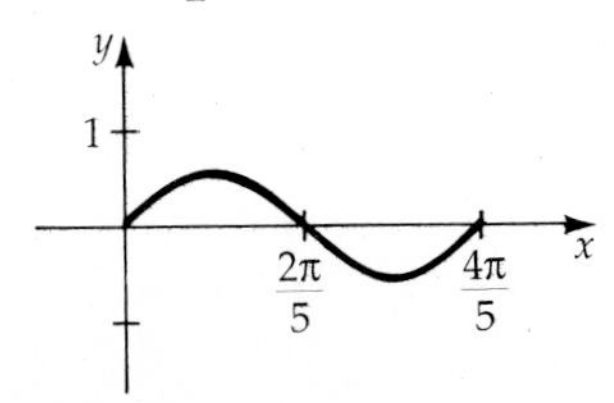

34. $f(x) = -\dfrac{3}{4}\cos 5x$, $a = \left|-\dfrac{3}{4}\right| = \dfrac{3}{4}$, $p = \dfrac{2\pi}{5}$.

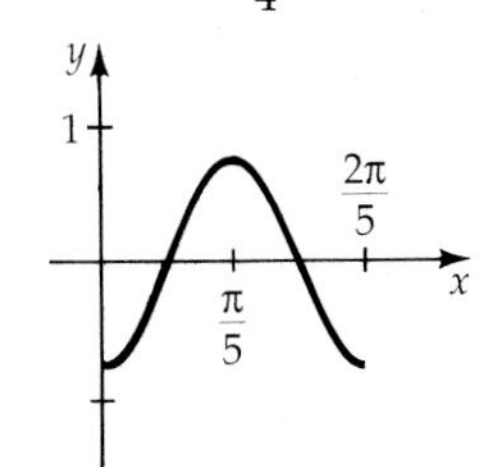

40. $f(x) = -\left|3\sin\dfrac{2}{3}x\right|$

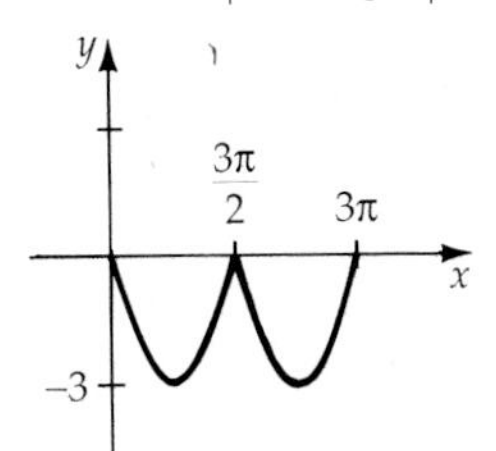

Exercise Set 2.6, page 95

30. $f(x) = -3 \tan 3x,\ p = \dfrac{2\pi}{3}.$

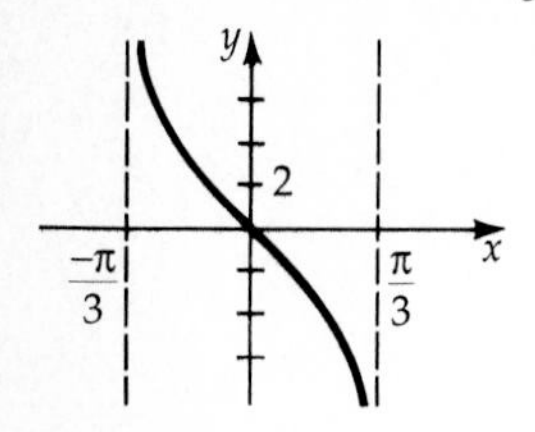

32. $f(x) = \dfrac{1}{2} \cot 2x,\ p = \dfrac{\pi}{2}.$

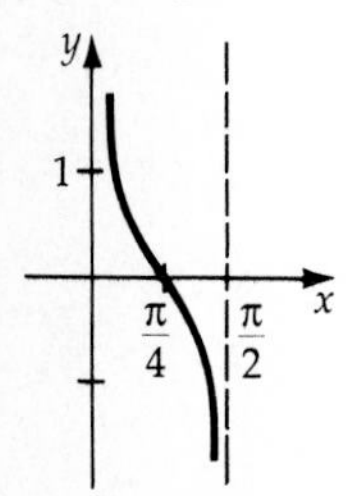

38. $f(x) = 3 \csc \dfrac{\pi x}{2},\ p = \dfrac{2\pi}{\pi/2} = 4.$

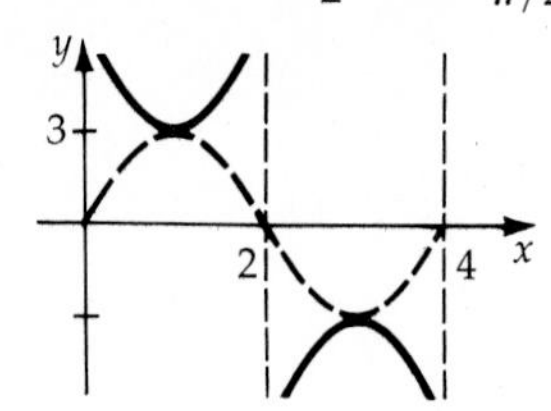

42. $f(x) = \sec \dfrac{x}{2},\ p = \dfrac{2\pi}{1/2} = 4\pi.$

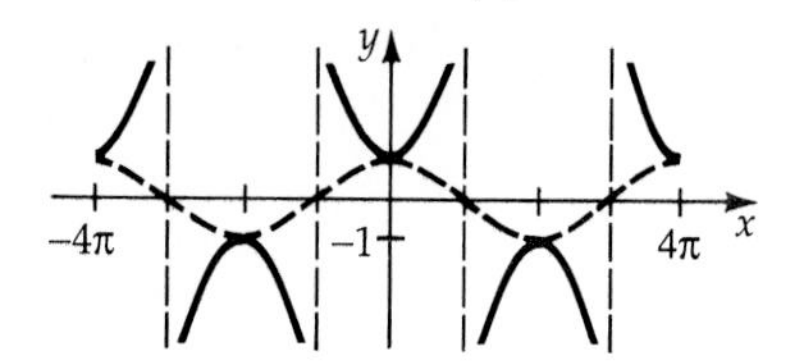

Exercise Set 2.7, page 101

20. $f(x) = \cos\left(2x - \dfrac{\pi}{3}\right)$

$$0 \le 2x - \frac{\pi}{3} \le 2\pi$$

$$\frac{\pi}{3} \le 2x \le \frac{7\pi}{3}$$

$$\frac{\pi}{6} \le x \le \frac{7\pi}{6}$$

Period $= \pi$, phase shift $= \dfrac{\pi}{6}$.

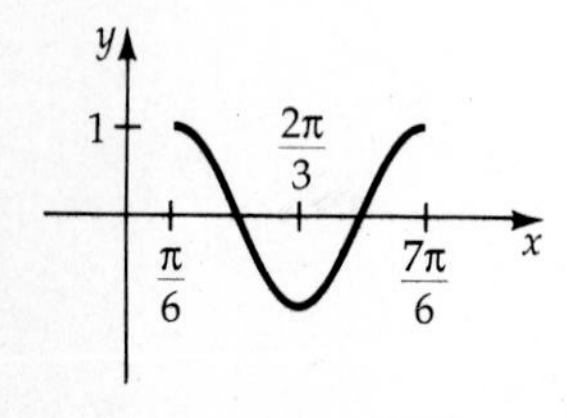

22. $f(x) = \tan(x - \pi)$

$$-\frac{\pi}{2} < x - \pi < \frac{\pi}{2}$$

$$\frac{\pi}{2} < x < \frac{3\pi}{2}$$

Period $= \pi$, phase shift $= \pi$.

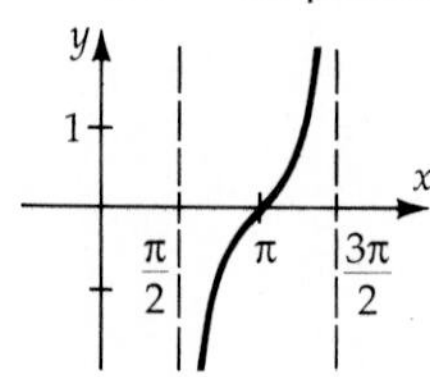

28. $f(x) = \sec\left(2x + \dfrac{\pi}{6}\right)$

$$0 \le 2x + \frac{\pi}{6} \le 2\pi$$

$$-\frac{\pi}{12} \le x \le \frac{11\pi}{12}$$

Period $= \pi$, phase shift $= -\dfrac{\pi}{12}$.

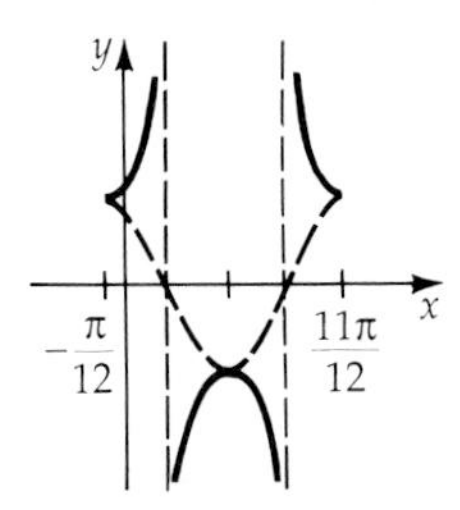

40. $f(x) = -2\cos\left(x + \dfrac{\pi}{3}\right) + 3, p = 2\pi$, phase angle $= -\dfrac{\pi}{3}$.

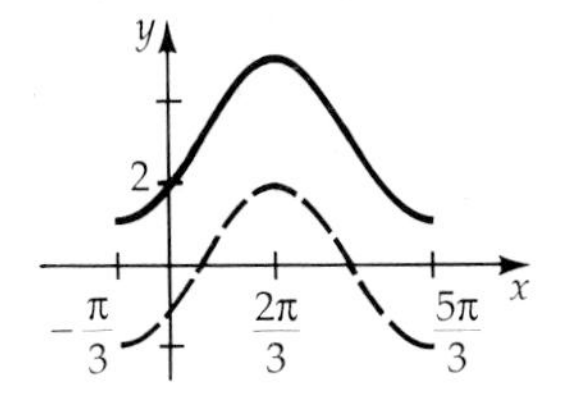

44. $f(x) = \cos\dfrac{x}{3} + 4,\ p = 6\pi$.

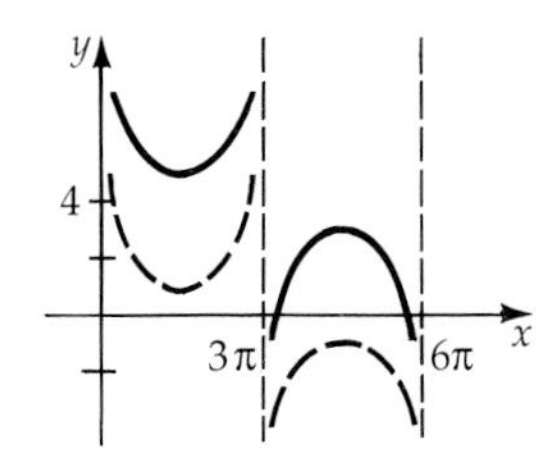

50. $f(x) = \dfrac{2}{3}x - \sin\dfrac{x}{2}$

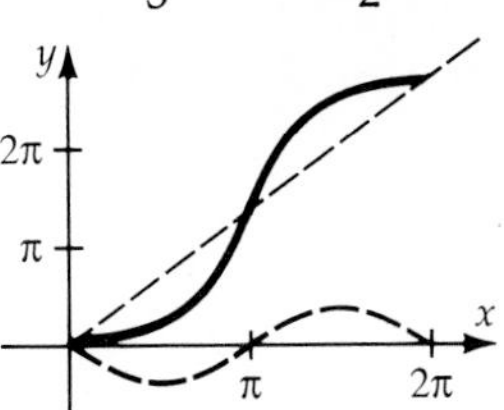

52. $f(x) = -\sin x + \cos x$

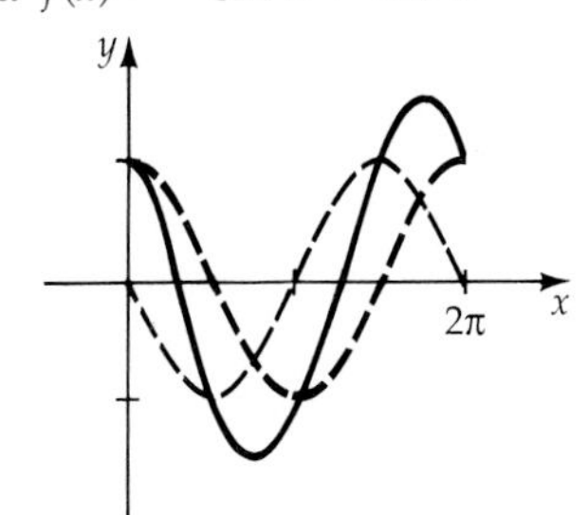

58. $f(x) = x\cos x$

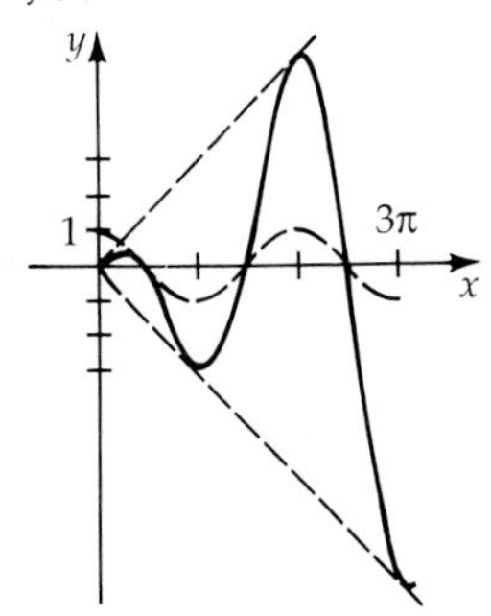

Exercise Set 2.8, page 107

20. Amplitude = 3, frequency = $1/\pi$, period = π. Since $2\pi/b = \pi$, we have $b = 2$. Thus

$$y = 3\cos 2t$$

28. Amplitude = 1.5.

$$f = \frac{1}{2\pi}\sqrt{\frac{k}{m}} = \frac{1}{2\pi}\sqrt{\frac{3}{27}} = \frac{1}{2\pi}\cdot\frac{1}{3} = \frac{1}{6\pi}, \qquad p = 6\pi$$

$$y = 1.5\cos 2\pi ft = 1.5\cos\left[2\pi\left(\frac{1}{6\pi}\right)t\right] = 1.5\cos\frac{1}{3}t$$

30. Amplitude = 4.

$$f = \frac{1}{2\pi}\sqrt{\frac{g}{l}} = \frac{1}{2\pi}\sqrt{\frac{32}{20}} = \frac{1}{2\pi}\cdot\frac{4}{\sqrt{10}} = \frac{\sqrt{10}}{5\pi}$$

$$p = \frac{5\pi}{\sqrt{10}} = \frac{\pi\sqrt{10}}{2}$$

$$y = 4\cos 2\pi ft = 4\cos\left[2\pi\left(\frac{\sqrt{10}}{5\pi}\right)t\right] = 4\cos\frac{2\sqrt{10}}{5}t$$

Exercise Set 3.1, page 116

30. $\sin x \cot x \sec x = \sin x \cdot \dfrac{\cos x}{\sin x} \cdot \dfrac{1}{\cos x} = 1$

40.
$$\begin{aligned}\sin^4 x - \cos^4 x &= (\sin^2 x + \cos^2 x)(\sin^2 x - \cos^2 x)\\ &= 1(\sin^2 x - \cos^2 x)\\ &= \sin^2 x - \cos^2 x\end{aligned}$$

52.
$$\begin{aligned}\frac{2 \sin x \cot x + \sin x - 4 \cot x - 2}{2 \cot x + 1} &= \frac{(\sin x)(2 \cot x + 1) - 2(2 \cot x + 1)}{2 \cot x + 1}\\ &= \frac{(2 \cot x + 1)(\sin x - 2)}{2 \cot x + 1}\\ &= \sin x - 2\end{aligned}$$

62.
$$\begin{aligned}\frac{\dfrac{1}{\sin x} + \dfrac{1}{\cos x}}{\dfrac{1}{\sin x} - \dfrac{1}{\cos x}} &= \frac{\dfrac{1}{\sin x} + \dfrac{1}{\cos x}}{\dfrac{1}{\sin x} - \dfrac{1}{\cos x}} \cdot \frac{\sin x \cos x}{\sin x \cos x}\\ &= \frac{\cos x + \sin x}{\cos x - \sin x}\\ &= \frac{\cos x + \sin x}{\cos x - \sin x} \cdot \frac{\cos x - \sin x}{\cos x - \sin x}\\ &= \frac{\cos^2 x - \sin^2 x}{\cos^2 x - 2 \sin x \cos x + \sin^2 x}\\ &= \frac{\cos^2 x - \sin^2 x}{1 - 2 \sin x \cos x}\end{aligned}$$

72.
$$\begin{aligned}\frac{\dfrac{1}{\tan x} + \cot x}{\dfrac{1}{\tan x} + \tan x} &= \frac{\dfrac{1}{\tan x} + \cot x}{\dfrac{1}{\tan x} + \tan x} \cdot \frac{\tan x}{\tan x}\\ &= \frac{1 + 1}{1 + \tan^2 x}\\ &= \frac{2}{\sec^2 x}\end{aligned}$$

Exercise Set 3.2, page 125

20.
$$\begin{aligned}\sin \frac{11\pi}{12} &= \sin\left(\frac{2\pi}{3} + \frac{\pi}{4}\right)\\ &= \sin \frac{2\pi}{3} \cos \frac{\pi}{4} + \cos \frac{2\pi}{3} \sin \frac{\pi}{4}\\ &= \frac{\sqrt{3}}{2} \cdot \frac{\sqrt{2}}{2} + \left(-\frac{1}{2} \cdot \frac{\sqrt{2}}{2}\right)\\ &= \frac{\sqrt{6}}{4} - \frac{\sqrt{2}}{4} = \frac{\sqrt{6} - \sqrt{2}}{4}\end{aligned}$$

22.
$$\begin{aligned}\cos \frac{\pi}{12} &= \cos\left(\frac{\pi}{3} - \frac{\pi}{4}\right)\\ &= \cos \frac{\pi}{3} \cos \frac{\pi}{4} + \sin \frac{\pi}{3} \sin \frac{\pi}{4}\\ &= \frac{1}{2} \cdot \frac{\sqrt{2}}{2} + \frac{\sqrt{3}}{2} \cdot \frac{\sqrt{2}}{2}\\ &= \frac{\sqrt{2}}{4} + \frac{\sqrt{6}}{4} = \frac{\sqrt{2} + \sqrt{6}}{4}\end{aligned}$$

24.
$$\begin{aligned}\tan \frac{7\pi}{12} &= \tan\left(\frac{\pi}{3} + \frac{\pi}{4}\right)\\ &= \frac{\tan \pi/3 + \tan \pi/4}{1 - \tan \pi/3 \tan \pi/4}\\ &= \frac{\sqrt{3} + 1}{1 - \sqrt{3}(1)} = \frac{1 + \sqrt{3}}{1 - \sqrt{3}} \cdot \frac{1 + \sqrt{3}}{1 + \sqrt{3}}\\ &= -2 - \sqrt{3}\end{aligned}$$

62. $\sin \alpha = \dfrac{24}{25}$, $\cos \alpha = -\dfrac{7}{25}$, $\cos \beta = -\dfrac{4}{5}$, $\sin \beta = -\dfrac{3}{5}$

a.
$$\begin{aligned}\cos(\beta - \alpha) &= \cos \beta \cos \alpha + \sin \beta \sin \alpha\\ &= -\frac{4}{5} \cdot \left(-\frac{7}{25}\right) + \left(-\frac{3}{5}\right) \cdot \frac{24}{25}\\ &= \frac{28}{125} - \frac{72}{125}\\ &= -\frac{44}{125}\end{aligned}$$

b.
$$\begin{aligned}\sin(\alpha+\beta) &= \sin\alpha\cos\beta+\cos\alpha\sin\beta\\ &= \frac{24}{25}\cdot\left(-\frac{4}{5}\right)+\left(-\frac{7}{25}\right)\cdot\left(-\frac{3}{5}\right)\\ &= -\frac{96}{125}+\frac{21}{125}\\ &= -\frac{75}{125}=-\frac{3}{5}\end{aligned}$$

70.
$$\begin{aligned}\cos(\theta+\pi) &= \cos\theta\cos\pi-\sin\theta\sin\pi\\ &= (\cos\theta)(-1)-(\sin\theta)(0)\\ &= -\cos\theta\end{aligned}$$

82.
$$\begin{aligned}\cos 5x\cos 3x+\sin 5x\sin 3x &= \cos(5x-3x)\\ &= \cos 2x\\ &= \cos(x+x)\\ &= \cos x\cos x-\sin x\sin x\\ &= \cos^2 x-\sin^2 x\end{aligned}$$

86.
$$\begin{aligned}&\sin(\alpha-\beta)-\sin(\alpha+\beta)\\ &= \sin\alpha\cos\beta-\cos\alpha\sin\beta-\sin\alpha\cos\beta-\cos\alpha\sin\beta\\ &= -2\cos\alpha\sin\beta\end{aligned}$$

Exercise Set 3.3, page 131

26.
$$\cos\theta=\frac{24}{25}\qquad \sin\theta=-\sqrt{1-\left(\frac{24}{25}\right)^2}=-\frac{7}{25}\qquad \tan\theta=\frac{-7/25}{24/25}=-\frac{7}{24}$$

$$\begin{aligned}\sin 2\theta &= 2\sin\theta\cos\theta & \cos 2\theta &= \cos^2\theta-\sin^2\theta\\ &= 2\left(\frac{-7}{25}\right)\left(\frac{24}{25}\right) & &= \left(\frac{24}{25}\right)^2-\left(-\frac{7}{25}\right)^2\\ &= -\frac{336}{625} & &= \frac{527}{625}\end{aligned}$$

$$\begin{aligned}\tan 2\theta &= \frac{2\tan\theta}{1-\tan^2\theta}\\ &= \frac{2(-7/24)}{1-(-7/24)^2}\\ &= \frac{-7/12}{1-49/576}\\ &= -\frac{336}{527}\end{aligned}$$

38.
$$\sin\alpha=-\frac{7}{25},\ \cos\alpha=-\sqrt{1-\left(-\frac{7}{25}\right)^2}=-\frac{24}{25}$$

$$\begin{aligned}\sin\frac{\alpha}{2} &= \sqrt{\frac{1-\cos\alpha}{2}} & \cos\frac{\alpha}{2} &= -\sqrt{\frac{1+\cos\alpha}{2}}\\ &= \sqrt{\frac{1+24/25}{2}} & &= \sqrt{\frac{1-24/25}{2}}\\ &= \frac{7\sqrt{2}}{10} & &= -\frac{\sqrt{2}}{10}\end{aligned}$$

$$\begin{aligned}\tan\frac{\alpha}{2} &= \frac{1-\cos\alpha}{\sin\alpha}\\ &= \frac{1+24/25}{-7/25}=-7\end{aligned}$$

54.
$$\begin{aligned}\frac{1}{1-\cos 2x} &= \frac{1}{1-1+2\sin^2 x}\\ &= \frac{1}{2\sin^2 x}=\frac{1}{2}\csc^2 x\end{aligned}$$

58.
$$\frac{2\sin x\cos x}{\cos^2 x-\sin^2 x}=\frac{\sin 2x}{\cos 2x}=\tan 2x$$

62.
$$\begin{aligned}\sin 4x &= 2\sin 2x\cos 2x\\ &= 2(2\sin x\cos x)(\cos^2 x-\sin^2 x)\\ &= 4\sin x\cos^3 x-4\sin^3 x\cos x\end{aligned}$$

68.
$$\begin{aligned}\sin 3x+\sin x &= \sin(2x+x)+\sin x\\ &= \sin 2x\cos x+\cos 2x\sin x+\sin x\\ &= (2\sin x\cos x)\cos x\\ &\quad +(1-2\sin^2 x)\sin x+\sin x\\ &= 2\sin x\cos^2 x+\sin x-2\sin^3 x\\ &\quad +\sin x\\ &= 2\sin x(1-\sin^2 x)+2\sin x-2\sin^3 x\\ &= 2\sin x-2\sin^3 x+2\sin x-2\sin^3 x\\ &= 4\sin x-4\sin^3 x\end{aligned}$$

72.
$$\begin{aligned}\cos^2\frac{x}{2} &= \left[\pm\sqrt{\frac{1+\cos x}{2}}\right]^2=\frac{1+\cos x}{2}\\ &= \frac{1+\cos x}{2}\cdot\frac{\sec x}{\sec x}=\frac{\sec x+1}{2\sec x}\end{aligned}$$

78.
$$\tan^2 \frac{x}{2} = \left(\frac{1 - \cos x}{\sin x}\right)^2$$
$$= \frac{(1 - \cos x)^2}{\sin^2 x}$$
$$= \frac{(1 - \cos x)^2}{1 - \cos^2 x}$$
$$= \frac{(1 - \cos x)^2}{(1 - \cos x)(1 + \cos x)}$$
$$= \frac{1 - \cos x}{1 + \cos x}$$
$$= \frac{\dfrac{1}{\cos x} - \dfrac{\cos x}{\cos x}}{\dfrac{1}{\cos x} + \dfrac{\cos x}{\cos x}}$$
$$= \frac{\sec x - 1}{\sec x + 1}$$

Exercise Set 3.4, page 139

12.
$$\sin 195° \cos 15° = \frac{1}{2}[\sin(195° + 15°) + \sin(195° - 15°)]$$
$$= \frac{1}{2}(\sin 210° + \sin 180°)$$
$$= \frac{1}{2}\left(-\frac{1}{2} + 0\right) = -\frac{1}{4}$$

22.
$$\cos 3\theta + \cos 5\theta = 2 \cos \frac{3\theta + 5\theta}{2} \cos \frac{3\theta - 5\theta}{2}$$
$$= 2 \cos 4\theta \cos(-\theta)$$
$$= 2 \cos 4\theta \cos \theta$$

36.
$$\sin 5x \cos 3x = \frac{1}{2}[\sin(5x + 3x) + \sin(5x - 3x)]$$
$$= \frac{1}{2}(\sin 8x + \sin 2x)$$
$$= \frac{1}{2}(2 \sin 4x \cos 4x + 2 \sin x \cos x)$$
$$= \sin 4x \cos 4x + \sin x \cos x$$

44.
$$\frac{\cos 5x - \cos 3x}{\sin 5x + \sin 3x} = \frac{-2 \sin \dfrac{5x + 3x}{2} \sin \dfrac{5x - 3x}{2}}{2 \sin \dfrac{5x + 3x}{2} \cos \dfrac{5x - 3x}{2}}$$
$$= -\frac{\sin 4x \sin x}{\sin 4x \cos x} = -\tan x$$

62. $a = 1$, $b = \sqrt{3}$, $k = \sqrt{(\sqrt{3})^2 + (1)^2} = 2$. Thus α is a first quadrant angle.
$$\sin \beta = \left|\frac{\sqrt{3}}{2}\right| = \frac{\sqrt{3}}{2}$$
$$\beta = \frac{\pi}{3}, \qquad \alpha = \frac{\pi}{3}$$
$$y = k \sin(x + \alpha)$$
$$y = 2 \sin\left(x + \frac{\pi}{3}\right)$$

70. From Exercise 62, we know
$$y = \sin x + \sqrt{3} \cos x$$
$$y = 2 \sin\left(x + \frac{\pi}{3}\right)$$

Exercise Set 3.5, page 151

2.
$$y = \sin^{-1} \frac{\sqrt{2}}{2}$$
$$\sin y = \frac{\sqrt{2}}{2} \quad \text{for} \quad -\frac{\pi}{2} \le y \le \frac{\pi}{2}$$
$$y = \frac{\pi}{4}$$

10.
$$y = \sec^{-1} \frac{2\sqrt{3}}{3}$$
$$\sec y = \frac{2\sqrt{3}}{3} \quad \text{for} \quad 0 \le y \le \pi$$
$$\cos y = \frac{\sqrt{3}}{2}$$
$$y = \frac{\pi}{6}$$

18.
$$y = \tan^{-1} 1$$
$$\tan y = 1 \quad \text{for} \quad -\frac{\pi}{2} \le y \le \frac{\pi}{2}$$
$$y = \frac{\pi}{4}$$

30. $y = \sin^{-1}(-0.9650)$
$y \approx -74.8°$

40. $y = \csc^{-1}(-10.9856)$
$y \approx -5.2°$

54. $y = \sin^{-1}\left(\cos \frac{7\pi}{6}\right) = \sin^{-1}\left(-\frac{\sqrt{3}}{2}\right)$
$y = -\frac{\pi}{3}$

58. $y = \cot^{-1}\left(\cos \frac{2\pi}{3}\right) = \cot^{-1}\left(-\frac{1}{2}\right) = \tan^{-1}(-2)$
$y \approx -1.1071$

70. Let $x = \cos^{-1} 3/5$. Thus $\cos x = 3/5$ and $\sin x = \sqrt{1 - (3/5)^2} = 4/5$.
$$y = \tan\left(\cos^{-1} \frac{3}{5}\right) = \tan x$$
$$y = \frac{\sin x}{\cos x} = \frac{4/5}{3/5} = \frac{4}{3}$$

84.
$$\tan^{-1} x = \sin^{-1} \frac{24}{25}$$
$$\tan(\tan^{-1} x) = \tan\left(\sin^{-1} \frac{24}{25}\right)$$
$$x = \tan\left(\sin^{-1} \frac{24}{25}\right)$$
$$x = \frac{24}{7}$$

94.
$$\sin^{-1} x + \cos^{-1} \frac{4}{5} = \frac{\pi}{6}$$
$$\sin^{-1} x = \frac{\pi}{6} - \cos^{-1} \frac{4}{5}$$
$$\sin(\sin^{-1} x) = \sin\left(\frac{\pi}{6} - \cos^{-1} \frac{4}{5}\right)$$
$$x = \sin \frac{\pi}{6} \cos\left(\cos^{-1} \frac{4}{5}\right) - \cos \frac{\pi}{6} \sin\left(\cos^{-1} \frac{4}{5}\right)$$
$$= \frac{1}{2} \cdot \frac{4}{5} - \frac{\sqrt{3}}{2} \cdot \frac{3}{5} = \frac{4 - 3\sqrt{3}}{10}$$

102. Let $\alpha = \cos^{-1} x$ and $\beta = \cos^{-1}(-x)$. Thus $\cos \alpha = x$ and $\cos \beta = -x$. We have $\sin \alpha = \sqrt{1 - x^2}$ and we have $\sin \beta = \sqrt{1 - x^2}$ because α is in quadrant I and β is in quadrant II.
$$\cos^{-1} x + \cos^{-1}(-x) = \alpha + \beta$$
$$= \cos^{-1}[\cos(\alpha + \beta)]$$
$$= \cos^{-1}(\cos \alpha \cos \beta - \sin \alpha \sin \beta)$$
$$= \cos^{-1}[x(-x) - \sqrt{1 - x^2} \cdot \sqrt{1 - x^2}]$$
$$= \cos^{-1}(-x^2 - 1 + x^2)$$
$$= \cos^{-1}(-1) = \pi$$

106.

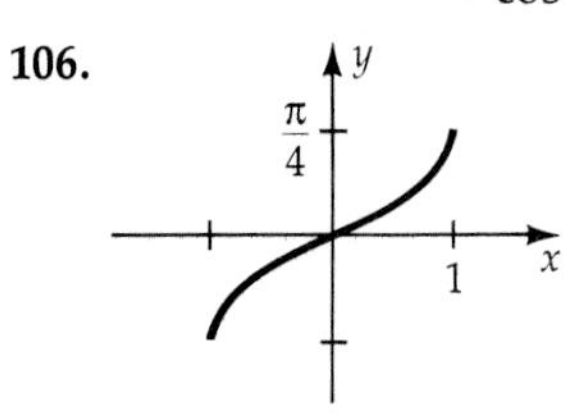

Exercise Set 3.6, page 157

44. $2 \sin^2 x + 5 \sin x + 3 = 0$
$$\sin x = \frac{-5 \pm \sqrt{5^2 - 4(2)(3)}}{2(2)} = \frac{-5 \pm 1}{4}$$
$$\sin x = -1 \qquad \sin x = -\frac{3}{2}$$
$$x = 270° \qquad \text{no solution}$$
The solution is 270°.

52.
$$\sin x + 2 \cos x = 1$$
$$\sin x = 1 - 2 \cos x$$
$$(\sin x)^2 = (1 - 2 \cos x)^2$$
$$\sin^2 x = 1 - 4 \cos x + 4 \cos^2 x$$
$$1 - \cos^2 x = 1 - 4 \cos x + 4 \cos^2 x$$
$$0 = 5 \cos^2 x - 4 \cos x$$
$$0 = \cos x(5 \cos x - 4)$$
$$\cos x = 0 \qquad 5 \cos x - 4 = 0$$
$$x = 90°, 270° \qquad \cos x = \frac{4}{5}$$
$$x \approx 36.9°, 323.1°$$
The solutions are 90°, 270°, 36.9°, and 323.1°.

56. $2\cos^2 x - 5\cos x - 5 = 0$

$$\cos x = \frac{5 \pm \sqrt{(-5)^2 - 4(2)(-5)}}{2(2)} = \frac{5 \pm \sqrt{65}}{4}$$

$\cos x \approx 3.26 \qquad \cos x \approx -0.7656$

no solution $\qquad x \approx 140.0°, 220.0°$

The solutions are 140.0° and 220.0°.

66. $\cos 2x = -\dfrac{\sqrt{3}}{2}$

$$2x = \frac{5\pi}{6} + 2k\pi \quad \text{or} \quad 2x = \frac{7\pi}{6} + 2k\pi, \quad k \text{ an integer}$$

$$x = \frac{5\pi}{12} + k\pi \quad \text{or} \quad x = \frac{7\pi}{12} + k\pi$$

84. $2\sin x\cos x - 2\sqrt{2}\sin x - \sqrt{3}\cos x + \sqrt{6} = 0$

$$2\sin x(\cos x - \sqrt{2}) - \sqrt{3}(\cos x - \sqrt{2}) = 0$$

$$(\cos x - \sqrt{2})(2\sin x - \sqrt{3}) = 0$$

$\cos x = \sqrt{2} \qquad \sin x = \dfrac{\sqrt{3}}{2}$

no solution $\qquad x = \dfrac{\pi}{3}, \dfrac{2\pi}{3}$

The solutions are $\dfrac{\pi}{3}$ and $\dfrac{2\pi}{3}$.

Exercise Set 4.1, page 168

14. $B = 90.00° - 45.89° = 44.11°$

$\tan 45.89° = \dfrac{a}{1.228} \qquad \cos 45.89° = \dfrac{1.228}{c}$

$a = 1.228 \tan 45.89° \qquad c = \dfrac{1.228}{\cos 45.89°}$

$\approx 1.267 \qquad \approx 1.764$

28. $A = 70°, b = 5.2$

$$\tan 70° = \frac{a}{5.2}$$

$$a = 5.2 \tan 70° \approx 14 \text{ meters}$$

32. $C = 33.8° + 56.2° = 90.0°$

$$c = \sqrt{453^2 + 1520^2} \approx 1590 \text{ yards}$$

34. $\alpha = 360° - 335.4° = 24.6° \qquad a = 2(480) = 960$

$C = 24.6° + 65.4° = 90° \qquad b = 2(215) = 430$

$$c = \sqrt{a^2 + b^2} = \sqrt{960^2 + 430^2} \approx 1050 \text{ mi}$$

Exercise Set 4.2, page 173

4. $A = 180° - 78° - 28° = 74°$

$\dfrac{b}{\sin B} = \dfrac{c}{\sin C} \qquad \dfrac{a}{\sin A} = \dfrac{c}{\sin C}$

$\dfrac{b}{\sin 28°} = \dfrac{44}{\sin 78°} \qquad \dfrac{a}{\sin 74°} = \dfrac{44}{\sin 78°}$

$b = \dfrac{44 \sin 28°}{\sin 78°} \approx 21 \qquad a = \dfrac{44 \sin 74°}{\sin 78°} \approx 43$

14. $\sin 22.6° = \dfrac{h}{13.8}$

$$h = 13.8 \sin 22.6 \approx 5.30$$

Since $h < 5.55$, two solutions exist.

$$\frac{a}{\sin A} = \frac{b}{\sin B}$$

$$\frac{13.8}{\sin A} = \frac{5.55}{\sin 22.6}$$

$$\sin A = 0.9555$$

$$A \approx 72.9° \quad \text{or} \quad 107.1°$$

$A = 72.9° \qquad A = 107.1°$

$C = 180° - 72.9° - 22.6° \qquad C = 180° - 107.1° - 22.6°$

$C = 84.5° \qquad C = 50.3°$

$\dfrac{c}{\sin 84.5°} = \dfrac{5.55}{\sin 22.6°} \qquad \dfrac{c}{\sin 50.3°} = \dfrac{5.55}{\sin 22.6°}$

$c = \dfrac{5.55 \sin 84.5°}{\sin 22.6°} \approx 14.4 \qquad c = \dfrac{5.55 \sin 50.3°}{\sin 22.6°} \approx 11.1$

20. $\alpha = 65°$

$B = 65° + 8° = 73°$

$A = 180° - 50° - 65° = 65°$

$C = 180° - 65° - 73° = 42°$

$$\frac{b}{\sin B} = \frac{c}{\sin C}$$

$$\frac{b}{\sin 73°} = \frac{20}{\sin 42°}$$

$$b = \frac{20 \sin 73°}{\sin 42°}$$

$$b \approx 29 \text{ mi}$$

24. $A = 56°, C = 56°, B = 35°,$

$$\theta = 56° - 35° = 21°$$

$$= 180° - 35° - 21° = 124°$$

$$\frac{AC}{\sin B} = \frac{BC}{\sin \alpha}$$

$$\frac{AC}{\sin 124°} = \frac{6}{\sin 35°}$$

$$AC \approx 8.7 \text{ m}$$

Exercise Set 4.3, page 180

12.

$$c^2 = a^2 + b^2 - 2ab \cos C$$

$$c^2 = 14.2^2 + 9.30^2 - 2(14.2)(9.30)\cos 9.20°$$

$$c^2 \approx 27.4$$

$$c \approx 5.24$$

18.

$$\cos A = \frac{b^2 + c^2 - a^2}{2bc}$$

$$\cos A = \frac{132^2 + 160^2 - 108^2}{2(132)(160)} \approx 0.7424$$

$$A \approx 42.1°$$

26.

$$K = \frac{1}{2}ac \sin B$$

$$K = \frac{1}{2}(32)(25)\sin 127° \approx 319$$

28.

$$A = 180° - 102° - 27° = 51°$$

$$K = \frac{a^2 \sin B \sin C}{2 \sin A}$$

$$K = \frac{8.5^2 \sin 102° \sin 27°}{2 \sin 51°} \approx 20.6$$

36.

$$s = \frac{1}{2}(a + b + c)$$

$$= \frac{1}{2}(10.2 + 13.3 + 15.4) = 19.45$$

$$K = \sqrt{s(s - a)(s - b)(s - c)}$$

$$= \sqrt{19.45(19.45 - 10.2)(19.45 - 13.3)(19.45 - 15.4)}$$

$$\approx 66.9$$

48.

$$\alpha = 90° - 74° = 16°$$

$$A = 16° + 90° + 32° = 138°$$

$$B = 4 \cdot 16 = 64$$

$$C = 3 \cdot 22 = 66$$

$$a^2 = b^2 + c^2 - 2bc \cos A$$

$$a^2 = 64^2 + 66^2 - 2(64)(66) \cos 138°$$

$$a^2 = 14{,}730$$

$$a \approx 121 \text{ mi}$$

56. $S = \frac{1}{2}(324 + 412 + 516) = 626$

$$K = \sqrt{626(626 - 324)(626 - 412)(626 - 516)}$$

$$K \approx 66{,}710$$

$$\text{cost} = 4.15(66{,}710) = \$276{,}848$$

Exercise Set 4.4, page 192

2. $|P_1P_2| = \sqrt{[-5-(-4)]^2 + [6-(-7)]^2}$

$|P_1P_2| = \sqrt{1 + 169}$

$P_1P_2 = \sqrt{170}$

$\alpha = \tan^{-1}\left|\frac{6-(-7)}{-5-(-4)}\right| = \tan^{-1}\frac{13}{1} = 85.6°$

$\theta = 360° - 85.6°$ θ is in quadrant IV

$\theta = 274.4°$

8.

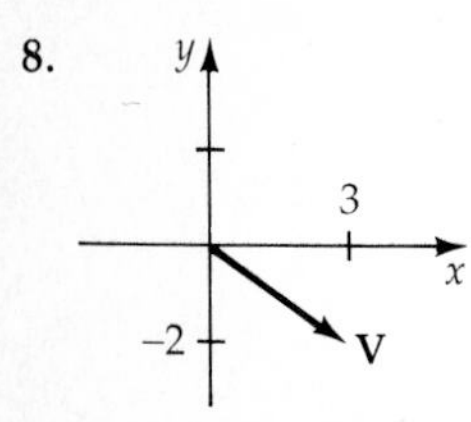

22. $x = -3,\ y = 4$

28. $x = 15\cos 140° \approx -11.5$

$y = 15\sin 140° \approx 9.6$

32. $(\mathbf{i} - 4\mathbf{j}) - (-3\mathbf{i} - 2\mathbf{j}) = [1-(-3)]\mathbf{i} + [-4-(-2)]\mathbf{j}$

$= 4\mathbf{i} - 2\mathbf{j}$

34. $-4\mathbf{V} = -4(-\mathbf{i} + 2\mathbf{j}) = 4\mathbf{i} - 8\mathbf{j}$

40. $\frac{3}{4}\mathbf{U} - \frac{1}{2}\mathbf{V} = \frac{3}{4}(2\mathbf{i} + 3\mathbf{j}) - \frac{1}{2}(-\mathbf{i} + 2\mathbf{j})$

$= \left(\frac{3}{2}\mathbf{i} + \frac{9}{4}\mathbf{j}\right) + \left(\frac{1}{2}\mathbf{i} - \mathbf{j}\right)$

$= \left(\frac{3}{2} + \frac{1}{2}\right)\mathbf{i} + \left(\frac{9}{4} - 1\right)\mathbf{j}$

$= 2\mathbf{i} + \frac{5}{4}\mathbf{j}$

56. $\mathbf{U}\cdot\mathbf{V} = (15\mathbf{i} - 10\mathbf{j})\cdot(18\mathbf{i} + 16\mathbf{j})$

$= (15)(18) + (-10)(16)$

$= 270 - 160 = 110$

62. $\cos\theta = \frac{\mathbf{U}\cdot\mathbf{V}}{|\mathbf{U}||\mathbf{V}|}$

$\cos\theta = \frac{(-1)(4) + (5)(7)}{\sqrt{(-1)^2 + 5^2}\sqrt{4^2 + 7^2}} = 0.7541$

$\theta \approx 41.1°$

68. $\alpha = \tan^{-1}\left|\frac{0.8}{2.6}\right| \approx 17.1°$

$\theta = 90° - 17.1° = 72.9°$

The heading is 72.9°.

70. $\alpha = \theta$

$\sin\alpha = \frac{120}{800}$

$\alpha = 8.6°$

72. $W = \mathbf{F}\cdot\mathbf{S}$

$W = |\mathbf{F}||\mathbf{S}|\cos 35°$

$W = (800\cos 35°)(45) = 29{,}489$ ft-lb

Exercise Set 5.1, page 203

2. $3 + \sqrt{-25} = 3 + i\sqrt{25} = 3 + 5i$

12. $(1 - 3i) + (6 + 2i) = 7 - i$

16. $(-3 + i) - (-8 + 2i) = -3 + i + 8 - 2i = 5 - i$

24. $(5 - 3i)(-2 - 4i) = -10 - 20i + 6i + 12i^2$

$= -10 - 14i - 12 = -22 - 14i$

34. $\frac{5-7i}{5+7i} = \frac{5-7i}{5+7i}\cdot\frac{5-7i}{5-7i} = \frac{25 - 35i - 35i + 49i^2}{25 - 49i^2}$

$= \frac{25 - 70i - 49}{25 + 49} = \frac{-24 - 70i}{74}$

$= -\frac{12}{37} - \frac{35}{37}i$

54. $i^{28} = (i^4)^7 = 1$

68. $\sqrt{-3}\sqrt{-121} = i\sqrt{3}\cdot i\sqrt{121} = i\sqrt{3}\cdot 11i$

$= 11i^2\sqrt{3} = -11\sqrt{3}$

Exercise Set 5.2, page 210

12. $r = \sqrt{1^2 + (\sqrt{3})^2} = 2$

$\alpha = \tan^{-1}\left|\frac{\sqrt{3}}{1}\right| = \tan^{-1}\sqrt{3} = 60°$

$\theta = 60°$

$z = 2(\cos 60° + i \sin 60°)$

26. $z = 4\left(\cos \frac{5\pi}{3} + i \sin \frac{5\pi}{3}\right)$

$= 4\left(\frac{1}{2} - \frac{\sqrt{3}}{2}i\right)$

$= 2 - 2i\sqrt{3}$

34. $z = 5(\cos 4 + i \sin 4)$

$= 5(-0.6536 - 0.7568i)$

$= -3.2682 - 3.7840i$

40. $z_1z_2 = 4(\cos 30° + i \sin 30°) \cdot 3(\cos 225° + i \sin 225°)$

$= 12[\cos(30° + 225°) + i \sin(30° + 225°)]$

$= 12(\cos 255° + i \sin 255°)$

50. $\frac{z_1}{z_2} = \frac{10(\cos \pi/3 + i \sin \pi/3)}{5(\cos \pi/4 + i \sin \pi/4)}$

$= 2\left[\cos\left(\frac{\pi}{3} - \frac{\pi}{4}\right) + i \sin\left(\frac{\pi}{3} - \frac{\pi}{4}\right)\right]$

$= 2\left(\cos \frac{\pi}{12} + i \sin \frac{\pi}{12}\right) = 1.93 + 0.518i$

62. $z_1 = 1 + i$ $\qquad z_2 = 1 - i$

$r_1 = \sqrt{1^2 + 1^2} = \sqrt{2}$ $\qquad r_2 = \sqrt{1^2 + (-1)^2} = \sqrt{2}$

$\alpha_1 = \tan^{-1}\left|\frac{1}{1}\right| = 45°$ $\qquad \alpha_2 = \tan^{-1}\left|\frac{-1}{1}\right| = 45°$

$\theta_1 = 45°$ $\qquad \theta_2 = 315°$

$z_1 = \sqrt{2}(\cos 45° + i \sin 45°)$

$z_2 = \sqrt{2}(\cos 315° + i \sin 315°)$

$\frac{z_1}{z_2} = \frac{\sqrt{2}(\cos 45° + i \sin 45°)}{\sqrt{2}(\cos 315° + i \sin 315°)}$

$= \cos(45° - 315°) + i \sin(45° - 315°)$

$= \cos 270° - i \sin 270°$

$= 0 - (-i) = i$

Exercise Set 5.3, page 216

10.

$z = 1 + i\sqrt{3}$

$r = \sqrt{1^2 + (\sqrt{3})^2} = 2$

$\alpha = 60°$

$\theta = 60°$

$z = 2(\cos 60° + i \sin 60°)$

$(1 + i\sqrt{3})^8 = [2(\cos 60° + i \sin 60°)]^8$

$= 2^8[\cos(8 \cdot 60°) + i \sin(8 \cdot 60°)]$

$= 256(\cos 480° + i \sin 480°)$

$= 256(\cos 120° + i \sin 120°)$

12.

$z = 2\sqrt{3} - 2i$

$r = \sqrt{(2\sqrt{3})^2 + (-2)^2} = 4$

$\alpha = 30°$

$\theta = 330°$

$z = 4(\cos 330° + i \sin 330°)$

$(2\sqrt{3} - 2i)^5 = [4(\cos 330° + i \sin 330°)]^5$

$= 4^5[\cos(5 \cdot 330°) + i \sin(5 \cdot 330°)]$

$= 1024(\cos 1650° + i \sin 1650°)$

$= 1024(\cos 210° + i \sin 210°)$

24. $-1 + i = \sqrt{2}(\cos 135° + i \sin 135°)$

$$w = \sqrt{2}^{1/5}\left(\cos \frac{135° + 360°k}{5} + i \sin \frac{135° + 360°k}{5}\right)$$

$k = 0, 1, 2, 3, 4$

$$k = 0, \quad w_1 = 2^{1/10}\left(\cos \frac{135°}{5} + i \sin \frac{135°}{5}\right)$$

$$\approx 0.955 + 0.487i$$

$$k = 1, \quad w_2 = 2^{1/10}\left(\cos \frac{135° + 360°}{5} + i \sin \frac{135° + 360°}{5}\right)$$

$$\approx -0.168 + 1.059i$$

$$k = 2, \quad w_3 = 2^{1/10}\left(\cos \frac{135° + 360° \cdot 2}{5} + i \sin \frac{135° + 360° \cdot 2}{5}\right)$$

$$\approx 1.052 + 0.168i$$

$$k = 3, \quad w_3 = 2^{1/10}\left(\cos \frac{135° + 360° \cdot 3}{5} + i \sin \frac{135° + 360° \cdot 3}{5}\right)$$

$$\approx 0.487 - 0.955i$$

$$k = 4, \quad w_4 = 2^{1/10}\left(\cos \frac{135° + 360° \cdot 4}{5} + i \sin \frac{135° + 360° \cdot 4}{5}\right)$$

$$\approx 0.758 - 0.756i$$

26. $-2 + 2i\sqrt{3} = 4(\cos 120° + i \sin 120°)$

$$w = 4^{1/3}\left(\cos \frac{120° + 360°k}{3} + i \sin \frac{120° + 360°k}{3}\right) \quad k = 0, 1, 2$$

$$k = 0, \quad w_1 = 4^{1/3}\left(\cos \frac{120°}{3} + i \sin \frac{120°}{3}\right) \approx 1.216 + 1.020i$$

$$k = 1, \quad w_2 = 4^{1/3}\left(\cos \frac{120° + 360°}{3} + i \sin \frac{120° + 360°}{3}\right)$$

$$\approx -1.492 + 0.543i$$

$$k = 2, \quad w_3 = 4^{1/3}\left(\cos \frac{120° + 360° \cdot 2}{3} + i \sin \frac{120° + 360° \cdot 2}{3}\right)$$

$$\approx 0.276 - 1.563i$$

34. $x^5 + 32i = 0$

$x^5 = -32i \quad -32i = 32(\cos 270° + i \sin 270°)$

$x = (-32i)^{1/5}$

$$x = 32^{1/5}\left(\cos \frac{270° + 360°k}{5} + i \sin \frac{270° + 360°k}{5}\right)$$

$k = 0, 1, 2, 3, 4$

$$k = 0, \quad x_1 = 2\left(\cos \frac{270°}{5} + i \sin \frac{270°}{5}\right)$$

$$= 2(\cos 54° + i \sin 54°)$$

$$k = 1, \quad x_2 = 2\left(\cos \frac{270° + 360°}{5} + i \sin \frac{270° + 360°}{5}\right)$$

$$= 2(\cos 126° + i \sin 126°)$$

$$k = 2, \quad x_3 = 2\left(\cos \frac{270° + 360° \cdot 2}{5} + i \sin \frac{270° + 360° \cdot 2}{5}\right)$$

$$= 2(\cos 198° + i \sin 198°)$$

$$k = 3, \quad x_4 = 2\left(\cos \frac{270° + 360° \cdot 3}{5} + i \sin \frac{270° + 360° \cdot 3}{5}\right)$$

$$= 2(\cos 270° + i \sin 270°)$$

$$k = 4, \quad x_5 = 2\left(\cos \frac{270° + 360° \cdot 4}{5} + i \sin \frac{270° + i \sin 360° \cdot 4}{5}\right)$$

$$= 2(\cos 342° + i \sin 342°)$$

Exercise Set 6.1, page 227

4.

$$x^2 = -\frac{1}{4}y$$

$$4p = -\frac{1}{4}$$

$$p = -\frac{1}{16}$$

vertex $(0, 0)$, focus $\left(0, -\frac{1}{16}\right)$, directrix $y = \frac{1}{16}$

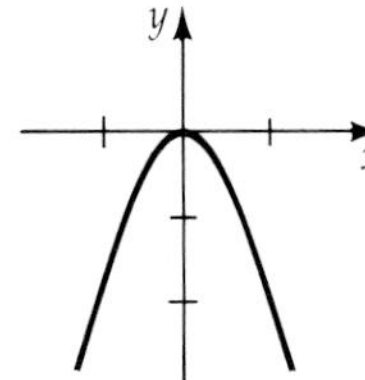

20.

$$x^2 + 5x - 4y - 1 = 0$$

$$x^2 + 5x = 4y + 1$$

$$x^2 + 5x + \frac{25}{4} = 4y + 1 + \frac{25}{4}$$

$$\left(x + \frac{5}{2}\right)^2 = 4\left(y + \frac{29}{16}\right)$$

$$4p = 4$$

$$p = 1$$

vertex $\left(-\frac{5}{2}, -\frac{29}{16}\right)$, focus $\left(-\frac{5}{2}, -\frac{13}{16}\right)$, directrix $y = -\frac{45}{16}$

28. vertex $(0, 0)$, focus $(5, 0)$, $p = 5$ since focus is $(p, 0)$

$$y^2 = 4px$$

$$y^2 = 4(5)x$$

$$y^2 = 20x$$

30. vertex $(2, -3)$, focus $(0, -3)$. $h = 2$ and $k = 3$ since the vertex is $(2, -3)$, and $p = -2$ since focus is $(h + p, k)$ and $h = 2$

$$(y - k)^2 = 4p(x - h)$$

$$(y + 3)^2 = 4(-2)(x - 2)$$

$$(y + 3)^2 = -8(x - 2)$$

Exercise 6.2, page 235

8. Write equation in standard form.

$$\frac{x^2}{12} + \frac{y^2}{25} = 1$$

Thus $a^2 = 25$, $b^2 = 12$, $c^2 = 25 - 12 = 13$.
Center $(0, 0)$, vertices $(0, 5)$, $(0, -5)$ and foci $(0, \sqrt{13})$, $(0, -\sqrt{13})$.

14.

$$9x^2 + 16y^2 + 36x - 16y - 104 = 0$$
$$9x^2 + 36x + 16y^2 - 16y - 104 = 0$$
$$9(x^2 + 4x) + 16(y^2 - y) = 104$$
$$9(x^2 + 4x + 4) + 16\left(y^2 - y + \frac{1}{4}\right) = 104 + 36 + 4$$
$$9(x + 2)^2 + 16\left(y - \frac{1}{2}\right)^2 = 144$$
$$\frac{(x + 2)^2}{16} + \frac{\left(y - \frac{1}{2}\right)^2}{9} = 1$$

center $(-2, \frac{1}{2})$, $a = 4$, $b = 3$, $c = \sqrt{4^2 - 3^2} = \sqrt{7}$
vertices $(2, \frac{1}{2})$ and $(-6, \frac{1}{2})$, foci $(-2 + \sqrt{7}, \frac{1}{2})$ and $(-2 - \sqrt{7}, \frac{1}{2})$

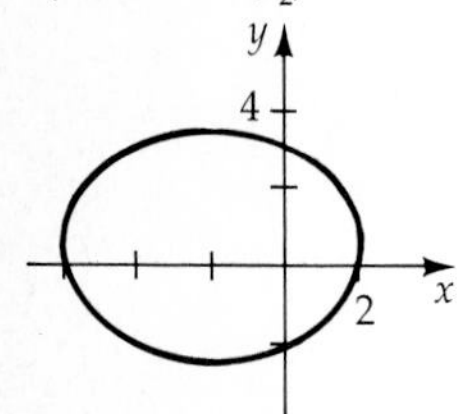

22.

$$2b = 6$$
$$b = 3$$

Since the foci are $(0, 4)$ and $(0, -4)$, $c = 4$, $a^2 = 4^2 + 3^2 = 25$.

$$\frac{x^2}{9} + \frac{y^2}{25} = 1$$

30.

$$2b = 8$$
$$b = 4$$
$$b^2 = 16$$

The equation of the ellipse is of the form

$$\frac{(x + 4)^2}{a^2} + \frac{(y - 1)^2}{16} = 1$$
$$\frac{(0 + 4)^2}{a^2} + \frac{(4 - 1)^2}{16} = 1$$
$$\frac{16}{a^2} + \frac{9}{16} = 1$$
$$\frac{16}{a^2} = \frac{7}{16}$$
$$a^2 = \frac{256}{7}$$

36. Because the foci are $(0, -3)$ and $(0, 3)$, $c = 3$ and center is $(0, 0)$.

$$e = \frac{c}{a}$$
$$\frac{1}{4} = \frac{3}{a}$$
$$a = 12$$
$$3^2 = 12^2 - b^2$$
$$b^2 = 144 - 9 = 135$$

The equation of the ellipse is $\frac{x^2}{135} + \frac{y^2}{144} = 1$.

Exercise Set 6.3, page 245

4.

$$a^2 = 25 \qquad b^2 = 36$$
$$a = 5 \qquad b = 6$$
$$c^2 = a^2 + b^2 = 25 + 36 = 61$$
$$c = \sqrt{61}$$

Transverse axis is on y-axis because y^2 term is positive. Center $(0, 0)$, foci $(0, \sqrt{61})$, $(0, -\sqrt{61})$, and asymptotes $y = \frac{5}{6}x$ and $y = -\frac{5}{6}x$.

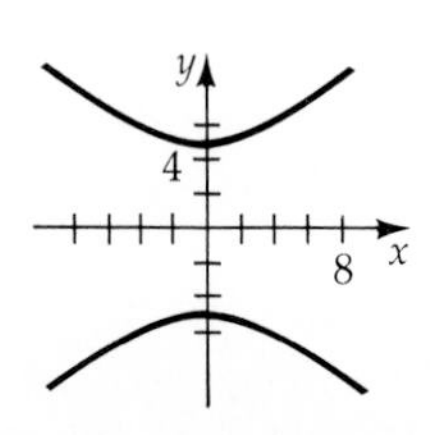

18. $16x^2 - 9y^2 - 32x - 54y + 79 = 0$

$$16(x^2 - 2x + 1) - 9(y^2 + 6y + 9) = -79 - 65 = -144$$

$$\frac{(y+3)^2}{16} - \frac{(x-1)^2}{9} = 1$$

Transverse axis is parallel to y-axis because y^2 term is positive. Center is at $(1, -3)$; because $a^2 = 16$ or $a = 4$ vertices are $(1, -3 + 4)$ and $(1, -3 - 4)$, or $(1, 1)$ and $(1, -7)$.

$$c^2 = a^2 + b^2 = 16 + 9 = 25$$

$$c = \sqrt{25} = 5$$

Foci are $(1, -3 + 5)$ and $(1, -3 - 5)$, or $(1, 2)$ and $(1, -8)$. Since $b^2 = 9$ or $b = 3$, asymptotes are $y + 3 = \frac{4}{3}(x - 1)$ and $y + 3 = -\frac{4}{3}(x - 1)$.

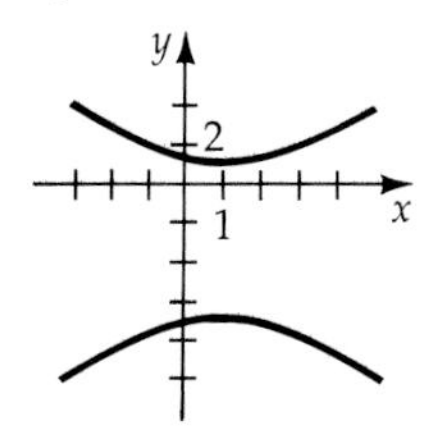

40. Because the vertices are $(2, 3)$ and $(-2, 3)$, $a = 2$ and center is $(0, 3)$.

$$e = \frac{c}{a}$$

$$\frac{5}{2} = \frac{c}{2}$$

$$c = 5$$

$$5^2 = 2^2 + b^2$$

$$b^2 = 25 - 4 = 21$$

The equation of the hyperbola is $\frac{x^2}{4} - \frac{(y-3)^2}{21} = 1$.

Exercise Set 6.4, page 257

14.

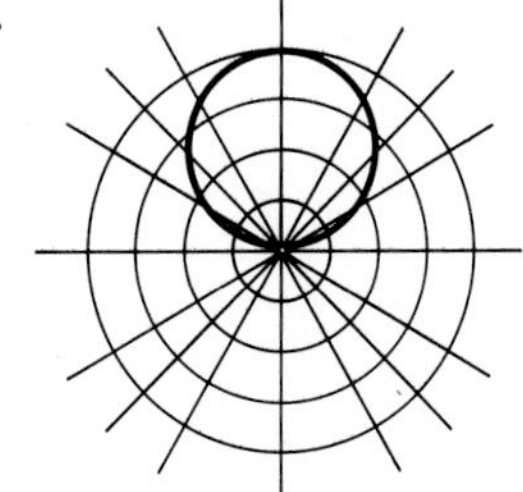

20.

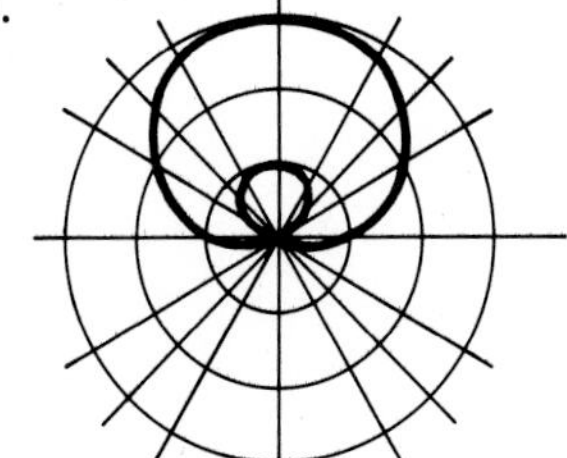

18.

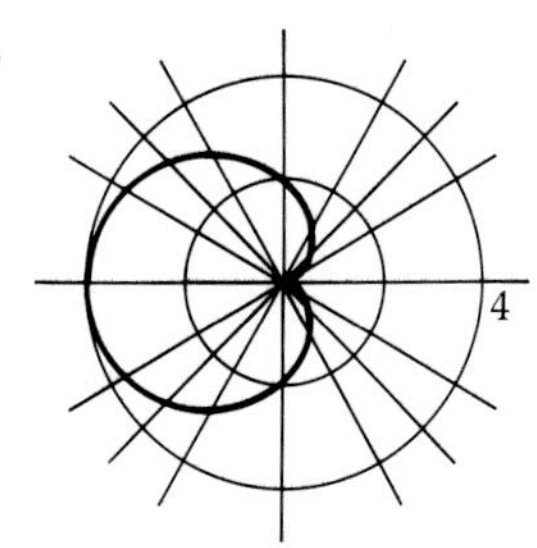

28.

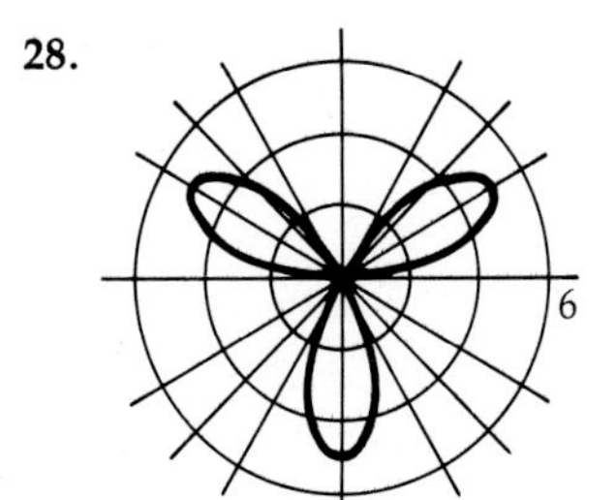

32.

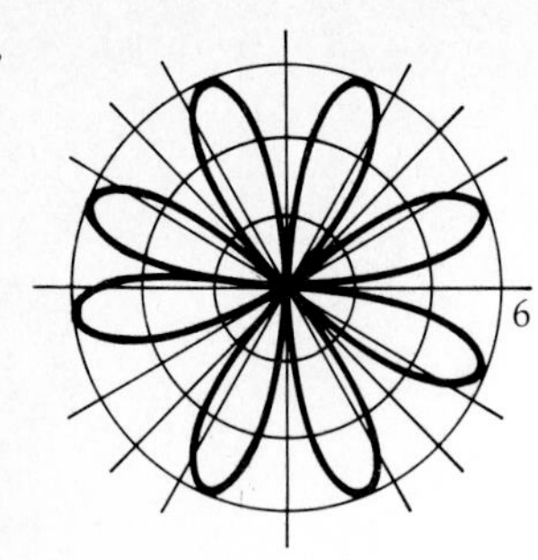

44.

$$\begin{aligned} x &= r\cos\theta & y &= r\sin\theta \\ &= 2\cos\left(-\frac{\pi}{3}\right) & &= 2\sin\left(-\frac{\pi}{3}\right) \\ &= (2)\left(\frac{1}{2}\right) = 1 & &= (2)\left(-\frac{\sqrt{3}}{2}\right) = -\sqrt{3} \end{aligned}$$

The rectangular coordinates are $(1, -\sqrt{3})$.

48.

$$\begin{aligned} r &= \sqrt{x^2 + y^2} & \theta &= \tan^{-1}\frac{y}{x} \\ &= \sqrt{(12)^2 + (-5)^2} & &= \tan^{-1}\left(-\frac{5}{12}\right) \\ &= \sqrt{144 + 25} = 13 & &\approx 337.4^\circ \end{aligned}$$

The polar coordinates are $(13, 337.4^\circ)$.

56. $r = \cot\theta = \dfrac{\cos\theta}{\sin\theta}$. Substitute $\sqrt{x^2 + y^2}$ for r, x/r for $\cos\theta$ and y/r for $\sin\theta$.

$$\begin{aligned} \sqrt{x^2 + y^2} &= \frac{x}{y} \\ x^2 + y^2 &= \frac{x^2}{y^2} \\ y^4 + x^2y^2 - x^2 &= 0 \end{aligned}$$

58.

$$\begin{aligned} r &= \frac{2}{1 - \sin\theta} \\ r - r\sin\theta &= 2 \\ \sqrt{x^2 + y^2} - y &= 2 \\ \sqrt{x^2 + y^2} &= 2 - y \\ x^2 + y^2 &= 4 - 4y + y^2 \\ x^2 + 4y - 4 &= 0 \end{aligned}$$

Exercise Set 7.1, page 268

26.

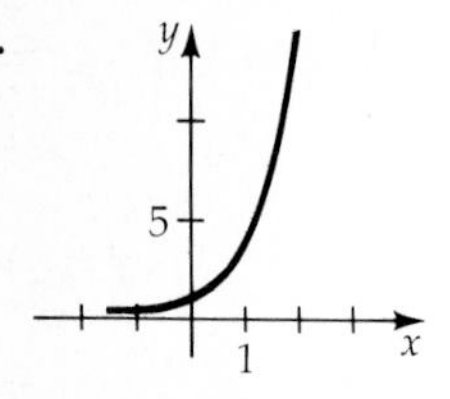

36.

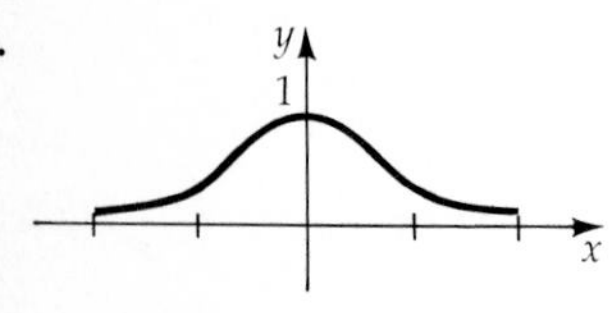

38.

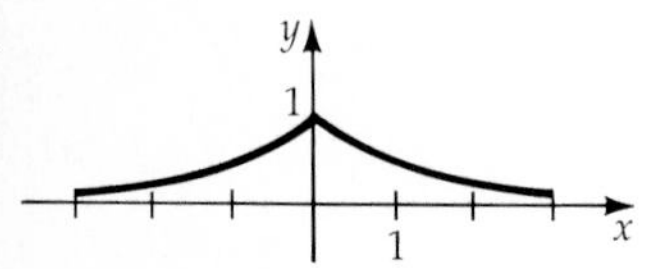

44.

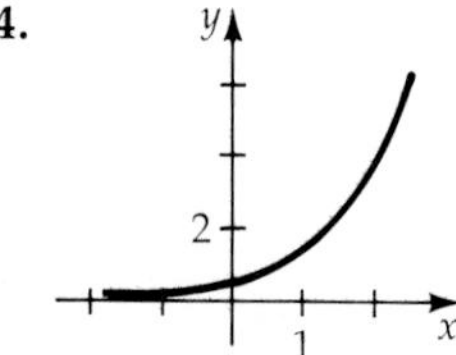

46.

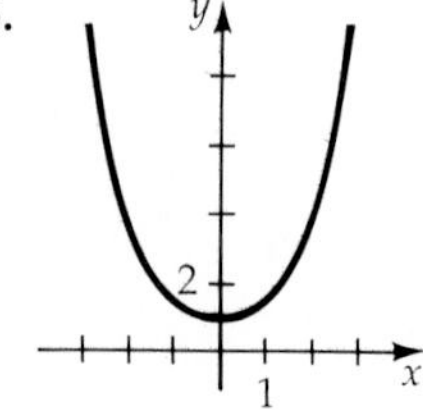

Exercise Set 7.2, page 277

2.
$$\log_{10} 1000 = 3$$
$$1000 = 10^3$$

12.
$$3^5 = 243$$
$$\log_3 243 = 5$$

22.
$$\log_{10} \frac{1}{1000} = n$$
$$10^n = \frac{1}{1000}$$
$$10^n = 10^{-3}$$
$$n = -3$$

32. $\log_b x^2y^3 = \log_b x^2 + \log_b y^3 = 2 \log_b x + 3 \log_b y$

42. $\log_b 20 = \log_b 2^2 \cdot 5 = 2 \log_b 2 + \log_b 5$
$= 2(0.3562) + (0.8271) = 1.5395$

52. $5 \log_3 x - 4 \log_3 y + 2 \log_3 z$
$= \log_3 x^5 - \log_3 y^4 + \log_3 z^2$
$= \log_3 \dfrac{x^5z^2}{y^4}$

62. $\log_5 37 = \dfrac{\log 37}{\log 5} \approx 2.243589$

Exercise Set 7.3, page 282

2.

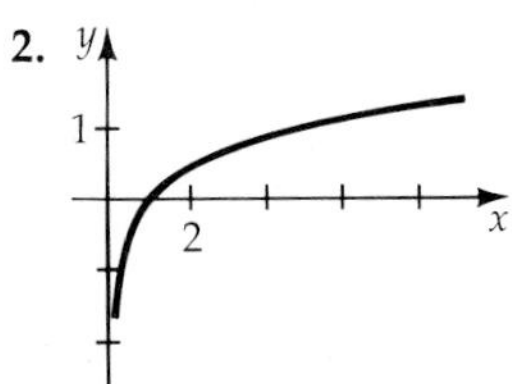

10.

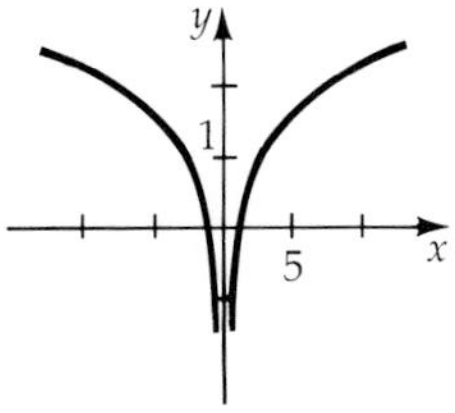

y 1 5 x

12.

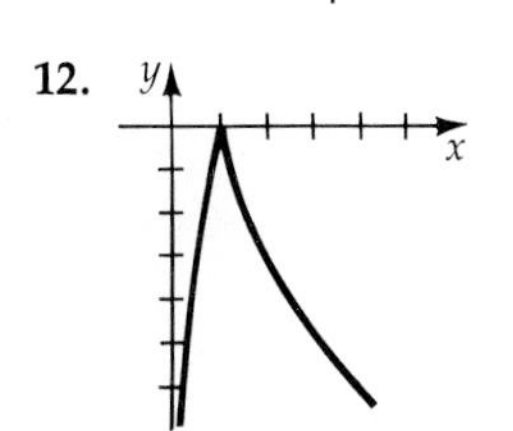

14.

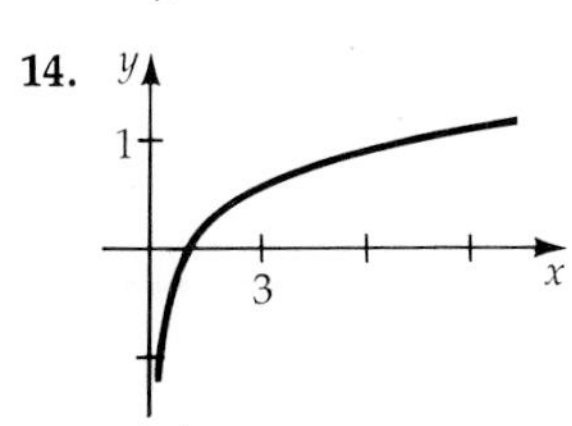

20.

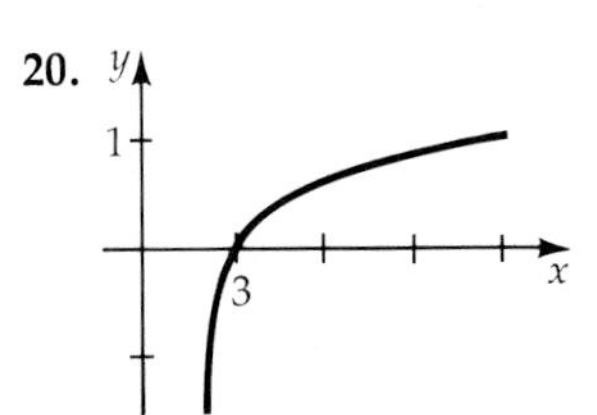

42.
$$I = 100{,}000I_0$$
$$R = \log \frac{100{,}000I_0}{I_0}$$
$$R = \log 100{,}000$$
$$R = 5$$

44. Let I = the intensity of the smaller earthquake.
Then $1000I$ = the intensity of the larger earthquake.

$$R_1 = \log \frac{I}{I_0} \qquad R_2 = \log \frac{1000I}{I_0} = \log 1000\frac{I}{I_0}$$

$$R_2 = \log 1000\frac{I}{I_0} = \log 1000 + \log \frac{I}{I_0}$$
$$= \log 10^3 + R_1 = 3 + R_1$$

The first earthquake has a Richter scale measure that is 3 more than the second earthquake.

Exercise Set 7.4, page 288

2.
$$3^x = 243$$
$$3^x = 3^5$$
$$x = 5$$

12.
$$6^x = 50$$
$$\log(6^x) = \log 50$$
$$x \log 6 = \log 50$$
$$x = \frac{\log 50}{\log 6} \approx 2.18$$

24.
$$\log_3 x + \log_3(x + 6) = 3$$
$$\log_3 x(x + 6) = 3$$
$$3^3 = x(x + 6)$$
$$27 = x^2 + 6x$$
$$x^2 + 6x - 27 = 0$$
$$(x + 9)(x - 3) = 0$$
$$x = 3 \quad \text{or} \quad -9$$

Since $\log_3 x$ is only defined for $x > 0$, the only solution is $x = 3$.

32.
$$\ln x = \frac{1}{2}\ln\left(2x + \frac{5}{2}\right) + \frac{1}{2}\ln 2$$
$$\ln x = \frac{1}{2}\ln 2\left(2x + \frac{5}{2}\right)$$
$$\ln x = \frac{1}{2}\ln(4x + 5)$$
$$\ln x = \ln(4x + 5)^{1/2}$$
$$x = \sqrt{4x + 5}$$
$$x^2 = 4x + 5$$
$$0 = x^2 - 4x - 5$$
$$0 = (x - 5)(x + 1)$$
$$x = 5, -1$$

Check: $\ln 5 = \frac{1}{2}\ln\left(10 + \frac{5}{2}\right) + \frac{1}{2}\ln 2$

$$1.6094 \approx 1.2629 + 0.3466$$

Since $\ln(-1)$ is not defined, -1 is not a solution. Thus the only solution is $x = 5$.

42.
$$\frac{10^x + 10^{-x}}{2} = 8$$
$$10^x(10^x + 10^{-x}) = (16)10^x$$
$$10^{2x} + 1 = 16(10^x)$$
$$10^{2x} - 16(10^x) + 1 = 0$$

Let $u = 10^x$.

$$u^2 - 16u + 1 = 0$$
$$u = \frac{16 \pm \sqrt{16^2 - 4(1)(1)}}{2} = 8 \pm 3\sqrt{7}$$
$$10^x = 8 \pm 3\sqrt{7}$$
$$x \log 10 = \log(8 \pm 3\sqrt{7})$$
$$x = \log(8 \pm 3\sqrt{7}) = \pm 1.20241$$

50. $\text{pH} = -\log(1.26 \times 10^{-12}) = 11.89$ basic solution

52.
$$\text{pH} = 3.1$$
$$-\log[H_3O^+] = 3.1$$
$$H_3O^+ = 10^{-3.1} \approx 7.9 \times 10^{-4}$$

Exercise Set 7.5, page 297

4. a. $P = 12{,}500,\ r = .08,\ t = 10,\ n = 1.$

$$B = 12{,}500\left(1 + \frac{.08}{1}\right)^{10} = \$26{,}986.56$$

b. $n = 365$

$$B = 12{,}500\left(1 + \frac{.08}{365}\right)^{3650} \approx \$27{,}816.82$$

c. $n = 8760$

$$B = 12{,}500\left(1 + \frac{.08}{8760}\right)^{87600} \approx \$27{,}819.16$$

6. $P = 32{,}000,\ r = .08,\ t = 3.$

$$B = Pe^{rt}$$

$$B = 32{,}000e^{3(.08)} \approx \$40{,}679.97$$

10.

$$N(t) = N_0e^{kt}$$

$$N(0) = 53{,}700 = N_0 \quad \text{(by definition)}$$

$$N(6) = 58{,}100$$

Use $t = 6$ and $N_0 = 53{,}700$.

$$58{,}100 = 53{,}700e^{k(6)}$$

$$\frac{58{,}100}{53{,}700} = e^{6k}$$

$$\ln\left(\frac{58{,}100}{53{,}700}\right) = 6k \ln e$$

$$\frac{\ln\left(\frac{58{,}100}{53{,}700}\right)}{6} = k$$

$$k \approx 0.0131$$

$$N(t) = 53{,}700e^{0.0131t}$$

12. $P(t) = 15{,}600(e^{0.09t})$

a. $P(10) = 15{,}600(e^{.9}) \approx 38{,}400$

b. We must determine the value of t when the population $N(t) = 15{,}600 \cdot 2 = 31{,}200$.

$$31{,}200 = 15{,}600e^{.09t}$$

$$\frac{31{,}200}{15{,}600} = e^{.09t}$$

$$t = \frac{\ln\left(\frac{31{,}200}{15{,}600}\right)}{.09} \approx 8$$

The population will double by 1992.

14.

$$N(138) = 100e^{138k}$$

$$50 = 100e^{138k}$$

$$.5 = e^{138k}$$

$$\ln .5 = 138k \ln e$$

$$k = \frac{\ln .5}{138} \approx -.005023$$

$$N(t) = 100e^{-.005023t}$$

16.

$$N(t) = N_0(0.5)^{t/138}$$

$$N(730) = N_0(0.5)^{730/138} \approx N_0(0.0256)$$

After 2 years (730 days), only 2.56% of the polonium remains.

18.

$$N(t) = N_0(0.5)^{t/5730}$$

$$N(t) = .65N_0$$

$$.65N_0 = N_0(0.5)^{t/5730}$$

$$t = 5730\,\frac{\ln .65}{\ln 0.5} \approx 3560$$

The bone is approximately 3560 years old.

22.

$$N(3.4 \times 10^{-5}) = 10 \log\left(\frac{3.4 \times 10^{-5}}{10^{-16}}\right)$$

$$= 10(\log 3.4 + \log 10^{11})$$

$$= 10 \log 3.4 + 110 \approx 115.3 \text{ decibels}$$

Answers to Odd-Numbered Exercises

Exercise Set 1.1, page 9

1. 4 **3.** -3 **5.** -9 **7.** -1 **9.** 0 **11.** $-2, 3$ **13.** $5, -2$ **15.** $\frac{1}{2}$ **17.** $-6 - \sqrt{7}, -6 + \sqrt{7}$ **19.** $1, -\frac{1}{3}$ **21.** $\sqrt{26}$
23. $\sqrt{85}$ **25.** $2\sqrt{61}$ **27.** $\sqrt{41}$ **29.** 1 **31.** $(1, 2)$ **33.** $\left(\frac{5}{2}, 1\right)$ **35.** $\left(1, -\frac{1}{2}\right)$ **37.** $\left(-2, -\frac{1}{2}\right)$ **39.** $(7, -4)$
41. a.

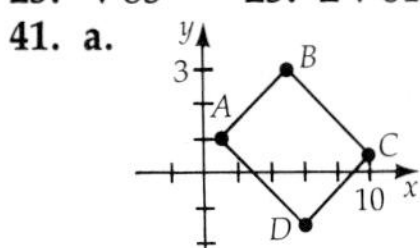

d. rectangle **43.** 40 **47. a.** 72 **c.** 36 **d.** 2

Exercise Set 1.2, page 15

1.

3.

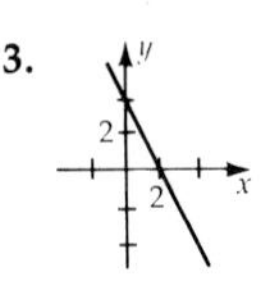

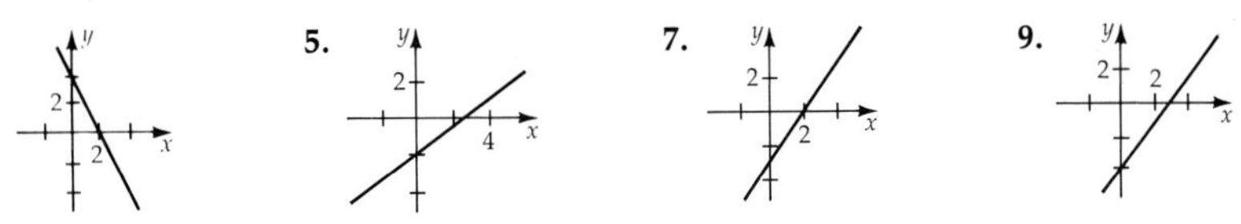

5.

7.

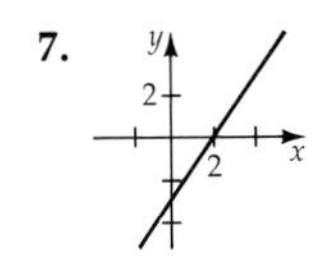

9.

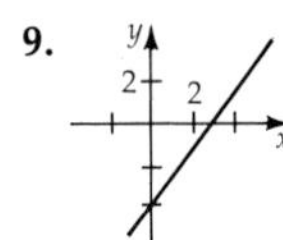

11.

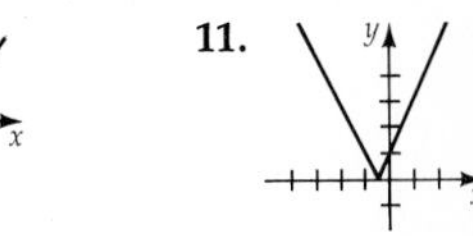

13.

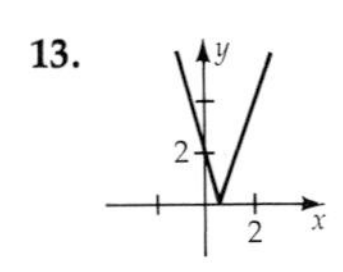

15.

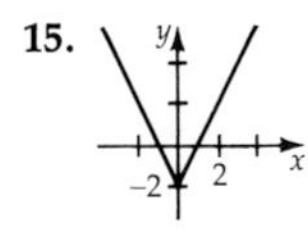

17.

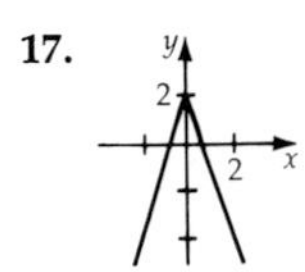

19.

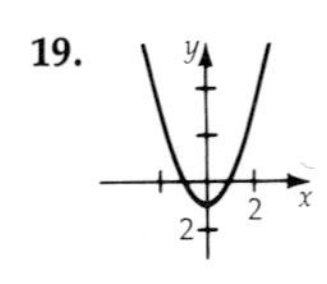

21.

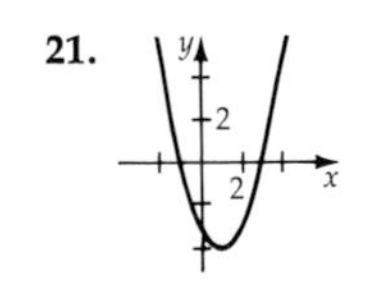

23.

25.

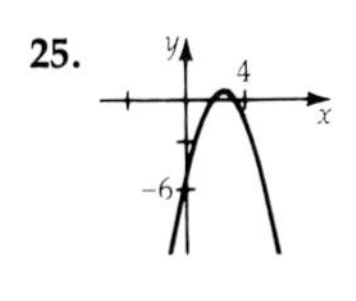

27. $(2, 0), (0, 5)$ **29.** $(6, 0), (0, -4)$ **31.** $(3, 0), \left(0, \frac{9}{5}\right)$ **33.** $(0, 0)$

35. $(-2, 0), (3, 0), (0, -6)$ **37.** $(-3, 0), \left(\frac{1}{2}, 0\right), (0, -3)$ **39.** $(1 - \sqrt{2}, 0), (1 + \sqrt{2}, 0), (0, -1)$ **41.** center $(0, 0)$, radius 1
43. center $(3, -1)$, radius 4 **45.** center $(-4, 2)$, radius 5 **47.** center $(5, 2)$, radius 1 **49.** center $(3, -4)$, radius 6
51. center $(-2, 2)$, radius 4 **53.** $\left(-\frac{5}{2}, 3\right)$, radius $\frac{\sqrt{57}}{2}$ **55.** the point $(2, 3)$ **57.** a degenerate circle **59.** $(x + 1)^2 + (y - 1)^2 = 25$

61. $x^2 + y^2 = 13$ **63.**

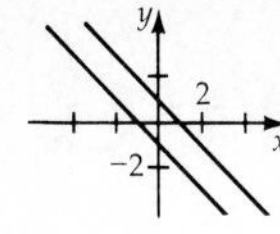

65.

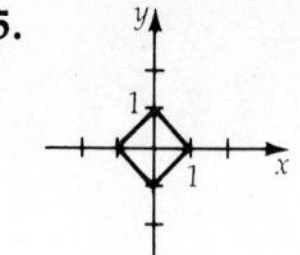

67. **69.**

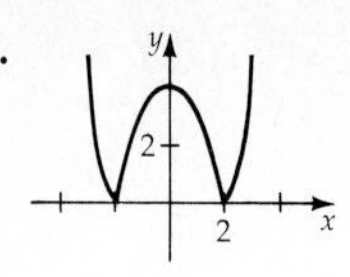

71. $(x-1)^2 + (y-1)^2 = 2$

Exercise Set 1.3, page 22

1. yes **3.** yes **5.** no **7.** the set of real numbers **9.** the set of real numbers **11.** the set of real numbers **13.** $\{x \mid x \geq 4\}$ **15.** $\{x \mid x \neq 3, x \neq -3\}$ **17.** the set of real numbers **19.** $\{z \mid z \neq 0\}$ **21.** 0 **23.** -24 **25.** $3a^2 - 8a + 6$ **27.** $3x^2 - 9x + 5$ **29.** $-\dfrac{2}{3}$ **31.** 2 **3.** ± 2 **35.** $9\pi/t$ **37.** $d(t) = \sqrt{1 + 5625t^2}$ **39.** $y = 300 - 4x$ **41.** yes **43.** no **47.** 3 **49.** $4 + h$

Exercise Set 1.4, page 31

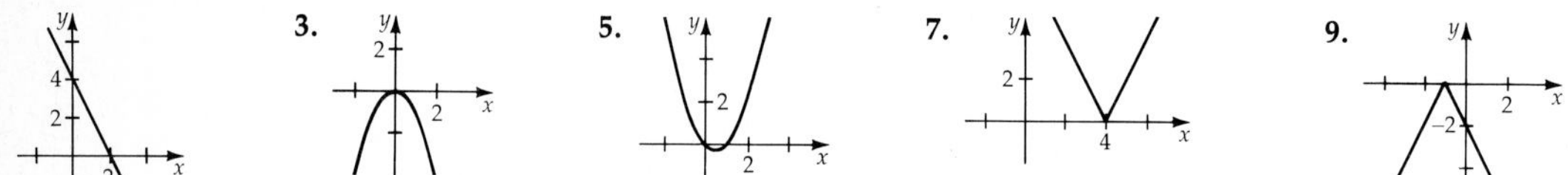

1.

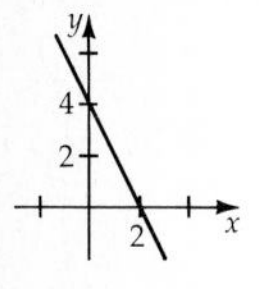

3.

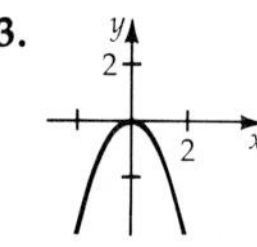

5.

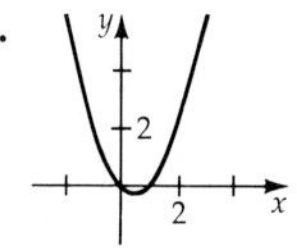

7. **9.**

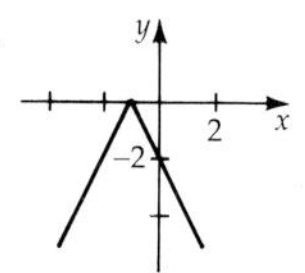

11. yes

13. yes **15.** no **17.** yes **19.** yes **21.** yes **23.** no **25.** no **27.** yes **29.** yes **31.** even **33.** odd **35.** neither **37.** odd **39.** even

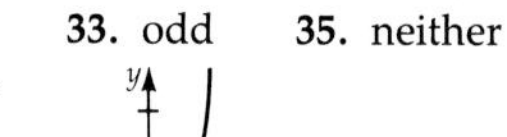

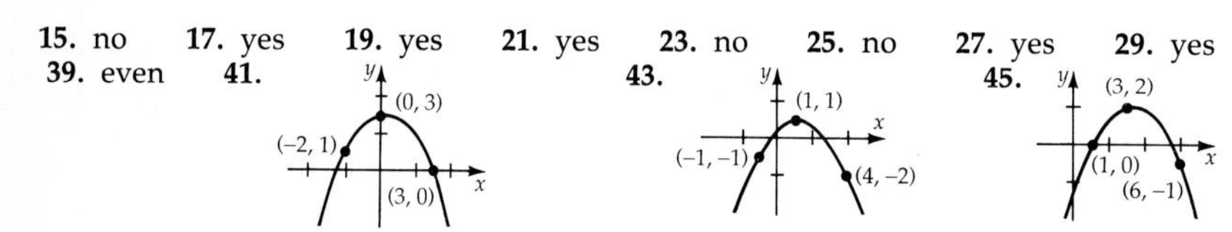

41.

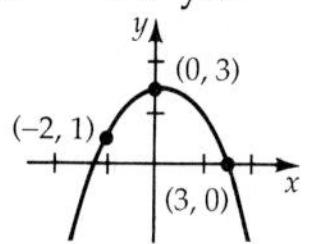

43.

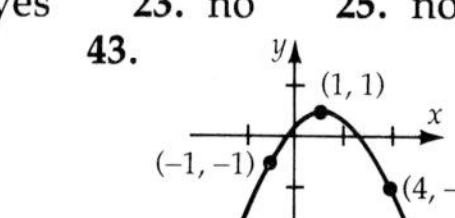

45.
(3, 2)
(1, 0)
(6, −1)

47.

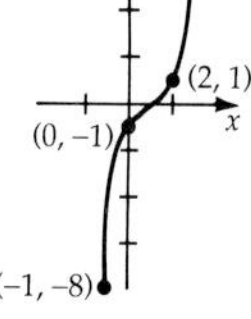

49.

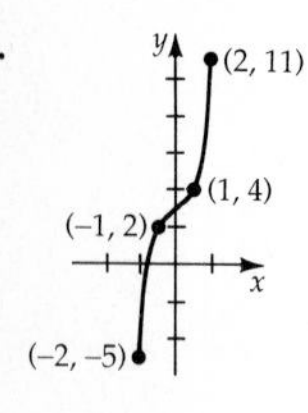

51.

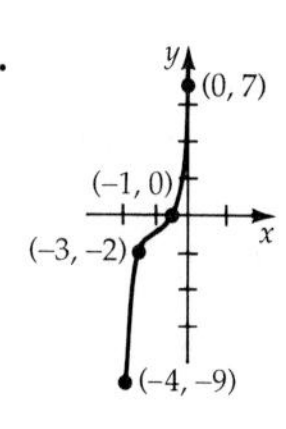

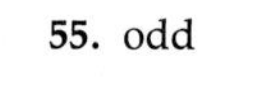

53. neither **55.** odd **57.**

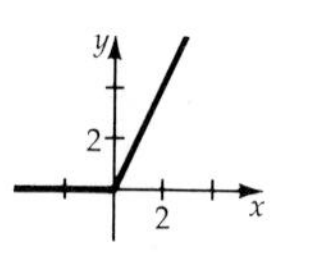

59.

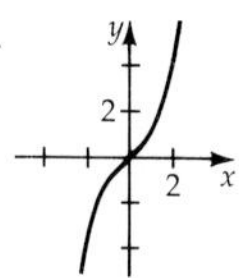

Exercise Set 1.5, page 39

1. 6 **3.** 1 **5.** -13 **7.** $\dfrac{1-a}{a^2}$ **9.** $4x^2 - 14x + 12$ **11.** $f^{-1}(x) = \dfrac{1}{2}x + 2$ **13.** $u^{-1}(v) = \dfrac{1-v}{3}$ **15.** $f^{-1}(z) = \sqrt{z},\ z \geq 0$

17. $p^{-1}(q) = \dfrac{1}{q}$ **19.** $r^{-1}(t) = t^2 - 1,\ t \geq 0$ **21.** $H^{-1}(s) = \sqrt{s+4},\ s \geq -4$ **23.** $P(v) = (v+1)^3$ **25.** $f^{-1}(x) = \dfrac{1}{x-1},\ x \neq 1$

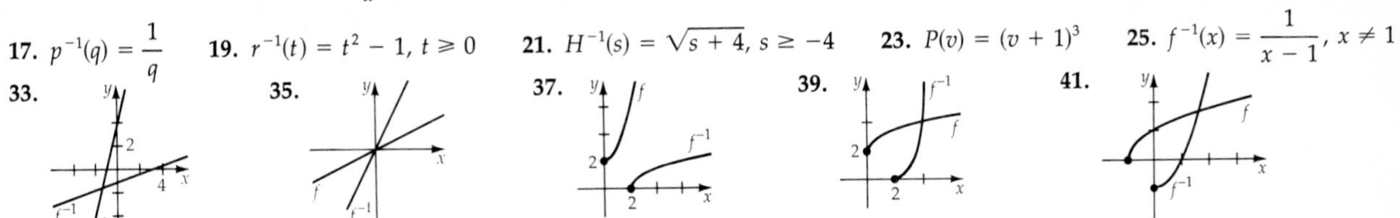

33.

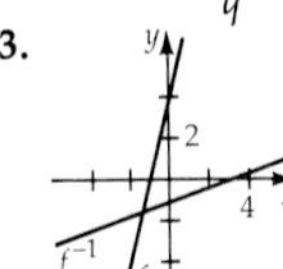

35.

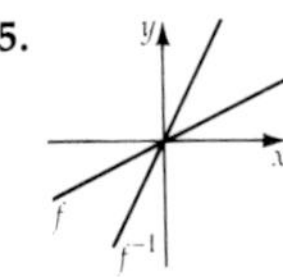

37.

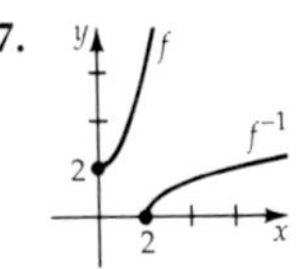

39.

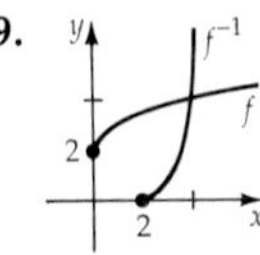

41.

43. $f^{-1}(x) = \sqrt{x+3} + 2,\ x \geq -3$ **45.** $f^{-1}(x) = \sqrt{x+25} - 4,\ x \geq -25$ **47.** $f^{-1}(x) = \dfrac{x+1}{1-x},\ x \neq 1$ **51.** 3 **53.** -9 **55.** $6x + 4$

Challenge Exercises, page 41

1. false, $f(x) = x^2$ does not have an inverse function **3.** true **5.** true **7.** false, consider $f(x) = x$ **9.** true

Review Exercises, page 42

1. $\sqrt{61},\ \left(1, \dfrac{1}{2}\right)$ **3.** $\sqrt{5x_1^2 - 12x_1 + 8}$

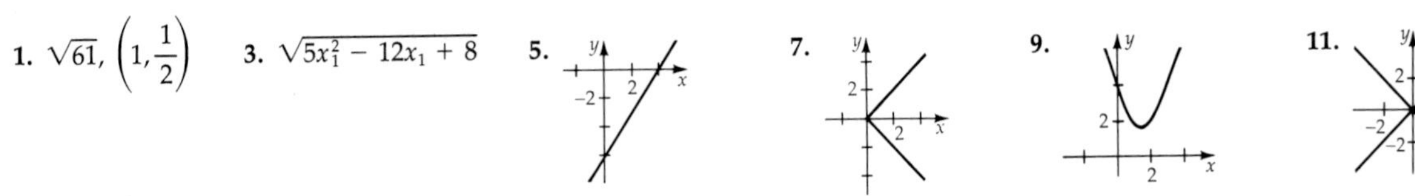

5.

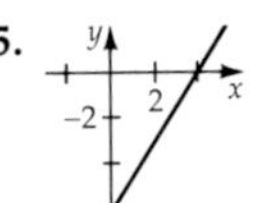

7.

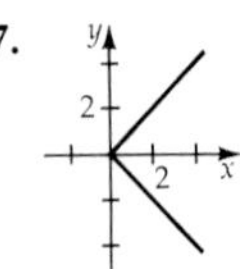

9.

11.

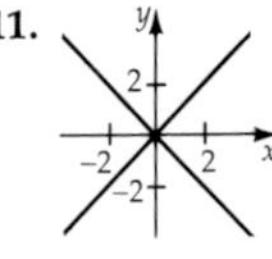

13.

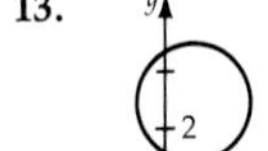

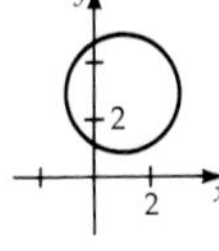

15. the point $(-4, 1)$

17.

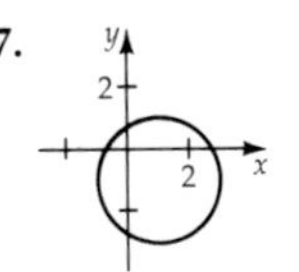

19.

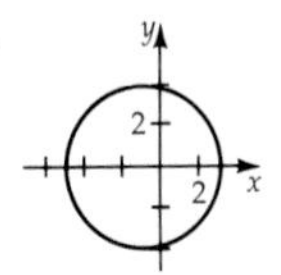

21.

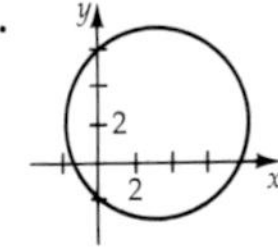

23. $(2, 0), (0, -5)$ **25.** $(-6, 0), (2, 0), (0, -12)$ **27.** $(0, 3)$ **29.** the set of real numbers **31.** $\{x \mid x = -1\}$

33. the set of real numbers **35.** $\left\{x \mid x \leq \dfrac{1}{2}\right\}$ **37.** $\{x \mid x \neq 3, x \neq -2\}$ **39.** 5 **41.** $a^2 + 2ab + b^2$ **43.** -20 **45.** 5

47. $-2x^2 + 4x + 7$ **49.** $-2x^3 + 5x^2 + 4x - 3$

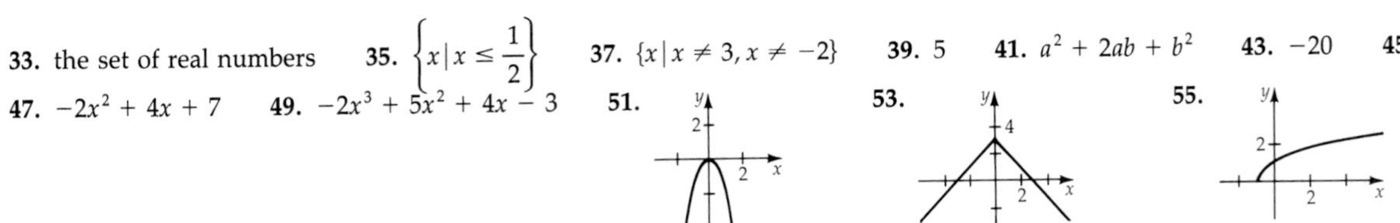

51.

53.

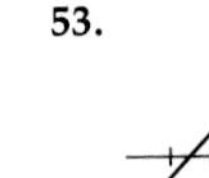

55.

57. 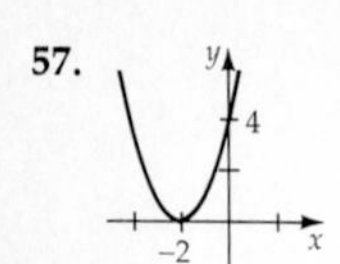
59. 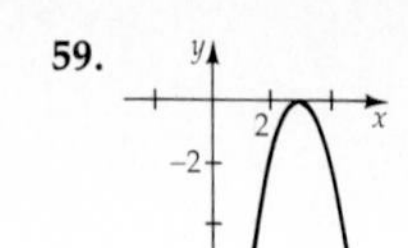
61. 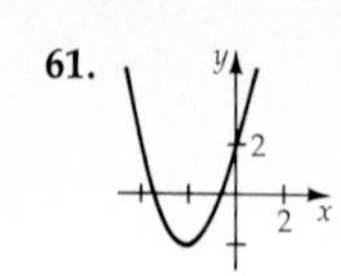
63. 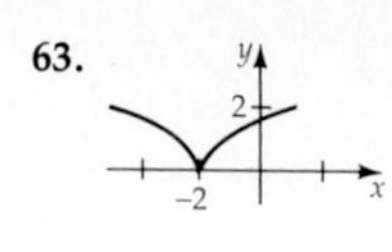
65. 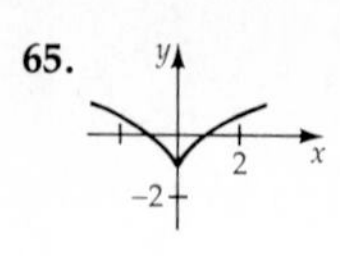

67. **69.** no **71.** no **73.** yes **75.** no **77.** no **79.** no **81.** $f^{-1}(x) = \frac{1}{2}x + 4$ **83.** $f^{-1}(x) = \frac{5 - 4x}{x}$

85. $f^{-1}(x) = -1 + \sqrt{x},\ x \geq 0$ **87.** $A(t) = 4\pi t^4$ **89.** $A(w) = 30w - 2w^2$

Exercise Set 2.1, page 54

1. 75° **3.** 3.7° **5.** 12.45° **7.** 43°51′ **9.** 79°4′25″ **11.** 113° **13.** 7.66° **15.** 56.2° **17.** 134.22° **19.** 78.133° **21.** 16.733° **23.** 47.333° **25.** 165.615° **27.** 95.469° **29.** 36°36′ **31.** 66°43′12″ **33.** 6°11′24″ **35.** 132°34′48″ **37.** $\pi/12$ **39.** $7\pi/4$ **41.** $7\pi/2$ **43.** 0.51 **45.** 2.90 **47.** 10.65 **49.** 30° **51.** 67.5° **53.** 660° **55.** 68.75° **57.** 36.67° **59.** 250.96° **61.** $\pi/3$ **63.** $\pi/12$ **65.** $\pi/2 - 1.22$ **67.** $\pi/4$ **69.** $\pi/8$ **71.** $\pi - 1.76$ **73.** 4, 229.2° **75.** 2.38, 136.6° **77.** 6.28 in **79.** 18.33 cm **81.** 3π **83.** $5\pi/12$ **85.** $\pi/30$ rad/s **87.** 100π rad/s **89.** 69.14 rad/s **91.** 53.55 mph **93.** 30.79 ft **95.** 486.95 ft^2 **97.** 1199.9 cm^2 **99.** 840,000 mi **101 a.** 6206 revolutions **b.** 5378 revolutions **103.** 157 cm/s **105. a.** 69.1 mi **b.** 1.15 mi **c.** 0.0192 mi

Exercise Set 2.2, page 61

1. $\sin\theta = 3\sqrt{10}/10$ $\csc\theta = \sqrt{10}/3$ $\cos\theta = \sqrt{10}/10$ $\sec\theta = \sqrt{10}$ $\tan\theta = 3$ $\cot\theta = 1/3$
3. $\sin\theta = \sqrt{2}/2$ $\csc\theta = \sqrt{2}$ $\cos\theta = \sqrt{2}/2$ $\sec\theta = \sqrt{2}$ $\tan\theta = 1$ $\cot\theta = 1$
5. $\sin\theta = \sqrt{2}/2$ $\csc\theta = \sqrt{2}$ $\cos\theta = \sqrt{2}/2$ $\sec\theta = \sqrt{2}$ $\tan\theta = 1$ $\cot\theta = 1$
7. $\sin\theta = 2\sqrt{13}/13$ $\csc\theta = \sqrt{13}/2$ $\cos\theta = 3\sqrt{13}/13$ $\sec\theta = \sqrt{13}/3$ $\tan\theta = 2/3$ $\cot\theta = 3/2$
9. $\sin\theta = \sqrt{5}/5$ $\csc\theta = \sqrt{5}$ $\cos\theta = 2\sqrt{5}/5$ $\sec\theta = \sqrt{5}/2$ $\tan\theta = 1/2$ $\cot\theta = 2$
11. $\sin\theta = \sqrt{2}/2$ $\csc\theta = \sqrt{2}$ $\cos\theta = \sqrt{2}/2$ $\sec\theta = \sqrt{2}$ $\tan\theta = 1$ $\cot\theta = 1$
13. 3/4 **15.** 4/5 **17.** 3/4 **19.** 12/13 **21.** 13/5 **23.** $\sqrt{10}/3$ **25.** 1 **27.** 2 **29.** $\sqrt{3}$ **31.** $\sqrt{2}/2$ **33.** 1/2 **35.** $\sqrt{3}$ **37.** 0.2079 **39.** 0.6249 **41.** 1.1190 **43.** 0.8221 **45.** 1.0053 **47.** 0.4816 **49.** 1.0729 **51.** 0.3153 **53.** 1.2331 **55.** $\sqrt{2}$ **57.** 0 **59.** $-3/4$ **61.** $\sqrt{3}/6$ **63.** 5/4 **65.** $\sqrt{3} - \sqrt{6}$ **67.** $\sqrt{3}$ **69.** $\frac{3\sqrt{2} + 2\sqrt{3}}{6}$ **71.** $\frac{3 - \sqrt{3}}{3}$ **73.** $2\sqrt{2} - \sqrt{3}$ **75.** 30°, $\pi/6$ **77.** 30°, $\pi/6$ **79.** 45°, $\pi/4$ **81.** 45°, $\pi/4$ **83.** 30°, $\pi/6$ **85.** 60°, $\pi/3$ **87.** 0.5932 **89.** 1.0051 **91.** 0.6327 **101.** $-3/4$ **103.** 34/9 **105.** $\sqrt{2} - 1$ **107.** 0 **109.** true **111.** true

Exercise Set 2.3, page 68

1. $\sin\theta = 3\sqrt{13}/13$ $\csc\theta = \sqrt{13}/3$ $\cos\theta = 2\sqrt{13}/13$ $\sec\theta = \sqrt{13}/2$ $\tan\theta = 3/2$ $\cot\theta = 2/3$
3. $\sin\theta = 3\sqrt{13}/13$ $\csc\theta = \sqrt{13}/3$ $\cos\theta = -2\sqrt{13}/13$ $\sec\theta = -\sqrt{13}/2$ $\tan\theta = -3/2$ $\cot\theta = -2/3$
5. $\sin\theta = -5\sqrt{89}/89$ $\csc\theta = -\sqrt{89}/5$ $\cos\theta = -8\sqrt{89}/89$ $\sec\theta = -\sqrt{89}/8$ $\tan\theta = 5/8$ $\cot\theta = 8/5$
7. $\sin\theta = 0$ $\csc\theta$ is undefined $\cos\theta = -1$ $\sec\theta = -1$ $\tan\theta = 0$ $\cot\theta$ is undefined
9. $\sin 330° = -1/2$ $\csc 330° = -2$ $\cos 330° = \sqrt{3}/2$ $\sec 330° = 2\sqrt{3}/3$ $\tan 330° = -\sqrt{3}/3$ $\cot 330° = -\sqrt{3}$
11. $\sin 210° = -1/2$ $\csc 210° = -2$ $\cos 210° = -\sqrt{3}/2$ $\sec 210° = -2\sqrt{3}/3$ $\tan 210° = \sqrt{3}/3$ $\cot 210° = \sqrt{3}$
13. $\sin\pi/3 = \sqrt{3}/2$ $\csc\pi/3 = 2\sqrt{3}/3$ $\cos\pi/3 = 1/2$ $\sec\pi/3 = 2$ $\tan\pi/3 = \sqrt{3}$ $\cot\pi/3 = \sqrt{3}/3$
15. $\sin 11\pi/6 = -1/2$ $\csc 11\pi/6 = -2$ $\cos 11\pi/6 = \sqrt{3}/2$ $\sec 11\pi/6 = 2\sqrt{3}/2$ $\tan 11\pi/6 = -\sqrt{3}/3$ $\cot 11\pi/6 = -\sqrt{3}$
17. 1/2 **19.** 0 **21.** 1 **23.** $\frac{3\sqrt{2} + 2\sqrt{3}}{6}$ **25.** $-3/2$ **27.** 1 **29.** quadrant I **31.** quadrant IV **33.** quadrant III **35.** $\tan\phi = \sqrt{3}/3$ **37.** $\sin\phi = -1/2$ **39.** $\cot\phi = -1$ **41.** $\tan\phi = -\sqrt{3}/3$ **43.** $\csc\phi = 2\sqrt{3}/3$ **45.** $\cot\phi = -\sqrt{3}/3$ **47.** $\cot\phi = -\sqrt{3}/3$ **49.** 0.7986 **51.** -0.4384 **53.** 0.1405 **55.** -0.7880 **57.** 0.5878 **59.** 0.8090 **61.** -0.8296 **63.** -0.2594 **65.** 30°, 150° **67.** 150°, 210° **69.** 225°, 315° **71.** 240°, 300° **73.** $3\pi/4, 7\pi/4$ **75.** $5\pi/6, 11\pi/6$ **77.** $\pi/3, 2\pi/3$ **79.** $\pi/6, 7\pi/6$ **95.** 0.5625 **97.** -1.2137 **103. a.** 0.479425532 **b.** 0.866021271

Exercise Set 2.4, page 76

1. $\pi/3$ **3.** $7\pi/4$ **5.** $2\pi/3$ **7.** $\pi/4$ **9.** $7\pi/4$ **11.** 2.4 **13.** 30° **15.** 330° **17.** 126° **19.** $(1/2, \sqrt{3}/2)$ **21.** $(-\sqrt{2}/2, \sqrt{2}/2)$ **23.** $(-\sqrt{3}/2, -1/2)$ **25.** $(1/2, -\sqrt{3}/2)$ **27.** $(\sqrt{3}/2, -1/2)$ **29.** $(0, -1)$ **31.** $(1/2, -\sqrt{3}/2)$ **33.** $(-1/2, -\sqrt{3}/2)$ **35.** 0.9391 **37.** −3.3805 **39.** −1.1528 **41.** −0.2679 **43.** 0.8090 **45.** 48.0889 **47.** $\sqrt{2}/2$ **49.** $\sqrt{3}/2$ **51.** $-\sqrt{3}/2$ **53.** 1/2 **55.** $-\sqrt{3}/2$ **97.** $-\sqrt{3}/2$ **99.** −1/2 **101.** −1 **103.** $\pm\sqrt{\sec^2\phi - 1}$ **105.** $\pm\sqrt{1 + \tan^2\phi}$ **107.** $\pm\sqrt{1 + \cot^2\phi}$ **109. a.** OF **b.** OB **c.** EF

Exercise Set 2.5, page 87

1. 2, 2π **3.** 1, π **5.** 1/2, π **7.** 2, 4π **9.** 1/2, 2π **11.** 1,8π **13.** 2,6π **15.** 3,3π

17. 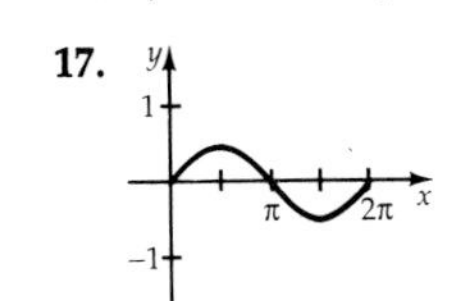

19. 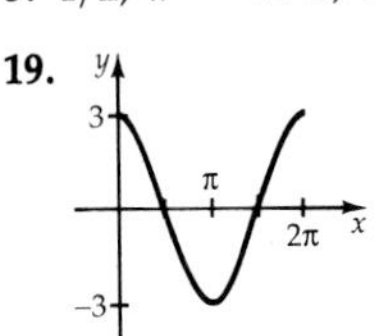

21. 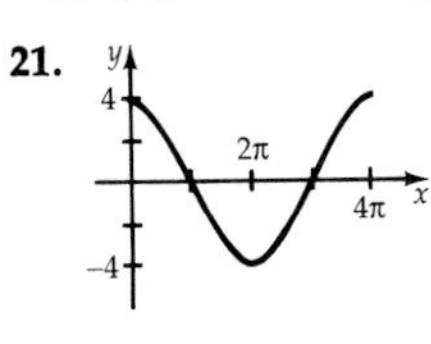

23. 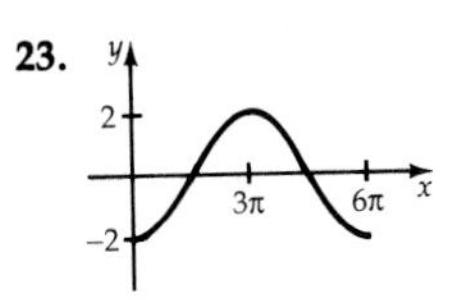

25. 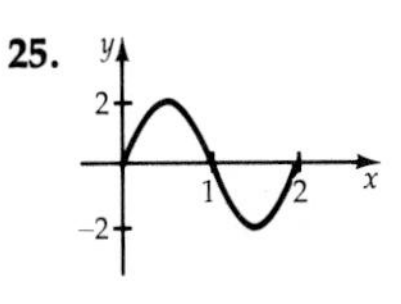

27. 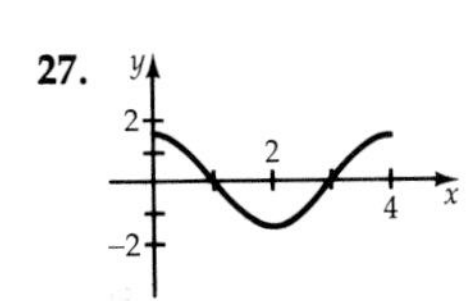

29. 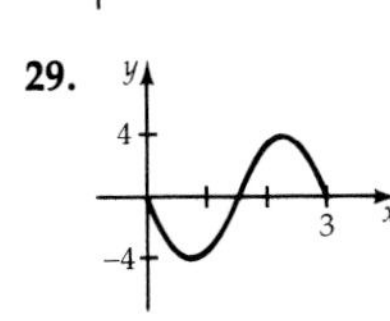

31. 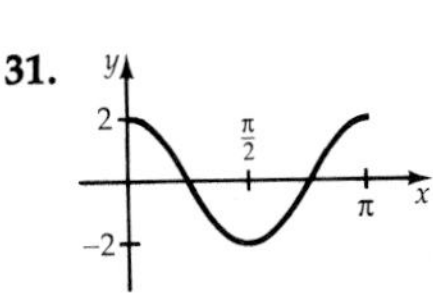

33. 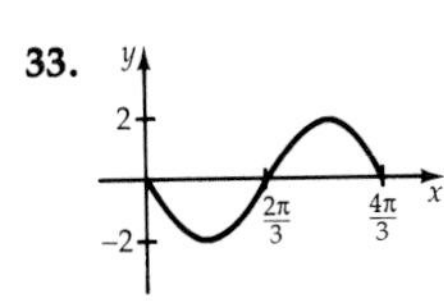

35. 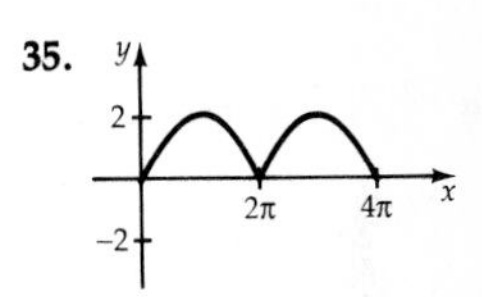

37. 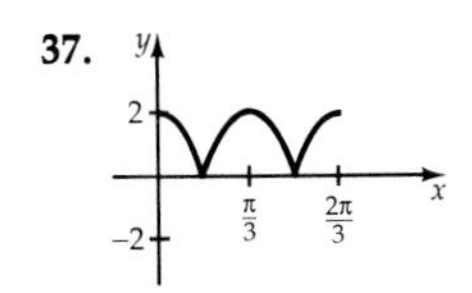

39. 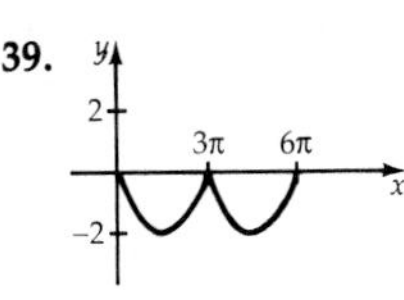

41. 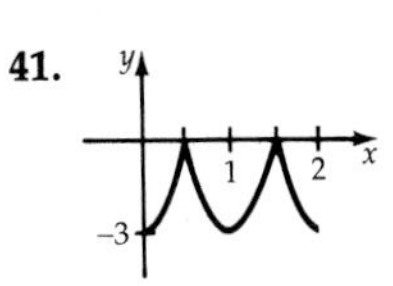

43. $y = \cos 2x$ **45.** $y = 3 \sin \frac{2}{3}x$ **47.** $y = -2 \cos \pi x$

49. 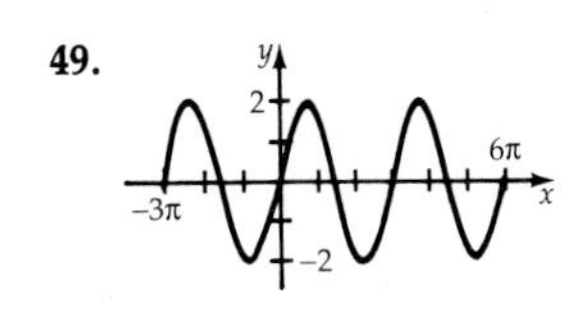

51. 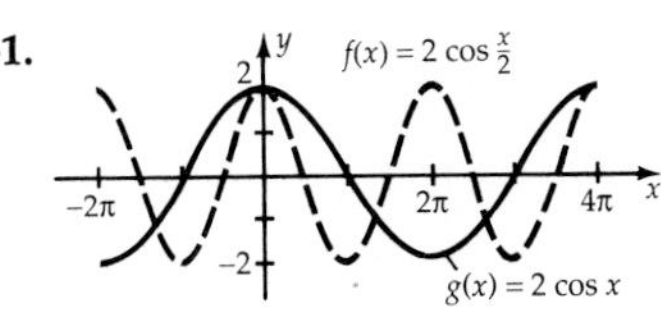

53. 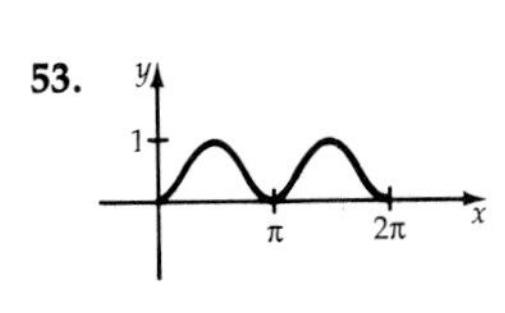

55. 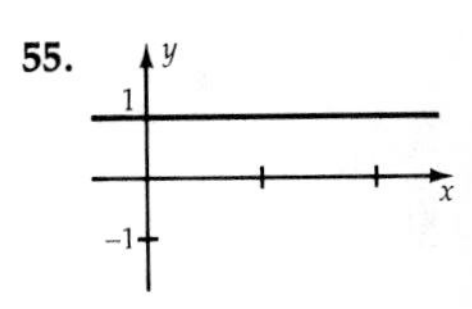

57. even function **59.** $y = 2 \sin \frac{2}{3}x$ **61.** $y = 0.5 \sin \frac{8}{5}x$ **63.** $y = 4 \sin \pi x$ **65.** $y = 3 \cos 4x$ **67.** $y = 1.8 \cos \frac{4}{3}x$

69. $y = 3 \cos \frac{4\pi}{5}x$ **71.** $f(t) = 60 \cos 20t$

period = $\pi/5$

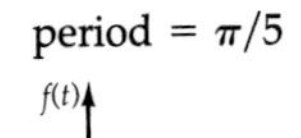

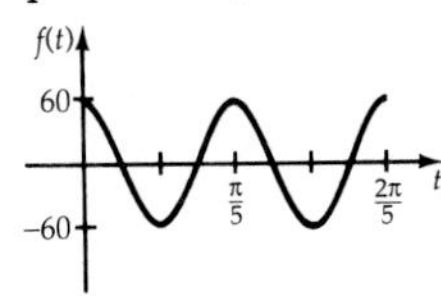

Exercise Set 2.6, page 95

1. $\pi/2 + k\pi$, k an integer **3.** $\pi/2 + k\pi$, k an integer **5.** 2π **7.** π **9.** 2π **11.** $2\pi/3$ **13.** $\pi/3$ **15.** 8π **17.** 1

19. 4 **21.**

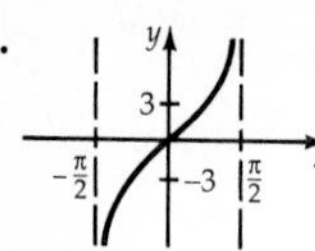

23.

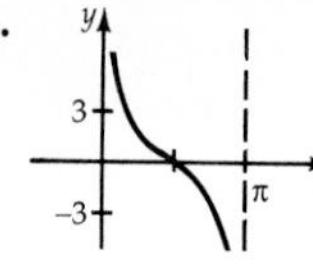

25.

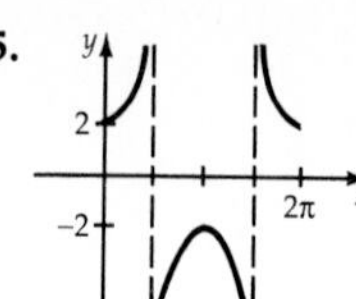

27.

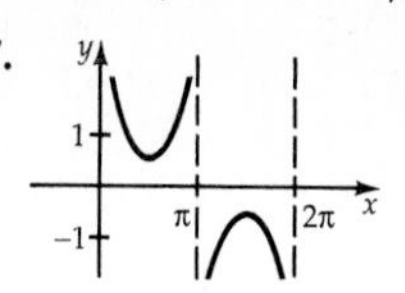

29.

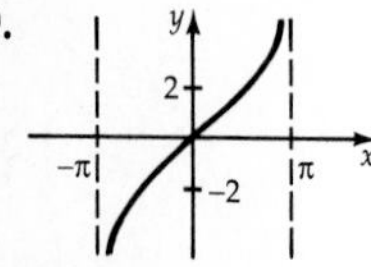

31.

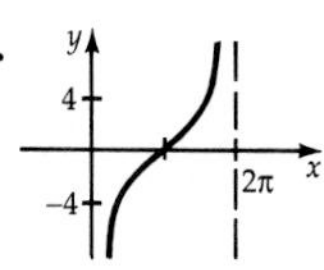

33.

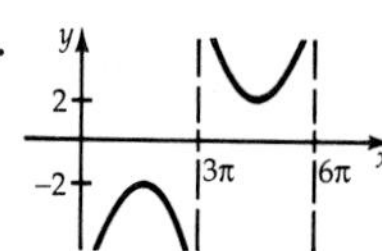

35.

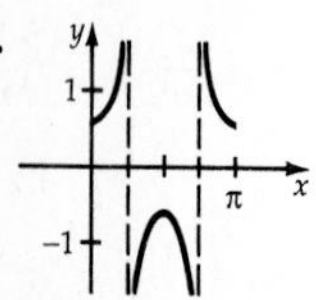

37.

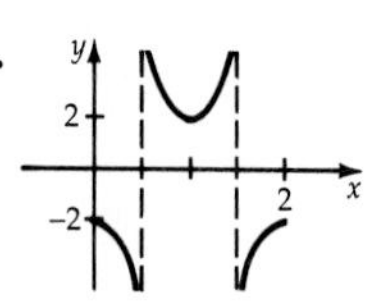

39.

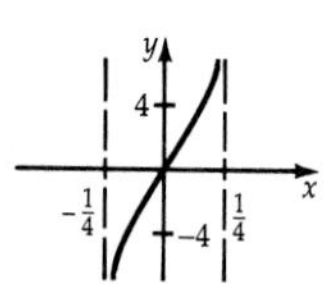

41.

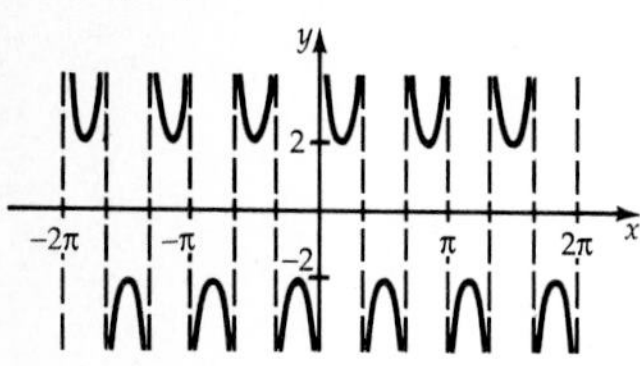

43.

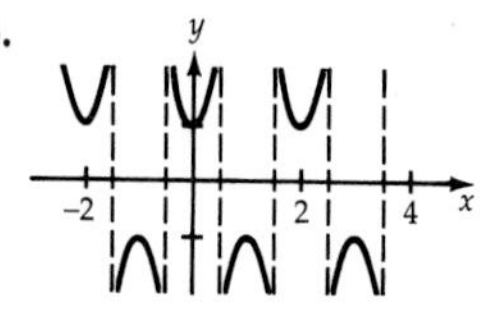

45.

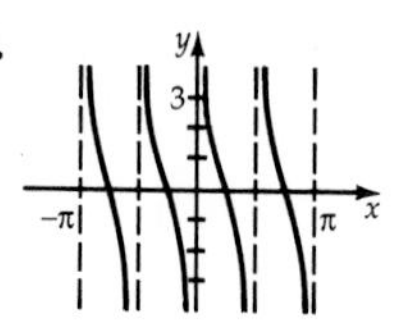

47.

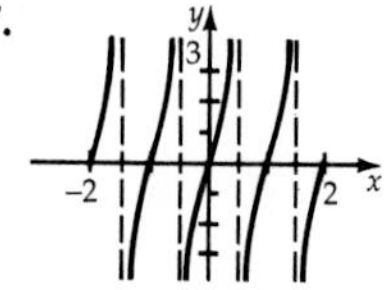

49. $f(x) = \frac{3}{2}x$ **51.** $f(x) = \csc \frac{2}{3}x$ **53.** $f(x) = \sec \frac{3}{4}x$ **55.** $y = \tan 3x$ **57.** $y = \sec \frac{8}{3}x$ **59.** $y = \cot \frac{\pi}{2}x$ **61.** $y = \csc \frac{4\pi}{3}x$

63.

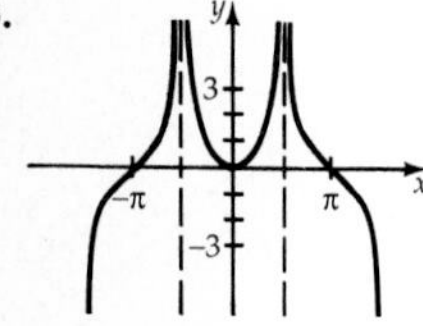

65.

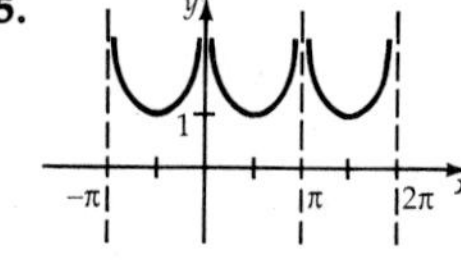

67.

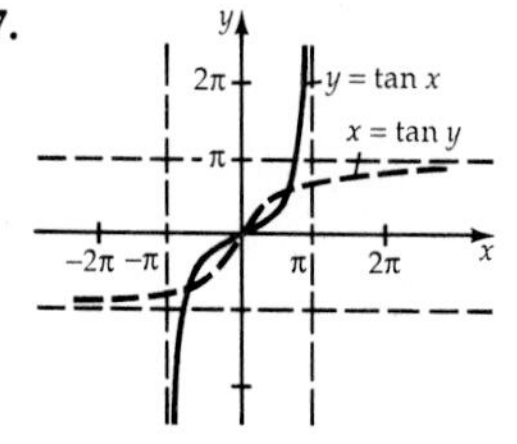

Exercise Set 2.7, page 101

1. 2, $\pi/2$, 2π **3.** 1, $\pi/8$, π **5.** 4, $-\pi/4$, 3π **7.** 5/4, $2\pi/3$, $2\pi/3$ **9.** $\pi/8$, $\pi/2$ **11.** -3π, 6π **13.** $\pi/16$, π **15.** -12π, 4π

17.

19.

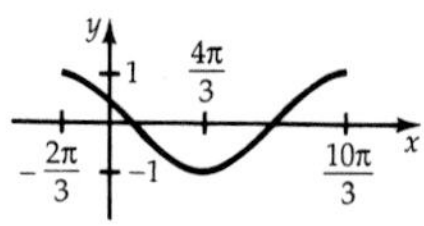

21.

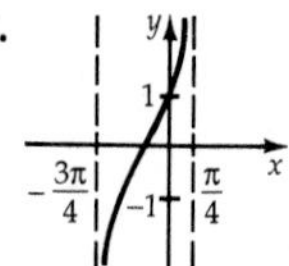

23.

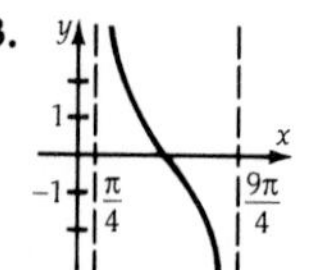

25.

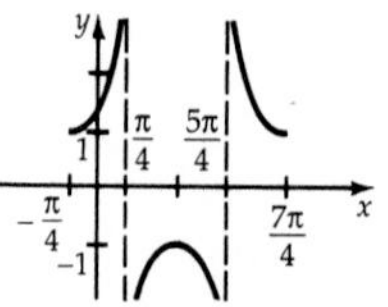

27.

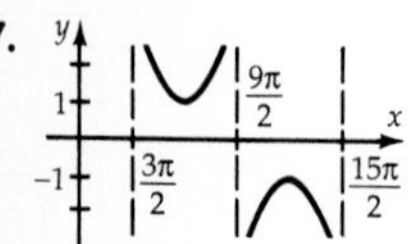

29.

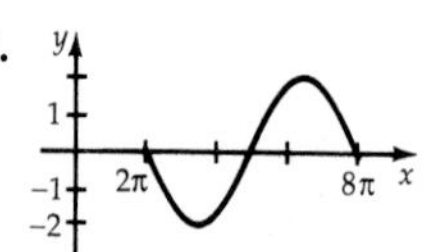

31.

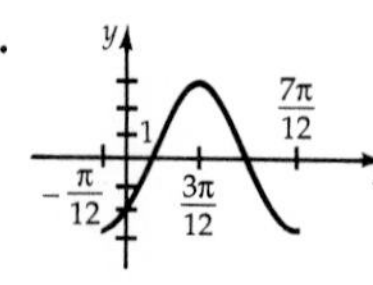

33.

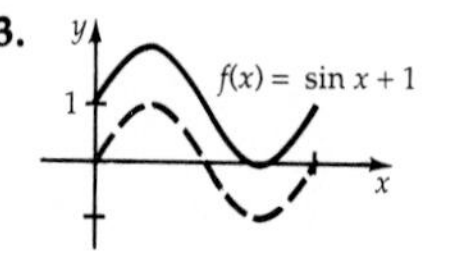

35.

37.

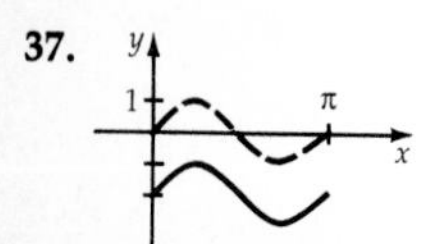

39.

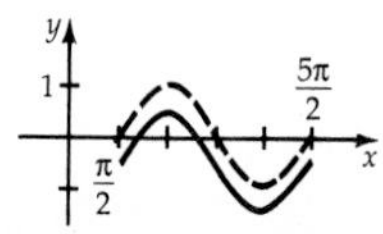

41.

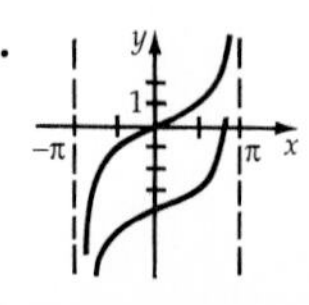

43.

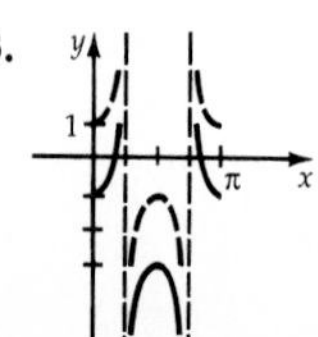

45.

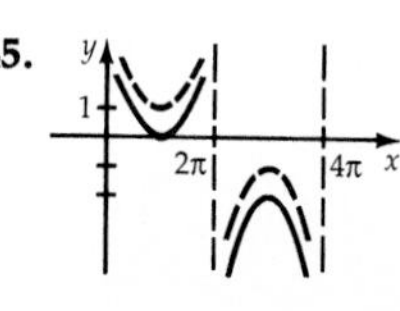

47. 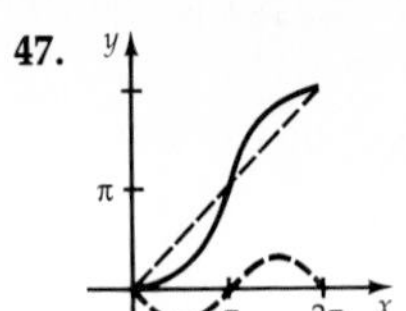

49. 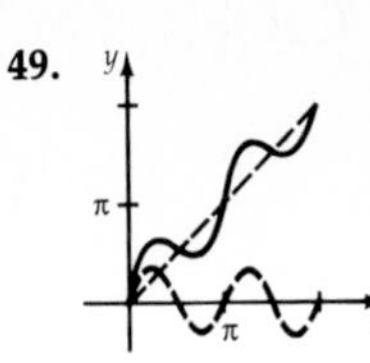

51. 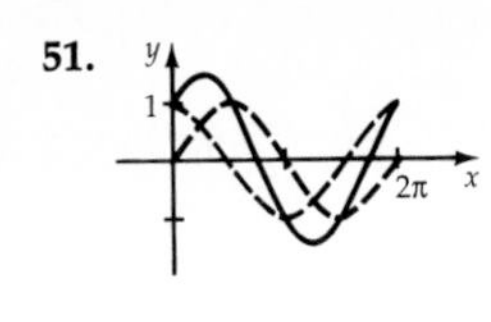

53. 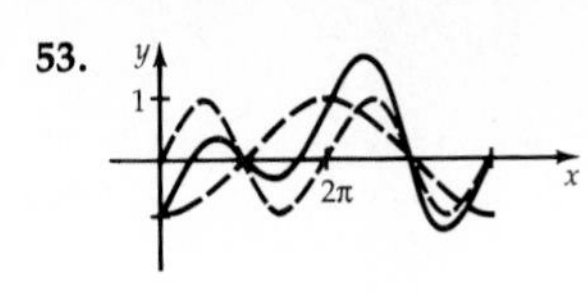

55. 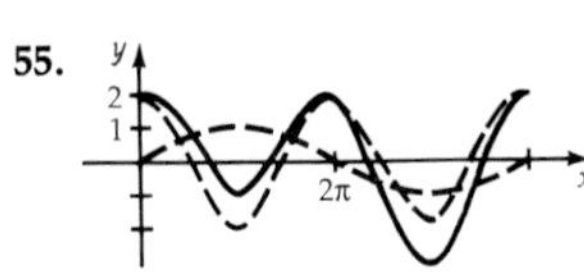

57. 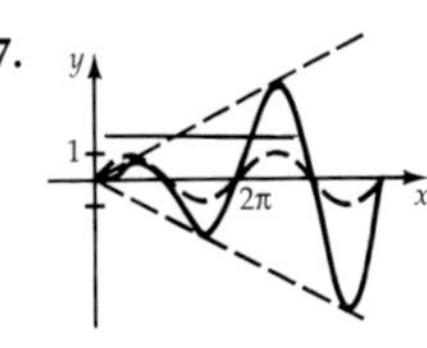

59. 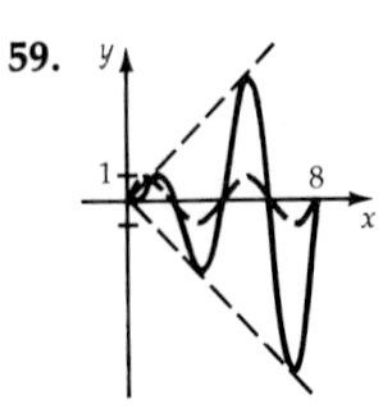

61. 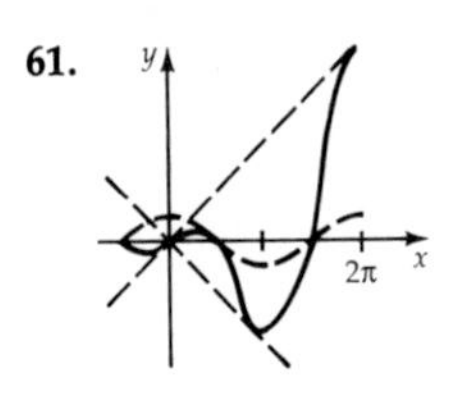

63. $f(x) = \sin\left(2x - \frac{\pi}{3}\right)$ **65.** $f(x) = \csc\left(\frac{x}{2} - \pi\right)$ **67.** $f(x) = \sec\left(x - \frac{\pi}{2}\right)$ **69.** $f(x) = 2\sin\left(2x - \frac{2\pi}{3}\right)$ **71.** $f(x) = \tan\left(\frac{x}{2} - \frac{\pi}{4}\right)$

73. $f(x) = \sec\left(\frac{x}{2} - \frac{3\pi}{8}\right)$ **75.** 1 **77.** $\cos^2 x + 2$ **79.**

81. 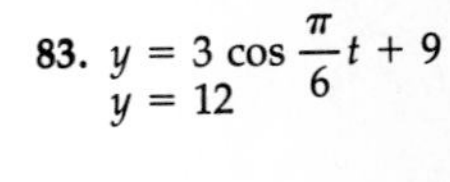

83. $y = 3\cos\frac{\pi}{6}t + 9$
$y = 12$

Exercise Set 2.8, page 107

1. $2, \pi, 1/\pi$ **3.** $3, 3\pi, 1/3\pi$ **5.** $4, 2, 1/2$ **7.** $3/4, 4, 1/4$ **9.** $y = 4\cos 3\pi t$

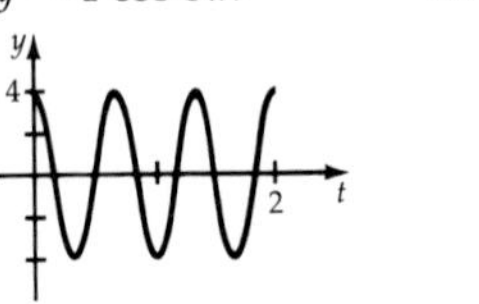

11. $y = \frac{3}{2}\cos\frac{4\pi}{3}t$

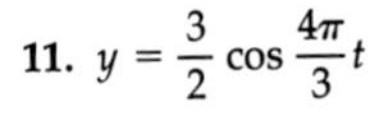

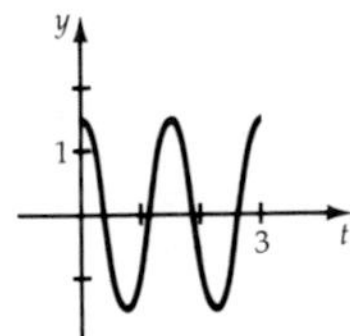

13. $y = 2\sin 2t$

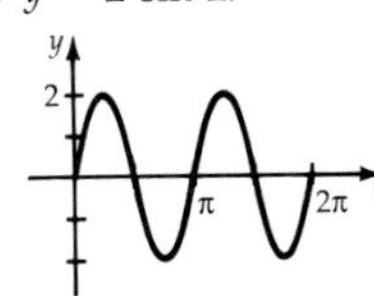

15. $y = \sin \pi t$

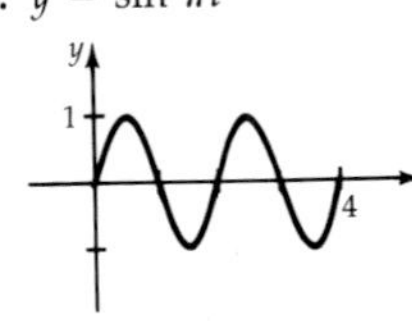

17. $y = 2\sin 2\pi t$

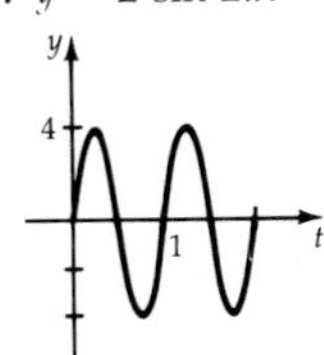

19. $y = \frac{1}{2}\cos 4t$ **21.** $y = 3\cos 2t$ **23.** $y = \frac{1}{2}\cos\frac{2\pi}{3}t$ **25.** $y = 4\cos 4t$ **27.** $4\pi, 1/4\pi, 2$
$y = 2\cos\frac{1}{2}t$ **29.** $y = \cos\frac{4\sqrt{3}}{3}t$
$\pi\sqrt{3}/2, 2\sqrt{3}/3\pi$

31. $\sqrt{10}\pi, \sqrt{10}/10\pi, 6$
$y = 6\cos\frac{\sqrt{10}}{5}t$ **33.** $y = \frac{1}{2}\cos 2\sqrt{2}\,t,$
$f = \sqrt{2}/\pi$ **35.** period = 2/3
$y = \frac{1}{2}\cos 3\pi t$ **37.** Period increases by a factor of $\sqrt{2}$

Challenge Exercises, page 109

1. false, initial site must be along the positive x-axis **3.** true **5.** false, $1 + \tan^2\theta = \sec^2\theta$ is an identity **7.** false, the period is 2π **9.** false, $\sin 45° + \cos(90 - 45°) \neq 1$ **11.** false, $\sin^2 30 \neq (\sin 30)^2$ **13.** false, 1 rad $\approx 57°$

Review Exercises, page 110

1. a. 120° **b.** −135°

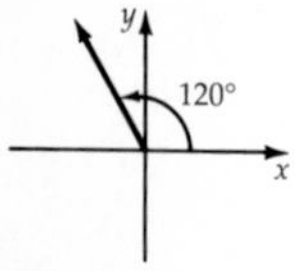

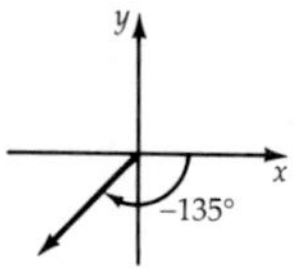

3. a. 114°48′ **b.** −38°22′48″ **5. a.** $7\pi/4$ **b.** 1.70

7. 3.67 in **9.** 14.32° **11.** 3.49 m^2 **13.** 53.78 rad/s **15.** $\sin\theta = -3\sqrt{10}/10$, $\cos\theta = \sqrt{10}/10$, $\tan\theta = -3$, $\csc\theta = -\sqrt{10}/3$, $\sec\theta = \sqrt{10}$, $\cot\theta = -1/3$ **17.** $\sqrt{5}/2$ **19.** $3\sqrt{5}/5$

21. a. −1 **b.** 0.5365 **23. a.** −3.2361 **b.** 0 **25. a.** $-2\sqrt{3}/3$ **b.** $-\sqrt{3}$ **27 a.** $\sqrt{2}$ **b.** −1 **29.** $-\dfrac{4+\sqrt{6}}{4}$ **31.** $\dfrac{\sqrt{3}-2\sqrt{2}}{2}$

33. −1/2 **35.** $-\sqrt{2}/2$ **37.** 1.955 **39.** 0.035 **41.** 2.6 **43.** 0.85

51. **53.** 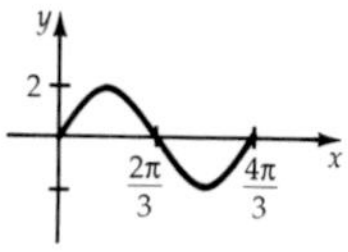 **55.** 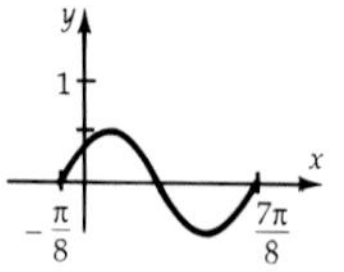 **57.** 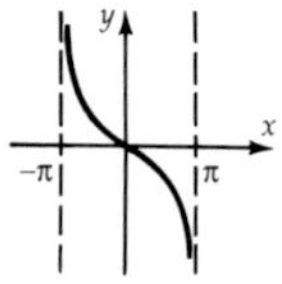 **59.**

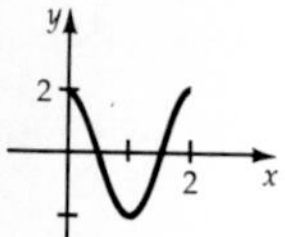

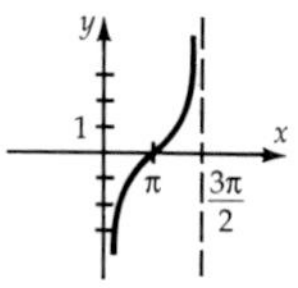

61. 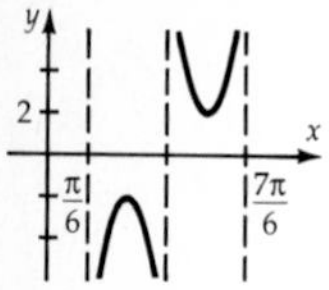 **63.** 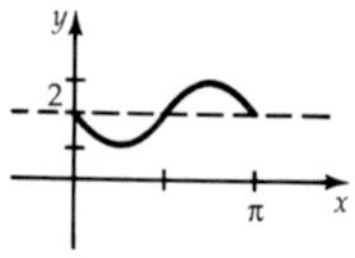 **65.** 2.5, $\pi/25$, $25/\pi$ **67.** π, $1/\pi$, 0.5, $y = 0.5\cos 2t$

Exercise Set 3.1, page 116

1. $\cos^2 x - 5\cos x + 6$ **3.** $\sin^4 x - 17\sin^2 x + 60$ **5.** $12\sin^2 x + 7\sin x - 10$ **7.** $\cos x \sin x + 7\cos x - 2\sin x - 14$ **9.** $3\sin^2 x - 2\sin x\cos x - 8\cos^2 x$ **11.** $(\sin x - 1)^2$ **13.** $(4\tan x - 3)(2\tan x - 5)$ **15.** $(2\sin x - 1)(2\sin x + 1)$

17. $(\sin x - 3)(\sin^2 x + 3\sin x + 9)$ **19.** $(\cos x + 1)(\sin x + 1)$ **21.** $\dfrac{\cos x + 3\sin x}{\sin x\cos x}$ **23.** $\dfrac{2\sin^2 x - \cos^2 x}{\sin x\cos x}$ **25.** $\dfrac{9\cos x + 2}{\cos x(3\cos x + 1)}$

27. $\dfrac{\sin x + 1}{\sin x - 1}$ **83.** $\pm\sqrt{1 - \sin^2 x}$ **85.** $\pm\dfrac{\sqrt{1 - \sin^2 x}}{1 - \sin^2 x}$

Exercise Set 3.2, page 125

1. $\dfrac{\sqrt{6}+\sqrt{2}}{4}$ **3.** $\dfrac{\sqrt{6}+\sqrt{2}}{4}$ **5.** 0 **7.** $\dfrac{-\sqrt{6}+\sqrt{2}}{4}$ **9.** $-\dfrac{\sqrt{6}+\sqrt{2}}{4}$ **11.** $2+\sqrt{3}$ **13.** $\dfrac{\sqrt{6}-\sqrt{2}}{4}$ **15.** $-\dfrac{\sqrt{6}+\sqrt{2}}{4}$

17. $2+\sqrt{3}$ **19.** $\dfrac{\sqrt{2}+\sqrt{6}}{4}$ **21.** $-\dfrac{\sqrt{6}+\sqrt{2}}{4}$ **23.** $-2-\sqrt{3}$ **25.** 0 **27.** $\sqrt{3}/2$ **29.** 1/2 **31.** 1/2 **33.** $\sqrt{3}$

35. cos 10° **37.** −sin 40° **39.** $\sin 5\pi/12$ **41.** $\cos 5\pi/12$ **43.** $\tan 7\pi/12$ **45.** $\sin 5x$ **47.** $\cos x$ **49.** $\sin 4x$ **51.** $\cos 2x$ **53.** $\sin x$ **55.** $\tan 7x$ **57.** −sin 1 **59.** cos 3 **61. a.** −63/65 **b.** −56/65 **63. a.** 63/65 **b.** 56/65 **c.** 33/56 **65. a.** −77/85 **b.** −84/85 **c.** −13/84 **67. a.** −33/65 **b.** −16/65 **c.** 63/16

Exercise Set 3.3, page 131

1. $\sin 4\alpha$ **3.** $\cos 4\beta$ **5.** $\cos 6\alpha$ **7.** $\tan 6\alpha$ **9.** $\dfrac{\sqrt{2-\sqrt{3}}}{2}$ **11.** $\sqrt{2}+1$ **13.** $-\dfrac{\sqrt{2+\sqrt{2}}}{2}$ **15.** $\dfrac{\sqrt{2-\sqrt{2}}}{2}$
17. $\dfrac{\sqrt{2-\sqrt{2}}}{2}$ **19.** $\dfrac{\sqrt{2-\sqrt{3}}}{2}$ **21.** $-2-\sqrt{3}$ **23.** $\dfrac{\sqrt{2+\sqrt{3}}}{2}$ **25.** $\sin 2\theta = -24/25$, $\cos 2\theta = 7/25$, $\tan 2\theta = -24/7$
27. $\sin 2\theta = -240/289$, $\cos 2\theta = 161/289$, $\tan 2\theta = -240/161$ **29.** $\sin 2\theta = -336/625$, $\cos 2\theta = -527/625$, $\tan 2\theta = 336/527$
31. $\sin \theta = 240/289$, $\cos 2\theta = -161/289$, $\tan 2\theta = 240/161$ **33.** $\sin 2\theta = -720/1681$, $\cos 2\theta = 1519/1681$, $\tan 2\theta = -720/1519$
35. $\sin 2\theta = 240/289$, $\cos 2\theta = -161/289$, $\tan 2\theta = -240/161$ **37.** $\sin \alpha/2 = \sqrt{26}/26$, $\cos \alpha/2 = 5\sqrt{26}/26$, $\tan \alpha/2 = 1/5$
39. $\sin \alpha/2 = 5\sqrt{34}/34$, $\cos \alpha/2 = -3\sqrt{34}/34$, $\tan \alpha/2 = -5/3$ **41.** $\sin \alpha/2 = \sqrt{5}/5$, $\cos \alpha/2 = 2\sqrt{5}/5$, $\tan \alpha/2 = 1/2$
43. $\sin \alpha/2 = \sqrt{2}/10$, $\cos \alpha/2 = -7\sqrt{2}/10$, $\tan \alpha/2 = -1/7$ **45.** $\sin \alpha/2 = \sqrt{17}/17$, $\cos \alpha/2 = 4\sqrt{17}/17$, $\tan \alpha/2 = 1/4$
47. $\sin \alpha/2 = 5\sqrt{34}/34$, $\cos \alpha/2 = -3\sqrt{34}/34$, $\tan \alpha/2 = -5/3$

Exercise Set 3.4, page 139

1. $\sin 3x - \sin x$ **3.** $\frac{1}{2}[\sin 8x - \sin 4x]$ **5.** $\sin 8x + \sin 2x$ **7.** $\frac{1}{2}[\cos 4x - \cos 6x]$ **9.** $1/4$ **11.** $-\sqrt{2}/4$ **13.** $-1/4$
15. $\dfrac{\sqrt{3}-2}{4}$ **17.** $2 \sin 3\theta \cos \theta$ **19.** $2 \cos 2\theta \cos \theta$ **21.** $-2 \sin 4\theta \sin 2\theta$ **23.** $2 \cos 2\theta \cos \theta$ **25.** $\sqrt{6}/2$ **27.** $-\sqrt{6}/2$
29. $\sqrt{2}/2$ **31.** $\sqrt{6}/2$ **49.** $y = \sqrt{2}\sin(x - 135°)$ **51.** $y = \sin(x - 60°)$ **53.** $y = \dfrac{\sqrt{2}}{2}\sin(x - 45°)$ **55.** $y = 17\sin(x + 61.9°)$
57. $y = 8.5\sin(x - 20.6°)$ **59.** $y = \sqrt{2}\sin(x + 3\pi/4)$ **61.** $y = \sin(x + \pi/6)$ **63.** $y = 9.8\sin(x + 1.99)$
65. $y = 5\sqrt{2}\sin(x + 3\pi/4)$ **67.** **69.** **71.**

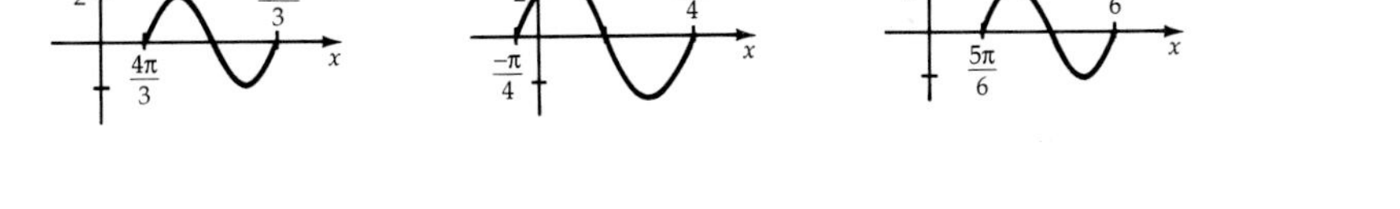

73. **75.** **91.** $\sqrt{2}$, 90°, 4π **93.** 2, $\pi/12$, π **95.** 2, 1/3, 2 **99.** 90°

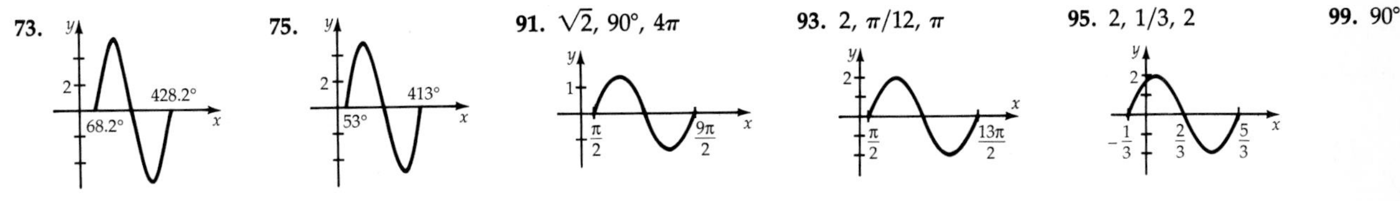

Exercise Set 3.5, page 151

1. $\pi/2$ **3.** $5\pi/6$ **5.** $-\pi/4$ **7.** $\pi/3$ **9.** $\pi/3$ **11.** $-\pi/4$ **13.** $-60°$ **15.** 120° **17.** 30° **19.** 30° **21.** 0.4729
23. 1.8087 **25.** -0.9620 **27.** 0.7943 **29.** $-16.1°$ **31.** 56.3° **33.** $-63.9°$ **35.** $-45.7°$ **37.** 106.7° **39.** 84.0° **41.** 1/2
43. 2 **45.** 3/5 **47.** 1 **49.** 1/2 **51.** $\pi/6$ **53.** $\pi/4$ **55.** not defined **57.** 0.4636 **59.** $-\pi/6$ **61.** $-\pi/3$ **63.** 2.1735
65. -0.7430 **67.** 0.6675 **69.** 24/25 **71.** 13/5 **73.** $\pi/3$ **75.** $\pi/6$ **77.** $-\pi/4$ **79.** 0 **81.** 24/25 **83.** 12/13 **85.** 2

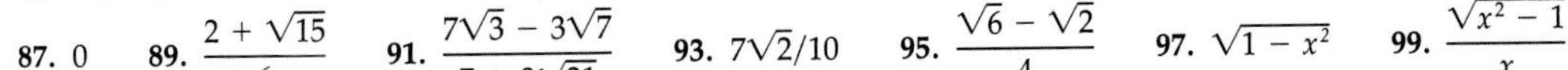

87. 0 **89.** $\dfrac{2+\sqrt{15}}{6}$ **91.** $\dfrac{7\sqrt{3}-3\sqrt{7}}{7+3\sqrt{21}}$ **93.** $7\sqrt{2}/10$ **95.** $\dfrac{\sqrt{6}-\sqrt{2}}{4}$ **97.** $\sqrt{1-x^2}$ **99.** $\dfrac{\sqrt{x^2-1}}{x}$

105. **107.** **109.**

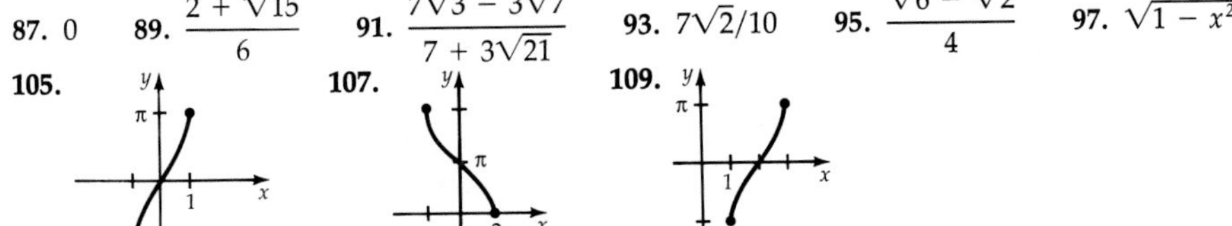

111. **113.** **115.**

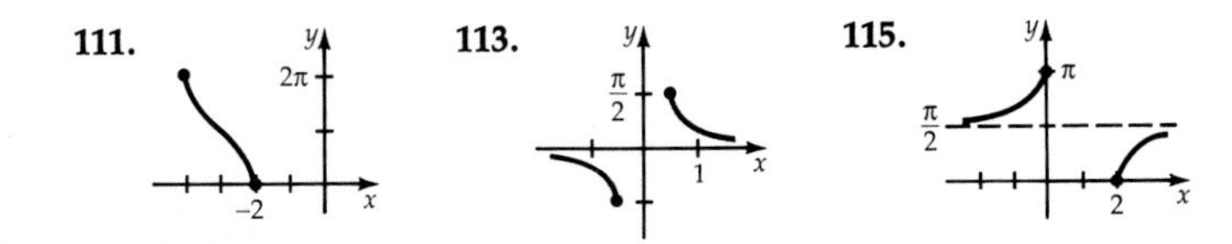

121. $y = \dfrac{1}{3}\tan 5x$ **123.** $y = 3 + \cos\left(x - \dfrac{\pi}{3}\right)$ **125.**

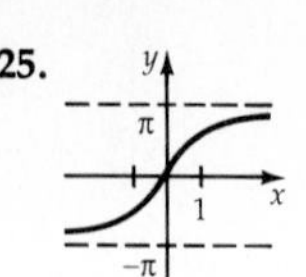

127.

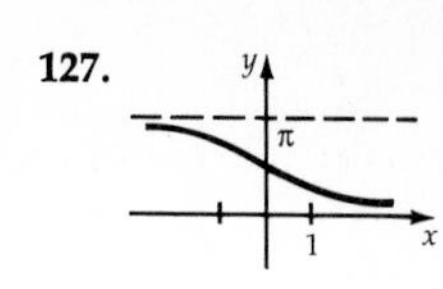

129. 0.1014

Exercise Set 3.6, page 157

1. $\pi/4, 7\pi/4$ **3.** $\pi/3, 4\pi/3$ **5.** $\pi/4, \pi/2, 3\pi/4, 3\pi/2$ **7.** $\pi/2, 3\pi/2$ **9.** $\pi/6, \pi/4, 3\pi/4, 11\pi/6$ **11.** $\pi/4, 3\pi/4$ **13.** $\pi/6, \pi/2, 5\pi/6$ **15.** $\pi/6, 5\pi/6, 7\pi/6, 11\pi/6$ **17.** $0, \pi/4, 3\pi/4, \pi, 5\pi/4, 7\pi/4$ **19.** $\pi/6, 5\pi/6, 4\pi/3, 5\pi/3$ **21.** $0, \pi/2, \pi, 3\pi/2$ **23.** 41.4°, 318.6° **25.** no solution **27.** 68.0°, 292° **29.** no solution **31.** 12.8°, 167.2° **33.** 15.5°, 164.5° **35.** 0°, 33.7°, 180°, 213.7° **37.** no solution **39.** no solution **41.** 0, 120°, 240° **43.** 70.5°, 289.5° **45.** 68.2°, 116.6°, 248.2°, 296.6° **47.** 19.5°, 90°, 160.5°, 270° **49.** 60°, 90°, 300° **51.** 53.1°, 180° **53.** 72.4°, 220.2° **55.** 50.1°, 129.9°, 205.7°, 334.3° **57.** no solution **59.** 22.5°, 157.5° **61.** $\pi/8 + k\pi/2$ or $5\pi/8 + k\pi/2$ where k is an integer **63.** $\pi/10 + 2k\pi/5$ where k is an integer **65.** $0 + 2k\pi, \pi/3 + 2\pi k, \pi + 2k\pi, 5\pi/3 + 2k\pi$ where k is an integer **67.** $\pi/2 + k\pi, 5\pi/6 + k\pi$ where k is an integer **69.** $0° + 2k\pi$ where k is an integer **71.** $0, \pi$ **73.** $0, \pi/6, 5\pi/6, \pi, 7\pi/6, 11\pi/6$ **75.** $0, \pi/2, 3\pi/2$ **77.** $0, \pi/3, 2\pi/3, \pi, 4\pi/3, 5\pi/3$ **79.** $4\pi/3, 5\pi/3$ **81.** $0, \pi/4, 3\pi/4, \pi, 5\pi/4, 7\pi/4$ **83.** $\pi/6, 5\pi/6, \pi$ **85.** $\pi/6, \pi/2$ **87.** $-\pi/3, 0$

89. $0, \pi/4, \pi/2, 3\pi/4, \pi, 5\pi/4, 3\pi/2, 7\pi/4$ **91.** $0, \pi/2, \pi, 3\pi/2$ **93.** $0, \pi/2, 5\pi/6, 7\pi/6, 3\pi/2, 11\pi/6$ **95.** $A = \dfrac{1}{2}r^2\theta$

97. $A = \dfrac{1}{2}r^2(\theta - \sin\theta)$

Challenge Exercises, page 161

1. false, $\dfrac{\tan 45°}{\tan 60°} \neq \dfrac{45°}{60°}$ **3.** false, $\sin^{-1}x \neq \dfrac{1}{\sin x}$ **5.** false, $\sin(30° + 60°) \neq \sin 30° + \sin 60°$

7. false, $\tan 45° = \tan 225°$ but $45° \neq 225°$ **9.** false, $\cos(\cos^{-1}(2)) \neq 2$ **11.** false, $\sin(180° - \theta) = \sin\theta$

Review Exercises, page 161

1. $\dfrac{\sqrt{6} - \sqrt{2}}{4}$ **3.** $\dfrac{\sqrt{6} - \sqrt{2}}{4}$ **5.** $-\dfrac{\sqrt{6} + \sqrt{2}}{4}$ **7.** $\dfrac{\sqrt{2 - \sqrt{2}}}{2}$ **9.** $\sqrt{2} + 1$ **11. a.** 0 **b.** $\sqrt{3}$ **c.** 1/2

13. a. $\sqrt{3}/2$ **b.** $-\sqrt{3}$ **c.** $-\dfrac{\sqrt{2 - \sqrt{3}}}{2}$ **15.** $\sin 6x$ **17.** $\sin 3x$ **19.** $\cos\beta$ **21.** 0.2740 **23.** $-\sqrt{3}/2$ **25.** $-2\sin 3\theta\cos 2\theta$

27. $2\sin 4\theta\sin 2\theta$ **47.** 13/5 **49.** 3/2 **51.** 4/5 **53.** 30°, 150°, 240°, 300°

55. $\pi/2 + 2k\pi$, $3.8713 + 2k\pi$, $5.553 + 2k\pi$ where k is an integer **57.** $\pi/12, 5\pi/12, 13\pi/12, 17\pi/12$ **59.**

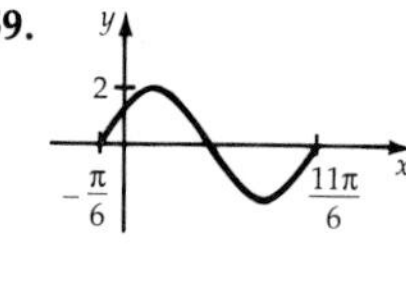

61.

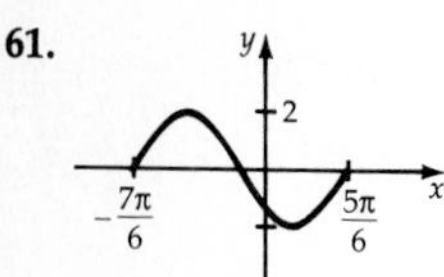

63.

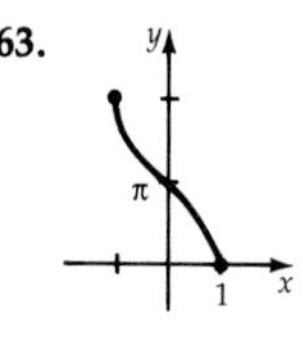

65.

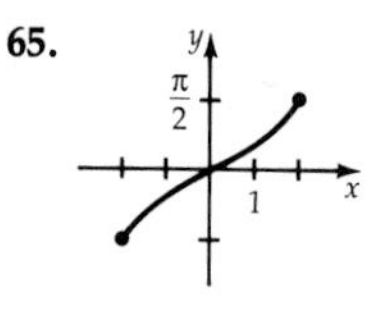

Exercise Set 4.1, page 168

1. 51° **3.** 24° **5.** 1570 **7.** 3.47 **9.** $B = 34°, b = 81, c = 140$ **11.** $B = 66.2°, a = 55.1, c = 137$ **13.** $A = 66.53°, b = 497.1, a = 1145$ **15.** $A = 68.61°, B = 21.39°, c = 34.74$ **17.** 14 **19.** $\sin A = 0.4472$ $\cot A = 0.8944$ $\tan A = 0.5$ **21.** 0.6 **23.** 0.625 **25.** $B = 48°, b = 11$ in, $c = 15$ in
$\csc A = 2.2361$ $\sec A = 1.1180$ $\cot A = 2$
27. $A = 41°, B = 49°, a = 4.0$ ft **29.** 17 ft **31.** 2230 ft **33.** 2.94 mi **35.** 38 mi **37.** 5.2 m **39.** 59 ft **41.** 6.8° **43.** 70 s

Exercise Set 4.2, page 173

1. $C = 77°, b = 16, c = 17$ **3.** $B = 38°, a = 18, c = 10$ **5.** $C = 32.6°, c = 21.6, a = 39.8$ **7.** $B = 47.7°, a = 57.4, b = 76.3$
9. $C = 59°, B = 84°, b = 46$ or $C = 121°, B = 22°, b = 17$ **11.** no solution **13.** no solution **15.** $C = 19.8°, B = 145.5°, b = 10.7$ $C = 160.3°, B = 4.9°, b = 1.6$ **17.** no solution
19. 8.1 mi **21.** 4.2 mi **23.** 193 yd **25.** 94 ft **27.** 33.5 ft **29.** 1170 mi

Exercise Set 4.3, page 180

1. 12.5 **3.** 150.5 **5.** 28.9 **7.** 9.5 **9.** 10.5 **11.** 40.1 **13.** 90.7 **15.** 38.7° **17.** 89.6° **19.** 47.9° **21.** 116.7°
23. 80.3° **25.** 139 **27.** 52.9 **29.** 58.8 **31.** 299 **33.** 36.3 **35.** 7.3 **37.** 709 mi **39.** 347 mi **41.** 40 cm
43. 9.17 in, 24.9 in **45.** 55.4 cm **47.** 2764 ft **49.** 47,520 m^2 **51.** 203 m^2 **53.** 162 in^2 **55.** $41,479 **57.** 6.23 acres
59. 12.5° **61.** 72.7° **63.** 52.0 cm **69.** 137 in^3

Exercise Set 4.4, page 192

1. $\sqrt{10}$, 161.6° **3.** 9, 0° **5.** 30, 180° **7, 9.**

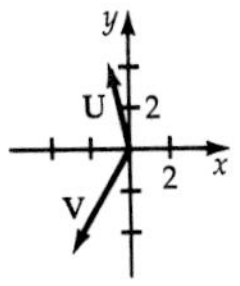

11. 5, 126.9° **13.** 44.7, 296.6° **15.** 4.5, 296.6° **17.** 45.7, 336.8° **19.** $x = -8, y = 5$ **21.** $x = -2, y = -6$ **23.** $x = -3, y = 4$
25. $x = 6, y = -4$ **27.** $x = 3.9, y = 4.6$ **29.** $6\mathbf{i} + \mathbf{j}$ **31.** $-5\mathbf{i} - 2\mathbf{j}$ **33.** $6\mathbf{i} + 9\mathbf{j}$ **35.** $5\mathbf{i} + 4\mathbf{j}$ **37.** $-6\mathbf{i} + \frac{17}{2}\mathbf{j}$ **39.** $\frac{5}{6}\mathbf{i} + 3\mathbf{j}$
41. $-3\mathbf{i} + 4\mathbf{j}$ **43.** $\mathbf{i} - 3\mathbf{j}$ **45.** $3.6\mathbf{i} + 3.5\mathbf{j}$ **47.** $6.1\mathbf{i} - 5.1\mathbf{j}$ **49.** -10 **51.** -10 **53.** 24 **55.** 186 **57.** 70.3° **59.** 63.4°
61. 47.7° **63.** 177.3° **65.** 186 lb, 11° **67.** 378 mph, 120° **69.** 293 lb **71.** 1192 ft-lb **77.** $3\mathbf{i} + \mathbf{j}$ **79.** $\frac{21\sqrt{10}}{5}\mathbf{i} - \frac{7\sqrt{10}}{5}\mathbf{j}$
81. $-\sqrt{2}\mathbf{i} + \sqrt{2}\mathbf{j}$

Challenge Exercises, page 194

1. false, cannot solve a triangle given the angle opposite one of the given sides **3.** true **5.** true **7.** true **9.** true
11. false, if $\mathbf{v} = i + j$ and $\mathbf{w} = i - j$, then $\mathbf{v} \cdot \mathbf{w} = 1 - 1 = 0$

Review Exercises, page 195

1. $B = 53°, a = 11, c = 18$ **3.** $B = 48°, C = 95°, A = 37°$ **5.** $c = 13, A = 55°, B = 90°$ **7.** No triangle is formed
9. $C = 45°, a = 29, b = 35$ **11.** 357 **13.** 917 **15.** 792 **17.** 167 **19.** 4.47, 333° **21.** 4.47, 153° **23.** 3.61, 124°
25. 5.10, 11° **27.** $x = -8, y = 5$ **29.** $x = 10, y = 6$ **31.** $-7\mathbf{i} - 3\mathbf{j}$ **33.** $-5\mathbf{i} - \frac{5}{2}\mathbf{j}$ **35.** -9 **37.** -9 **39.** 86° **41.** 125°

Exercise Set 5.1, page 203

1. $2 + 3i$ **3.** $4 - 11i$ **5.** $8 + i\sqrt{3}$ **7.** $7 + 4i$ **9.** $0.9i$ **11.** $5 + 12i$ **13.** $4 - 3i$ **15.** $-2 - 5i$ **17.** $12 - 2i$
19. $-2 + 11i$ **21.** $16 + 16i$ **23.** $23 + 2i$ **25.** $74 + 0i$ **27.** $-117 - i$ **29.** $\frac{1}{2} - \frac{i}{2}$ **31.** $\frac{7}{58} + \frac{3}{58}i$ **33.** $\frac{5}{13} + \frac{12}{13}i$
35. $\frac{1}{61} + \frac{11}{61}i$ **37.** $1 - 6i$ **39.** $-16 - 30i$ **41.** $-29 - 17i$ **43.** $-2 - 2i$ **45.** $-16 + 0i$ **47.** $0 + 75i$ **49.** $-i$ **51.** i
53. -1 **55.** -1 **57.** $-i$ **59.** i **61.** -1 **63.** $-i$ **65.** -2 **67.** $-8\sqrt{5}$ **69.** 11 **71.** $9 + 40i$ **73.** $\frac{1}{2} \pm \frac{\sqrt{3}}{2}i$
75. $-1 \pm i$ **77.** $-\frac{3}{2} \pm \frac{\sqrt{3}}{2}i$ **79.** $-\frac{1}{4} \pm \frac{\sqrt{23}}{4}i$ **81.** 5 **83.** $\sqrt{29}$ **85.** $\sqrt{65}$ **87.** 3 **95.** no **97.** $66 + 6\sqrt{5} - 14\sqrt{3}$
99. 1 **101.** 0 **103.** $(x + 3i)(x - 3i)$ **105.** $(2x + 9i)(2x - 9i)$ **107.** 0 **109.** 0

Exercise Set 5.2, page 210

1–7. i, x, $\sqrt{3} - i$, $-2 - 2i$, $-2i$, $3 - 5i$

9. $z = 2(\cos 315° + i \sin 315°)$ **11.** $2(\cos 330° + i \sin 330°)$ **13.** $3(\cos 90° + i \sin 90°)$

15. $5(\cos 180° + i \sin 180°)$ **17.** $\sqrt{2} + i\sqrt{2}$ **19.** $\frac{\sqrt{2}}{2} - \frac{i\sqrt{2}}{2}$ **21.** $-4.60 + 3.86i$ **23.** 8 **25.** $-\sqrt{3} + i$ **27.** $-3i$

29. $-4\sqrt{2} + 4i\sqrt{2}$ **31.** $\frac{9\sqrt{3}}{2} - \frac{9}{2}i$ **33.** $-0.832 + 1.820i$ **35.** $-4\sqrt{3} - 4i$ **37.** $-4\sqrt{3} - 4i$ **39.** $6(\cos 225° + i \sin 225°)$

41. $12(\cos 335° + i \sin 335°)$ **43.** $10\left(\cos \frac{7\pi}{12} + i \sin \frac{7\pi}{12}\right)$ **45.** $15(\cos 165° + i \sin 165°)$ **47.** $-4 + 4i\sqrt{3}$ **49.** $-\frac{3\sqrt{3}}{2} + \frac{3}{2}i$

51. $3i$ **53.** $-\frac{5}{2} + \frac{5\sqrt{3}}{2}i$ **55.** $\frac{5}{3}\left(\cos \frac{5\pi}{12} - i \sin \frac{5\pi}{12}\right)$ **57.** $2.7321 - 0.7321i$ **59.** 6 **61.** $-\frac{1}{2} - \frac{\sqrt{3}}{2}i$ **63.** $-i\sqrt{2}$

65. $8\sqrt{2} - 8i\sqrt{2}$ **67.** $-\frac{3}{8} - \frac{\sqrt{3}}{8}i$ **69.** $59.1 + 42.9i$

Exercise Set 5.3, page 216

1. $256(\cos 240° + i \sin 240°)$ **3.** $32(\cos 120° + i \sin 120°)$ **5.** $32(\cos 45° + i \sin 45°)$ **7.** $64(\cos 0° + i \sin 0°)$
9. $32(\cos 270° + i \sin 270°)$ **11.** $1024\sqrt{2}(\cos 315° + i \sin 315°)$ **13.** $\cos 270° + i \sin 270°$ **15.** $1.148 + 2.772i$, $-1.148 - 2.772i$

17. $1.879 + 0.684i$, $0.347 + 1.970i$, $-1.532 + 1.286i$, $-1.879 - 0.684i$, $-0.347 - 1.970i$, $1.532 - 1.286i$

19. $0.951 + 0.309i$, $0 + i$, $-0.951 + 0.309i$, $-0.588 - 0.809i$, $0.588 - 0.809i$

21. $1 + 0i$, $-0.5 + 0.866i$, $-0.5 - 0.866i$

23. $1.070 + 0.213i$, $-0.213 + 1.070i$, $-1.070 - 0.213i$, $0.213 - 1.070i$

25. $-0.276 + 1.563i$, $-1.216 - 1.020i$, $1.492 - 0.543i$

27. $2\sqrt{2} + 2i\sqrt{6}$, $-2\sqrt{2} - 2i\sqrt{6}$

29. $2(\cos 60° + i \sin 60°)$, $2(\cos 180° + i \sin 180°)$, $2(\cos 300° + \text{i} \sin 300°)$

31. $\cos 67.5° + i \sin 67.5°$, $\cos 157.5° + i \sin 157.5°$, $\cos 247.5° + i \sin 247.5°$, $\cos 337.5° + i \sin 337.5°$

33. $3(\cos 0° + i \sin 0°)$, $3(\cos 120° + i \sin 120°)$, $3(\cos 240° + i \sin 240°)$

35. $3(\cos 45° + i \sin 45°)$, $3(\cos 135° + i \sin 135°)$, $3(\cos 225° + i \sin 225°)$, $3(\cos 315° + i \sin 315°)$

37. $1.19(\cos 75° + i \sin 75°)$, $1.19(\cos 165° + i \sin 165°)$, $1.19(\cos 225° + i \sin 255°)$, $1.19(\cos 345° + i \sin 345°)$

39. $1.26(\cos 80° + i \sin 80°)$, $1.26(\cos 200° + i \sin 200°)$, $1.26(\cos 320° + i \sin 320°)$

Challenge Exercises, page 218

1. false, $\sqrt{(-1)^2} = \sqrt{1} = 1$ **3.** true **5.** false, $\frac{1 + i}{1 - i} = i$ **7.** true **9.** true **11.** false, $\cos \pi + i \sin \pi = -1$

Review Exercises, page 218

1. $3 - 8i$, $3 + 8i$ **3.** $-2 + i\sqrt{5}$, $-2 - i\sqrt{5}$ **5.** -4 **7.** $5 + 2i$ **9.** $-3 + 3i$ **11.** $25 - 19i$ **13.** $\frac{8}{25} - \frac{6}{25}i$ **15.** $-2 - 2i$
17. $6 + 6i$ **19.** 7 **21.** $-i$ **23.** 1 **25.** 8 **27.** $\sqrt{41}$ **29.** $2\sqrt{2}(\cos 315° + i \sin 315°)$ **31.** $\sqrt{13}(\cos 146.3° + i \sin 146.3°)$
33. $3.54 - 3.54i$ **35.** $-0.832 + 1.818i$ **37.** $-30i$ **39.** $-8.918 + 8.030i$ **41.** $-6.01 - 13.74i$ **43.** $3(\cos 100° - i \sin 100°)$

45. $5(\cos 59° - i \sin 59°)$ **47.** $\frac{5}{3}(\cos 1.9 + i \sin 1.9)$ **49.** $-171.8 - 171.8i$ **51.** $64 - 110.9i$ **53.** $-22.6 + 22.6i$

55. $3(\cos 30° + i \sin 30°)$, $3(\cos 150° + i \sin 150°)$, $3(\cos 270° + i \sin 270°)$

57. $4(\cos 30° + i \sin 30°)$, $4(\cos 120° + i \sin 120°)$, $4(\cos 210° + i \sin 210°)$, $4(\cos 300° + i \sin 300°)$

59. $3(\cos 0° + i \sin 0°)$, $3(\cos 90° + i \sin 90°)$, $3(\cos 180° + i \sin 180°)$, $3(\cos 270° + i \sin 270°)$

Exercise Set 6.1, page 227

1. vertex: $(0, 0)$
focus: $(0, -1)$
directrix: $y = 1$

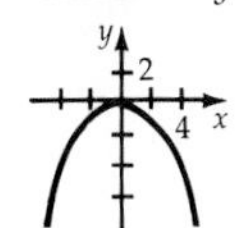

3. vertex: $(0, 0)$
focus: $(1/12, 0)$
directrix: $x = -1/12$

5. vertex: $(2, -3)$
focus: $(2, -1)$
directrix: $y = -5$

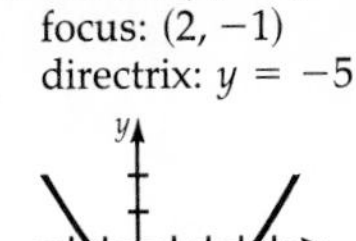

7. vertex: $(2, -4)$
focus: $(1, -4)$
directrix: $x = 3$

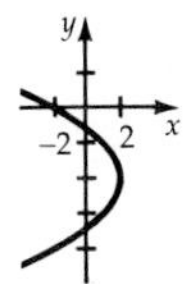

9. vertex: $(-4, 1)$
focus: $(-7/2, 1)$
directrix: $x = -9/2$

11. vertex: $(2, 2)$
focus: $(2, 5/2)$
directrix: $y = 3/2$

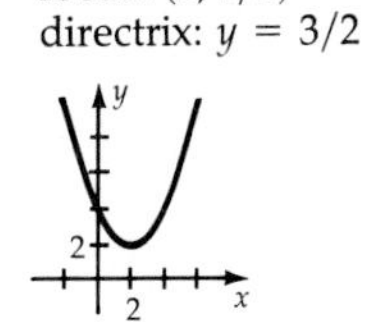

13. vertex: $(-4, -10)$
focus: $(-4, -39/4)$
directrix: $y = -41/4$

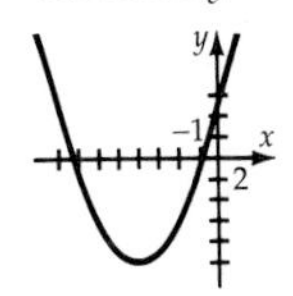

15. vertex: $(-7/4, 3/2)$
focus: $(-2, 3/2)$
directrix: $x = -3/2$

17. vertex: $(-5, -3)$
focus: $(-9/2, -3)$
directrix: $x = -11/2$

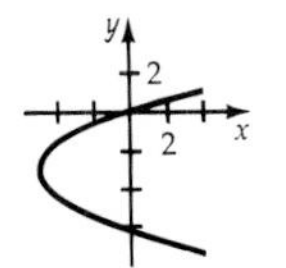

19. vertex: $(-3/2, 13/12)$
focus: $(-3/2, 1/3)$
directrix: $y = 11/6$

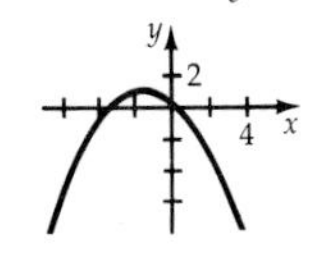

21. vertex: $(2, -5/4)$
focus: $(2, -3/4)$
directrix: $y = -7/4$

23. vertex: $(9/2, -1)$
focus: $(35/8, -1)$
directrix: $x = 37/8$

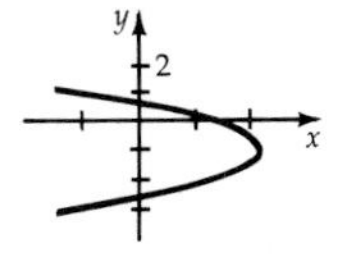

25. vertex: $(1, 1/9)$
focus: $(1, 31/36)$
directrix: $y = -23/36$

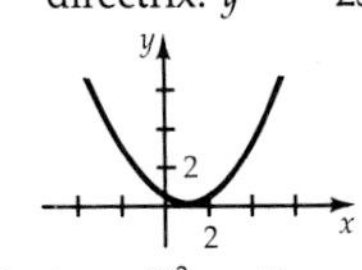

27. $x^2 = -16y$

29. $(x + 1)^2 = 4(y - 2)$ **31.** $(x - 3)^2 = 4(y + 4)$ **33.** $(x + 4)^2 = 4(y - 1)$ **35.** 4 **37.** $4|p|$ **39.**

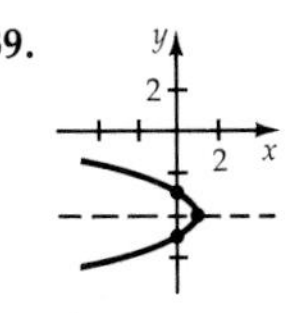

43.

45. $x^2 + y^2 - 8x - 8y - 2xy = 0$

Exercise Set 6.2, page 235

1. vertices: $(0, 5)$, $(0, -5)$
foci: $(0, 3)$, $(0, -3)$

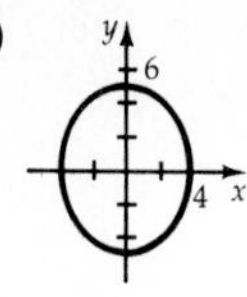

3. vertices: $(3, 0)$, $(-3, 0)$
foci: $(\sqrt{5}, 0)$, $(-\sqrt{5}, 0)$

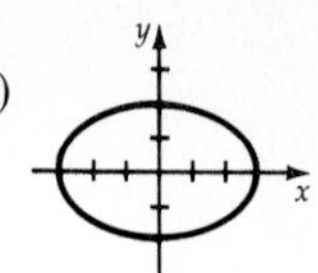

5. vertices $(2, 0)$, $(-2, 0)$
foci: $(1, 0)$, $(-1, 0)$

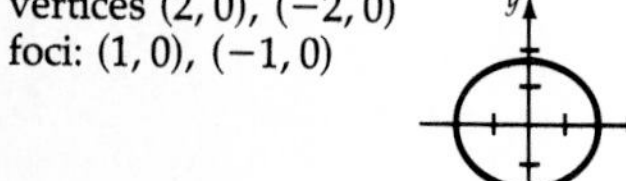

7. vertices: $(0, 5)$, $(0, -5)$
foci: $(0, 3)$, $(0, -3)$

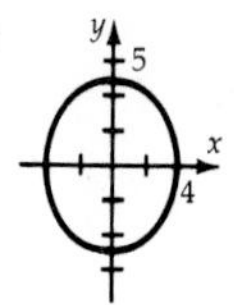

9. vertices $(0, 4)$, $(0, -4)$
foci: $(0, \sqrt{39}/2)$, $(0, -\sqrt{39}/2)$

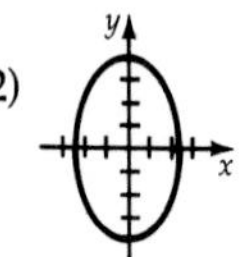

11. vertices: $(3, 6)$, $(3, 2)$
foci: $(3, 4 + \sqrt{3})$, $(3, 4 - \sqrt{3})$

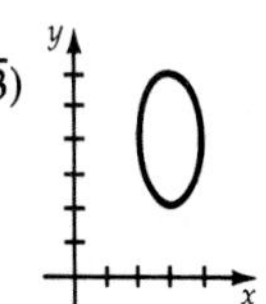

13. vertices: $(-1, -3)$, $(5, -3)$
foci: $(0, -3)$, $(4, -3)$

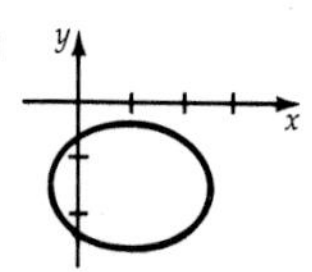

15. vertices: $(2, 4)$, $(2, -4)$
foci $(2, \sqrt{7})$, $(2, -\sqrt{7})$

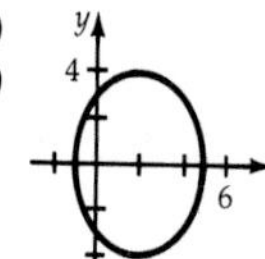

17. vertices: $(-1, 6)$, $(-1, -4)$
foci: $(-1, 4)$, $(-1, -2)$

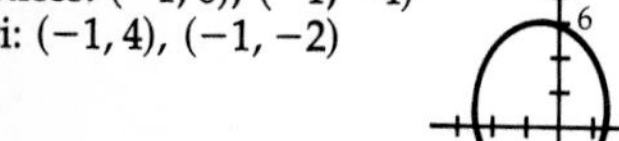

19. vertices: $(11/2, -1)\,(1/2, -1)$
foci: $(3 + \sqrt{17}/2, -1)$, $(3 - \sqrt{17}/2, -1)$

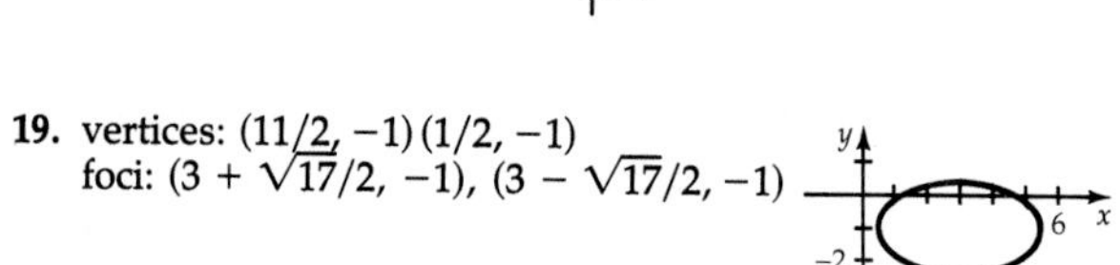

21. $\frac{x^2}{25} + \frac{y^2}{9} = 1$ **23.** $\frac{x^2}{36} + \frac{y^2}{16} = 1$ **25.** $\frac{x^2}{36} + \frac{y^2}{81/8} = 1$ **27.** $\frac{(x+2)^2}{16} + \frac{(y-4)^2}{7} = 1$ **29.** $\frac{(x-2)^2}{25/24} + \frac{(y-4)^2}{25} = 1$

31. $\frac{(x-5)^2}{16} + \frac{(y-1)^2}{25} = 1$ **33.** $\frac{x^2}{25} + \frac{y^2}{21} = 1$ **35.** $\frac{x^2}{20} + \frac{y^2}{36} = 1$ **37.** $\frac{(x-1)^2}{25} + \frac{(y-3)^2}{21} = 1$ **39.** $\frac{x^2}{80} + \frac{y^2}{144} = 1$

41. The information does not describe an ellipse. **43.** $\frac{x^2}{36} + \frac{y^2}{27} = 1$ **45.** $\frac{(x-1)^2}{16} + \frac{(y-2)^2}{12} = 1$ **47.** $9/2$ **51.** $x = \pm\frac{9\sqrt{5}}{5}$

Exercise Set 6.3, page 245

1. center: $(0, 0)$
vertices: $(\pm 4, 0)$
foci: $(\pm\sqrt{41}, 0)$
asymptotes: $y = \pm 5x/4$

3. center: $(0,0)$
vertices: $(0, \pm 2)$
foci: $(0, \pm\sqrt{29})$
asymptotes: $y = \pm 2x/5$

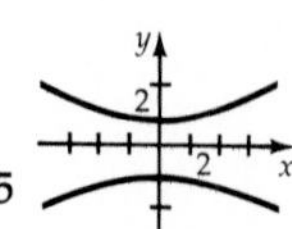

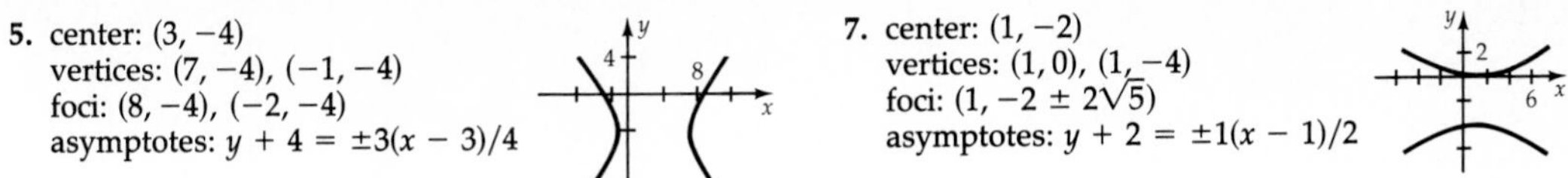

5. center: $(3, -4)$
vertices: $(7, -4)$, $(-1, -4)$
foci: $(8, -4)$, $(-2, -4)$
asymptotes: $y + 4 = \pm 3(x - 3)/4$

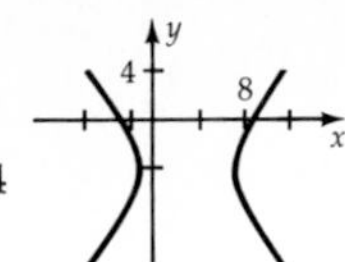

7. center: $(1, -2)$
vertices: $(1, 0)$, $(1, -4)$
foci: $(1, -2 \pm 2\sqrt{5})$
asymptotes: $y + 2 = \pm 1(x - 1)/2$

9. center: $(0, 0)$
vertices: $(\pm 3, 0)$
foci: $(\pm 3\sqrt{2}, 0)$
asymptotes: $y = \pm x$

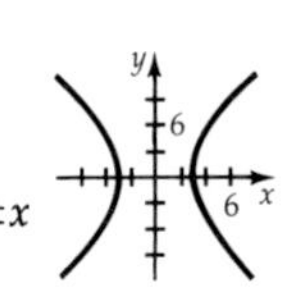

11. center: $(0, 0)$
vertices: $(0, \pm 3)$
foci: $(0, \pm 5)$
asymptotes: $y = \pm 3x/4$

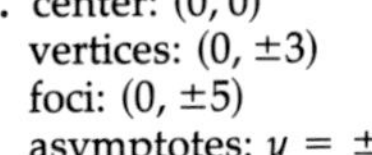

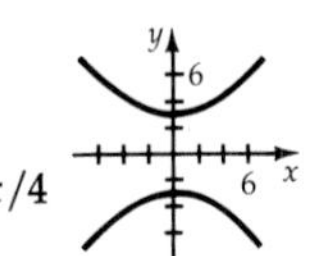

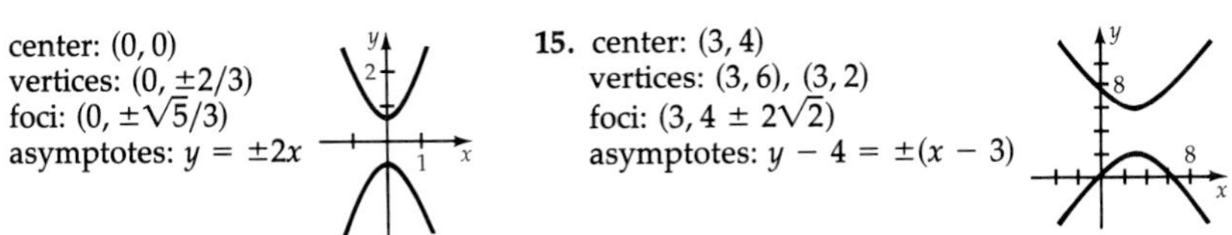

13. center: $(0, 0)$
vertices: $(0, \pm 2/3)$
foci: $(0, \pm\sqrt{5}/3)$
asymptotes: $y = \pm 2x$

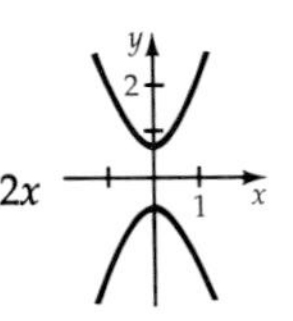

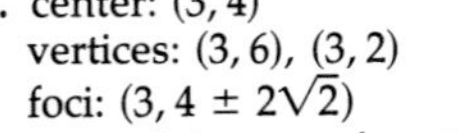

15. center: $(3, 4)$
vertices: $(3, 6)$, $(3, 2)$
foci: $(3, 4 \pm 2\sqrt{2})$
asymptotes: $y - 4 = \pm(x - 3)$

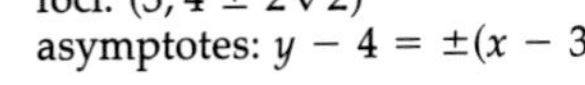

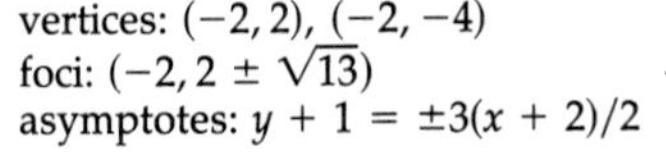

17. center: $(-2, -1)$
vertices: $(-2, 2)$, $(-2, -4)$
foci: $(-2, 2 \pm \sqrt{13})$
asymptotes: $y + 1 = \pm 3(x + 2)/2$

19. center: $(-4, 3)$
vertices: $(-2, 3)$, $(-6, 3)$
foci: $(-4 \pm 2\sqrt{5}, 3)$
asymptotes: $y - 3 = \pm 2(x + 4)$

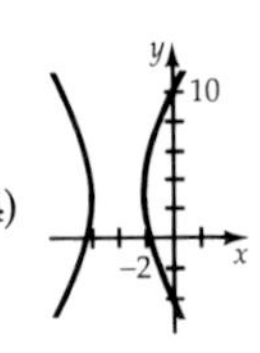

21. center: $(2, -2)$
vertices: $(2, 1)$, $(2, -5)$
foci: $(2, 3)$, $(2, -7)$
asymptote: $y + 2 = \pm 3(x - 2)/4$

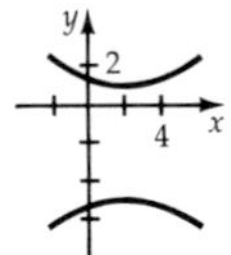

23. center: $(-1, -1)$
vertices: $(-1/2, -1)$, $(-3/2, -1)$
foci: $((-6 \pm \sqrt{13})/6, -1)$
asymptotes: $y + 1 = \pm 2(x + 1)/3$

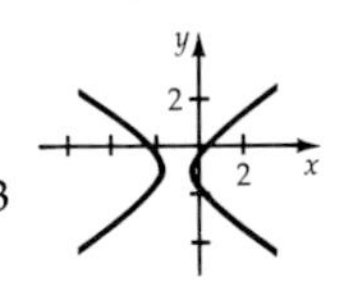

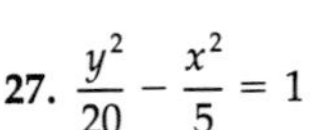

25. $\dfrac{x^2}{9} - \dfrac{y^2}{7} = 1$ **27.** $\dfrac{y^2}{20} - \dfrac{x^2}{5} = 1$ **29.** $\dfrac{y^2}{9} - \dfrac{x^2}{36/7} = 1$ **31.** $\dfrac{y^2}{16} - \dfrac{x^2}{64} = 1$ **33.** $\dfrac{(x-4)^2}{4} - \dfrac{(y-3)^2}{5} = 1$

35. $\dfrac{41(x-4)^2}{144} - \dfrac{41(y+2)^2}{225} = 1$ **37.** $\dfrac{(y-2)^2}{3} - \dfrac{(x-7)^2}{12} = 1$ **39.** $\dfrac{(y-7)^2}{1} - \dfrac{(x-1)^2}{3} = 1$ **41.** $\dfrac{x^2}{4} - \dfrac{y^2}{12} = 1$

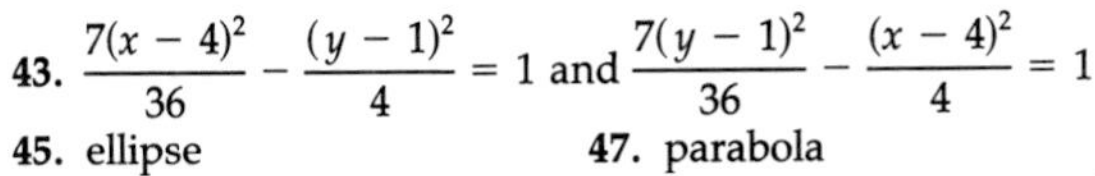

43. $\dfrac{7(x-4)^2}{36} - \dfrac{(y-1)^2}{4} = 1$ and $\dfrac{7(y-1)^2}{36} - \dfrac{(x-4)^2}{4} = 1$

45. ellipse

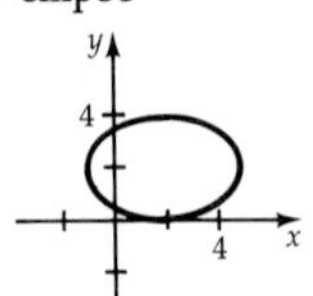

47. parabola

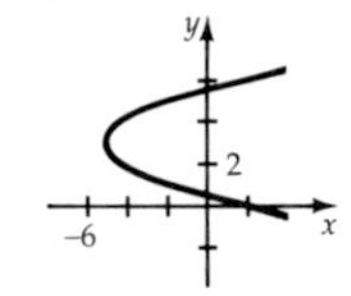

49. parabola

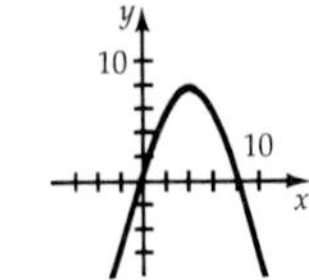

51. ellipse

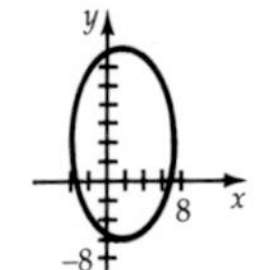

53. $\dfrac{(x - \sqrt{1215}/2)^2}{324} + \dfrac{y^2}{81/4} = 1$ **55.** $\dfrac{x^2}{2500} - \dfrac{y^2}{7500} = 1$ **57.** $\dfrac{x^2}{1} - \dfrac{y^2}{3} = 1$ **59.** $\dfrac{y^2}{9} - \dfrac{x^2}{7} = 1$ **61.** $y + 2 = \pm 9/5$ **67.**

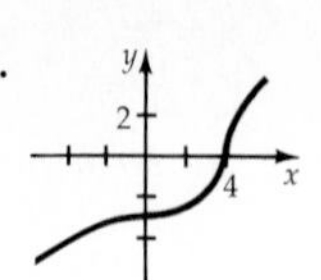

Exercise 6.4, page 257

1–7.

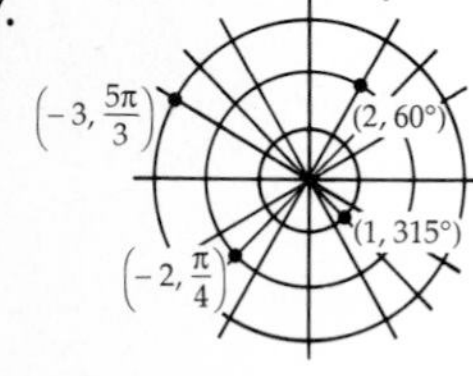

9.

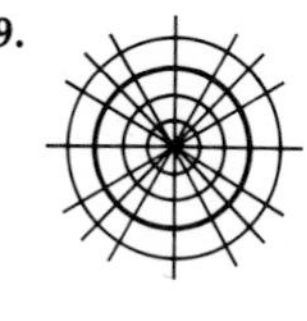

11.

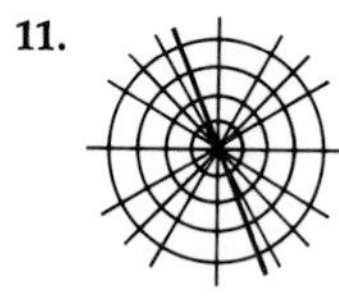

13.

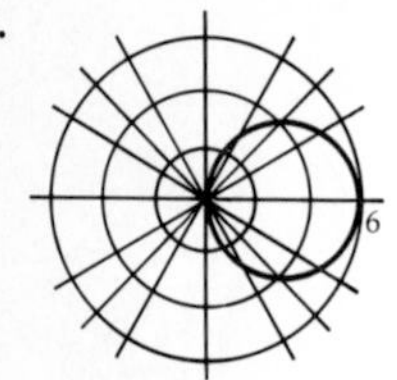

15.

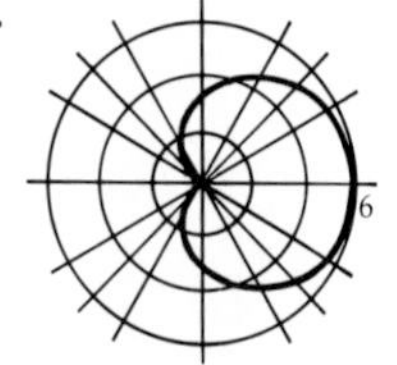

17.

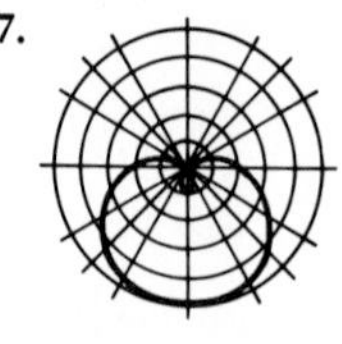

19.

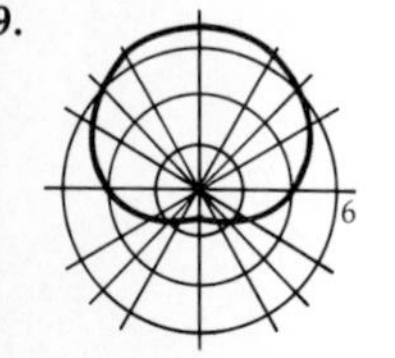

21.

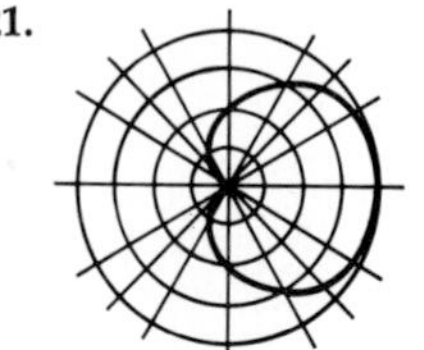

23.

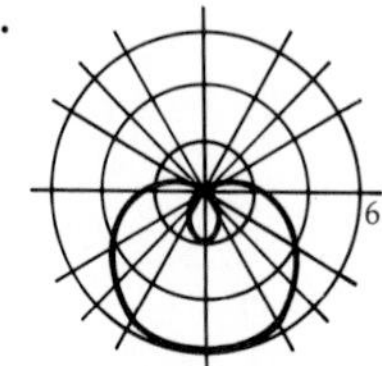

25.

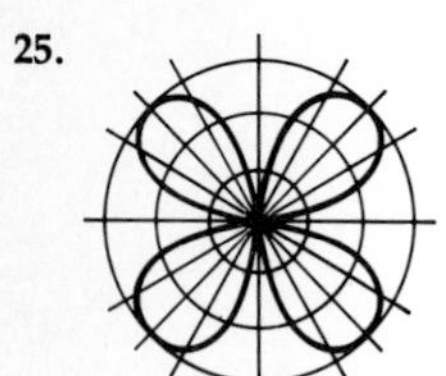

27.

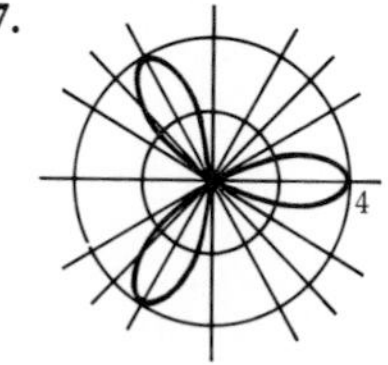

29.

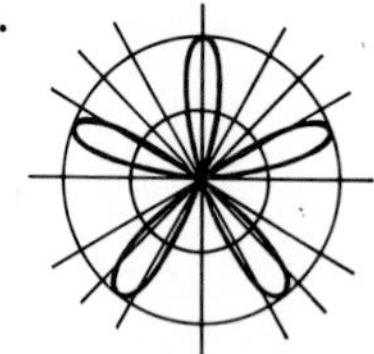

31.

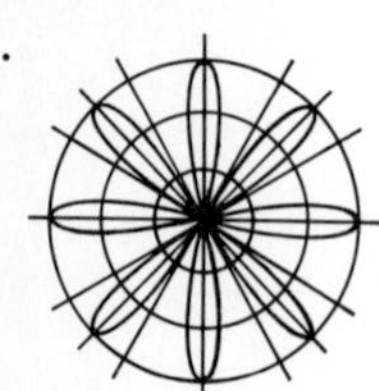

33.

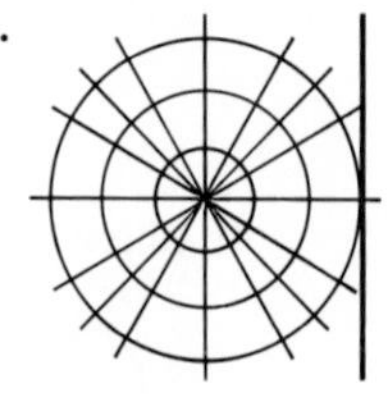

35.

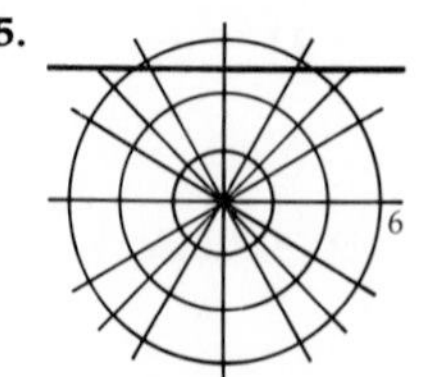

37.

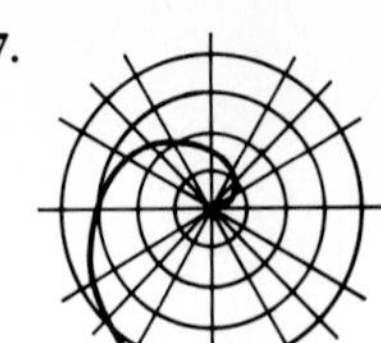

39.

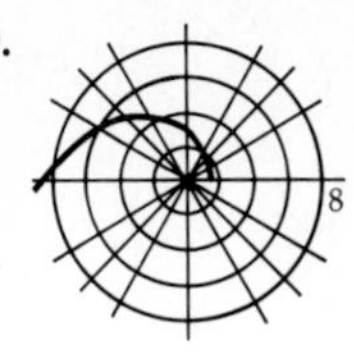

41. $(2, -60°)$

43. $(3/2, -3\sqrt{3}/2)$ **45.** $(0, 0)$ **47.** $(5, 53.1°)$ **49.** $x^2 + y^2 - 3x = 0$ **51.** $x = 3$ **53.** $x^2 + y^2 = 16$ **55.** $x^4 - y^2 + x^2y^2 = 0$
57. $y^2 + 4x - 4 = 0$ **59.** $y = 2x + 6$ **61.** $r = 2 \csc \theta$ **63.** $r = 2$ **65.** $r \cos^2 \theta = 8 \sin \theta$ **67.** $r^2 \cos 2\theta = 25$

69.

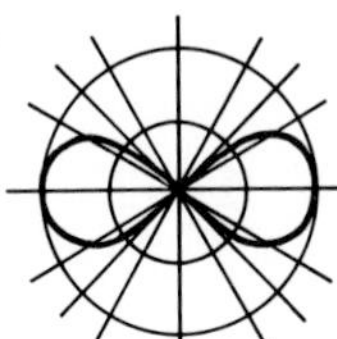

71.

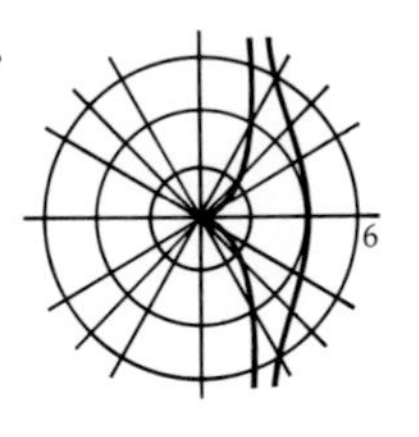

73.

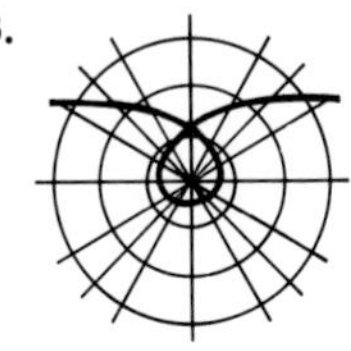

75.

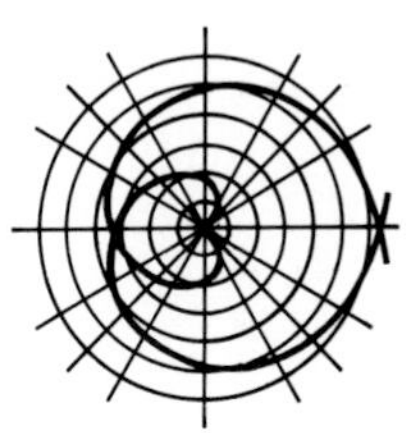

Challenge Exercises, page 259

1. false, a parabola has no asymptotes
3. false, by keeping foci fixed and varying asymptotes, we can make conjugate axis any size needed.
5. false, parabolas have no asymptotes **7.** false, a parabola can be a function **9.** true

Review Exercises, page 260

1. vertices: $(\pm 2, 0)$
foci: $(\pm 2\sqrt{2}, 0)$
asymptotes: $y = \pm x$

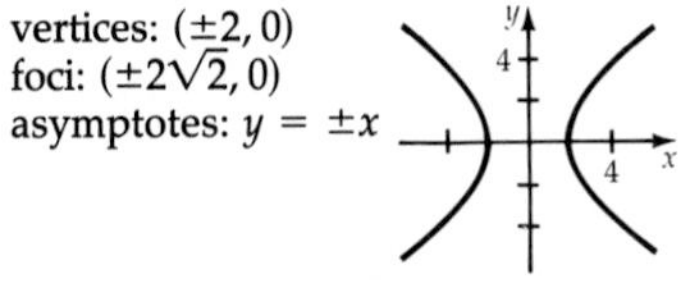

3. vertices: $(-1, -1)$, $(7, -1)$
foci: $(3 \pm 2\sqrt{3}, -1)$

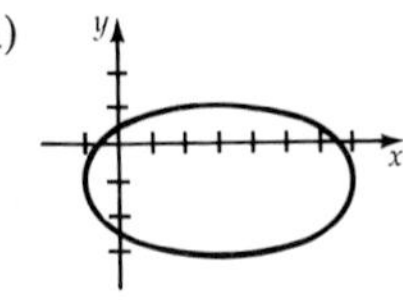

5. vertex: $(22, -4)$
foci: $(349/16, -4)$

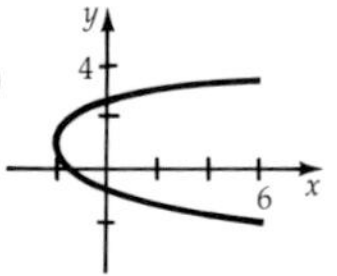

7. vertices: $(-2, -2)$, $(-2, 4)$
foci: $(-2, 1 \pm \sqrt{5})$

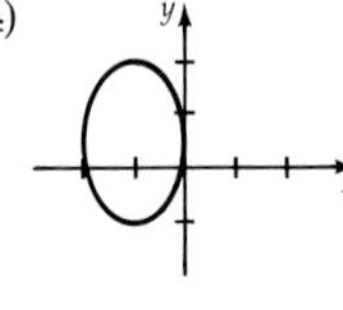

9. vertices: $(-5, 2/3)$, $(7, 2/3)$
foci: $(1 \pm 2\sqrt{13}, 2/3)$
asymptotes: $y - 2/3 = \pm 2(x - 1)/3$

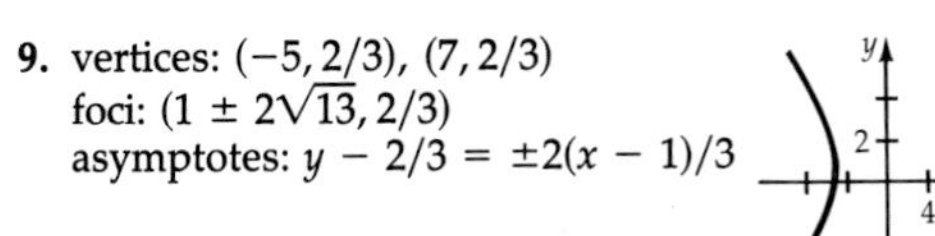

11. vertex: $(-7/2, -1)$
focus: $(-7/2, -3)$

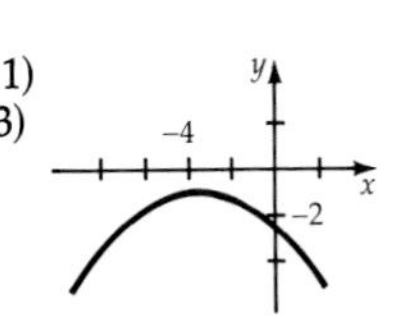

13. $\dfrac{(x-2)^2}{25} + \dfrac{(y-3)^2}{16} = 1$ **15.** $\dfrac{(x+2)^2}{16} - \dfrac{(y-2)^2}{20} = 1$ **17.** $x^2 = 3(y+2)/2$ or $(y+2)^2 = 12x$ **19.** $\dfrac{x^2}{36} - \dfrac{y^2}{4/9} = 1$

21. $(y - 3)^2 = -8x$ **23.** $\dfrac{(x-1)^2}{25} + \dfrac{(y-1)^2}{9} = 1$

25.

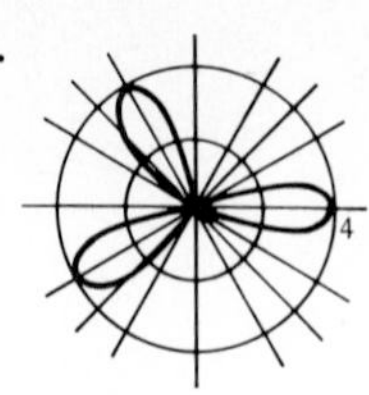

27.

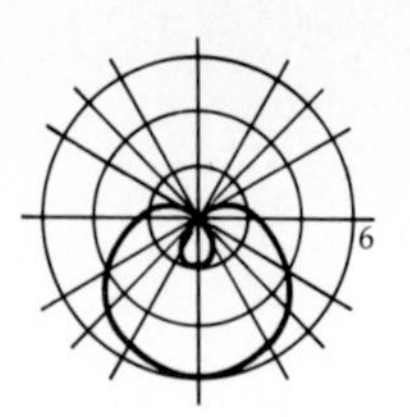

29.

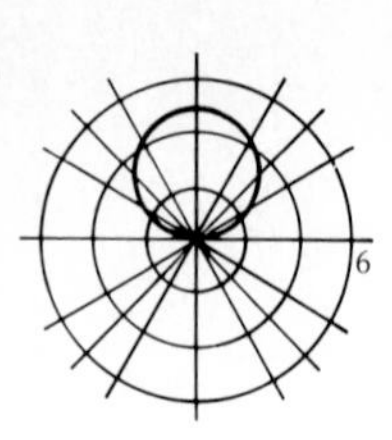

31.

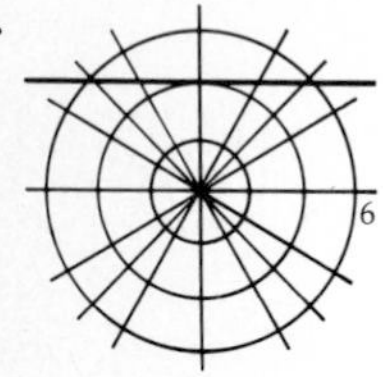

33.

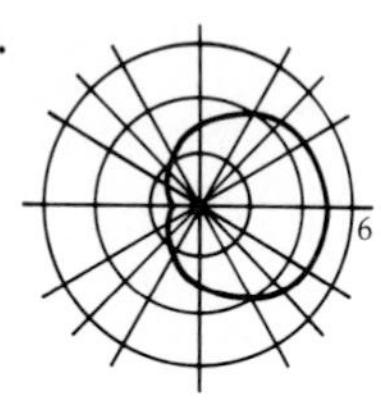

35. $r \sin^2 \theta = 16 \cos \theta$ **37.** $3r \cos \theta - 2r \sin \theta = 6$ **39.** $y^2 = 8x + 16$

41. $x^4 + y^4 + 2x^2y^2 - x^2 + y^2 = 0$

Exercise Set 7.1, page 268

1. 4.72880 **3.** 442.335 **5.** 2.17458 **7.** 164.022 **9.** 5.65223 **11.** 0.969476 **13.** 70.4503 **15.** 19.8130 **17.** 14.0940 **19.** 15.1543 **21.** 3353.33 **23.** 8103.08

25.

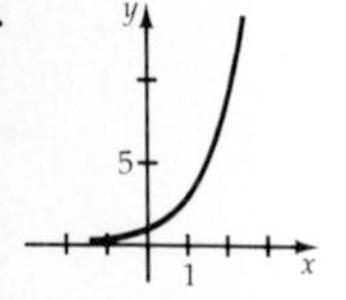

27.

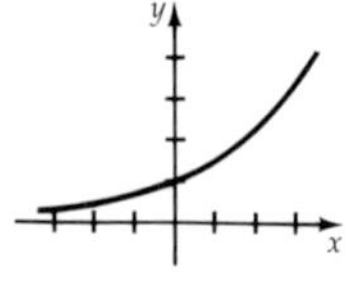

29.

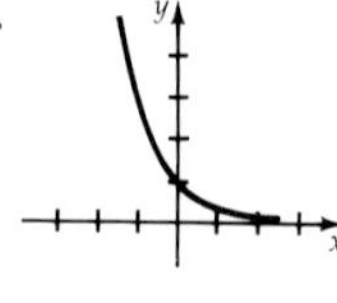

31.

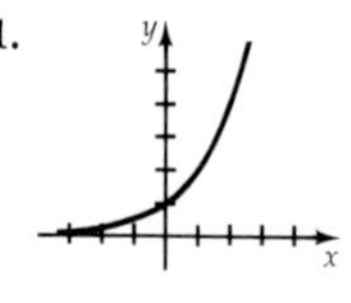

33.

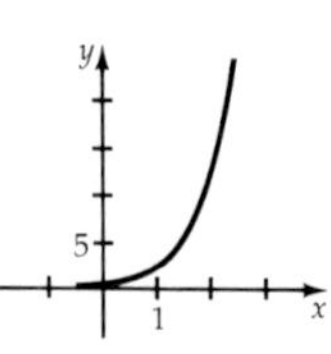

35.

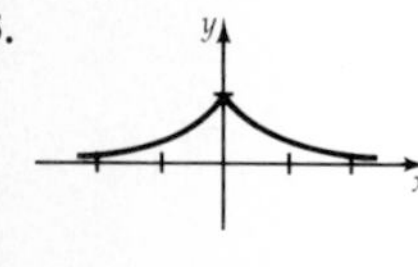

37.

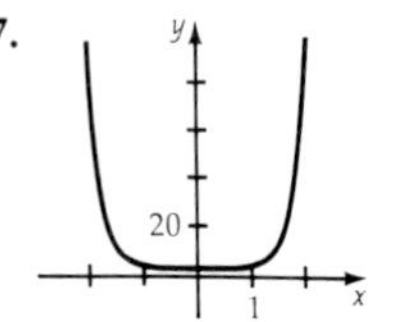

39.

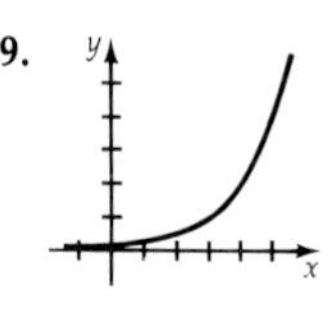

41.

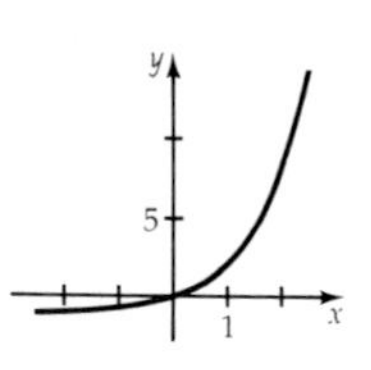

43.

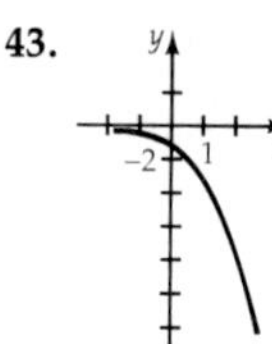

45.

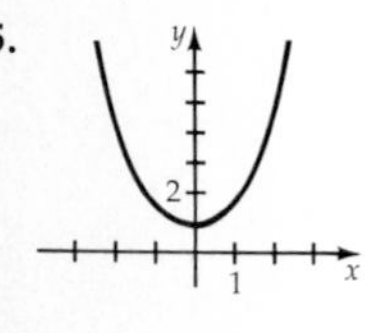

47.

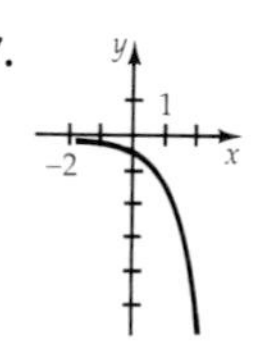

49.

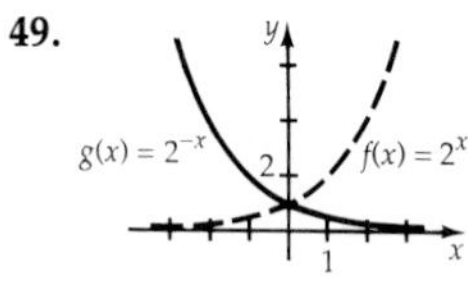

51.

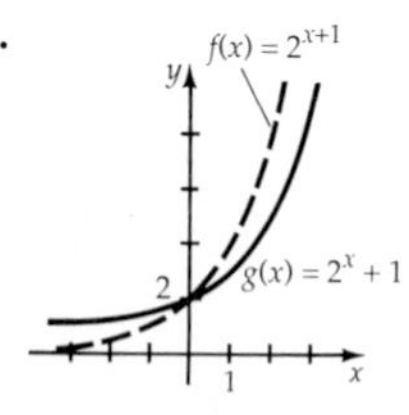

53.

Value of n	Value of $\left(1 + \dfrac{1}{n}\right)^n$
5	2.48832
50	2.691588029
500	2.715568520
5000	2.718010041
50,000	2.718254646
500,000	2.718280469
5,000,000	2.718281828

55. **a.** -2 **b.** 4 **c.** does not exist
h is not an exponential function because b is not a positive constant.

57. 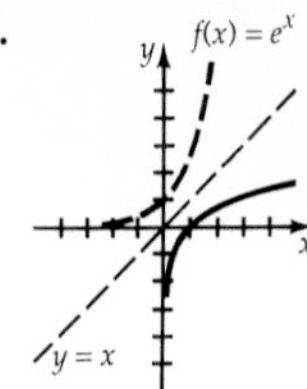

61. a. 20,000 **b.** 40,000 **c.** 320,000 **63.** e^{π} **65.** 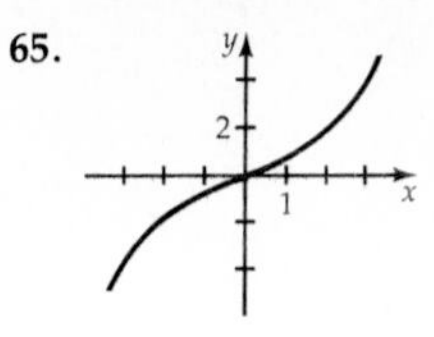

Exercise Set 7.2, page 277

1. $10^2 = 100$ **3.** $5^3 = 125$ **5.** $3^4 = 81$ **7.** $b^t = r$ **9.** $3^{-3} = \frac{1}{27}$ **11.** $\log_2 16 = 4$ **13.** $\log_7 343 = 3$ **15.** $\log_{10} 10{,}000 = 4$ **17.** $\log_b j = k$ **19.** $\log_b b = 1$ **21.** 6 **23.** 5 **25.** 3 **27.** -2 **29.** 0 **31.** $\log_b x + \log_b y + \log_b z$ **33.** $\log_3 x - 4 \log_3 z$ **35.** $\frac{1}{2} \log_b x - 3 \log_b y$ **37.** $\log_b x + \frac{2}{3} \log_b y - \frac{1}{3} \log_b z$ **39.** $\frac{1}{2} \log_7(x + z^2) - \log_7(x^2 - y)$ **41.** 0.9208 **43.** 1.1292 **45.** -0.4709 **47.** 1.7479 **49.** 1.3562 **51.** $\log_{10} x^2(x + 5)$ **53.** $\log_b \sqrt{\dfrac{(x - y)^3(x + y)}{z}}$ **55.** $\log_8(x + y)$ **57.** $\ln x^2(x - 3)^4$ **59.** $\ln \dfrac{xz}{y}$ **61.** 1.5395 **63.** 0.86719 **65.** -1.7308 **67.** -2.3219 **69.** 0.87357 **71.** 3.06 **73.** 3190 **75.** 0.00334 **77.** 7.40 **79.** 2.00 **81.** 0.300 **83.** $[1, 10^{1000}]$ **85.** $[e^e, e^{e^3}]$ **87.** $(0, 1)$ **91.** reflexive property of equality, definition of $\log_b x = n$

Exercise Set 7.3, page 282

1. **3.** 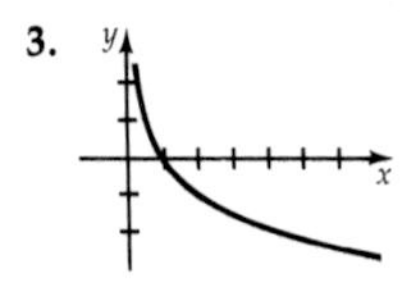 **5.** 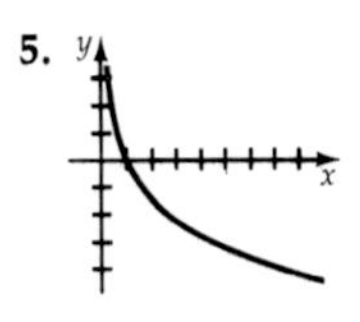 **7.** 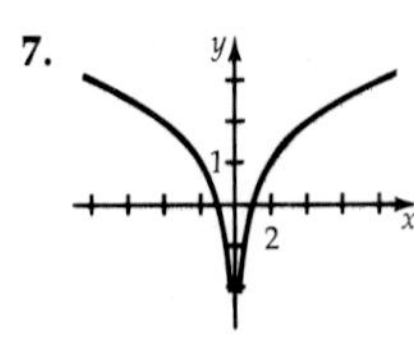 **9.** 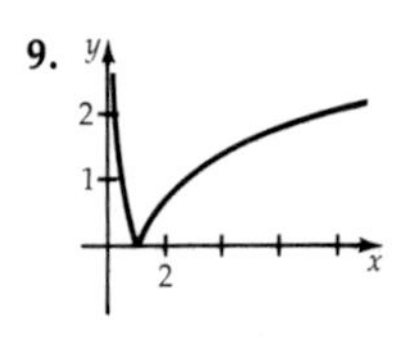

11. **13.** 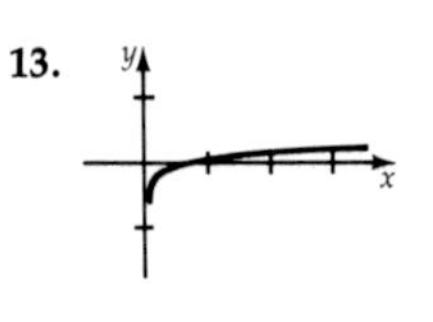 **15.** 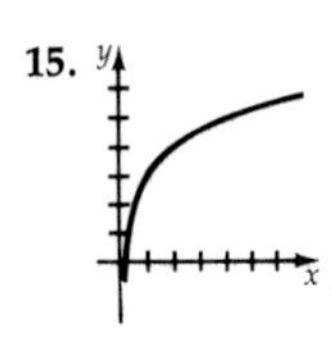 **17.** 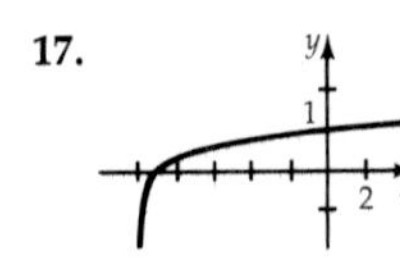 **19.** 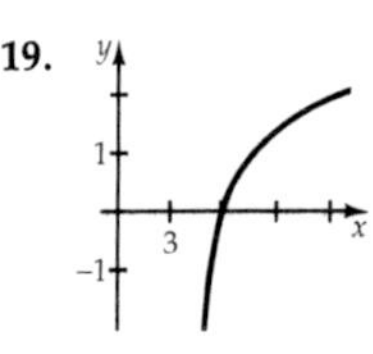

21. **23.** 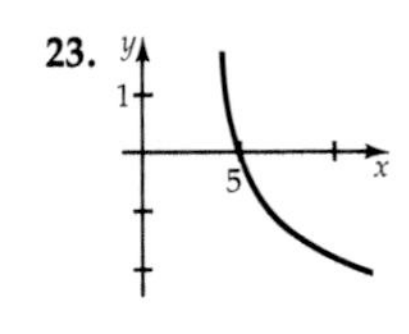 **25.** 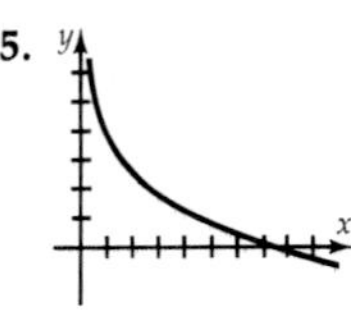 **27.** 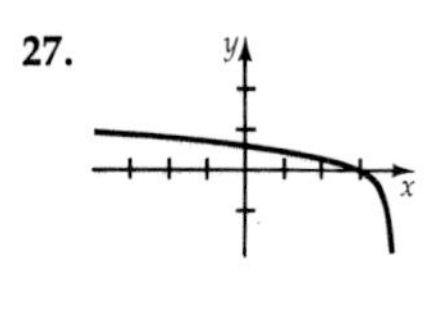

29. $\{x \mid x > 0, x \neq 1\}$ **31.** $\{x \mid x < -3 \text{ or } x > 3\}$ **33.** $\{x \mid x \neq 0\}$ **35.** $\{y \mid y \neq 0\}$ **37.** all real numbers **39.** all real numbers **41. a.** 15 **b.** 54 **c.** \$6,565,990,000 **43.** 8.6 **45. a.** 3 **b.** 1.386 **c.** 3.296

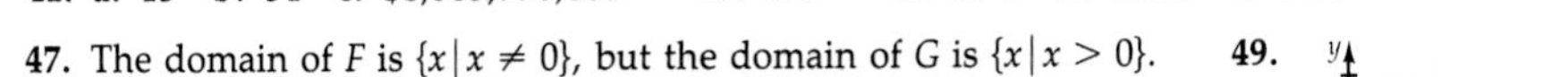

47. The domain of F is $\{x \mid x \neq 0\}$, but the domain of G is $\{x \mid x > 0\}$.

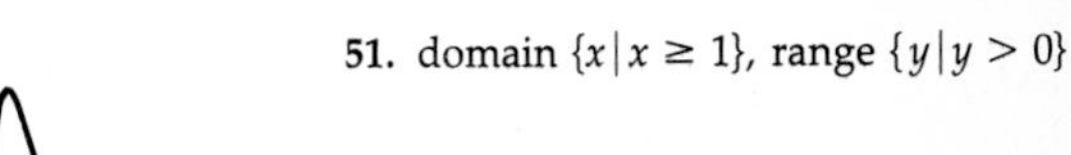

49. **51.** domain $\{x \mid x \geq 1\}$, range $\{y \mid y > 0\}$

53. domain $\{x \mid -1 < x < 1\}$, range $\{y \mid y \geq 100\}$ **55.** domain $\{x \mid x > 1\}$, range all real numbers **57.** 8.8

Exercise Set 7.4, page 288

1. 6 **3.** 3 **5.** $-3/2$ **7.** $-6/5$ **9.** 3 **11.** 2.64 **13.** -4.36 **15.** 0.163 **17.** -0.25 **19.** 2.3 **21.** 5.0852
23. $2 + 2\sqrt{2}$ **25.** 199/95 **27.** 3 **29.** 10^{10} **31.** 25 **33.** 2 **35.** 3, -3 **37.** no solution **39.** 5 **41.** $\log(20 + \sqrt{401})$
43. $\frac{1}{2}\log\frac{3}{2}$ **45.** $\ln(15 \pm 4\sqrt{14})$ **47.** $\ln(1 + \sqrt{65}) - \ln 8$ **49.** 2.2 **51.** 3.2×10^{-5}
53. **a.** 8500, 10,285 **b.** in 6 years **55.** **a.** 60°F **b.** in 27 minutes
57. The second step, because $\log 0.5 < 0$, thus the inequality sign must be reversed.

59. no solutions

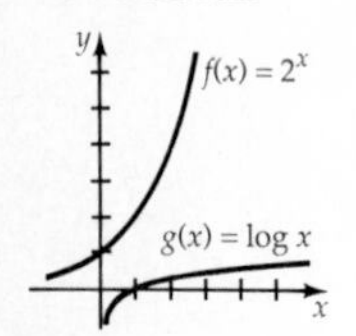

61. 2 solutions

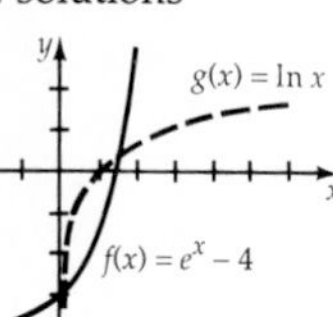

63. 2 solutions

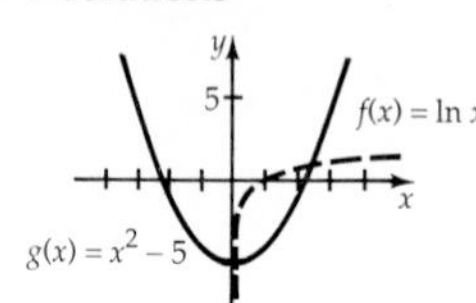

65. 4 solutions

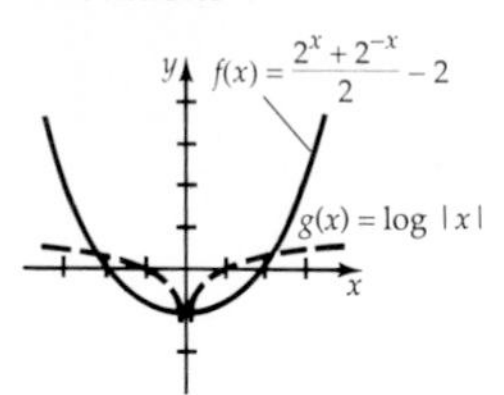

67. $x = \dfrac{y}{y - 1}$

69. $e^{0.336} \approx 1.4$ **71.** $\left(0, \dfrac{1}{\ln 2}\right)$

Exercise Set 7.5, page 297

1. **a.** \$9724.05 **b.** \$11,256.80 **3.** **a.** \$48,885.72 **b.** \$49,282.20 **c.** \$49,283.29 **5.** \$24,730.82 **7.** **a.** 2200 **b.** 17,600
9. $N(t) = 22{,}600(e^{.01368t})$ **11.** **a.** 18,400 **b.** 1994 **13.** $N(t) = 100e^{-.000418t}$ **15.** 39.1% **17.** 2161 years old
19. **a.** 0.056 **b.** 42° **c.** 54 minutes **d.** never **21.** 10 times more powerful **23.** 3.01 decibels
25. **a.** 211 hours **b.** 1386 hours **27.** 3.1 years **29.** **a.** 21.7, 0.87 **b.** 1086, .88 **c.** 72,382, .92
31. **a.** 0.023639 **b.** 6.8 billion **c.** 80 billion **33.** **a.** 16,000 **b.** 10,500 (nearest 500) **35.** $\dfrac{1}{k}\ln\left[\dfrac{P(t)(m - P_0)}{P_0(m - P(t))}\right]$
37. **a.** 286,500 gallons (nearest 500) **b.** 250,500 gallons (nearest 500) **c.** 34.5 hours **39.** 15.8 times stronger **41.** 6.93%

Challenge Exercises, page 301

1. true **3.** true **5.** false, $h(x) = \left(\frac{1}{2}\right)^x$ is a decreasing function **7.** true **9.** true
11. false, $\log(1 + 1) = \log 2$, but $\log 1 + \log 1 = 0$ **13.** true

Review Exercises, page 302

1. 2 **3.** 3 **5.** -2 **7.** -3 **9.** 1000 **11.** 7 **13.** 15.6729 **15.** 5.47395 **17.** 13.6458

19.

21.

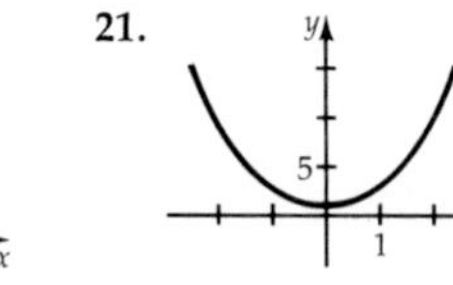

23.

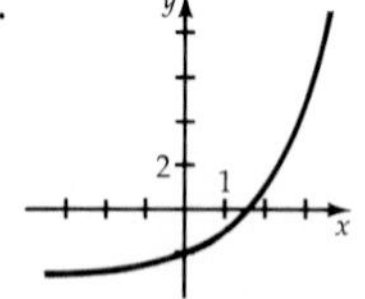

25.

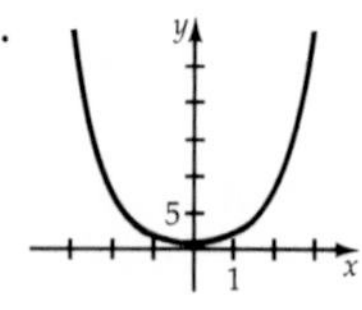

27. 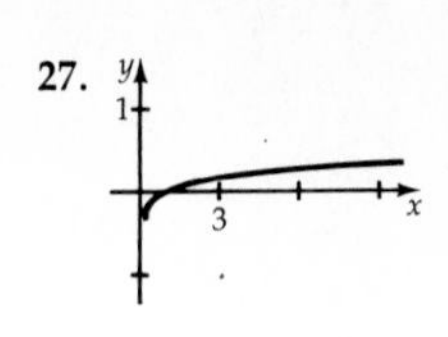

29. 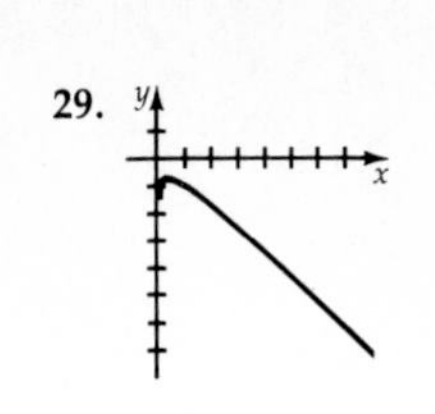

31. 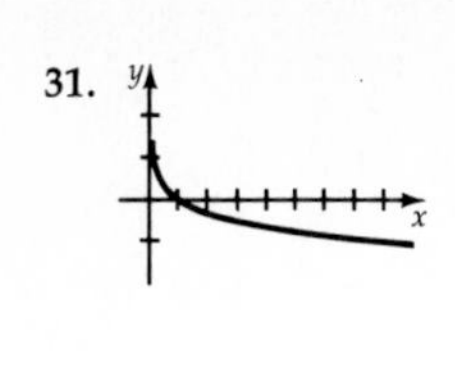

33. $4^3 = 64$ **35.** $\sqrt{2}^4 = 4$ **37.** $\log_5 125 = 3$ **39.** $\log_{10} 1 = 0$

41. $\log_b \dfrac{x^2y^3}{z}$ **43.** $\ln x + 3 \ln y$ **45.** $\log x^2 \sqrt[3]{x + 1}$ **47.** $\ln\left(\dfrac{\sqrt{2xy}}{z^3}\right)$ **49.** 2.86754 **51.** −0.117233 **53.** 295 **55.** 1.41×10^{22}

57. $\dfrac{\ln 30}{\ln 4}$ **59.** 4 **61.** 4 **63.** $\dfrac{\ln 3}{2 \ln 4}$ **65.** e^{e^2} **67.** 1,000,005 **69.** 81 **71.** 4 **73.** 2.2 **75. a.** \$20,323.79 **b.** \$20,339.99

77. \$4438.10 **79.** $N(t) = e^{0.3219t}$ **81.** $N(t) = 2.899e^{0.3219t}$

Index